MEDICAL
INTELLIGENCE
UNIT 35

DNA Vaccines

Hildegund C. J. Ertl, M.D.
The Wistar Institute
Philadelphia, Pennsylvania, U.S.A.

SPRINGER SCIENCE+BUSINESS MEDIA, LLC

DNA Vaccines

Medical Intelligence Unit 35

Copyright ©2003 Springer Science+Business Media New York
Originally published by Kluwer Academic Publishers in 2003

http://www.wkap.nl/

Please address all inquiries to Landes Bioscience / Eurekah.com:
Landes Bioscience / Eurekah.com, 810 South Church Street, Georgetown, Texas, U.S.A. 78626
Phone: 512/ 863 7762; FAX: 512/ 863 0081; www.Eurekah.com; www.landesbioscience.com
Landes tracking number: 1-58706-057-4

DNA Vaccines edited by Hildegund Ertl, Landes / Kluwer dual imprint/ Landes series: Medical Intelligence Unit 35
    ISBN 978-0-306-47444-6   ISBN 978-1-4615-0105-3 (eBook)
    DOI 10.1007/978-1-4615-0105-3

Library of Congress Cataloging-in-Publication Data

CIP applied for but not received at time of publication.

CONTENTS

EDITOR

Hildegund C. J. Ertl, M.D.
The Wistar Institute
Philadelphia, Pennsylvania, U.S.A
Chapters 3, 16

CONTRIBUTORS

Constantin Bona
Department of Microbiology
Mount Sinai School of Medicine
New York, New York, U.S.A.
Chapter 2

Robert K. Bright
Robert W. Franz Cancer
   Research Center
Portland, Oregon, U.S.A.
Chapter 11

M. A. Chambers
Veterinary Laboratories Agency
   Weybridge
TB Research Group
Department of Bacterial Diseases
Addlestone, Surrey, England, U.K.
Chapter 10

Kaw Yan Chua
Department of Pediatrics
National University of Singapore
Republic of Singapore
Chapter 12

Joachim Fensterle
Max-Planck Institute for Infection
   Biology, Berlin
Institute of Medical Radiology
   and Cell Research
University of Würzburg
Würzburg, Germany
Chapter 15

Julie Fitzgerald
Wistar Institute
Philadelphia, Pennsylvania, U.S.A.
Chapter 3

R. G. Hewinson
Veterinary Laboratories Agency
   Weybridge
TB Research Group
Department of Bacterial Diseases
Addlestone, Surrey, England, U.K.
Chapter 10

Katherine A. High
Children's Hospital of Philadelphia
Philadelphia, Pennsylvania, U.S.A.
Chapter 13

Maurice R. Hilleman
Merck Institute for Vaccinology
West Point, Pennsylvania, U.S.A.
Chapter 1

Patrick G. Holt
TVW Telethon Institute for Child
   Health Research
Centre for Child Health Research
The University of Western Australia
Perth, Western Australia, Australia
Chapter 12

Kayo Inaba
Laboratory of Immunobiology
Graduate School of Biostudies
Kyoto University
Kyoto, Japan
Chapter 2

Stefan H. E. Kaufmann
Max-Planck Institute for Infection
   Biology
Berlin, Germany
Chapter 15

Garnett Kelsoe
Department of Immunology
Duke University Medical Center
Durham, North Carolina, U.S.A.
Chapter 4

Ronald C. Kennedy
Department of Microbiology
  and Immunology
Texas Technical University of the Health
  Science Center
Lubbock, Texas, U.S.A.
Chapter 11

Dennis M. Klinman
Center for Biologics Evaluation
  and Research
Food and Drug Administration
Bethesda, Maryland, U.S.A.
Chapter 18

K. Kilpatrick
TriPath Oncology
Durham, North Carolina, U.S.A.
Chapter 4

Jiri Kovarik
World Health Organization
  Collaborating Centre for Vaccinology
  and Neonatal Immunology
Department of Pathology and Pediatrics
University of Geneva
Geneva, Switzerland
Chapter 14

Jens Leifert
Department of Neuropharmacology
The Scripps Research Institute
La Jolla, California, U.S.A.
Chapter 5

Karl Ljungberg
Department of Microbiology
  and Tumorbiology
Karolinska Institute
Swedish Institute for Infectious Disease
  Control
Stockholm, Sweden
Chapter 9

D. B. Lowrie
National Institute for Medical Research
Mill Hill, London, England, U.K.
Chapter 10

Richard T. Mahoney
Institutional Development
International Vaccine Institute
Seoul National University Campus
Seoul, Korea
Chapter 19

Xavier Martinez
World Health Organization
  Collaborating Centre for Vaccinology
  and Neonatal Immunology
Department of Pathology and Pediatrics
University of Geneva
Geneva, Switzerland
Chapter 14

Philip M. Murphy
Laboratory of Host Defenses
National Institute of Allergy
  and Infectious Diseases
National Institutes of Health
Bethesda, Maryland, U.S.A.
Chapter 17

Christopher Pack
Department of Microbiology
The University of Tennessee
Knoxville, Tennessee, U.S.A.
Chapter 8

Barry T. Rouse
Department of Microbiology
The University of Tennessee
Knoxville, Tennessee, U.S.A.
Chapter 8

Denise E. Sabatino
Children's Hospital of Philadelphia
Philadelphia, Pennsylvania, U.S.A.
Chapter 13

M. Sarzotti
Department of Immunology
Duke University Medical Center
Durham, North Carolina, U.S.A.
Chapter 4

Michael H. Shearer
Department of Microbiology
   and Immunology
Texas Technical University of the Health
   Science Center
Lubbock, Texas, U.S.A.
Chapter 11

Claire-Anne Siegrist
World Health Organization
   Collaborating Centre for Vaccinology
   and Neonatal Immunology
Department of Pathology and Pediatrics
University of Geneva
Geneva, Switzerland
Chapter 14

Herbert A. Smith
Center for Biologics Evaluation
   and Research
Food and Drug Administration
Bethesda, Maryland, U.S.A.
Chapter 18

Ralph M. Steinman
Laboratory of Cellular Physiology
   and Immunology
Rockefeller University
New York, New York, U.S.A.
Chapter 2

Jeffrey B. Ulmer
Vaccines Research
Chiron Corporation
Emeryville, California, U.S.A.
Chapter 7

H. M. Vordermeier
Veterinary Laboratories Agency
   Weybridge
TB Research Group
Department of Bacterial Diseases
Addlestone, Surrey, England, U.K.
Chapter 10

Britta Wahren
Department of Microbiology
   and Tumorbiology
Karolinska Institute
Swedish Institute for Infectious Disease
   Control
Stockholm, Sweden
Chapter 9

Yu-Mei Wen
Department of Molecular Virology
Medical Center Fudan University
Shanghai, China
Chapter 19

J. Lindsay Whitton
Department of Neuropharmacology
The Scripps Research Institute
La Jolla, California, U.S.A.
Chapter 5

Henry Wilde
Chulalongkorn University
Thai Red Cross Society
Bangkok, Thailand
Chapter 19

Betina Wolfowicz
Department of Pediatrics
National University of Singapore
Republic of Singapore
Chapter 12

Zhi-Yi Xu
International Vaccine Institute
Seoul National University Campus
Seoul, Korea
Chapter 19

Jonathan W. Yewdell
Laboratory of Viral Diseases
National Institute for Allergy
    and Infectious Diseases
National Institutes of Health
Bethesda, Maryland, U.S.A.
Chapter 6

Anne Kjerrström Zuber
Department of Microbiology
    and Tumorbiology
Karolinska Institute
Swedish Institute for Infectious Disease
    Control
Stockholm, Sweden
Chapter 9

Bartek Zuber
Department of Microbiology
    and Tumorbiology
Karolinska Institute
Swedish Institute for Infectious Disease
    Control
Stockholm, Sweden
Chapter 9

PREFACE

DNA vaccines, first described less than 10 years ago, have shown efficacy in experimental animals against a number of afflictions caused by infectious agents, cancer or misguided immunological responses. Results obtained in clinical trials have been disappointing thus far. The 'DNA Vaccine' book details in the initial chapters the immunological mechanisms that govern immune responses to vector encoded antigens. The remaining chapters illustrate the use of DNA vaccines for prevention or therapy of infectious diseases, cancer, autoimmune reactions, allergies and rejection of proteins delivered within the realm of gene therapy.

Hildegund C. J. Ertl

CHAPTER 1

Overview of Vaccinology in Historic and Future Perspective:
The Whence and Whither of a Dynamic Science with Complex Dimensions

Maurice R. Hilleman

Abstract

U nderstanding of the present and future can be aided immensely by acquaintance with a breadth of knowledge of the past. The history of the world records its dread experiences with diseases, and also the degree to which ignorance, authoritarian dogma, and institutionalized group-thinking restricted those individuals who sought to understand and to seek remedies.

The history of vaccines and vaccinology lends itself to discussion of its progress in terms of periods or eras, in which new advances were made. This recounting of the dynamic history of vaccines has been written to capture the big picture as it covers the span of time from its ancient past to the present, with an added view of the possibilities and projections for the future.

In projecting what lies ahead, it is necessary to take into account what needs vaccines fulfill and, importantly, the degree to which they are being utilized. Further to this is the requirement that the vaccine enterprise approach its targets with greatest energy and efficiency while providing assurance that its role in the contract between science and society is being fulfilled, especially in providing an equitable return to the public for public funding. There may be no more cogent example for this than in the brilliant but mostly pedestrian pursuit of recombinant DNA vectorology for which there is urgent need, if safe and effective, to bring to practical application in immunoprophylactic preventive medicine.

It is hoped that this historic and futuristic review will convey information of value to researchers of the future who may be encouraged to continue the legacy of vaccines of the past.

History of Disease, Science and Vaccines to 1875 (Table 1)

Among the many quests pursued by mankind, none can be more noble than that of preventing infectious diseases by procedures which derive from the science of vaccinology. In such endeavor, the roots of its past and present provide guidelines for future possibilities and realities. Immunoprophylaxis can be understood best in the perspective of the historic evolution of the sciences of medicine in general and of infectious diseases and vaccinology in particular,[1-8] together with a consideration of all the elements which play a role in bringing them to practical application. Figure 1 provides a diagrammatic overview of the past and future of science and vaccinology.

Ancient

Ancient peoples were in possession of much rudimentary knowledge about infectious diseases that included infectious transmission by contact, the association of disease with food and

Table 1. History of change and advance—by arbitrary periods

Approximate Period	Changes
Ancient (500 AD)	Rudiments of understanding disease and infectious transmission. Hygienic rules codified in laws and taboos.
Dark Ages (500-900)	Barbarian invasions and downfall of West Roman Empire. Small kingdoms replace empires, Christianity replaces paganism. Clerical dogma of disease caused by sin. Galen's writings adopted as authoritarian.
Middle Ages (900-1400)	Holy Roman Empire: power struggle between church and state. Building of towns, cities, universities, institutions. Greek and Arabic recordings dominate medical practice.
(1200-1300)	Formation of European states, reform of church fails. Hundred Years War, plagues of disease.
Renaissance (1400-1600) and Reformation (1500-1600)	Revival of individuality and humanism, Hellenic classicism. Copernicus describes solar system, challenges religious dogma. Age of voyages and discovery. Luther leads reformation with breakaway from Roman church. Diseases and especially syphilis rampant. Hellenic theory of internal harmony replaces sin as cause of disease. Fracastoro describes disease contagion and relates disease to putrefaction.
Liberation and Rise of Science (1600-1700)	Democratic principles and scientific experimentation, but with religious opposition. Galileo evolves scientific methods. Kepler shows planetary movement around the sun. Hooke discovers cellular structure; Leeuwenhoek perfects microscope and discovers microscopic life. Redi refutes spontaneous generation. Epidemics rampant, hygienic and epidemiologic control begins. Universities and academies become the venue for scientific endeavor.
Industrialization and Revolution (1700-1800)	Logic, trial and error, and development of systems in scientific experiments. Spallanzani attacks spontaneous generation, discovers cell fusion in fertilization. Philadelphia is cradle for American medicine. Rosa presents germ theory for disease. Lady Montagu introduces variation against smallpox. Jennerian prophylaxis against smallpox. Agricultural and industrial advances. American and French revolutions create republics.
Nineteenth Century (1800-1975)	Realism and materialism replace idealism and humanitarianism. Machine age powered by machines, transport by powered ships and railways. Schleiden and Schwann establish cellular basis for living structures. Cell nucleus discovered. Darwinian concepts for evolution and origin of species gives a secular view to life. Mendelian genetics lay fallow until "rediscovery" in 1900.

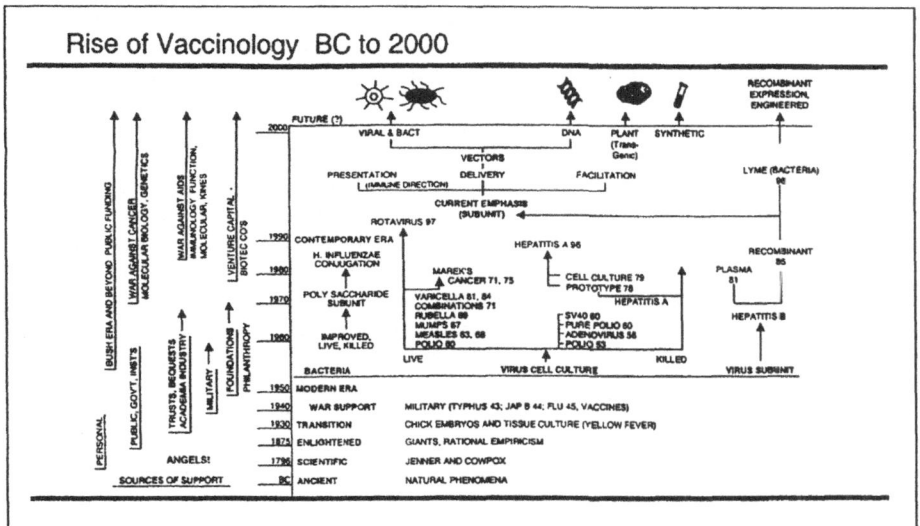

Figure 1. Rise of vaccinology BC-2000. (Reprinted from Hilleman, MR. Vaccines in historic evolution and perspective: A narrative of vaccine discoveries, In: Vaccine. Vol. 18. 2000:1436-1447, with permission from Elsevier Science).

water, the application of ill-defined public and private hygienic practices, and the development of resistance to second occurrence which may follow certain clinically definable diseases. Such empirical observations led to cause-and-effect associations and rules for prevention that were codified in ancient laws and in religious taboos.

Dark Ages

Such knowledge of the past largely was lost to the peoples of Europe following the barbarian invasions and the fall of the West Roman Empire near the end of the sixth century. Small Germanic kingdoms replaced empire and Christianity replaced Paganism. This marked the onset of the Dark Ages that led to the evolution of Feudalism. Ecclesiastical dogma replaced reason and held that disease was caused by sin. Mystic therapies were formulated to dispel the demons of disease in an atmosphere of terror and ignorance. Plagues and warfare abounded. In a feeble effort to resurrect Hellenic medicine of the past, the anatomic writings of Galen (AD 138-201), though laced with gross inaccuracies, were adopted as authoritarian by the Church and were actively enforced for many centuries that followed.

Middle Ages

With the dawn of the Middle Ages in the late 900's, a federation of states of the Holy Roman Empire was formed which strengthened the power of the Church and led to great struggles between Church and State. Among the more important events during the next two hundred years were the conquest of the Holy Land by the Crusaders, the building of towns and cities, the founding of great universities, and the collection and teaching of ancient medical knowledge in the monasteries. The practice of medicine by the monks, however, was not allowed outside the monastery and this brought about secular practice of medicine in the hands of the laity, especially in the universities where ancient Greek and Arabic recordings were being studied and followed.

The 1200's and 1300's were notable for the formation of the individual European states, reducing the previous leading roles of empire and papacy. The attempts to reform the church by Wyclif in England and by Hus in Bohemia ended in failure. The Hundred Years' War began in 1337 and bubonic plague, leprosy, and influenza were rampant. The plague, called the Black Death, was ascribed to supernatural and other spurious causes, even though a contagious nature was recognized. The treatments prescribed were irrelevant, including the pursuit of ineffective herbal remedies.

Renaissance and Reformation

The early 1400's signaled the beginning of the Renaissance, to be followed a century later by the Reformation. The Renaissance was a time of revival of human individuality and humanity together with discard of slavery to theology that had dominated the medieval period. There was revival of Hellenistic classicism and a new freedom for individuals. Appreciation for enjoyment of life prevailed. Copernicus challenged religious dogma by determining that the planets revolved around the sun, and new horizons were opened by the voyages of discovery to the West Coast of Africa, of India, and especially to the New World by Christopher Columbus. In the early 1500's, Martin Luther led a rebellion against the authority and improprieties of the Roman church, and the protestant reformation, which followed, gave rise to a century of religious warfare.

It was recorded that, during this period, smallpox, measles, varicella and influenza were of increased severity and frequency. The true origin of the pandemic of the new disease called syphilis was never established, and there was no specific therapy for the disease until Ehrlich's Salvarsan in the early 1900's and penicillin during World War II.

The theistic idea that disease was the result of sin was replaced by the equally spurious Hellenic belief that it was derived from lack of harmony of the humors in the body which nature itself could cure. Fracastoro classified contagion in epidemic diseases to those arising from direct contact, from indirect contact (fomite), or from a distance with outside dissemination. He noted similarities between certain infectious diseases and putrefaction.

Liberation and the Rise of Science

The dawn of the seventeenth century heralded the presence of democratic principles that replaced the political structures of the past. Science, now unshackled, initiated a new era of experimentation that penetrated the unanswered secrets of nature, albeit in the presence of continuing religious opposition, some of which persists to the present. The emphasis in science of the time was to discover and to define phenomena.

The dominant scientific figure in this new era was Galileo who introduced experimental methods for scientific study and who invented the telescope that was used by Kepler to define the movements of the planets around the sun. Hooke observed that plant tissue was comprised of individual units that he called cells. The existence of microbes was first revealed by Leeuwenhoek[8] in Delft who perfected the microscope and discovered the world of "animalcules" which, hitherto, were totally unknown.

There was a concept, prevalent for centuries, that life could generate spontaneously from inert matter, and the appearance of maggots and flies in putrefying meat was commonly given as support for the phenomenon. This belief was attacked by Redi who showed that exclusion of existing flies precluded the generation of new flies. Redi's demonstration was credible and convincing but the prevailing judgment was that Redi's observations applied only to primitive single-cell organisms called infusoria. Common belief in spontaneous generation continued.

The epidemics of the century were among the most severe in history. They included malaria, typhus, bubonic plague, smallpox, and diphtheria. Studies and observations pointed the way for control by hygienic and epidemiologic procedures that were started near the end of the century. There was no recognition of linkage between these diseases and the world of microbes discovered by Leeuwenhoek. The century ended with the establishment of cooperation between the medical and natural sciences that set the stage for experimental medical research of the future. The principles for hygiene and sanitation were established as the universities and academies became the principal venue for scientific endeavors.

The Eighteenth Century

The eighteenth century was a time when experimentation by trial and error was encouraged, when logic aided understanding, and when systems were constructed to explain individual phenomena. A new attack on the theory of spontaneous generation was made by Spallanzani, who showed that the joining of two preexisting cells was required to give rise to new creatures in the fertilization process. Even more convincing, he isolated single cell microbes that he found to reproduce by fission and did not generate spontaneously. Factual data were presented to show that life comes only from previous life, but the belief in spontaneous generation was not to end until the nineteenth century.

Philadelphia was the cradle of American Medicine and was centered in the University of Pennsylvania where the first medical school was founded by John Morgan and the first degrees in medicine were conferred in 1768. Pennsylvania Hospital and the University itself had been founded by Benjamin Franklin who was a noted publisher, intellectual, scientist and politician. He invented bifocal glasses, demonstrated the electrical nature of lightening, and described other scientific phenomena. He founded the American Philosophical Society, dedicated to learning and to useful knowledge, that is alive and well today. In its organization and purpose, the Society was the precursor of the U.S. National Academy of Sciences that was chartered by President Lincoln. Benjamin Rush was the best-known physician of the period who was made famous by his leadership in medicine and by his descriptions of epidemics of yellow fever and dengue. His modes of treatment of diseases centered on the dubious practices of bloodletting and purgation of the intestine by calomel. Phillip Syng Physick of the University of Pennsylvania was the father of American surgery.

The pestilences and epidemics of the seventeenth century continued in the eighteenth, but with less intensity. During this time, Rosa presented the view that disease could be caused by germs, by air, or by emanations from the ground.

The greatest triumph of medicine in the eighteenth century was the creation by Edward Jenner,[9,10] in England, of the sciences of vaccines and immunology. The ancient Chinese practice of reducing the severity of smallpox by inoculation of infected pus (variolation) from a human case of the disease had been introduced into England by Lady Mary Wortly Montagu[11] in the early 1700's. Folklore had revealed that human smallpox did not occur in persons who had been infected previously with the related virus of cowpox. Jenner made the first scientific observations of the phenomenon and demonstrated that dermal inoculation of cowpox pus induced immunity against smallpox on subsequent challenge with virulent human virus. His findings of 1796 were published in 1798[9] and were the beginnings of the sciences of vaccines and immunology. The practice of vaccination spread rapidly around the world and smallpox vaccine was the only immunogen discovered until the late 1800's.

The agricultural and industrial revolutions of the period came with the application of science to the tasks of production. Breeding programs to improve plants and animals were carried out and improved tools for farming were invented. The changes in industrial production were centered on the replacement of handcraft by power-driven machinery that was boosted by the invention of the steam engine and was applied initially to the production of textiles. Political and economic changes were brought about in America by revolution against English sovereignty, and in Europe by the French revolution which overthrew the monarchy.

The triumphs of the eighteenth century can be considered a Golden Age for science and medicine that derived from primordial beginnings during the Renaissance and that paved the way for the great advances which followed. The practices of science and medicine were divided into many branches and were pursued mainly in the Universities.

Nineteenth Century to 1875: New Application of Science

The nineteenth century was marked by realism and materialism that replaced the idealism and humanitarianism of the near past. It was a century in which invention and ingenuity fueled the industrial revolution and contributed immensely to the development of new waterways and to means for travel and transport by railways and machine-powered ships. International settlement of opposing economic interests continued to be made by warfare.

In biology, Schleiden and Schwann[40] established the cellular basis of living structures. The cell nucleus was discovered, and studies of altered structure and function of abnormal cells founded the science of histopathology.

Perhaps the most significant upheaval for society for all recorded history came in mid-century with the creation by Charles Darwin of his theories of evolution and the origin of species[12] that provided a secular view to life. Darwin's thesis was that there was continuous variation in plants and animals and that their natural selection through survival of the fittest was the basis for evolution of species. These new revelations freed the mind and thinking from what had been dominated by creationist dogma. Institutionalized beliefs were replaced by knowledge supported by evidence.

A tragedy of the period was the failure to recognize the contemporary works of Gregor Mendel[13] who elucidated the mechanisms of inheritance and the mathematical explanations by which mutations could be introduced and transmitted in subsequent generations. Mendel's works lay fallow until 1900 when they were "discovered" and became the founding concepts for the science of genetics. One can speculate that knowledge of the laws of Mendelian inheritance could have contributed significantly to understanding the mechanisms for biological evolution during the time of Darwin.

Enlightened Empiricism, 1875-1930 (Table 2)

The last quarter of the nineteenth century was marked by one of the most outstanding series of breakthroughs in all of science and medicine up to that time. The principal architects for the new science[4,11] were Louis Pasteur[8,14] in France, and Robert Koch,[8,15] Emil von Behring,[8,11,16] and Paul Ehrlich[8,11,17] in Germany. This was a period in which enlightened empiricism came to dominate the field and firmly established the sciences of bacteriology, vaccines, passive immunotherapy, and chemotherapy of infectious diseases. Vaccinology and immunology had lain dormant since the time of Jenner.

At the beginning of the period, the theories of spontaneous generation needed to be put to rest and the germ theory of disease needed to be established. The former was accomplished in large measure by the chemist, Louis Pasteur,[14] and the latter by both Pasteur and Koch.[8,15] Pasteur likened fermentation to putrefaction and was able to save the French wine and beer industries through cleaning of fermentation vessels and by heating to eliminate contaminating microorganisms. Application of the principle to milk gave the process of Pasteurization. Through simple experiments, Pasteur was able to allay the theory of spontaneous generation and to reveal the role of microorganisms in disease. He discovered the fowl plague bacterium and showed that its virulence could be attenuated in the laboratory and used as a safe and protective vaccine. The same idea of attenuation or "fixation" gave rise to rabies vaccine. The transition of Pasteur from the brilliant chemist to the knowledgeable vaccinologist may have been facilitated by the contributions of Auzias-Turenne who was not cited for his pioneering concepts[41] on vaccines and immunology.

Robert Koch[8,15] was the architect of the critical technologies for propagation and selective pure culture of bacteria. He discovered the bacteria, which cause tuberculosis and cholera, and provided rigid criteria needed to establish an etiologic relationship between a pathogen and a disease. His discovery of the phenomenon of hypersensitivity ranked in importance with Metchnikov's discovery of phagocytosis.

Emil von Behring[8,11,16] discovered antibodies. He detoxified the toxins of diphtheria and tetanus and immunized animals to prepare antisera that he used for passive therapy of infections in man. Induction of passive immunity through administration of preformed antibody became the principal form of specific treatment of infectious diseases for decades that followed.

Paul Ehrlich's[8,11,17] earliest works focused on his discovery of selective binding of aniline dyes to proteins that he used in his seminal experiments on staining tissues and tuberculosis

Table 2. **Enlightened empiricism and the dawn of immunology and vaccines (1875-1930)**

Period	Changes - Discoveries
1875-1910	New sciences of microbiology, vaccinology, passive immunotherapy and chemotherapy are pioneered and established; principal architects were: **Louis Pasteur** (French). Allays theory of spontaneous generation, establishes procedures for specific microbial fermentation, establishes a link between putrefaction and disease, establishes germ theory of disease, discovers microbes, and develops live attenuated and killed vaccines. **Robert Koch** (German). Develops precise methods for selective bacterial isolation and propagation, creates postulates to relate microbe to disease, discovers hypersensitivity, discovers microbes of tuberculosis and cholera. **Emil von Behring** (German). Discovers antibodies, develops passive immunotherapy for diphtheria. **Paul Ehrlich** (German). Discovers specificity in binding of chemicals to cells and proteins (now called receptor/ligand binding); selective staining in bacteriology and histopathology; establishes prime/boosting in immunization; pioneers and develops world's first synthetic drug (Salvarsan) against syphilis.
1910-1930	New bacterial vaccines and therapeutic antisera created in early 1900's. League of Nations, post World War I, establishes principles for regulatory standardization and control of biologicals.

bacilli. Ehrlich was an expert immunologist who developed the means for precise standardization of potency of antisera. He developed the concept of attachment of chemicals to cells in the framework of a lock-and-key specificity and developed the world's first synthetic drug, that of Salvarsan against syphilis. Ehrlich's concept for the lock-and-key phenomenon is now known as ligand and receptor binding and is the basis for specificity of reactions observed in modern cell biology, immunology, and pharmaceutical drugs.

During the first three decades of the twentieth century, the science of vaccines spread rapidly throughout the world and vaccines were made in individual nations. New bacterial vaccines and antisera were developed. The end of World War I (1918) was followed by formation of the League of Nations and the creation of the Permanent Commission on Biological Standardization that would develop tests to assure potency and safety of biological preparations.[18] One of its principal achievements was the establishment of the valuable concept that each nation should form its own national regulatory control authority, but it also promoted a flawed concept that proper control of biological products could be achieved by examination of the end product alone without in-process testing.

Many new bacterial vaccines and therapeutic antisera were developed following 1900 but the development of a new viral vaccine needed to await the period of the 1930's.

Premodern Era: Transition, War and Recovery (Table 3)

Transition

The two decades between 1930 and 1950, which included World War II, was a time of transition that preceded the start of the modern era of vaccines when cell culture was introduced. The breakthroughs of the early time were Goodpasture's discovery,[19] in 1931, of growth of viruses in embryonated hen's eggs that permitted an alternative to use of tissues of infected mammals as the source for viruses and viral antigens. Theiler's safe and effective live 17D yellow fever vaccine[20] grown in minced chick embryo tissue was the first viral vaccine to follow that against rabies.

Wartime and Post War Efforts

E.R. Squibb and Sons Research Laboratories (1942-1948)

The author entered the vaccine enterprise on leaving the University of Chicago in 1944.[2] At that time, several of the pharmaceutical companies in the USA were engaged to develop and manufacture vaccines that were critical to the needed support for the military efforts of World War II.

A killed vaccine against typhus was developed at the Squibb Laboratories, using Rickettsiae grown in embryonated hen's eggs. In cooperation with Dr. Wendell Stanley,[1,2] we developed the commercial process for mass production of Sharples' centrifuge-purified influenza vaccine using virus grown in embryonated hen's eggs. In addition, my colleagues and I developed and produced a crude formalin-killed mouse brain-derived vaccine against Japanese B encephalitis[1,2] that was used in 1944 and thereafter to protect the American troops in the Pacific Offensive. During this time, the bacteriological laboratories at E.R. Squibb & Sons developed and were licensed to produce a six-valent pneumococcal polysaccharide vaccine.

Walter Reed Army Institute of Research, with Inclusion of the Early Modern Era (1948-1957)

Having joined the laboratories at Walter Reed in 1948,[2] my prime assignment was to study respiratory diseases, to develop means for laboratory analysis of influenza viruses and to discover the means to avert the next Pandemic of influenza. Early work in my laboratory included the breakthrough discovery[21-23] of a pattern of annual progressive minor changes in the antigenic form of influenza A virus, interspersed on infrequent occasions by very major changes, rendering the existent vaccine totally noneffective. This phenomenon is now called drift (minor) and shift (major)[24] and relates to the surface antigens against which host immunity is directed. Early detection[25] in our laboratories of the 1957 antigenic shift in Asian influenza A virus (from H_1N_1 to H_2N_2) permitted the production of 40 million doses of the vaccine before the pandemic reached its peak during November.

In 1952, we discovered[26,27] a new family of viruses that caused epidemic acute respiratory disease and primary atypical pneumonia in newly recruited soldiers in the military. Covert viruses of similar kind but of different serotypes were isolated from tonsils and adenoids of normal children in Dr. Huebner's laboratory[28] at the National Institutes of Health. The collective serotypes of these agents were joined in a new family and were named the Adenoviruses.[29] We developed a highly effective formalin-killed adenovirus vaccine[30] that was made using viruses grown in monkey kidney cell cultures. We evaluated the vaccine in clinical studies at Fort Dix, New Jersey, in 1956,[31] and at Fort Leonard Wood in 1957.[32] It showed near 100% efficacy. Killed adenovirus vaccine was licensed for commercial distribution in 1958 but was discontinued a few years later.

Modern Era Vaccines, Review and Highlights (Table 4)

The year 1950 may reasonably be considered the beginning of the Modern Era of Vaccinology.[1,2] It gave the early soundings for a remarkable era of breakthroughs and triumphs in vaccines, which were marred only by minor adverse experiences that might have been expected for a fledgling science, which grew from a primitive database. It brought the continued development of new and improved bacterial vaccines, propelled especially by the revival of subunit pneumococcal polysaccharide vaccine that had been licensed during World War II and that disappeared following the introduction of the antibacterial sulfonamide and penicillin drugs. The new era of viral vaccines was fueled mostly by the renaissance of cell culture for viral propagation by Enders and associates[33] and opened the door to a vast array of vaccines to

Table 3. Premodern era: Transition, war and recovery (1930-1957)

Period	Changes and Accomplishments
1930-1942	1931 Goodpasture cultivates viruses in embryonated hens' eggs. Theiler develops live attenuated 17D yellow fever vaccine using virus grown in minced chick embryo tissue cultures.
1942-1948	**Squibb Laboratories** and vaccines Formalin-killed Epidemic Typhus (Rickettsia) vaccine. Wendel Stanley's Sharples centrifuge purified influenza vaccine. Formalin-killed Japanese B encephalitis vaccine.
1949-1955	(See below. Developments of cell culture and killed poliovirus vaccine).
1948-1958	**Walter Reed Army Institute of Research** 1950 Discovery of the phenomenon now called drift and shift in influenza virus surface antigens. 1953 Codiscovery of the Adenoviruses at Walter Reed and the NIH. 1956 Killed Adenovirus vaccine developed and proved highly effective in military troops. 1957 Asian Pandemic influenza detected and virus analyzed. Killed vaccine produced and delivered. 1958 Civilian Adenovirus vaccine licensed by Parke-Davis Laboratories.

come. The modern era of vaccines was brought to realistic accomplishment by major entry of the pharmaceutical industry into the world of basic and applied research on vaccines.

The first viral vaccine of the Modern Era was that of killed poliovaccine[34] (see below). Although licensed in 1955, it was faced with immediate problems that were not fully resolved until the 1960's.

The New Department of Virus and Cell Biology Research in the Merck Research Laboratories (Table 4)

Origin

In the late 1950's, Dr. Vannevar Bush,[2,35,36] who had been the U.S. wartime head of the Civilian Office of Scientific Research and Development (OSRD) and the President of the Carnegie Institute, became the Chairman of Merck & Co. following the death of George W. Merck. Bush, who was an inventive genius in science and engineering, presented a view that, one day, viruses would be of major significance to science and medicine. He requested that Dr. Max Tishler, then President of the Merck Research Laboratories, establish a major new laboratory of virology that would take its place among the leading laboratories of the world. It did become the focal point for the pioneering of the great majority of new vaccines of the modern era. I was persuaded to leave Walter Reed and joined Merck, in late 1957,[2] as the Director of the new Department of Virus and Cell Biology Research and assumed duties on December 31.

Important to the new venture was the Merck corporate policy supporting the pursuit of basic research as the basis for its commercial initiatives to bring products of value to people.

Table 4. Review and highlights of modern era vaccines

1950	Beginning of modern era of vaccines.
1957	Merck & Co. establishes new initiative in a new department of Virus and Cell Biology. George Merck, Vannevar Bush, Max Tishler.

Corporate policy promotes basic research pursuits. Merck holds concept that "medicine is for the patient, not for profit."

Unique organizational structure provides central authority with responsibility and accountability.
    Collective inclusion, direction, and responsibility for all the sciences spanning the distance from clinic to clinic, plus development and product licensure.

Far reaching and productive relations with Dr. Joseph Stokes, Jr. of the University of Pennsylvania, and Dr. Victor Villarejos of Louisiana State University Center in San Jose, Costa Rica.

Programs undertaken to develop bacterial, viral and plasma-derived and recombinant hepatitis B vaccines.

Mr. George Merck, long-time Chairman, had held that "Medicine is for the patient, not for the profits." The biological enterprise was funded and initiated at a time when pertinent knowledge was limited, when applicable precedents were few in number, when the ability to acquire significant patent protection was limited, and when there was little interest on the part of knowledgeable academic scientists to enter industry. The initiative was driven by abiding optimism and faith for the future.

Efficiency and effectiveness of the new freestanding departmental operation[2] was facilitated greatly by a corporate mandate for strong central authority in return for my assumption of total responsibility and accountability. Through such organization, that was novel to industry, decisions could be made rapidly in defining problems and in providing likely solutions, with deployment of critical numbers of scientists of diverse disciplines needed for problem solving. Staff morale was high, driven by participation in successful research at the cutting edge of science. The departmental venture embraced all the pertinent basic sciences plus development, engineering, data analysis, governmental liaison, pursuit of licensure, and the overseeing of production. Very important, also, was the responsibility for and direction of the planned collaborative clinical-laboratory studies that ranged from engagement in etiologic discovery, field epidemiology, and clinical evaluation of vaccines for immunogenicity, reactogenicity, and protective efficacy. Far-reaching and productive relationships[2] were established with Dr. Joseph Stokes, Jr., then physician-in-chief of Children's Hospital of Philadelphia and Chief of Pediatrics of the Medical School of the University of Pennsylvania. A second partnering in clinical and epidemiological investigations was with Dr. Victor Villarejos, then Director of the Louisiana State University International Center for Medical Research and Training (LSU-ICMRT) in San Jose, Costa Rica.

The year 1958 was a time of beginning in our laboratories in which concepts were based largely on perceived possibilities rather than a previously established knowledge base. Further discussion in this section will be focused on three lines of endeavor: Bacterial vaccines, killed and live viral vaccines, and the breakthrough first vaccine (hepatitis B) made by recombinant expression technology.

In 1958, knowledge, concepts, and precedents for new vaccine research and development were restricted (Table 5) and new ground needed to be turned for the vaccines that were to come. There were common elements that, collectively, included etiologic discovery, determination of the live, killed, or subunit approach, development of means for attenuation or inactivation, formulation, preclinical assessment, proof of safety and efficacy plus creation and compliance with regulatory procedures.

Table 5. Common elements in development of vaccines

- Etiologic discovery and definition.
- Propagation of organisms in cell culture systems acceptable for human use.
- Selective choice of live attenuated, whole, or subunit killed approach.
- Appropriate attenuation or inactivation to achieve acceptable reactogenicity and immunogenicity.
- Formulation to achieve adjuvantation, appropriate immune response, stability and long-term immunity.
- Preclinical assessment for safety, reactogenicity, immunogenicity.
- Proof of protective efficacy and retained immunity in controlled clinical trials.
- Development of surrogate markers governing acceptability for product release.
- Engineered scale-up, licensure, manufacturing, and distribution.

Bacterial Vaccines (Table 6)

The first engagement of our laboratories with bacterial vaccines was in response to a U.S. military request that we develop meningococcal vaccines for use in the Armed Forces based on the subunit capsular polysaccharide vaccination studies of Artenstein and Gotschlich.[37] Our task was to evolve satisfactory fermentation of bacteria, to extract, to purify and to characterize the capsular polysaccharides, to formulate into single and multivalent combinations,[38] and to prove clinical safety, immunogenicity, and efficacy. The controlled studies to establish protective efficacy were carried out by Mäkela et al in Finland.[39] Vaccines were distributed in monovalent and combined formulations following licensures in 1974, 1975, and 1982. The ultimate vaccine[38] was quadrivalent and was comprised of A, C, WI-135, and Y serogroup polysaccharides. Lack of predictability of epidemics and hence of market needs led eventually to discontinuance of commercial distribution of the product by the company.

The pneumococcal vaccines, as noted above, represented an expansion of the six-valent vaccine that had been licensed to Squibb in the 1940's (see above). The further scientific initiative was kept alive largely by Dr. Robert Austrian of the University of Pennsylvania who established continued need for the vaccine, defined the worldwide prevalence of the various capsular serotypes, determined which deserved inclusion in a polyvalent vaccine, and did clinical evaluations as well.[42,43] We pioneered and pursued the commercial development and clinical studies of polytypic pneumococcal vaccines and the 14- and 23-valent vaccines that were the first to be licensed in 1977 and 1983, respectively.[44-46]

The capsular polysaccharide vaccine against disease caused by *Haemophilus influenzae* proved to be of very limited usefulness in protecting infants who are vulnerable to serious disease in early life at a time before the immune response could be raised. All this changed with the seminal observation by a now unidentified scientist that the coupling[2] of polysaccharide with protein elicited T cell immune responses and immunized babies in early infancy.[47,48] Several pharmaceutical companies, including our own, pursued the observation and the conjugate vaccine is now being distributed widely with excellent results.

Live Viral Vaccines

Pediatric

Live attenuated pediatric vaccines against measles, mumps, rubella, and varicella were conceived by us[2] in 1957 as only theoretical future possibilities. They were to become the realities of the 1960's and 1970's. Collectively (Table 7), they shared common hurdles that included the need to characterize the individual vaccine viruses and to achieve an optimal attenuation of each whereby protection could be established with acceptable reactogenicity. Further, there were no meaningful animal models and no markers of attenuation. This created the necessity

Table 6. Modern era—Bacterial capsular polysaccharide vaccines

Meningococcus
1963	Sulfonamide resistance.
1969-1972	Vaccine feasibility and protective efficacy shown at Walter Reed Laboratories.
1974-1982	Merck develops and licenses Groups A, C, A-C, W135 and Y and quadrivalent vaccine.
	Sporadic occurrence of infection restricts commercial interest.

Pneumococcus
1945	Type-specific antibody is marker of protection.
1946	Hexavalent vaccine licensed (Squibb).
1965-1968	Austrian—effective chemotherapy may not prevent death. Vaccine needed.
1977	14-valent licensed (Merck).
1984	23-valent licensed (Merck).

Hemophilus influenzae
1985	Polysaccharide vaccine licensed for older children.
	Ineffective in children <2 years.
	Mid 1970's, conjugation with protein gives vaccine that protects newborn animals.
1987-1990	Licensure of diverse Hemophilus influenzae vaccines.

Table 7. Modern era—Hurdles for live virus pediatric vaccines

Target:
  Live attenuated vaccines: measles, mumps, rubella, varicella and combinations of them.
Common hurdles:
  Discover, grow virus in cell cultures acceptable for human use.
  No animal models and no markers of attenuation.
  All tests in children with decisions made by judgment.
  Preparation of numerous passage-level vaccines of commercial quality for clinical testing.
  Testing by judgment starting with likely most attenuated and working backward to finding acceptable balance between reactogeniaty and immunogenicity.
  Retention of acceptance of research by scientific and regulatory committees (overcoming objections, e.g., varicella vaccine).
  Elimination and avoidance of viral contaminants.
  Huge and long-term vaccine preparation and testing time.
  Protective efficacy in two placebo-controlled studies for each vaccine.
  Safety validation in 10-20,000 susceptibles.
  Safety for susceptible contacts.
  Retained protection for long term.
  Vaccine stable on storage and distribution.
  Combined vaccines having no increase in reactogenicity, and with formulation adjustment to prevent interference between live viruses.

to define all attributes of the candidate virus strains in tests in children, which involved the preparation and assay of vaccines of numerous virus attenuation levels and the making of decisions that were guided by judgment alone aided by review of the previous clinical and laboratory data as accumulated. Prior to licensure, each vaccine required validation for safety in

Table 8. Modern era—Individual problems and solutions for pediatric live virus vaccines

Measles—chick embryo cell culture.
    Reduce reactions:
        Coadministration of immune globulin initially.
        Further attenuation (no globulin needed).
        Eliminate avian leukemia virus from cultures – development of experimental
            leukemia-free flocks.
        High-level potency and safety.

Mumps—chick embryo cell culture.
    Jeryl Lynn strain lacks neurovirulence.
    High-level potency with nonreactogenicity.

Rubella—discovery, propagation in duck cells.
    Rapid and reliable attenuation.
    Nontransmission to susceptible adult contacts.

Bivalent and trivalent formulations.
    Acceptable potency and reactogenicity.
    Very successful clinical acceptance.
    Flagship for pediatric immunogens.

Varicella—all aspects were worked out by 1981 with KMcC strain.
    Virus refused to achieve acceptable balance of retained reactogenicity vs.
        immunogenicity on application of attenuation procedures.
    OKA strain substituted.

10-20,000 susceptible persons and protective efficacy needed to be established in at least two placebo-controlled clinical trials. Each vaccine virus needed to be propagated in acceptable cell cultures, to establish long-term immunity and to comply with all the general and individual tests for safety.

    Individually (Table 8), solutions to notable problems of each of the vaccines needed to be achieved. Measles vaccine,[49-51] derived from the Enders Edmonston virus,[52] was propagated in chick embryo cell culture and was laden with contaminating viruses of avian leukosis. The solution to the problem lay in the development of leukemia-free chicken flocks.[53] The excess clinical virulence of the Edmonston measles virus for babies was solved initially by coadministration of measles immune globulin[2] and later by development of the more attenuated measles virus, Moraten.[54] The Jeryl Lynn strain of mumps virus was isolated and attenuated in our laboratories and easily achieved high-level potency and efficacy without untoward clinical reactions, especially that of residual neurotropism.[55-58] Rubella vaccine development awaited discovery of the virus by Weller[59] and by Parkman[60] and the breakthrough finding by us of propagation and reliable attenuation of the virus in duck embryo cell culture.[61] An important initial problem related to achieving assurance for nontransmission of vaccine virus from a vaccinee to susceptible pregnant females who were in the first trimester prior to the time when fetal safety of attenuated virus was not known. The vaccine was licensed and showed exemplary performance.[62-66] Varicella vaccine was pioneered in our laboratories in all aspects including propagation, preparation, general safety, and protective efficacy.[67,68] There was initial inability of the virus to reach an acceptable potency:toxicity ratio but was resolved later by substituting the Japanese OKA[69] virus strain in the vaccine. Combinations of the measles, mumps, and rubella vaccines in bivalent and trivalent combinations were pioneered in our

Table 9. Modern era—Marek's chicken cancer vaccine

Marek's disease, a neural or visceral lymphomatosis of chickens (range paralysis) caused by a herpes virus.

Economic loss to poultry industry is large, through lowered productivity (eggs) and condemnations at slaughter (meat).

1970   Burmester discovers turkey herpesvirus antigenically related to chicken Marek's.
        Burmester virus causes no disease in chickens but protects against Marek's.
        After long and complicated research to establish safety and efficacy for chickens
            and safety for man in food consumption, two highly effective vaccines
            were developed in our laboratories.

1971   Frozen infected cell vaccine licensed.

1975   Purified free virus vaccine, dried, licensed.

This was the world's first licensed vaccine against any cancer. It revolutionized the economics of the poultry industry.

laboratories[70-72] and facilitated vaccine administration. The triple combination called MMR[70-72] has been the flagship for disease prevention in pediatric populations for three decades. It is being readied for addition of varicella vaccine in a quadrivalent formulation. Measles vaccine now saves one million lives per year and has saved many millions of lives as the world now looks forward to eventual measles elimination and eradication of the virus.[73] The economic value of application of MMR in the USA, as revealed in estimates[74] by the Batelle organization for the Centers for Disease Control, amounted to a 5.1 billion-dollar savings in 1994 alone.

Avian Cancer

Mareks' disease (Table 9) is a neural and visceral lymphomatosis of chickens that is of extreme economic importance to the poultry industry because of reduced productivity of laying hens, and rejection of the carcass at slaughter because of the visible presence of tumors. Mareks' disease is caused by a virus of the herpesvirus family and Burmester and colleagues[75] isolated a related virus from turkeys that did not cause disease in chickens but immunized chickens against Mareks' Disease. Our laboratories pioneered and developed both frozen infected cell vaccine and purified live virus vaccine that were licensed in 1971 and 1975, respectively.[76] The Mareks' vaccine represents the world's first licensed vaccine against a viral cancer. It revolutionized the economics of the poultry industry.

Killed Viral Vaccines

Poliomyelitis (Table 10)

The first of the Modern Era vaccines was that against poliomyelitis which came as a fulfillment of initiatives of the National Foundation for Infantile Paralysis[77] which gathered financial support from the public through its March of Dimes campaign which honored President Franklin Delano Roosevelt. The first of the National Foundation's efforts was toward support of the previously established Warm Springs Poliomyelitis Foundation—a poliomyelitis rehabilitation center. The program of the National Foundation led gradually to initiation and support of studies of the etiology, pathogenesis, and epidemiology of poliomyelitis. The findings disclosed the feasibility for poliomyelitis prevention by a vaccine and opened the way to a formal and focused vaccine initiative. Technical planning and implementation of efforts were

Table 10. Modern era: Viral vaccines against poliomyelitis

National Foundation for Infantile Paralysis supports establishment of basics in virus and pathogenesis of disease. Antibody is marker for protection.

Killed Vaccine:
Macacus renal cell culture grown, and formaldehyde-inactivated with first order kinetics. Licensure in 1955 by several manufacturers.

Problems:
Incomplete poliovirus inactivation caused paralysis in children. Process changes made and vaccine rereleased.
Variability and less than acceptable potency.
Indigenous wild viruses in monkey kidneys – requiring assurance of inactivation of all agents present in cultures.

1960: Merck's Purivax licensed. Purified standardized potency vaccine. Discontinued for commercial reasons.
1960: Hitherto undetectable monkey polyomavirus, SV_{40}, discovered. Resists total inactivation by formaldehyde.
1962: SV_{40} virus eliminated

Live Vaccine: (nonneurotropic, attenuated).
Pioneered initially by Koprowski, followed by Sabin.
Licensure in 1960.

Problems:
Retained or reversible neurovirulence.
SV_{40} contamination, initially present, was removed.
Live vaccine became paradigm for poliomyelitis prevention and worldwide eradication of virus.

carried out by committees comprised of groups of the most outstanding and capable clinical and laboratory scientists of the time.

The technical effort of the programs were successful and this led to licensure, in 1955, of the Salk killed poliomyelitis vaccine by several pharmaceutical companies. What had been hailed as a scientific triumph rapidly became a disaster[77,78] when the vaccine that was manufactured by the Cutter Laboratories was released for general use and caused poliomyelitis in at least 260 persons, of whom 94 were primary vaccinees, 126 were family contacts, and 40 occurred in community contacts.[77,78] The vaccine was prepared using formaldehyde-treated virus that had been grown in cell cultures of *Macacus* monkey kidneys. There was incomplete inactivation of the virus in some lots of commercially produced vaccine that caused poliomyelitis in children and this led to removal of the vaccine from the market. Extensive investigations of the technical process were carried out. Flaws were detected and corrected and there was return of the vaccine to the market.

Improvements and Problems in Poliovaccines

Entry of the new department of Virus & Cell Biology into poliovaccine research came with its endeavor to correct the variable potency of the commercial vaccine (Table 10). A vaccine comprised of highly purified virus in which each viral component (serotype) was precisely standardized in viral antigen content was brought to market by us under the label of Purivax in 1960.[79,80] It was ultimately discontinued for commercial reasons.

A major problem in poliovaccine production during its first five years following licensure was the presence of a large number of different indigenous monkey viruses[81] that were commonly present in the monkey kidneys and were amplified during cell cultivation. Presence of such contaminating viruses was permitted so long as they were inactivated by formaldehyde in the vaccine production process. Because of our interest in live viral vaccines, we were led to seek another source for monkey kidneys free of contaminants, and the African Green (*Cercopithecus aethiops*) was substituted for the *Macacus* monkey on the advice sought from Dr. William Mann who was then the Director of the National Zoological Park in Washington, DC. Renal cells from the *Cercopithecus* species were found to be highly permissive to viral replication with very evident cytopathogenic changes that allowed us to detect the presence of hitherto undetectable viruses covertly present in *Macacus* cells. Of special importance, use of these cells permitted our discovery of SV_{40} virus,[82,83] a *Macacus* monkey polyomavirus that we later found to be oncogenic[84] for baby hamsters.

There was major disruption in the killed poliovaccine manufacture when a miniscule portion of SV_{40} virus was found to have resisted inactivation. The problem was rapidly solved by the substitution of *Cercopithecus* for *Macacus* monkey kidneys that were nearly free of any wild virus, including SV_{40}. SV_{40} became of great importance in studies of cancer viruses and molecular biology[85] and of vaccines and immune responses against cancer.[86]

Live Poliovirus Vaccine (Table 10), Summarized Here Together with Killed Poliovaccine

The evolution of concepts and the search for a nonneurovirulent live poliovirus vaccine were pioneered by Dr. Hilary Koprowski, and also the first clinical tests of a live virus vaccine were made by him.[87-89] The live attenuated poliovirus vaccine, that was developed by Albert Sabin,[90] was licensed in 1960, having been made possible by the huge clinical trials of the vaccine that were conducted by Prof. Mikhail Chumakov in the Soviet Union. Like the killed vaccine, research on the yet unlicensed Sabin product was suspended until it was freed of SV_{40} virus[82,83] by clean up of seed stocks and the use of *Cercopithecus* monkey cells.

The live virus vaccine,[90] to this day, retains low level neurovirulence and causes poliomyelitis with very low frequency in recipients and in susceptible contacts to recipients of Sabin vaccine. Because of its effectiveness and convenience of administration by oral feeding, Sabin vaccine has become the gold standard for worldwide poliovirus immunization and ultimate eradication of the virus itself. The killed poliovirus vaccine is now being substituted by the live vaccine in some populations in which the numbers of cases of poliomyelitis associated with vaccine exceed the number of cases caused by wild virus.

Hepatitis Viral Vaccines

Hepatitis A Vaccine (Table 11)

We initiated large-scale laboratory and field studies in the early 1960's to discover the viruses related etiologically to type A (epidemic) and type B (endemic, serum) hepatitis. In 1973, we published on our earlier isolation of an enterovirus-like agent, CR326,[91,92] recovered in marmoset monkeys from the feces of a Costa Rican patient by the methods of Deinhardt. It was found recently[93] that the GB virus that Deinhardt isolated and thought to be the cause for hepatitis A was not. Instead, it was a Flavivirus that is related to hepatitis G and represents the first isolation of a Flavivirus hepatitis virus, predating that of both hepatitis C and hepatitis G.

Studies in our laboratories[94-96] using hepatitis A virus purified from infected marmoset liver provided a supply of virus for use in developing assays, opened the door to tests for viral antigen, for antibody, and for diagnosis, and allowed us to write a whole new chapter on the etiology, biology, clinical picture and epidemiology of hepatitis A. In addition, we prepared a prototype formalin-killed hepatitis A vaccine,[97] using virus purified from marmoset liver that was proved safe and highly protective in challenge studies carried out in marmosets. Application to man, however, needed to await a suitable means to cultivate the virus in vitro. The

Table 11. Modern era: Hepatitis A virus vaccine

Early 1960's Program initiated at Merck to discover viruses causing viral hepatitis.
  1973 Recover CR326 Hepatitis A virus in marmoset monkeys.
  1978 Prototype, highly effective, formalin-killed, infected marmoset liver-derived vaccine prepared and proved safe and effective in marmoset and chimpanzee studies.
  1979 Breakthrough propagation of virus in cell culture opens door to vaccine for man.
  1991 Cell culture propagated, formalin-killed vaccine for man prepared using procedures of prototype marmoset liver vaccine.
  1992 Safety and efficacy for man are proved.
  1994 Licensure.

major breakthrough to resolve that problem was the successful cultivation of the virus in cell culture in our laboratories reported in 1979.[98] A highly safe and effective cell culture-derived vaccine against hepatitis A for use in people was prepared by Lewis et al[99] in the Merck Research Laboratories using the virus we had attenuated[100] and the procedure developed for preparing the prototype killed vaccine in which we had used virus purified from infected marmoset liver. Protective efficacy was established[101] in controlled field studies in 1992 and licensure was granted in 1994. The vaccine is now used widely throughout the world to prevent the disease.

Plasma-Derived Subunit Human Hepatitis B Vaccine[102-108] (Table 12)

The availability of a source for antigen of an important virus quickly elicits the interest of the research vaccinologist as a new vaccine possibility. The discovery by Blumberg in 1965[109] of the surface antigen of hepatitis B virus in the blood of human carriers provided such a moment and raised the question of whether viral antigen purified from human blood might be used to make a safe and effective vaccine. We began vaccine work in our laboratories in 1968 at a time when the database of knowledge of hepatitis B virus was near zero. Antigen purification, with yields in practical amounts, was rapidly accomplished employing a three-step sequential purification/inactivation process involving digestion with pepsin, followed by urea denaturation/renaturation, and treatment with formaldehyde. Work on the vaccine was greatly hampered by failure of growth of the virus in cell culture. Surrogate tests for inactivation of possible carryover viral contaminants from human blood were carried out in which examples of viruses representing a wide diversity of virus families were performed. It was shown that all the surrogate viruses were destroyed readily by each treatment procedure individually and gave assurance that the sequential application of the three steps in the process would suffice to destroy all possible life forms. The agent of scrapie was shown to be destroyed by urea treatment, and while only anticipatory at the time, it did relieve the fears of possible transmission of Creutzfeld-Jakob Disease (C-JD) as was raised with the British outbreak of Mad Cow Disease[110] (variant C-JD agent) and when the presence of prion proteins (conformationally misfolded)[111] in human blood was shown. Recent reports suggest the PrP^{sc} protein of prion disease may be present in splenic dendritic cells of the immune system.[112]

The safety and high-level protective efficacy of the alum adjuvanted vaccine were shown in challenge studies in chimpanzees and in placebo-controlled field studies[108] in high-risk populations. The product was licensed in 1981, just 13 years following the start of studies in our laboratories. This vaccine was a major breakthrough in vaccine science and was both the world's first viral subunit vaccine and the first licensed viral vaccine to prevent cancer in human beings. The vaccine is now used routinely throughout the world to immunize children and susceptible adults and to break the maternal carrier/infant transmission cycle.

Table 12. Modern era: Hepatitis B virus vaccines

Early 1960's	Programs initiated at Merck to discover viruses causing hepatitis.

Human carrier plasma-derived vaccine.

1965	Blumberg and Prince discover Hepatitis B virus surface antigen.
1968	Probes initiated at Merck for purification, inactivation, efficacy and safety of vaccine having possible blood-derived contaminants including prion (Creutzfeldt-Jacobs) disease-agents.
1975	Safety for man is proved.
1980	Efficacy proved in controlled clinical studies.
1981	Vaccine licensed after 13 years of intensive research.

Recombinant Hepatitis B vaccine (antigen expressed in yeast).

1975	Collaborative studies initiated between Drs. Rutter and Hall and our laboratories to develop a recombinant expression system for antigen.
1982	Expression system developed in yeast. Recombinant antigen substituted for plasma-derived antigen in the licensed vaccine.
1986	Recombinant Hepatitis B vaccine is licensed.

Hepatitis B vaccines represent:
World's first subunit virus vaccine.
World's first vaccine against human cancer.
World's first recombinant-expressed vaccine.

Recombinant-Expressed Subunit Human Hepatitis B Vaccine (Table 12)

It was evident in 1975 that the available supply of acceptable carrier plasma would not suffice to supply the world requirement for hepatitis B vaccine. The new science of recombinant expression of proteins had recently been born and a collaborative investigation was established between Drs. Rutter and Hall of the Universities of California and Washington, respectively, and our laboratory at West Point, Pennsylvania, to develop an expression system for hepatitis B surface antigen. Effective recombinant expression of surface antigen was achieved in baker's yeast.[113] The process was optimized and the purified yeast recombinant antigen replaced the plasma-derived antigen in the existing product.[114,115] The plasma-derived and yeast-derived vaccines performed the same and yeast recombinant vaccine was licensed in 1986, just 11 years after the recombinant studies were initiated. This was the world's first recombinant-expressed vaccine and provided all the benefits of the plasma-derived product.

Contemporary Era (in Retrospect, 1990-2000): Vaccine Quiesence During a Time of Explosion in New Scientific Knowledge

By 1984, all of the Modern Era vaccines had been pioneered and developed except for final scale-up and licensure of three products (hepatitis A, varicella, *H. influenzae*) by our laboratories. During the Contemporary Period, only four new vaccines were licensed. These were the reassortant Rotavirus (Wyeth-Lederle, 1997) and the recombinant Lyme disease (SmithKline Beecham, 1996) vaccines. Both of these vaccines were based on technologies developed and established during the Modern Era. In the most recent time, conjugate pneumococcal (Wyeth Lederle 2000) and meningococcal (Chiron 2000) vaccines have been licensed, using the *H. influenzae* conjugate vaccine technology of the Modern Era.

The Future Will Be Driven by New Science and Technology (Table 13)

Public Funding of Basic Research

Even though there has been relative quiescence in pioneering new vaccine technologies and developments since 1984, a new Golden Era was dawning in the creation of new and sophisticated basic knowledge in the sciences of immunology, microbiology, and vaccinology. These represent a pay-off to the public for the many years of publicly funded science that were driven by the Bush plan for publicly supported basic research,[36] by the War on Cancer (molecular biology), by the AIDS initiatives (immunology) and, most recently, by increased appropriations of public monies by the Congress to the National Institutes of Health for support of programs in broad basic research in biology and the medical sciences. The players in the vaccine initiatives of the past decade were academia that was supported by NIH grants for basic research, the biotechnology companies that were funded mainly by venture capital, the large pharmaceutical companies that were funded with private resources, and, in addition, the internal and external activities of the NIH itself.

The sciences of greatest significance in driving the future of vaccines are those of immunology, vaccinology, and molecular genetics, as discussed below. The literature is replete with reviews that may be easily accessed.

Immunology

Immunology,[116,117] as it relates to prevention and control of infectious disease agents, deals directly with host immune responses. Advances in understanding immunologic functions have been a dynamic process, greatly accelerated in recent years, converting them from primitive understandings to a productive rational science. One may draw special attention to progress in

1. definition, characterization and function of all the classes and subsets of the cells of the innate and adaptive immune systems, including the scavenger, processing, and antigen-presenting cells of bone marrow derivation;
2. the processes for immune activation, suppression and depletion through the dynamics of recognition, specific binding, costimulation, and signal transduction;
3. regulation by cytokines and chemokines, apoptosis, and anergy;
4. antigen processing and presentation, with bias toward humoral or cellular immune responses introduced by the class of MHC presentation to T helper cells, and by allelic restriction through receptor binding affinities;
5. definition and direction of humoral vs. cell-mediated effector responses and the need for achieving optimal and appropriate balance between them; and
6. increased understanding of participation of immune function and dysfunction in allergic and in autoimmune and neoplastic diseases.

The long-standing black box of mystery in immune function has opened widely, and immunology has achieved an importance second to none in the medical sciences. It has become increasingly clear that efficient pursuit of vaccinology relies heavily on understanding immunology and suggests that vaccinology, in a sense, may be considered to be a branch of applied immunology.

Vaccinology and Microbiology

The science that deals with the parasite and its engagement with the host is called pathogenesis. The creation of substances suitable for immunologic prevention or mitigation of infectious diseases lies within the science of vaccinology.

Vaccinology,[116-118] as applied to cellular microbes and to viruses, has been aided by remarkable advances in new or improved technologies that may be applied successfully in the foreseeable future. Important themes have focused on a judicious identification and assembly of substances needed to evoke effective immune responses and on the definition of appropriate means

Table 13. Science and technologies that will drive future vaccinology

Public funding drove the basic research of the past.
    Bush plan: "Science: The Endless Frontier."
    War on Cancer (molecular biology).
    War against AIDS (Immunology).
Immunology
    Complex of immune responses of the host.
Microbiology
    Bacterial, viral, and parasitic invaders.
    Pathogenesis – engagement of pathogen with host.
Vaccinology
    Designing, creating, assaying, and providing vaccines.
Genetics, genomics, proteomics, and informatics
    Chemical definition of genes and proteins encoded, together with rapid throughput and
        computer analysis, may guide the composition of future vaccines.

for delivery and presentation of antigenic determinants that will be suitable to proper engagement of a highly complex and capricious immune system. Worthy of notation are:

1. focus on antigen mapping and definition of what antigens—whole, subunit or epitope, linear or folded conformational—are to be included in a vaccine which will have sufficient breadth of antigenic specificity to provide practical coverage of the diversity of microbial organisms and capability for effective engagement in overcoming allelic restriction on presentation in the individual host;

2. determination of optimal dosage and regimen for prime/boosting to evoke protective levels of both humoral and cellular immune responses, with memory and retention for an adequate time period. Antibody is recognized as important for preventing infection or in reducing viral load, and cell-mediated immunity is seminal to suppression or elimination of infection in cells by soluble factors (e.g., cytokines) or by deletion of infected cells themselves;

3. adjuvantation (enhancement) of the immune response through formulation, inclusion of immune modulators or cytokines, sustained or intermittent antigen release from repositories or encapsulation, or by pulsing of antigens onto antigen-presenting cells;

4. formulation to achieve appropriate and adequate responses against antigens given parentally, on mucosal or on transdermal application, or when fed orally;

5. development of recombinant live attenuated microbial and DNA plasmid vectors that will express foreign antigens in host somatic cells, especially precursor or mature dendritic cells; and

6. antigen production by recombinant expression in plants for use in conventional vaccines or for possible oral feeding.

Genetics

The new breakthroughs of genetics, genomics, proteomics, automated throughput assays, and computer-linked bio-information processes portend great applications to the sciences of immunology and vaccinology. Genetic sequencing of microbes, now completed for at least 30 organisms,[118,119] permits predictions of what gene segments may evoke humoral or cellular immune responses, and what shared antigens of immunopreventive significance may be present in a diversity of related microbes, such as described recently for the five subgroups of meningococci.[120,121] The continued creation of knockout and transgenic mice, guided by application of genetic information, may provide useful animal models for vaccine studies.

Future Utilization of Vaccines in a World of Change (Table 14)

Reality of One World in Public Health

The multitude of the world's transmissible infectious diseases may now afflict the peoples of any nation. Emerging new infections that may travel with the speed of aircraft, escaping geographic and political boundaries, give reason to caution that the systems for control in an effective past are becoming increasingly archaic and unable to cope. AIDS, tuberculosis, and malaria are classic examples of diseases out of control, which may include viruses such as those of epidemic and pandemic influenza, and the agents of measles and poliomyelitis when reintroduced into areas from which they had been eliminated. Campaigns to control the sexually transmitted and blood transmissible (injection) diseases through public education, especially for AIDS, have failed miserably to solve the problems in the world at large. Therapeutic drugs, hampered by development of resistance, excessive toxicity, cost, and practicality in administration have provided no real solutions to most of the world problems in infectious disease and offer no substantive evidence that they can or will prevail in the future. All this has dramatically enforced the reality that the only credible hope for resolution lies with immunoprophylaxis by vaccines that exist or remain to be developed. They are unique in providing the most feasible and cost-effective solutions to the problems of infectious diseases.

Politicoeconomic Imperatives, Developments, and Opportunities in Disease Control by Vaccines

Losses to mankind from infectious disease may be social, political, and economic. Social and political obligations and reality, long understood and practiced by the developed nations, are now being voiced in the World Health Assembly of the United Nations as declarations of basic human rights that include access to clean water, nutrition, health care, and prevention of disease by vaccination. Responses[122] to such proclamations fall within the mandated responsibilities of the World Bank (WB), United Nations Infants and Children's Emergency Fund (UNICEF), United Nations Development Program (UNDP), and World Health Organization (WHO) aided by governments and private institutions such as the Rockefeller Foundation. Importantly, it has not been lost to the world community that loss of people to disease or premature death, brings economic upsets, destabilization, and lack of balance reflected within the entire world community.

It seems worthy of note that a new pro-activity in the world's vaccination enterprise[122,123] began with the 1990 Declaration of New York,[122] a document describing desirable attributes and feasibilities for new and improved vaccines delineated by a group of leading technical specialists under the auspices of UNICEF. The Declaration was adopted at the World Summit for Children attended by chief executives of the world's nations. Application was pursued in a new organization, called the Children's Vaccine Initiative (CVI), that was formed under the auspices of WHO, UNICEF, WB, UNDP and the Rockefeller Foundation. Working under the general umbrella of WHO, it assumed responsibility for the WHO Expanded Program for Immunization (EPI) and endeavored to implement successful development of new vaccines of the kinds prescribed in the Declaration of New York. After nearly a decade of productive existence, the CVI was dissolved and replaced in 1999 by the Global Alliance for Vaccines and Immunization (GAVI),[123] under the authority of WHO, UNICEF, and WB that now administers the EPI and, in addition, pursues matters relative to finances, advisory policies, country coordination, and research and development. GAVI is now functional and active and invites voluntary participation of the world community to assist in its endeavors.

Consistent with the objectives of GAVI, the Bill and Melinda Gates Foundation appropriated monies and established a Children's Vaccine Program (CVP) implemented by the Program for Appropriate Technology in Health (PATH in Seattle, Washington) for purpose of introducing new vaccines into developing countries through disease assessment, clinical trials,

Table 14. Creation, delivery and utilization of vaccines in a changing world

Infectious diseases have no boundaries and are a collective problem of the whole world.
Vaccines are the tools for cheap and effective prevention.
Principal players in politics and economics of preventative disease control in underdeveloped
nations.
- World Health Assembly of the United Nations.
- World Bank.
- United Nations Development Program.
- United Nations Infants and Children's Emergency Fund.
- World Health Organization.
- Governments of developed nations.
- Private Institutions such as the Rockefeller Foundation and the Bill and Melinda
  Gates Foundation.
- Global Alliance for Vaccines and Immunization. A Coalition of United
  Nations agencies, together with national governments and worldwide
  voluntary participation. Proactive in finance, promotion, coordination and
  delivery of vaccines to the world population.
Incentives to Industry for vaccine R&D and production.
Push-Research & Development grants and loans.
Pull-Guaranteeing of market and profitability.
Special focus, based on 1990 Declaration of New York, on vaccines requiring fewest doses,
low cost, long-term immunity, safety, efficacy, and simplicity in delivery.

feasibility studies, identification of roadblocks, and advocacy fattened by special appropriations to support of research on a malaria and other vaccines. Of pivotal importance, the Gates Foundation recently established a very large Global Vaccine Fund within UNICEF for purchase of nontraditional vaccines by small nations having a Gross Domestic Product of less than $1,000 per person.

Incentives to Industry

In the overall, it is the intent of GAVI to limit its purchases of vaccines for poorly developed nations by a sliding formula that requires partial monetary contribution by governments, in accord with the ability of the national state to pay. The world's large vaccine companies sometimes make existing vaccines available for purchase at less than customary market price, though the implications of the practice of two-tiered pricing have yet to be settled. It is not economically feasible for private industry to allocate expenditure of huge amounts of money and resources for new vaccines in which there would not be a domestic market that would assure recapture of money spent in high-risk ventures. This might be changed, however, in the long term as developing nations achieve developed status with added capability for purchase of its vaccines. In the meantime, push and pull mechanisms are being studied by GAVI and UN agencies to encourage engagement of industry in vaccines for use in third world nations. *Push* mechanisms relate to direct support for research and development grants and loans while *pull* relates to guaranteeing an adequate market and profit for the finished product. As an incentive, the latter would seem the more appropriate.

Objectives of the Declaration of New York

The players in the vaccine enterprise in the USA are government (both the Congress and the Executive), academia, and the pharmaceutical industry, including biotechnology companies. The charge to the coalition is to develop useful knowledge that can and will be made available for disease prevention using new and improved vaccines. The mandate for future vaccines for use in all peoples are consistent with the intents expressed in the Declaration of New York.[122] Points to be heeded are:

1. simplicity in delivery as by administration onto mucosal surfaces, transdermal application, or oral feeding;
2. fewest number of doses implying high potency or combinations of different vaccines given in single shot preparations;
3. safe and nontoxic vaccines implying purity, detoxification, or use of highly attenuated vectors;
4. immunity of long duration implying immunologic memory;
5. cheap, implying a low cost and a high benefit ratio; and
6. stability, implying retention of potency during storage or in transit outside the cold chain.

The huge and bountiful toolbox of technologies for pursuing these objectives has been recorded above and need not be discussed here.

Science and the Social Contract (Table 15)

Entry into the twenty-first century invites careful review of the social contract between science and society. In the end, it is society itself, which chooses, empowers and holds accountable. As part of science in general, it may be time for the vaccine enterprise to ascertain whether it gives reasonable and acceptable fulfillment of its obligations to society. In his brilliant expose of December 1999, Gibbons[124,125] points out that in the prevailing contract between science and society, "science has been expected to produce 'reliable' knowledge, provided merely that it communicates its discoveries to society." He goes on to say that "a *new* contract must now ensure that scientific knowledge is 'socially robust', and that its production is seen by society to be both transparent and participative." Society can impose "appropriate sanctions if these expectations are not met."

In expanding his discourse, Gibbons points out that there are three individual elements in the social contract that relate to

1. government and society,
2. industry and society, and
3. higher education and society.

Universities, with substantial autonomy, have provided research and teaching in return for public funding, and with added need to provide fundamental knowledge and to train the manpower essential to an advanced industrial society.

Industrial research and development (R & D), through its own laboratory activities, carries basic knowledge into product and process innovations.

Government, through use of research establishments, is expected to fill the gaps of knowledge between university science and industrial R & D to bring products to fruition. The state is directly responsible to carry out research related to national needs such as defense, energy, public health and the like.

Since World War II, the laissez-faire of the day has shifted national priorities from security and military readiness to the also important maintenance of international competitiveness and enhancement of the quality of life. In the new contract, Gibbons[124] sees a loss of the demarcation between basic science and applied industrial science with a blurring of professional identities and career patterns. He also notes an arrangement in which the autonomy of science gave a flow of speech from science to the public. This is altered now to a situation in which society speaks back to science. Society speaks out on subjects such as national objectives, new regulatory regimens, and user-producer interfaces with special respect to safety. He sees a transition of science from one of "searching for truth" to provisional understandings of an empirical world that works, thus, replicability coupled with surrogate validity through formation of consensus within a relevant peer group. Social robustness and acceptance is not achieved until the concerns of a much broader section of the total community are considered. To have recognized the importance of this earlier might have forestalled the objections, concerns, and debates for health in using genetically modified organisms (GMA) for food.

Table 15. Science and the contract with society

Present Contract

    Science is expected to produce reliable knowledge and to communicate its discoveries to society.

    Gibbons calls for a new contract to create knowledge that is socially robust, with transparency, and participative with the public in its pursuit.

    Academia, with autonomy and public funding, provides basic research, teaching, and training of manpower needed for an industrialized society.

    Industry, with private funding, conducts R&D and converts basic knowledge into products and processes.

    Government, with public funding, needs to fill the gap between academic and industrial research to bring products to fruition. Also conducts research in defense, energy, public health, and the like.

    Call for a close association and continuity between Industry and Academia.

    Peter Drucker notes that the new century will be driven by knowledge and the conversion of knowledge to practical value.

    Society at large now owns the majority of businesses, financial enterprises, and financial obligations in the U.S. Public criticisms of our institutions may amount to no more than criticism of self.

    Society itself assumes an increasingly proactive role in monitoring science and in voicing its opinions and objections.

Peter Drucker,[126] in his comments on the post-capitalistic society, pointed out that its initiatives would be driven by knowledge and information, which must be harnessed to create value. This takes place at a time when the real ownership of the majority of businesses and financial enterprises in the USA no longer resides with the few but rather by society at large which is represented by vested interests in pension funds, in mutual funds, and other investment vehicles dedicated both to debt and equity. Because of this, it is difficult to distinguish between society at large and institutions, including the science enterprise. It follows, then,[125] that public criticism of the conduct of business may amount to no more than criticism of self. Fairness and allocation, therefore, may only be identifiable with managers and policies of both publicly and privately funded institutions.

Supporting Publicly Funded Science (Table 16)

The written, spoken, and electronic literature of our time is replete with expressions, which concur in the observations made by Gibbons[124] for the tenuous relationship of science and society. One marker for measuring the closeness of the relationship is that of the level of public funding for science. Overall, the science of the past has enjoyed a favored status. A most cogent time of public dissatisfaction came in the mid-1990's at the end of Vannevar Bush[36,125,127,128] era when budgets for science were deeply slashed and when considerations were made to end the public science-supported initiatives. This ended quickly with the appearance of a more robust economy.

A very good bellwether for the contemporary present is provided by examining the public funding appropriated for research by the National Institutes of Health, where the interests of the infectious disease and vaccine enterprises are focused. In the contemporary period, the life sciences receive about 50% of the U.S. national science budget. In this, the NIH has been immensely favored at the expense of other important scientific programs. For two years, the NIH appropriations by Congress grew by about 15% per year to $17.9 billion for fiscal year 2000 with intent for such increases for three more years to doubling that of 1998. For 2001, the President, perhaps in responding to serious objections by the nonmedical science establishment, chose to begin a correction of the imbalances. He proposed to expand research spending

Table 16. Budgetary support for publicly funded science

Federal budget appropriations for science reflect the relationship between science and society.
Science of the past has enjoyed a favored status, except for the mid-1990's when public
dissatisfaction and societal initiatives resulted in a deep cut in appropriations for science.
The life sciences receive about 50% of the U.S. national science budget at the expense of other
important sciences.
The NIH has been immensely favored, with stated intent to double the budget of 1998 by 2003
through 15% annual increases.
The NIH budget for 2001 gives 15% increase to NIH, but with lesser increases in other basic
science support. Questions are raised as to whether NIH is growing too fast to be effective.
The NIH extramural funding is the primary source for basic research on infectious disease and
technologies essential to pursuit of research on vaccines.

by 7% (to $43 billion) with provisions for substantial increases for basic sciences in nearly all the concerned departments and agencies, but with only 6% increase to NIH. In the Congressional action, however, NIH was given a 14.6% increase over year 2000[129] and significant increases were given also to nearly all publicly funded R & D organizations.

Enthusiasm for the NIH budgeting increases are not shared by all members of Congress, some of whom wonder whether the NIH budget is growing too fast and whether it can spend all the money effectively.[130]

Monitoring Performance of Public Funded Science (Table 17)

Perhaps an early warning that all may not have been well in federal science and technology support, came with the U.S. National Academy of Sciences' report of 1995 on Allocating Federal Funds for Science & Technology.[131] While very supportive, the report said that there is need for change. Among its thirteen recommendations, it was made clear that there must be accountability and evaluation of the Federal programs in research and technology, consistent with the Government Performance and Results Act of 1993. In response to this, the Academy's Committee on Science, Engineering, and Public Policy (COSEPUP) now issues an annual report that started[132] with assessment of the 1999 budget. They point out that the report is intended to be useful for government, the science establishment, and all of society interested in our national research investment. Greater sharing of research tools and materials within the science enterprise has been promoted recently.[133]

The U.S. government also may be able to demand more accountability and successful accomplishments in its own R & D laboratories. This problem is very complex since benefits from basic research may only come years later. Papadakis,[134] in a treatise on the subject declares that the economic impacts of public science can be measured. If a workable system for government R & D can be developed, then such systems for appraisal may be carried to its external funding for research.

Evaluating Grant Applications

In keeping with reform and improvement of the extramural grant evaluation system, the NIH Center for Scientific Review (CSR) convened a Blue Ribbon Panel chaired by Dr. Bruce Alberts, President of the U.S. National Academy of Sciences, to create a new system that "can be continuously evaluated by outside experts".[135,136] The Alberts Committee would eliminate the fractured organization of grant requests into twenty-one integrated review groups, with sixteen centered on diseases and organ systems and five on basic research areas in which the application cannot be predicted. Not all would agree. AIDS researchers, for example, believe that the disease is too important to be folded into the broader research grant system. Credibility for such concern may lie in the past practice of equating the spending of research dollars

Table 17. Monitoring public science

U.S. National Academy of Sciences report of 1995 calls for change, with special emphasis on accountability in evaluation of federally funded programs.

Academy now monitors government-supported sciences and issues an annual "COSEPUP" report. Accountability is considered essential for both external and internal funding by NIH.

The Blue Ribbon Panel on grants review (2000 draft 1) calls for drastic change for grant review with 16 evaluation groups focused on diseases and organ systems plus 5 on basic research areas.

The Institute of Medicine, 1998 report, calls for linking costs for specific diseases to allocation of money for study, and for bringing the public into research policy and into the grant review mechanisms.

The research grant review process is presently under study and drastic changes may be recommended.

with the burden of the individual disease which may have appeal to the public. In fact, the 1998 Institute of Medicine Report (commissioned by Congress) on Scientific Opportunities and Public Needs: Improving Priority Setting and Public Input at the National Institutes of Health,[137,138] recommends that the NIH should develop data comparing costs of specific diseases against the monies allocated to them. In addition, the public itself should have more to say. Most recently, lay members have been added to the membership of some grant review study sections.[139] While the participation of lay public in the grant review process remains undecided, Dr. Ruth Kirschstein (former acting director of the NIH) recently stated that the Council of Public Representatives (COPR) has been successful in helping to assess issues of importance.

Improving Science Education in the USA (Table 18)

Of all the initiatives being undertaken in the United States National Academies of Sciences,[140,141] none can be more important than that calling for improved education in science and mathematics from Kindergarten through grade 12. Without basic education, institutions for higher learning will not have the number of trained minds to educate and to fulfill its obligations under the social contract to provide a cadre of scientists needed to create fundamental knowledge and to give the manpower needed for an advanced industrial society. Science education needs to be lifelong. Without such preparation, the public itself will not be able to understand and to judge the merits of what science gives to society and this applies equally well to those who work in the media and are empowered to provide the communication between science and society.

Review of the current situation reveals that the majority of our public education in science is abysmal,[142] even after a decade of effort to make it right. A major problem lies with the lack of qualified teachers[143-147] who are needed to impart such instruction now. Funded by the Rockefeller Foundation and the Carnegie Corporation, the National Commission on Teaching and America's Future reported in 1996 that 23% of all secondary teachers lack even a minor in their principal teaching field. More than 25% of U.S. teachers have no license or only a provisional license, and 56% of high school students taking physics are taught by out-of-field teachers. Education in biology is especially poor and is increased by the misunderstood conflict between evolution and creationism[148] and the recent prohibition of teaching of evolution in Kansas, since corrected.

Science itself may be partly to blame for the declarations against it, because of its own failure to communicate its position to the press and to the public and to do so simply and clearly. The press and the media, looking for a good story, are often the means for conveying bad information to the public. This gives anti-science zealots a level of credibility in which the

Table 18. Improving science education in the USA

The U.S. National Academics of Science Initiative on K-12 public education is of critical importance in improving education and training in science and mathematics in primary level education. A strong cadre of well-informed and qualified teachers, themselves, are required.

The current and foreseeable lack of qualified teachers is judged abysmal.

Basic science education is needed to direct and train the people required for our advanced industrial society.

Education needs to be extended to the media, that conveys information, and to the public itself.

Science and scientists, themselves, need to provide simple and understable communications to the public and to the media, giving special attention not to overstate the implications of their work.

Good communication from the science establishment can be of very great importance in allaying public mistrust and in providing evidence-based knowledge that will counter belief systems which can inhibit freedom for practice of needed medical interventions in a healthy society.

anti-vaccine organizations deny the children their rights for protection against serious diseases by vaccines through fictitious representations of purported side effects.[149,150] In communicating with public and press, it is critically important not to overstate the projections[151,152] from its work as is so oft reported with the tired quote "scientists say." Creation of mistrust in science[153-155] is an ever-present danger when a full disclosure of any new product or procedure has not been made. It finds its best remedy in full education and disclosure during the time that the science is being pursued. Finally, only adequate educational background and free communication of information can protect the lay individual from developing belief systems[149] that are not based on facts and that can greatly inhibit the free practice of medical interventions so essential to maintaining a healthy society.

Cooperation between Academia and Industry in Creating New Products for the Public (Table 19)

Vannevar Bush's plan[36] for post-war public funding of science accounted for great scientific developments during the past five decades. In his plan, Bush recognized the role of academia in creating the basic knowledge needed by industry to develop new products, but he stated that the two should be separate. This concept is undergoing change. In the late years, driven by the need for industrial participation in the AIDS Initiative, there have been continuing calls for close collaboration between academia, government, and industry.

Congressman Ehlers, in the report of the House Scientific Committee in 1998[127,157] called for a continuity in a collaborative new model which creates a continuum of research from basic to applied to industrial development without separation between them. In a new interpretation of the contract between science and society, it is stated, as already related above,[124] that the government is responsible for filling the gap between the basic and product feasibility to be pursued by industry. This has *not* been accomplished to the present, though the Innovation Grants Program[158] now empowers the individual entrepreneurial investigator with collaboration and services to allow him/her to create a virtual company to achieve practical results. Industry, because of need for profitability, cannot undertake commitment of huge resources to undertake product development before feasibility has been established.

In the present era of intrusion by competition and by government, it will be necessary for industry to hold valid patents and other protections to preclude diminution of its ownership of intellectual property.

Table 19. Cooperation between academia and industry

In creating a new contract for public science, the Ehler's report calls for continuity in a
     collaborative new model, which creates a continuum between basic and applied
     research.

As stated, government is responsible for filling the gap between the basic and that required to
     establish product feasibility.

The Innovative Grants Program of the NIH now empowers entrepreneurial investigators to create
     collaborative ventures which are comparable to a virtual company and which can
     achieve practical objectives.

Industry, because of need for profitability, cannot undertake commitment of huge development
     resources before feasibility has been established.

Industry needs to hold valid patents and other protections to preclude intrusions into its ownership
     of intellectual property.

A Call to the Vaccine Research Enterprise (Table 20)

Those workers who wish to pursue immunologic intervention for prophylaxis of infectious
diseases and cancer may now do so with reasonable assurance for a large pool of public money
to support research. Funding will reward well thought out research proposals that emphasize
innovation and avoid the redundancy that is already present in overabundance.

Individual workers may select different targets to pursue. But the still dismal outlook for
meaningful answers to biological control of AIDS,[117,118,125,158-161] malaria,[117,162-164] and tuberculosis[117,165-168] gives a special opportunity for the science establishment to engage in serious and significant attempts to create the new scientific knowledge needed to bring vaccines
against these three diseases to a point of feasibility and practicality for industrial development.
Technical problems that apply to all these diseases are the elements of immunologic heterogeneity and variability of the causal agents that need to be overcome. Added, are the uncanny
capabilities of the pathogens for subversion, evasion, and destruction of the immune response,
giving means for escape from immunologic control. The problems of multiplicity of subtypes
and the mutational hypervariability in antigenic specificities of the HIV agent that causes
AIDS are unique and have never been experienced for any other important infection of
mankind.

Efforts that can achieve only small and incremental improvements in now futile endeavors
may be foolish to continue at a time when bold new and creative approaches are to be encouraged. The more recent efforts taken to achieve both humoral and cell-mediated immune responses should be worthy of reaching for attainment. Simultaneously, there is need for exercise
of creativity with the development and testing of new concepts and approaches of rational
possibility in the search for group-reactive antivirals.

One such approach is based on yet to be confirmed findings by J. Nunberg, R.A. LaCasse,
and coworkers[169] that viral infection of the cell at the point of binding of gp120 with CD4,
and the engagement with gp41 and the chemokine receptor, may create a complex in HIV
infection which may be conserved and immunologically group-specific. When "frozen in time"
by treatment of infected cells with formaldehyde, group specificity was captured and preserved.
This vision, assuming independent confirmation, presents a hope for a single entity vaccine
against all subtype and variant HIV-1 viruses.

Following on earlier leads, Fouts et al[170] prepared soluble, stable, covalently bound, single
chain chimeras of CD4 and gp120 proteins in which a highly conserved binding site for the
chemokine coreceptor is exposed. Even more pertinent may be the studies of Kim and coworkers[171] which disclose a coiled coil trimeric complex of gp41 that brings about fusion of the HIV
envelope with the cell membrane and permits entry of the viral machinery needed for infection. Such coiled coil motifs may be common to many viral membrane-fusion proteins.[172]

Table 20. The call for successes in the vaccine research enterprise

-Public monies for support of research are now abundant.
-Investigators who pursue innovation and who eliminate redundancy will be rewarded.
-Targeting for small and incremental improvements, in now futile endeavors, are foolish at a time when bold new approaches are needed.
-The outlook for meaningful biological control of AIDS, Malaria, and Tuberculosis is abysmal and should provide attractive challenges for the innovative investigator.
-The common theme for all vaccines is to determine **WHAT** antigens or immunologic determinants need to be included.
-Collectively, the most broadly group-specific antigens need to be sought and appropriate humoral and cellular responses need to be achieved.
-Identification of **WHAT** to deliver in a vaccine may be aided by application of genomics, proteomics, rapid throughput assays and informational sciences.
-It is possible that linear linkage and scaffold presentations of all the needed antigens may be successful in achieving the appropriate specific humoral and cellular immune responses to polytopic vaccines.
-Knowledge relating to the **HOW** to present has already been described in seemingly endless kind and with tiresome redundancy.

New concepts and approaches, such as those of Kim, Nunberg and Fouts, irrespective of validity or success, must be taken as equivalent to a breath of fresh air in "a smoke-filled room!"

In malaria, with multi-stage events in the life cycle, there is also antigenic diversity, especially among blood stage merozoites. By contrast, there is substantive conservation in intracellular liver stage parasites that may signal cytotoxic T lymphocytes and, which, if in sufficient number, might be capable of eliminating all or most of the infected liver cells. In tuberculosis, it has been shown that there is diversity in expression of the different antigenic proteins by individual strains of the organism. Efforts to identify and to assure presentation of all the critically important antigen epitopes that provide the trail to a single effective vaccine are highly needed.

These examples clearly emphasize the need for research emphasis on *what* to deliver to match the *how* to deliver in the vaccine.[158] It opens credibility for the approach that engages in the identification of short immunologic determinants or epitopes that can be linked in tandem to activate cellular immune responses. Some linear epitopes may be significant also in activating of B cells while other antigens may need to be presented in appropriate constructs to preserve molecular conformation. Judicious selection of epitopes must be in sufficient variety to engage the MHC receptors of sufficient allelic diversity. Success in application of genomic and proteomic analyses to identify common specificity for all five serotypes of meningococcus[120,121] has been achieved but the meaningfulness for a simple multi-group protein vaccine remains to be shown. It is noteworthy that sequencing of the genomes of at least 30 microbial parasites have already been completed[118,119] as stated above.

Amid all the complexities resident in an overwhelming database, it is evident that judicious selective choices must be made. Economically effective initiatives for elimination and eradication of polioviruses and measles virus may be achieved in the present decade. This is made increasingly imperative in the face of the new and emerging or reemerging infectious agents through the world.

As described above, the largest single problem for vaccine development lies with the 'what' and 'how' to deliver[158] vaccines of high level antigenic diversity. The technologies for 'how' to deliver antigens in vaccines have been described in seemingly endless forms and with tiresome redundancy. The determination of 'what' to deliver may be difficult to ascertain but may be aided materially by application of genomics, proteomics, automated throughput assays and informatics in genetic approaches[173-175] to aid in identifying appropriate and optimal antigens. It may become necessary or expedient to prepare vaccines consisting only of identified linear and conformational epitopes in broad-spectrum complex display so as to achieve appropriate humoral and cellular immune responses and to overcome individual allelic restriction of the host.

Alternatively, it may be fruitful to discover subdominant group-specific epitopes that are sequestered and go unseen in the response to natural infection. Emphasis may also be placed on the identification of group-specific antigens in the virus core where conservation of specificity is needed to retain species identity.

In view of the dire emergency, there is no time to waste on despair. Rather, to go for it and to do it in the spirit that what needs to be accomplished will be accomplished!

Recombinant DNA Vaccines

The title of the book in which this review appears is *DNA VACCINES* and it seems fitting to end this paper with an overview relating to the DNA vector approach. The question to be addressed is whether DNA vaccines really represent practicable advances at the cutting edge of vaccine research or whether they are misguided hopes in a field of dreams. More simply, is DNA really a new era in vaccinology?

Facing hundreds of papers on DNA vaccines that were recently reviewed, I was taken by the uniquely brilliant innovation and discovery that was represented. There is no good purpose, in this summary, either to duplicate or preempt what has already been recorded in the exceptional papers that comprise the book.

Seeking to reduce what we know to simplest terms, I was struck by a simile that was made in a popular television advertisement about the merits of a competitive brand of hamburgers. The statement made, in effect, was that "you've got nice buns, but where's the beef?" The bun in this simile represents the good and relevant publications about the basics of DNA vaccines. The beef is whether we now have the makings for a real and licensable vaccine which is the meat of the program.

Looking at the buns of the matter, it is evident that there is redundant redundancy in several definable areas. As summarized in Table 21, it is clear that the basics which ought to have been done actually have been done, viz., concerning recombinant plasmid composition, plasmid carriers, guided perturbation of the immune system, vaccine delivery systems, animal models, and animal experimentation to date. Most importantly, it is evident that a choice means for priming the immune system for a cytotoxic T cell response is by use of a recombinant DNA vaccine, and this can be amplified by boosting using expressed protein or by use of another recombinant vector. It is sobering, however, in that only eight among 171 studies summarized in two publications[176-177] dealt with tests in man, and these were limited to only the most superficial probes of HIV, hepatitis B, malaria, and cancer, which are prime targets for trials of DNA vaccines in man. Clearly, there is a barrier, which greatly limits progress from basic pursuits to the defining goals of tests in man.

Possible/probable deterrents to human experimentation with DNA plasmid vaccines are listed in Table 22. Safety issues are defined by regulatory control authorities[178] for purpose of giving definition to products for eventual licensure and commercial distribution. It is not the duty of the control authorities to obtain answers to the omissions by the research establishment to provide definitive answers, favorable or unfavorable, because of limited or nonpursuit of definable end objectives. It may be noted, however, that the problem of cellular integration of plasmid DNA may have been resolved, as well as it can be resolved,[178-180] and, in addition, there is no compelling evidence to support the likelihood of danger from antibody responses against DNA.[178] The remaining questions are for resident basic research scientists to answer, using staff and resources that otherwise might be dedicated to less priority purpose.

Appropriate and inappropriate immune perturbation, and the achievement of appropriate immunologic balance, are clear and definable dangers that need to be studied extensively in nonhuman primate models and eventually in the nonexpendable human volunteer. A cogent

Table 21. Examples of brilliant and often redundant research in the quest for licensable DNA vaccines

Recombinant plasmid composition
>   Codon usage and optimal signaling for transcription.
>   Enhancers of expression and immune response by substances such as CpG and encoded immunomodulators such as cytokines.
>   Coding for appropriate and needed antigens/epitopes.

Plasmid carriers
>   Naked, lipid encased, or ligand directed to cell receptors.
>   Delivery and expression in attenuated bacteria.

Perturbation of the immune system
>   Lymphoid and dendritic cell targeting.
>   Prime/boosting systems.
>   Durable immunity with anamnestic memory.
>   Appropriate engagement and balance in humoral and cellular immune responses.
>   Approach to elimination/eradication of persistent viral infections.

Vaccine delivery systems
>   Syringe, biojectors, transcutaneous passage, mucosal immunization with microparticles, oral feeding.
>   Targeting dendritic cells or precursors and lymph nodes.

Animal models
>   Rodents, ungulates, sub-human primates, birds.
>   Expression of antigens of wide diversity including pathogenic prokaryotes, viruses, and parasitic infections with demonstration of protection on challenge.

Animal and human experimentation to date
>   Among 171 experiments listed (with some duplication) in two publications only 8 dealt with initial probes in man (HIV, hepatitis B, malaria, cancer).
>   Projected clinical tests of DNA HIV vaccines in man in 2001-2002 may light the way to advances with DNA vaccines against other agents.

example is provided by the inadvertent experiences in man with viruses of the Paramyxoviridae family[181] that have been serious and sobering.

Practical considerations to weigh the use of alternative vectors is clearly a way to make an alternative choice. There are many different and superior or inferior vectors for which comparative appraisal needs to be made. There is no basis for a love affair with one particular vector which may stand in the way of another in achieving a licensed vaccine goal. An unbiased response to make such a determination is a charge to the basic research community.

Finally, a laissez-faire attitude or a resident fear to tread into the unknowns of human experimentation is hardly tenable in the face of the huge and unsolved problems from diseases in the Twenty-first Century! It must be recalled, as stated above, that the public increasingly and rightfully is asserting itself in publicly supported science and desires valued solutions.[182] The abundant annual federal appropriation increases being made to NIH may end by 2003 or before if societal needs are seen to exceed the need for science such as happened in 1995.[182] Additionally, Federal budget excesses may become budget deficiencies in a world with unpredictable economic certainties. For the good of science, for the good of the nation, and for the good of the scientific enterprise, there is need to plan efficiently and effectively to give valuable returns to the public for their public support. In this respect, there is no more cogent example than resides with DNA vaccines. Recent good news from Dr. Harriet Robinson, R.R. Amara, and colleagues[183] casts portents for more meaningful progress toward clinical suppression if not viral eradication in tests in monkeys which were primed with multi-protein recombinant DNA HIV vaccine and followed by boosting with corresponding recombinant poxvirus

Table 22.   Deterrents to human experimentation with recombinant DNA plasmid vaccines

Safety
>   DNA integration (theoretical cancer issue).
>   Autoimmunity – antibody against DNA.
>   Immune tolerance from long-term vector presence and expression.
>   Reactions to cytokines encoded in the DNA.

Appropriate and Inappropriate Immune Perturbation
>   Appropriate and effective balance in cellular and humoral responses against proper antigens/epitopes.
>   Adverse experiences encountered with previous vaccines.
>>   Killed whole measles or subunit H measles vaccine.
>>>   Short-term immunity with appearance of severe atypical measles on subsequent wild measles virus infection.
>>   Lethal disease in young infants given killed respiratory syncytial virus vaccine or in animals given killed SV5 vaccine.
>>   Severe immunosuppression following administration of high-titer measles virus vaccine and death from opportunistic infections.
>   Practical considerations and alternative vectors.
>>   DNA vaccines may require large doses in man and may not be cheap.
>>   Size for insertion of foreign gene sequences may not be large enough to include the necessary B and T cell antigens and to overcome allelic restriction.
>>   Alternative vectors such as alphavirus, stripped adenovirus, canarypox and herpesvirus vectors may be superior in important aspects.
>   Lack of will to pursue vaccines to definitive endpoints in man.
>>   Lack of determination to come to determinations and definitive agreements with national licensing authorities.
>>   Fear or lack of drive to engage in human experimentation.
>>   Practice of academic apathy against pursuit of basic research to practical endpoints.
>   Note: Public tolerance for public support of research that does not achieve achievable practical endpoints may not last forever.

vaccine prior to intra-rectal challenge with SHIV. The human HIV counterpart vaccine is planned to be in trial in 2002 (personal communication, Harriet Robinson). Added to this are the remaining seven trials with DNA HIV vaccines (see ref. 184) that are planned to begin testing in man in 2001-2002. Such promising initiatives with HIV may light the way to similar approaches in the pursuit of vaccines against malaria, tuberculosis, hepatitis C, cancer, and many others that are in need of development.

References

1. Hilleman MR. Vaccines in historic evolution and perspective: A narrative of vaccine discoveries. Vaccine 2000; 18:1436-1447.
2. Hilleman MR. Personal historical chronicle of six decades of basic and applied research in virology, immunology, and vaccinology. Immunol Rev 1999; 170:7-27.
3. Hilleman MR. Six decades of vaccine development—A personal history. Nature Medicine 1998; (Suppl.) 4:507-514.
4. Plotkin SL, Plotkin SA. A Short History of Vaccination in Vaccines. 3rd ed. Plotkin SL and Orenstein WA, eds. Philadelphia: W. B. Saunders, 1999:1-27.
5. Castiglioni A. A History of Medicine. 2nd ed. New York: Alfred A. Knopf, 1958.

6. Reither J. World History at a Glance. A Record of History of Events from Earliest Civilizations to the Present. Revised ed. Everyday Handbook Series. New York: Barnes & Noble, 1952.
7. McNeill WH. The Rise of the West. A History of the Human Community. Chicago: University of Chicago Press, 1963.
8. de Kruif P. Microbe Hunters. 31st printing, 1965. New York: Harcourt Brace, 1926.
9. Jenner E. An Inquiry into the Causes and Effects of the Variolae Vaccinae. London: Low, 1798.
10. Bailey I. Edward Jenner, benefactor to mankind. In: Plotkin SA and Fantini B, eds. Vaccinia, Vaccination, Vaccinology. Jenner, Pasteur and Their Successors. New York: Elsevier, 1995.
11. Bellanti JA, Kadlec JV. Introduction to immunology. In: Bellanti JA, ed. Immunology III. Philadelphia, W. B. Saunders, 1985:1-15.
12. Darwin C. On the Origin of Species by Means of Natural Selection. London: John Murray, 1859.
13. Henig RM.. The Monk in the Garden. The Lost and Found Genius of Gregor Mendel, the Father of Genetics. Boston: Houghton Mifflin, 2000.
14. Gerson GL. The Private Science of Louis Pasteur. Princeton: Princeton University Press, 1995.
15. Brock TD. Robert Koch: A Life in Medicine and Bacteriology. New York: Springer, 1988.
16. Gronski P, Seiler FR, Schwick HG. Discovery of antitoxins and development of antibody preparations for clinical uses from 1890 to 1990. Molec Immunol 1991; 28:1321-1332.
17. Witkop B. Paul Ehrlich and his magic bullets—revisited. Proc Am Philos Soc 1999; 143:540-556.
18. Hilleman MR. International biological standardization in historic and contemporary perspective. In A Celebration of 50 Years of Progress in Biological Standardization and Control at WHO. Brown F, Griffiths E, Horaud F, Schild GC, eds. Dev Biol Stand. Basel: Karger, 1999; 100:19-30.
19. Woodruff AM, Goodpasture EW. The susceptibility of the chorioallantoic membrane of chick embryos to infection with the fowl-pox virus. Am J Pathol 1931; 7:209-222.
20. Theiler M, Smith HH. The use of yellow fever virus modified by in vitro cultivation for human immunization. J Exp Med 1937; 65:787-800.
21. Hilleman MR, Mason RP, Rogers NG. Laboratory studies on the 1950 outbreak of influenza. Pub Hlth Repts 1950; 65:771-777.
22. Hilleman MR., Mason RP, Buescher EL. Antigenic pattern of strains of influenza A and B. Proc Soc Exp Biol Med 1950; 75:829-835.
23. Hilleman MR, Werner JH, Gauld RL. Serological studies of influenza antibodies in the population of the United States. An epidemiological investigation. Bull WHO 1953; 8:613-631.
24. Webster RG. Influenza viruses. In: Webster RG and Granoff A, eds. Encyclopedia of Virology. Vol. 2. New York: Academic Press, 1994:709-727.
25. Meyer HM Jr. et al. New antigenic variant in Far East influenza epidemic. Proc Soc Exp Biol Med 1957; 95:609-616.
26. Hilleman MR, Werner JH. Recovery of new agent from patients with acute respiratory illness. Proc Soc ExpBiol Med 1954; 85:183-188.
27. Hilleman MR. et al. Epidemiology of RI (RI-67) group respiratory virus infections in recruit populations. Am J Hygiene 1955; 62:29-43.
28. Rowe WP et al. Isolation of a cytopathic agent from human adenoids undergoing spontaneous degeneration in tissue culture. Proc Soc Exp Biol Med 1953; 84:570-573.
29. Enders JF et al. "Adenoviruses." Group name proposed for new respiratory tract viruses. Science 1957; 124:119-120.
30. Hilleman MR et al. Prevention of acute respiratory illness in recruits by adenovirus (RI-APC-ARD) vaccine. Proc Soc Exp Biol Med 1956; 92:377-383.
31. Stallones RA et al. Adenovirus (RI-APC-ARD) vaccine for prevention of acute respiratory illness. 2. Field evaluation. J Amer Med Assoc 1957; 163:9-15.
32. Hilleman MR. et al. Second field evaluation of bivalent type 4 & 7 adenovirus vaccine for prevention of acute respiratory disease in military recruits. AMA Arch Int Med 1958; 102:428-436.
33. Enders JF, Weller TH, Robbins FC. Cultivation of the Lansing strain of poliomyelitis virus in cultures of various embryonic tissues. Science 1949; 109:85-87.
34. Salk J, Drucker J. Noninfectious poliovirus vaccine. In: Plotkin SA and Mortimer EA, eds.Vaccines. Philadelphia: W. B. Saunders, 1988:158-181.
35. Zachary GP. Endless Frontier. Vannevar Bush, Engineer of the American Century. New York: Free Press, 1997.
36. Bush V. Science: The Endless Frontier. A Report to the President on a Program for Post-war Scientific Research, July 1945. Reprinted by the National Science Foundation, Washington, DC, 1990.
37. Gotschlich EC, Liu TY, Artenstein MS. Human immunity to the meningococcus. III. Preparation and immunochemical properties of the group A group B and group C meningococcal polysaccharides. J Exp Med 1965; 129:1349-1365.

38. Weibel RE et al. Clinical and laboratory investigations of monovalent and combined meningococcal polysaccharide vaccines Groups A and C. Proc. Soc Exp Biol Med 1976; 153:436-440.
39. Lepow ML. Meningococcal vaccines. In: Plotkin SA and Mortimer EA, Jr., eds.Vaccines. 2nd ed. Philadelphia: W. B. Saunders, 1994:503-515.
40. Nurse P. The incredible life and times of biological cells. Science 2000; 289:1711-1716.
41. Burke DS. Auzias-Turenne, J-A.. Louis Pasteur and early concepts of virulence attenuation and vaccination. Perspect Biol Med 1996; 39:171-186.
42. Austrian R. Pneumococcus the first 100 years. Rev Infect Dis 1981; 3:183-189.
43. Austrian R et al. Prevention of pneumococcal pneumonia by vaccination. Trans Assoc Amer Physicians 1976; 89:184-189.
44. Smit P et al. Protective efficacy of pneumococcal vaccines. J Amer Med Assoc 1977; 238:2613-2616.
45. Hilleman MR et al. Vaccination against pneumococcal infections. In: Lambert HP and Caldwell ADS, eds. Pneumonia and Pneumococcal Infections. Royal Soc Med Intl Congr, Symp. Series 27. London: Academic Press, 1980:67-85.
46. Hilleman MR et al. Polyvalent pneumococcal polysaccharide vaccines. J Infection 1979; 1:1-16.
47. Tai JY et al. *Haemophilus influenzae* type b polysaccharide-protein conjugate vaccine. Proc Soc Exp Biol Med 1987; 184:154-161.
48. Santosham M et al. The efficacy in Navajo infants of a conjugate vaccine consisting of *Haemophilus influenzae* type b polysaccharide and *Neisseria meningitidis* outer-membrane protein complex. N Engl J Med 1991; 324:1767-1772.
49. Stokes J Jr. et al. Efficacy of live attenuated measles virus vaccine given with human immune globulin. A preliminary report. N Engl J Med 1961; 265:507-513.
50. Stokes J Jr. et al. Studies of live attenuated measles virus vaccine in man. 1. Clinical aspects. Am J Publ Hlth 1962; 52:29-43.
51. Hilleman MR. et al. Studies of live attenuated measles virus in man. 2. Appraisal of efficacy. Am J Publ Hlth 1962; 52:44-56.
52. Enders JF, Katz SL, Holloway A.. Studies on an attenuated measles-virus vaccine. I. Development and preparation of the vaccine: Technics for assay of effects of vaccination. N Engl J Med 1960; 263:153-159.
53. Hughes WF, Watanabe DH, Rubin H. The development of a chicken flock apparently free of leukosis virus. Avian Dis 1963; 7:154-165.
54. Hilleman MR. et al. Development and evaluation of the Moraten measles virus vaccine. J Amer Med Assoc 1968; 206:587-590.
55. Buynak EB, Hilleman MR. Live attenuated mumps virus vaccine 1 Vaccine development. Proc Soc Exp Biol Med 1966; 123:768-775.
56. Stokes J Jr. et al. Live attenuated mumps virus vaccine. 2. Early clinical studies. Pediatrics 1967; 39:363-371.
57. Weibel RE, et al. Live attenuated mumps virus vaccine. 3. Clinical and serologic aspects in field evaluation. N Engl J Med 1967; 276:245-251.
58. Hilleman MR. et al. Live attenuated mumps virus vaccine. 4. Protective efficacy measured in field evaluation. N Engl J Med 1967; 276:252-258.
59. Weller TH, Neva FA. Propagation in tissue culture of cytopathic agents from patients with rubella-like illness. Proc Soc Exp Biol Med 1962; 111:215-225.
60. Parkman PD, Buescher EL, Artenstein MS. Recovery of rubella virus from Army recruits. Proc Soc Exp Biol Med 1962; 111:225-230.
61. Buynak EB et al. Preparation and testing of duck embryo cell culture rubella vaccine. Amer J Dis Child 1969; 118:347-354.
62. Weibel RE et al. Live attenuated rubella virus vaccines prepared in duck embryo cell culture: II. Clinical tests in families and in an institution. J Amer Med Assoc 1968; 205:554-558.
63. Buynak EB, Hilleman MR, Weibel RE, Stokes J Jr. Live attenuated rubella virus vaccines prepared in duck embryo cell culture. I. Development and clinical testing. J Amer Med Assoc 1968; 204:195-200.
64. Stokes J Jr., Weibel RE, Buynak EB, Hilleman MR. Protective efficacy of duck embryo rubella vaccines. Pediatrics 1969; 44:217-224.
65. Weibel RE, Stokes J Jr., Buynak EB, Hilleman MR. Live rubella vaccines in adults and children. HPV-77 and Merck-Benoit strains. Amer J Dis Child 1969; 118:226-229.
66. Hilleman MR et al. Live attenuated rubella virus vaccines. Experiences with duck embryo cell preparations. Amer J Dis Child 1969; 118:166-171.
67. Neff BJ et al. Clinical and laboratory studies of KMcC strain live attenuated varicella virus. Proc Soc Exp Biol Med 1981; 166:339-347.

68. Weibel RE et al. Live attenuated varicella virus vaccine. Efficacy trial in healthy children. N Engl J Med 1984; 310:1409-1415.
69. Takahashi M, Asano Y, Kamiya H, Baba K. Varicella vaccine: case studies. Microbiol Sciences 1985; 2:249-254.
70. Buyna, EB et al. Combined live measles mumps and rubella virus vaccines. J Amer Med Assoc 1969; 207:2259-2262.
71 Stokes J Jr. et al. Trivalent combined measles-mumps-rubella (M-M-R) vaccine. Findings in clinical-laboratory studies. J Amer Med Assoc 1971; 218:57-61.
72. Hilleman MR et al. Combined live virus vaccines. PAHO Sci Publ 1971; 226:397-400.
73. Report. Advances in global measles control and elimination. MMWR 1998; 47(Suppl.):1-25.
74. Report to the Centers for Disease Control and Prevention. (May 14, 1994). Cost benefit analysis of the measles-mumps-rubella (MMR) vaccine. Battelle, Arlington VA.
75. Okazaki W, Purchase HG, Burmester BR. Protection against Marek's disease by vaccination with a herpesvirus of turkeys. Avian Dis 1970;14:413-429.
76. Hilleman MR. Marek's Disease Vaccine. Its implication in biology and medicine. Avian Dis 1972; 16:191-199.
77. Robbins FC. Polio—Historical. In: Vaccines. Plotkin SA and Mortimer EA Jr., eds. Philadelphia: W. B. Saunders, 1998:98-114.
78. Nathanson N, Langmuir AD. The Cutter Incident. Poliomyelitis following formaldehyde inactivated poliovirus vaccination in the United States during the Spring of 1955. 1. Background. (reprinted from Am J Hyg 1963; 78:16-28). Rev Med Virol 1995; 5:126-131.
79. Hilleman MR et al. Investigation into the development and clinical testing of a poliomyelitis vaccine containing standardized amounts of purified poliomyelitis virus antigens. Acad Med NJ Special Bull 1960; 6:1-31.
80. Hilleman MR. Development of a purified poliomyelitis virus vaccine. J Amer Med Assoc 1961; 177:591-595.
81. Hul RN. The Simian viruses. In: Virology Monographs. Gard S, Hallauer C, and Meyer KF, eds. New York: Springer-Verlag, 1968:2-66.
82. Sweet BH, Hilleman MR. 6. Detection of a "nondetectable" simian virus (vacuolating agent) present in rhesus and cynomolgus monkey-kidney cell culture material. A preliminary report. Second International Conference on Live Poliovirus Vaccines. Pan American Health Organization and the World Health Organization. Washington DC, 6-7 June 1960:79-85.
83. Sweet BH, Hilleman MR. The Vacuolating Virus, S.V.40. Proc Soc Exp Biol Med 1960; 105:420-427.
84. Girardi AJ, Sweet BH, Slotnick VB, Hilleman MR. Development of tumors in hamsters inoculated in the neonatal period with vacuolating virus SV_{40}. Proc Soc Exp Biol Med 1962; 109:649-660.
85. Levine A. The origins of the small DNA tumor viruses. Adv Cancer Res 1994; 65:141-148.
86. Goldner H, Girardi AJ, Larson VM, Hilleman MR. Interruption of SV_{40} virus tumorigenesis using irradiated homologous tumor antigen. Proc Soc Exp Biol Med 1964; 117:851-857.
87. Koprowski H. Proceedings of the Round Table Conference on Immunization in Poliomyelitis. Hershey, PA. Sponsored by the National Foundation for Infantile Paralysis. March 15-19, 1951.
88. Koprowski H et al. Immunization of children by the feeding of living attenuated type 1 and type 2 poliomyelitis virus and the intramuscular injection of immune serum globulin. Amer J Med Sci 1956; 232:378-388.
89. Koprowski H, Plotkin S. History of Koprowski vaccine against poliomyelitis. In: Plotkin S and Fantini B, eds. Vaccinia, Vaccination and Vaccinology: Jenner, Pasteur and Their Successors. Paris: Elsevier, 1996:229-240.
90. Melnick JL. Live attenuated poliovaccines. In: Plotkin SA and Mortimer EA Jr., eds. Vaccines. Philadelphia: W. B. Saunders, 1988:115-157.
91. Mascoli CC et al. Recovery of hepatitis agents in the marmoset from human cases occurring in Costa Rica. Proc Soc Exp Biol Med 1973; 142:276-282.
92. Provost PJ et al. Etiologic relationship of marmoset propagated CR326 hepatitis A virus to hepatitis in man. Proc Soc Exp Biol Med 1973; 142:1257-1267.
93. Simons JN et al. Identification of two flavivirus-like genomes in the GB hepatitis agent. Proc Natl Acad Sci USA 1995; 92:3401-3405.
94. Hilleman MR. Immune adherence and complement-fixation tests for human hepatitis A. Diagnostic and epidemiologic investigations. Dev Biol Stand 1975; 30:383-389.
95. Hilleman MR. Characterization of CR326 human hepatitis A virus, a probable enterovirus. Dev Biol Stand 1975; 30:418-424.
96. Hilleman MR. Hepatitis and hepatitis A vaccine: A glimpse of history. J Hepatol 1993; 18:S5-S10.

97. Provost P, Hilleman MR. An inactivated hepatitis A virus vaccine prepared from infected marmoset liver. Proc Soc Exp Biol Med 1978; 159:201-203.
98. Provost PJ, Hilleman MR. Propagation of human hepatitis A virus in cell culture in vitro. Proc Soc Exp Biol Med 1979; 160:213-221.
99. Lewis JA et al. Use of a live attenuated hepatitis A vaccine to prepare a highly purified formalin inactivated hepatitis A vaccine. In: Hollinger FB, Lemon SM, and Margolis H, eds. Viral Hepatitis and Liver Disease. Baltimore: Williams & Wilkins, 1991:94-97.
100. Provost PJ et al. New findings in live attenuated hepatitis A vaccine development. J Med Virol 1986; 20:165-175.
101. Werzberger A. et al. A controlled trial of formalin-inactivated hepatitis A vaccine in healthy children. N Engl J Med 1992; 327:453-457.
102. Hilleman MR. Purified and inactivated human hepatitis B vaccine. Progress report. Amer J Med Sci 1975; 270:401-404.
103. Buynak EB et al. Vaccine against human hepatitis B. J Amer Med Assoc 1976; 235:2832-2834.
104. Hilleman MR. Hepatitis A and hepatitis B vaccines. In: Szmuness W, Alter HJ, and Maynard J, eds. Viral Hepatitis 1981 International Symposium. Philadelphia: The Franklin Institute Press, 1982:385-397.
105. Hilleman MR. Plasma-derived hepatitis B vaccine—A breakthrough in preventive medicine. In: Ellis RW, ed. Hepatitis B Vaccines in Clinical Practice. New York: Marcel Dekker, 1979:17-39.
106. Hilleman MR. Immunology vaccinology and pathogenesis of hepatitis B. In: KoprowskiH, Oldstone MBA, eds. Microbe Hunters Then and Now. Bloomington: Medi-Ed Press, 1996:221-233.
107. Hilleman MR. Three decades of hepatitis vaccinology in historic perspective. A paradigm of successful pursuits. In: Plotkin SA, Fantini B, eds. Vaccinia, Vaccination, Vaccinology: Jenner, Pasteur and Their Successors. New York: Elsevier, 1996:199-209.
108. Szmuness W et al. Controlled clinical trial of the efficacy of the hepatitis B vaccine (Heptavax B): A final report. Hepatology 1991; 1:377-385.
109. Blumberg BS, Alter HJ, Visnich S. A "new" antigen in leukemia sera. J Amer Med Assoc 1965; 191:541-546.
110. Minor PD, Will RG, Salisbury D. Vaccines and variant CJD. Vaccine 2001; 19:409-410.
111. Kopito R, Ron D. Conformational disease. Nature Cell Biol 2000; 2:E207-E209.
112. Sy M-S, Gambetti P. Prion replication—Once again blaming the dendritic cell. Nature Med 1999; 5:1235-1236.
113. Valenzuela P et al. Synthesis and assembly of hepatitis B virus surface antigen particles in yeast. Nature 1982; 298:347-350.
114. McAleer WJ et al. Human hepatitis B vaccine from recombinant yeast. Nature 1984; 307:178-180.
115. Hilleman MR, Weibel RE, Scolnick EM. Recombinant yeast human hepatitis B vaccine. J Hong Kong Med Assoc 1985; 37:75-85.
116. Hilleman MR. Overview of vaccinology with special reference to papillomavirus vaccine. J Clin Virol 2000; 19:79-90.
117. National Institute of Allergy and Infectious Disease. The Jordan Report 2000. Accelerated Development of Vaccines. National Institutes of Health, Bethesda, MD, 2000.
118. Hilleman MR. A simplified vaccinologists' vaccinology and the pursuit of a vaccine against AIDS. Vaccine 1998; 16:778-793.
119. Chakravati DN, Fiske MJ, Fletcher LD, Zagursky RJ. Application of genomics and proteomics for identification of bacterial gene products as potential vaccine candidates. Vaccine 2001; 19:601-612.
120. Tettelin H et al. Complete genome sequence of *Neisseria meningitidis* serogroup B strain MC58. Science 2000; 287:1809-1815.
121. Pizza M et al. Identification of vaccine candidates against serogroup B meningococcus by whole-genome sequencing. Science 2000; 287:1816-1820.
122. Basch PF. Vaccines and World Health. Science, Policy and Practice. New York: Oxford University Press, 1994:181-199.
123. Nossal GJV. The Global Alliance for Vaccines and Immunization—A millennial challenge. Nature Immunol 2000; 1:5-8.
124. Gibbons M. Science's new social contract with society. Nature 1999; 402(Suppl.):C81-C84.
125. Hilleman MR. The business of science and the science of business in the quest for an AIDS vaccine. Vaccine 1999; 17:1211-1222.
126. Drucker PF. Post-capitalist society. New York: Harper Collins, 1994.
127. Ehler's Report. A Report to Congress by the House Committee in Science. Unlocking Our Future toward a New National Science Policy. Sept. 24, 1998.
128. Press F. Science and Technology Policy for a New Era. Presidential Address, National Academy of Sciences, April 27, 1992, Washington, DC.

129. A Preview Report for Congressional Action on Research and Development in the Fiscal Year 2001 Budget. Final Fiscal Year 2001 Appropriations. AAAS Research and Development Funding Update, December 19, 2000.
130. News. Senator seeks reassurance on NIH budget. Nature 1999; 397:94.
131. U.S. National Academy of Sciences Committee on Criteria for Federal Support of Research and Development. Allocating Federal Funds for Science and Technology. Washington, DC: National Academy Press, 1995.
132. U.S. National Academy of Sciences Committee on Science, Engineering, and Public Policy. Observations on the President's Fiscal Year 1999 Federal Science and Technology Budget. Washington, DC: National Academy Press, 1998.
133. Marshall E. New NIH rules promote greater sharing of tools and materials. Science 1999; 286:2430-2431.
134. Papadakis M. The economic impacts of public science can be measured. The Scientist October 27, 1997:8.
135. Alberts BM et al. Proposed changes for NIH's Center for Scientific Review. Science 1999; 285:666-667.
136. Dove A. NIH proceeds with overhaul of grant system. Nature 1999; 5:1219.
137. Committee on the NIH Priority-Setting Process. Scientific Opportunities and Public Needs. Washington, DC: National Academy Press, 1998.
138. Gross CP, Anderson GF, Powe NR. The relation between funding by the National Institutes of Health and the burden of disease. N Eng J Med 1999; 340:1881-1887.
139. Agnew B. NIH invites activists into the inner sanctum. Science 1999; 283:1999-2001.
140. President's Committee of Advisors on Science and Technology, Panel on Educational Technology. Report to the President on the Use of Technology to Strengthen K-12 Education in the United States. Executive Office of the President of the United States, March, 1997.
141. Moreno NP. K-12 science education reform—A primer for scientists. BioScience 1999; 49:569.
142. Wheeler G. The wake-up call we dare not ignore. Science 1998; 279:1611-1612.
143. Gottesman MM. Math and science education: Training of teachers. Science 2000; 290:273.
144. Lerner LS. Good and bad science in U.S. schools. Nature 2000; 407:287-298.
145. Allen EE, Hood L. Biotechnology, inquiry and public education. TIBTECH 2000; 180:329-330.
146. MacIlwain C. U.S. universities under fire for antiquated teaching. Nature 1998; 392:746.
147. Hervis J. Ehler's bill wins bipartisan backing. Science 2000; 289:713.
148. U.S. National Academy of Sciences Report. Science and Creationism. 2nd Ed. Washington, DC: National Academy Press, 1999.
149. Griffiths PD. Defending vaccines from the enemy within. Rev Med Virol 1999; 9:143-146.
150. Editorial. Science wars and the need for respect and vigour. Nature 1997; 385:373.
151. Salmon S. Scientists, media should not overstate 'breakthroughs.' Ann Oncol 1998; 9:794-795.
152. Jensen P. Scientists must bridge the communication gap. Nature 1999; 399:406.
153. Haerlin B, Parr D. How to restore public trust in science. Nature 1999; 400:499.
154. Dickson D. Wellcome survey reveals public mistrust of scientists. Nature Med 1999; 5:10.
155. Reichhardt T. Biotech panel set up in U.S. may help allay public fears. Nature 1999; 399:508.
156. MacIlwain C. U.S. public puts faith in science, but still lacks understanding. Nature 1998; 394:107.
157. Lawler A. House study tackles new era in R and D. Science 1977; 277:28.
158. Hilleman MR. The frustrating journey toward an AIDS vaccine. Proc Amer Philos Soc 2000; 144:349-360.
159. Rousseau MC, Moreau J, Delmont J. Vaccination and HIV: A review of the literature. Vaccine 2000; 18:825-831.
160. Ruprecht RM. Live AIDS viruses as vaccines: Promise or peril? Immunol Revs 1999; 170:135-149.
161. Fauci AS. The AIDS epidemic. Considerations for the 21st century. N Eng J Med 1999; 341:1046-1060.
162. Russell PK, Howson CP, eds. Committee on Malaria Vaccines. Hope in a Gathering Storm. Institute of Medicine. Washington, DC: National Academy Press, 1996.
163. Anders RF, Saul A. Malaria vaccine. Parasitol Today 2000; 16:444-447.
164. Perlmann P, Björkman A. Malaria research: host-parasite interactions and new developments in chemotherapy, immunology, and vaccinology. Curr Opin Infect Dis 2000; 13:431-443.
165. Keith PWJ, McAdam FRCP. Recent progress in bacterial vaccines: tuberculosis. Int J Infect Dis 1997; 1:172-178.
166. Ginsberg AM. A proposed national strategy for tuberculosis vaccine development. Clin Infect Dis 2000; 30(Suppl.):5233-5242.
167. Kaufmann SHE. Is the development of a new tuberculosis vaccine possible? Nature Med 2000; 6:955-960.

168. Cole ST et al. Deciphering the biology of *Mycobacterium tuberculosis* from the complete genome sequence. Nature 1998; 393:537-544.
169. LaCasse RA et al. Fusion-competent vaccines: Broad neutralization of primary isolates of HIV. Science 1999; 283:357-362.
170. Fouts TR et al. Expression and characterization of a single-chain polypeptide analogue of the human immunodeficiency virus type 1 gp 120-CD4 receptor complex. J Virol 2000; 74:11427-11436.
171. Root MJ, Kay MS, Kim PS. Protein design of an HIV—1 entry inhibitor. Science 2001; 291:884-888.
172. Singh M, Berger B, Kim PS. Learn coil VMF: Computational evidence for coiled-coil-like motifs in many viral membrane-fusion proteins. J Molec Biol 1999; 290:1031-1041.
173. Williams KL. Genomes and proteomes: Towards a multidimensional view of biology. Electrophoresis 1999; 20:678-688.
174. Service RF. Structural genomics offers high-speed look at proteins. Science 2000; 287:1954-1956.
175. Anderson NL, Matheson AD, Steiner S. Proteomics: Applications in basic and applied biology. Curr Opin Biotech 2000; 11:408-412.
176. Hasan UA, Abai AM, Harper DR, Wren BW, Morrow WJW. Nucleic acid immunization: Concepts and techniques associated with third generation vaccines. J Immunol Methods 1999; 229:1-22.
177. Schultz J, Dollenmaier G, Molling K. Update on antiviral DNA vaccine research (1998-2000). Intervirology 2000; 43:187-217.
178. Cichutek K. DNA vaccines: Development, standardization and regulation. Intervirology 2000; 43:331-338.
179. Ledwith BJ et al. Plasmid DNA vaccines: Investigation of integration into host cellular DNA following intramuscular injection into mice. Intervirology 2000; 43:258-272.
180. Manam SJ et al. Plasmid DNA vaccines: Tissue distribution and effects of DNA sequence, adjuvants and delivery method on integration into host DNA. Intervirology 2000; 43:273-281.
181. Lamb RA, Kolakofsky D. Paramyxoviridae: The viruses and their replication. In: Fields BN, Knipe DM, Howley PM, eds. Fields Virology. Vol 1. Philadelphia: Lippincott-Raven, 1996:1177-1351.
182. Hilleman MR. The business of science and the science of business in the quest for an AIDS vaccine. Vaccine 1999; 17:1211-1222.
183. Amara RR. Control of a mucosal challenge and prevention of AIDS in rhesus macaques by a multiprotein DNA/MVA vaccine. Science 2001; 292:69-74.
184. Cohen J. AIDS vaccines show promise after years of frustration. Science 2001; 291:1686-1688.

CHAPTER 2

Dendritic Cells: Important Adjuvants During DNA Vaccination

Ralph M. Steinman, Constantin Bona and Kayo Inaba

Abstract

Vaccine design focuses on the identification of safe forms of antigen that elicit protective immunity. Adjuvants are also critical for efficacy, especially for inducing strong T cell-mediated responses. Dendritic cells (DCs) are nature's adjuvants, specialized to capture and process antigens and exert several costimulatory functions that expand Th1 helper and cytolytic T lymphocytes. Antigen-bearing DCs, in the absence of additional adjuvants, immunize mice to develop antimicrobial and anti-tumor immunity.

This chapter concentrates on the features and mechanisms that allow DCs to control immunity. So-called immature DCs can capture antigens by many routes, including uptake of other cells transfected by a DNA vaccine. Maturing DCs then 1) generate large amounts of MHC-peptide complexes, 2) produce chemokines that recruit other DCs and T cells, 3) reshape their chemokine receptors, e.g., upregulate CCR7, to increase homing and function in lymph nodes, 4) release cytokines like IL-12 that activate natural killer cells and polarize T cells to the protective Th1 phenotype, and 5) express numerous T cell adhesion and costimulatory products. The latter include C-type lectins such as DC-SIGN, several TNF/TNF-R family members such as CD40 and TRANCE-R, and many B7 family molecules such as CD86. By mobilizing DCs in the setting of DNA vaccination, one therefore exploits specialized antigen presenting cells with many mechanisms for enhancing immunity. This adjuvant role has been demonstrated recently in humans who have been vaccinated with autologous, antigen-bearing DCs.

In the context of DNA vaccines, three valuable potentials of DCs are evident. 1) DCs are directly transduced with vaccine DNA, leading to presentation of vaccine antigens on MHC class I and II products. In mice, DCs are the main white blood cells that are directly transfected. Since it is known that T cells are primed to bone marrow derived cells, rather than nonhematopoietic cells, the transfected DCs likely account for the initial immune priming by DNA vaccines. However with current methods, the frequency of transfected DCs is small, and the cells are short-lived. 2) DCs can capture antigens from other cells, presumably including DNA-transfected muscle and skin cells that die during normal cell turnover. The resulting "cross presentation" of cellular antigens may allow DCs to expand and sustain vaccine memory in the CD4$^+$ helper and CD8$^+$ killer compartments. 3) DCs also respond directly to DNA and to specific CpG oligodeoxynucleotides. This means that DNA vaccines, distinct from their capacity to encode specific antigens, stimulate DCs to mature and become powerful stimulators of T cell immunity. As a corollary, if a DNA vaccine fails to stimulate DC maturation, then the cells might be able to induce different forms of tolerance and suppress immunity. DNA vaccination therefore brings into focus important areas of DC physiology. Reciprocally DC physiology should provide useful guidelines for improving the efficacy of DNA vaccination.

DNA Vaccines, edited by Hildegund C. J. Ertl. ©2003 Eurekah.com
and Kluwer Academic / Plenum Publishers.

Dendritic Cells as Effective Initiators of T Cell-Immunity

When a vaccine is being designed to elicit T cell-mediated immunity, especially Th1 helper cells and cytolytic T lymphocytes (CTL),[1] a central challenge is to deliver the vaccine to appropriate antigen presenting cells. In the case of DNA or other vaccines, the expression and delivery of a foreign protein by itself is insufficient. If simple expression of vaccine antigens would suffice to elicit immunity, we might have more candidate HIV-1 vaccines because the genes, proteins, and processed peptides of this virus have been known for some time. In this section, we consider some of the early work that revealed a role for DCs as nature's adjuvants for initiating T cell immunity. In the next section, we point out that DCs must mature to exert their adjuvant roles, and that maturation is stimulated by select infectious agents and DNA vaccines.

Microbial and Cellular Extracts Are by Themselves Poor Initiators of T Cell-Mediated Immunity

At the time that DCs were discovered in the 1970's, the standard experimental models of cellular immunity to microbial infection involved mycobacteria and Listeria monocytogenes. Killed bacteria and protein rich microbial extracts would elicit recall immune responses termed delayed type hypersensitivity. Yet these same proteins were insufficient as vaccines.[2] Likewise, people who are infected with *M. Tuberculosis* or the BCG vaccine readily express delayed type hypersensitivity during the PPD skin test. Yet this same skin test does not prime uninfected individuals, even when given on an annual basis for dozens of years. The interpretation, that one needs live organisms to elicit strong immunity, more or less restates the finding but does not explain why foreign antigens alone often are not immunogenic. An analogous situation held true for the most powerful T cell response known, graft rejection. Medawar, who discovered the immune basis of graft rejection, spent years trying to extract active transplantation antigens but without success,[3] even though his extracts very likely contained MHC products. In both these infection and transplantation systems, it was not known that strong cell-mediated immunity occurs when antigens are presented on viable DCs. Although it is not yet established that DCs are absolutely essential for the induction of various T cell responses in vivo, it is clear that DCs are effective adjuvants and express several underlying mechanisms to carry out this role.[4,5]

Antigen Presenting Cells in the Afferent and Efferent Limbs of T-Dependent Immunity in Culture

The starting point for experimentation on DCs was a system for studying primary immune responses to antigen, the Mishell-Dutton culture. In this system, mouse splenocytes formed IgM antibody to sheep red blood cells in a T cell dependent manner. In addition, nonlymphocytic, radioresistant "accessory cells" were needed to elicit the primary response. It was not known what the cells were or how they worked. Nevertheless, Mishell-Dutton cultures were the available model to ask questions about primary immune responses, to determine the requirements for converting a foreign antigen into an immunogen.

An analysis of the accessory cells revealed the presence of a new cell type, the DC, which had to be distinguished from macrophages to be isolated and characterized.[6-8] The DCs were purified on the basis of several distinctive properties such as a paucity of Fc receptors, presence of the 33D1 antigen, and poor adherence to plastic (in each case, using their unusual shape and motility as additional markers).[6-8] Human blood and tonsil also contained comparably distinct DCs.[9,10] The identification of DCs did not require the use of antibodies to MHC II products, the alternative means that was being used at that time to identify accessory cells. These antibodies initially were called anti-Ia or anti "I region associated" antibodies, because the I region (later, MHC II) was involved in immune responsiveness. The use of anti-MHC II antibodies in retrospect was a circular way to select active accessory cells, since MHC II later proved to be essential for antigen presentation to CD4$^+$ helper T lymphocytes.

After purification as distinct leukocytes, DCs were found to express very high levels of MHC II[7,9,10] and to act as remarkably potent stimulators of many different T cell responses. Small numbers of DCs mediated the antibody response to red blood cells and to hapten-carrier conjugates.[11,12] Potent direct stimulation of T cells was observed in the responses to carrier proteins[12] and to alloantigens in the mixed leukocyte reaction.[13] Prior work had used 1:1 ratios of antigen presenting cells to T cells, but enriched DCs were active at 1:100 and even lower ratios. When depleted of DCs, other antigen presenting populations, such as MHC class II positive B cells and macrophages were weak or inactive in initiating immunity in culture. While DCs initiated responses in culture, other antigen presenting cells proved to be critical for the effector limb of immunity. For example, in T-dependent antibody formation, DCs first present antigens to expand and differentiate CD4+ helper T cells. Then the activated helper cells respond very efficiently to antigens presented by B cells to bring about antibody formation.[12,14] Likewise, DCs activate T cells that induce macrophages in an MHC-restricted fashion to produce inflammatory cytokines.[15] DCs also stimulate CD8+ T cells in the afferent or first stage of the cytotoxic T cell response, whereupon these T cells kill targets presenting antigen in the efferent limb.[16,17] Therefore, many different types of antigen presenting cells work together to generate T cell-dependent immune reactions.

The Costimulatory Properties of Dendritic Cells

In the early days of research on DCs, their capacity to vigorously stimulate T cell growth was noted in many systems in which antigen processing was not required. The systems included the mixed leukocyte reaction to major transplantation antigens,[13] and T cell proliferation to mitogens[18,19] and later superantigens.[20] Therefore DCs were not simply antigen presenting cells, but expressed additional "accessory" functions or if one wishes "second signals" and "costimulatory" effects. This was especially evident in situations where the required T cell stimulus, e.g., a superantigen or anti-CD3 antibody, was estimated to be very small. As few as 100-200 T cell ligands per DC were sufficient to activate polyclonal populations of resting T cells,[20,21] again suggesting that DCs had well developed accessory functions. Ironically (below), DCs are proving to have special mechanisms for forming MHC-peptide complexes or "signal one."

The Potency of DCs in Initiating MHC-Restricted Immunity

An unusual feature of DCs was their potency. Small numbers of DCs resulted in strong T cell growth, CTL differentiation, and lymphokine production. Simultaneously, DCs controlled the MHC restriction of the response. If DCs of MHC-A were used to prime T cells of MHC-B in the afferent limb of the MLR, the activated T cells helped B cells of MHC-A, not MHC-B, to grow and produce antibody in the efferent limb of the response.[14] If DCs of MHC-A primed MHC-A helper T cells to the KLH carrier protein, the helpers only triggered antibody responses to a hapten-KLH complex if the lymphocytes were also MHC-A.[12] DCs in the afferent limb therefore control the MHC restriction observed in the efferent limb of immunity. Current research on the mechanisms underlying DC function are considered below, after we first discuss DC maturation and the quality of the T cell response induced by DCs.

Dendritic Cell Maturation: A Control Point for Initiating Immunity in Tissue Culture

The studies above indicate that vaccine antigens gain efficacy when presented on DCs. Yet the targeting of proteins or preprocessed peptides to DCs is not enough. DCs also must differentiate or mature into potent stimulators of T cell immunity.

Maturation of Epidermal Langerhans Cells in Culture

After DCs had been described in lymphoid tissues, the sites for generating primary immune responses, it was natural to turn to peripheral tissues, the sites for antigen entry. Skin[22] and lung[23] were the first examples. In skin, MHC class II rich cells (Langerhans cells) had been shown to be antigen presenting cells for recall responses.[24] However in epidermal cell suspensions, typical MHC II positive DCs could only be identified after the cells had undergone a series of major changes in culture. These cultured DCs expressed high levels of surface MHC products and potent T cell stimulatory function, the strongest that had been observed. The term "maturation" was used to describe the terminal differentiation of Langerhans cells to powerful DC stimulators. It was proposed that DC maturation was critical for converting antigens into strong immunogens.[22] Similar events were apparent in the lung[25] and spleen.[26]

Antigen Uptake and T Cell Stimulation by Dendritic Cells Can Be Separated in Time

A further surprise came when potent DCs, cultured from skin or spleen, were tested for their capacity to present protein antigens.[26,27] Presentation of proteins was actually weak or nondetectable, even though the same cells were potent presenters of alloantigens and preprocessed peptides. Instead, the protein had to be administered when the DCs were immature; otherwise the antigen seemed to be ignored. It later became evident that immature DCs could endocytose antigens by a number of routes: macropinocytosis, phagocytosis, and adsorptive or receptor mediated uptake.[28-31] Mature DCs in contrast had weak endocytic activity for many soluble tracers and particulates. Some underlying mechanisms are discussed below. Nevertheless, antigen capture and T cell stimulation were separate components of immunogenicity, and DC maturation encompassed both sequentially in time.

Dendritic Cell Maturation Stimuli, Including Select CpG Oligodeoxynucleotides

Maturation occurred "spontaneously" in the initial experiments in cultured mouse epidermis,[22] mouse spleen,[26,32] and human blood.[33] When the immature DCs were purified, cytokine requirements became evident including GM-CSF and other factors in monocyte conditioned media.[34,35] Subsequent studies identified a combination of TNFα, IL-1β, and the prostaglandin PGE2 as effective inflammatory mediators of maturation.[36,37] CD40L and TRANCE (RANK ligand) are additional, cell-associated, TNF family members that control DC maturation and function.[38,39] There likely will be more, including TNF/TNF-R molecules that DCs use to influence other cells. Certain necrotic cells can induce maturation,[40,41] possibly through heat shock proteins.[42]

Importantly, many microbial products stimulate DC maturation in culture, including LPS,[43,44] double stranded RNA,[45,46] and CpG oligodeoxynucleotide (ODN) sequences.[47,48] CpG ODN's additionally expand the number of mature DCs in lymphoid organs in mice.[49] Many receptors for maturation (IL-1R, TNF R like CD40 and CD120, and Toll receptors) transduce signals for NF-κB activation via the TNF receptor associated factor, TRAF 6. Toll receptors signal via the MyD88 adapter, but a MyD88 independent pathway is also apparent in DCs.[50] The implication of these experiments is that a vaccine for strong T cell-mediated immunity will require two major components. First the vaccine and/or the encoded antigens need to be captured and processed by DCs, and second, the vaccine must comprise a maturation stimulus to ensure NF-κB activation and the potent T cell stimulatory function of these cells.

Dendritic Cells as Nature's Adjuvant

The above tissue culture studies were extended by three major lines of evidence that DCs were major antigen presenting cells for initiating immunity in situ: 1) The distribution of DCs in vivo was consistent with their role as physiologic adjuvants (Figure 1); 2) DCs efficiently captured antigens administered to animals and were the main cell type expressing the antigen in a form that is immunogenic to T cells; and 3) DCs could be used in adoptive transfer experiments to actively immunize mice and rats (Figure 2). Therefore, mature DCs serve as powerful and natural adjuvants. This has been extended to humans (next section).

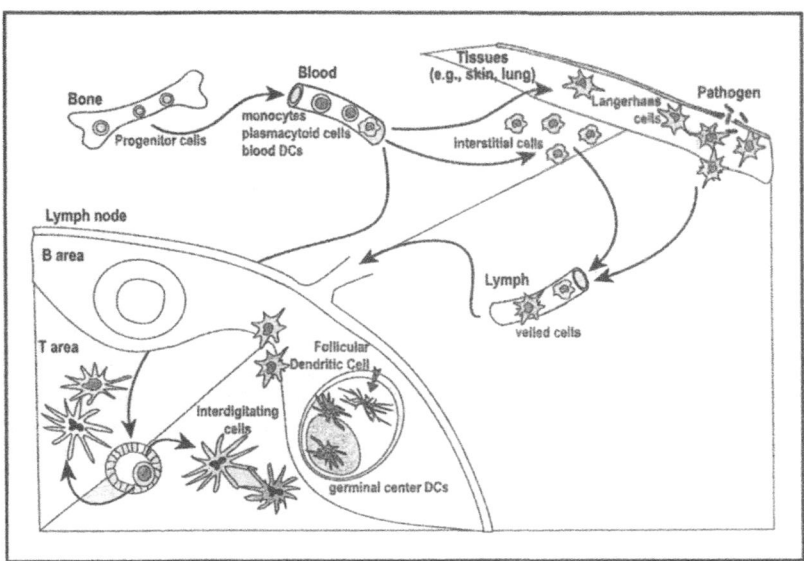

Figure 1. Distribution of dendritic cells in vivo. DCs are derived from proliferating progenitors in the bone marrow. The progenitors are responsive to flt-3L and G-CSF. The nonproliferating progeny or DC precursors are found primarily in blood as monocytes and plasmacytoid cells. Precursors typically require days in culture and several differentiation stimuli (CD40L, many cytokines) to become DCs. In contrast, immature DCs rapidly (< 1 day) mature upon encountering microbial stimuli and inflammatory cytokines. Immature DCs are found in blood, in skin (Langerhans cells) and other epithelia such as the airways, and the interstitial spaces of many organs. DCs, termed "veiled cells", move from peripheral tissues through afferent lymphatics to the T cell areas of lymphoid organs. DCs also can be found in the germinal center, where they are different from the stromal "follicular dendritic cells." The latter retain native antigens as immune complexes for presentation to B cells, whereas DCs present processed antigens to T cells.

Distribution of Dendritic Cells in vivo

The distinctive tissue distribution of DCs (Figure 1) was outlined using a common set of criteria, especially the isolation of MHC II rich, stellate, nonadherent, nonphagocytic cells. The cells were abundant in the T cell area of peripheral lymphoid organs: spleen, lymph node, and Peyer's patch,[51] where they formed an extensive network of MHC II rich processes. Even though DCs were outnumbered by B cells in suspensions of lymphoid cells, 50 to 1 approximately, their abundant MHC II products and distribution dominated the T cell area. DCs were also identified in peripheral tissues, like skin or airways, as well as afferent lymphatics[52-56] and blood.[9,57] Therefore DCs are distributed in a way that facilitates antigen capture and transport to lymphoid organs.

It is known that immune responses begin in peripheral lymphoid tissues, and that immunization leads to substantial but temporary depletion of antigen-reactive cells from the circulation. DCs help to explain this finding. DCs are designed to take antigen from wherever it is deposited, migrate to the lymphoid organs, and then efficiently select relevant T cell clones from the recirculating pool to start the immune response, as has been observed directly in situ.[58,59] Following clonal expansion and differentiation in the T cell area, T cells can reenter the circulation, returning as effectors to the site of antigen deposition. DCs in contrast are not found in efferent lymphatics. In sum, when antigens in the periphery are captured by DCs, they can gain access to the rare, antigen-specific, naïve T cells in the recirculating pool. The DCs then induce numerous activated effector T cells, which return to the initial sites of inflammation.

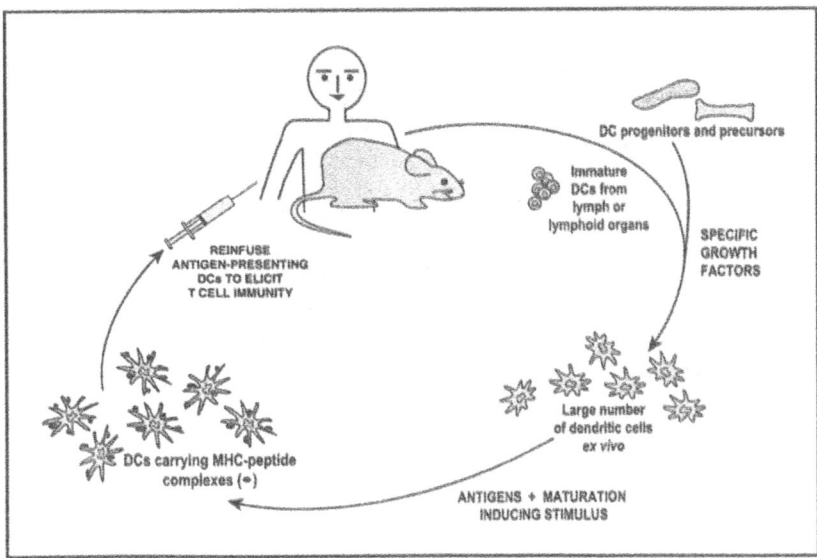

Figure 2. Demonstrating the role of DCs as adjuvants in vivo. If DCs from lymphoid tissues, or DCs derived from progenitors, are charged ex vivo with antigens, exposed to a maturation stimulus, and then injected into syngeneic animals or autologous humans, the DCs induce specific, CD4+ and CD8+ T cell immunity.

Therefore the distribution of immune cells in situ is such that events in the afferent limb of an immune response are orchestrated by DCs, while events in the efferent limb are the proviso of many other antigen presenting cells.

Efficient Capture of Antigens by DCs in vivo

If antigens were administered in vivo by different routes (skin, airway, muscle, gut, blood stream), and then the corresponding depots of DCs were isolated (Figure 2), the DCs presented antigen to specific T cells in culture.[55,56] In fact, when a complex organ like spleen was analyzed, DCs were the main cell type capturing antigens in vivo in a form immunogenic for T cells.[60] This does not imply that DCs are the only cells to capture antigens, or even to form MHC-peptide complexes, but instead that DCs are the main reservoirs of "immunogen".

Antigen Pulsed DCs Prime T-Dependent Immunity in Rodents

DCs were used as adjuvants to actively immunize or prime rodents. The DCs were obtained from lymphoid organs and lymph, or generated from progenitors in marrow. The cells were pulsed with antigens during their maturation ex vivo and used to prime mice or rats in an antigen-specific manner (Figure 2). The induced T cell immunity was specific for the antigens and MHC of the injected DCs.[26,56] When tested, mature DCs were more effective in priming T cells.[25,61-63] It was also feasible to elicit protective anti-microbial and anti-tumor immunity using antigen-pulsed DCs without other adjuvants.[64-66] The DCs in these experiments had undergone maturation ex vivo prior to injection. Therefore these adjuvant findings need not represent DC function in the steady state in vivo, where a role in peripheral tolerance is becoming apparent (next section).

Human DCs Control the Quality of the Immune Response: New Findings from Studies in Humans

It has been difficult to identify adjuvants for amplifying strong T cell-mediated immunity in humans. Once it became possible to prepare large numbers of DCs from different precursors, it was exciting to test their capacity to elicit immunity in humans, as observed in experimental animals above (Figure 2). In fact, strong immunity has now been induced with DCs in humans, both Th1 type CD4+ helpers and CD8+ cytolytic T lymphocytes.

Initial Use of DCs to Actively Immunize Patients Against Cancer

Cancer has been the first setting in which autologous DCs have been charged with antigens ex vivo, exposed to maturation stimuli, and then reinfused to try to elicit immunity. In all cases, the vaccinations have been nontoxic. Although this approach is in its early stages in terms of methodology, DC vaccination already has expanded T cell immunity, detectable in fresh blood samples from the vaccinees.[67,68] To date, when other adjuvants have been assessed in humans, it is typically necessary to measure immunity after prolonged restimulation of the blood sample in culture. Furthermore, some striking regressions in the setting of metastatic disease have been reported following vaccination,[69-71] again using methods that only begin to exploit DC physiology.

Antigen-Bearing DC Reliably Expand T Cell Immunity in Healthy Volunteers

The apparent safety of DCs as adjuvants led to studies in healthy volunteers, to determine immune efficacy in a vaccine vs. therapeutic setting. DCs were pulsed with model antigens. These were keyhole limpet hemocyanin, KLH, as a priming protein; tetanus toxoid as a recall protein; and influenza matrix peptide as a recall CD8+ T cell epitope. The injected DCs rapidly expanded T cell immunity.[72] The conditions of these new DC vaccinations experiments have been difficult to predetermine. It was elected to inject 2-4 x 10^6 mature DCs subcutaneously. However, much needs to be learned about the dose, frequency and route of DC administration. Many features of DC biology (below) also remain to be manipulated.

Improving the Quality and Affinity of the T Cell Response with Mature DCs

Subsequent studies in humans revealed two critical ways in which DCs improved the quality of the T cell response (Table 1). First, mature DCs rapidly polarized the CD4+ T cell response to the Th1 type. After a single injection of KLH-pulsed DCs, the antigen-dependent cytokine producing cells in blood made IFNγ but little or no IL-4.[73] Th1 IFNγ-secreting cells are known to be more protective and lead to better memory in experimental models of tumor growth and virus infection.[74-77] Second, mature DCs pulsed with an MHC class I restricted peptide, are able to improve the quality of T cell memory. After a second booster dose of peptide-pulsed DCs, the antigen-specific CD8+ T cells recognized antigen at 10-100 fold lower doses than observed initially.[78] Likewise, when booster doses of DCs were given with tetanus toxoid protein in cancer patients, strong delayed type hypersensitivity reactions were induced by the 3rd dose.[79] In other words, DCs not only control T cell priming, but they also influence the quality of the T cell response and can boost T cell memory (Table 1).

Silencing and Regulating Immune Responses with Immature DCs, One Mechanism for DC-Based Tolerance

The biggest surprise came when immature DCs were tested as APCs. It had been assumed from several studies in mice[25,61-63] that immature DCs were simply ignored, because they lacked several accessory properties to be outlined in the next section. However the immature DCs actually silenced the CD8+ IFNγ producing cells that were present prior to vaccination.[72] The silencing was specific for the antigen that had been given on the immature DCs, and it was

Table 1. Classical and new functions for DCs

Classical	New
Adaptive immunity	Innate immunity and tolerance
Priming	Th1/Th2; T regulatory cells
T cells	B, NK-T, NK cells

accompanied by the appearance of IL-10 producing T cells. In one of the individuals vaccinated with immature DCs, sufficient blood was available to show that functional regulatory T cells were induced (unpublished). More detailed in vitro experiments have shown that immature DCs elicit IL-10 producing regulatory T cells,[80] which markedly reduce the function of preformed Th1 type cells. Therefore, DCs can both enhance and dampen immunity in an antigen-specific way, and their state of maturation has a critical role in these distinct outcomes (Table 1).

Induction of regulatory T cells is one of the ways that DCs could mediate peripheral tolerance, an emerging area of DC function. The importance of DCs in tolerance was stimulated by the observations that DCs could process antigens from dying influenza-infected cells.[81,82] During influenza infection, there is extensive cell death, e.g., the airway epithelium is almost entirely killed. It is reasoned[83] that DCs have no way to distinguish microbial from self antigens in the dying airway epithelium, and likewise, to distinguish microbial proteins from the many environmental nonself proteins in the airway. Yet recovery from influenza or other infections is generally not associated with chronic immunity to the airway cells or its lumenal proteins. It has therefore been proposed that the critical function of DCs in the steady state is to induce tolerance, either through the induction of T regulatory cells, or the deletion and anergy of antigen-responsive T cells. Both of these functions have now been demonstrated for DCs.[80] Therefore, the immune response is not solely controlled by DC maturation[22] or danger[84] or microbial pattern recognition.[85] Instead, DCs in the steady state must first induce peripheral tolerance to those peptides from environmental proteins and self tissues that could later be presented when infection induces DC maturation. In this way, the immune response can safely focus on microbial antigens.

Other Types of Immune Responses and Subsets of DCs

Dendritic Cells Influence Other Classes of Lymphocytes, not Just T Cells

DCs are able to control other parts of the immune system. These other classes of lymphocytes have been the subject of only a few studies, primarily in culture. DCs have direct effects on human B cells inducing growth and high-level Ig secretion, even switching to the IgA isotype.[86] Interestingly, monocyte-derived DCs but not Langerhans cells have this B cell stimulatory function.[87] DCs present glycolipids to NK-T cells, resulting in high levels of IL-12 production.[88] Through this IL-12, and by direct interactions,[89] DCs have the potential to recruit NK cells.[90] Therefore, with the appropriate stimuli and ligands, DCs can control several types of lymphocytes, not just T cells (Table 1).

DCs Link Innate and Active Immunity

DCs have an innate capacity to respond rapidly to microbial stimuli, and in so doing, set in motion the adaptive immune response. Microbial extracts can selectively induce high-level production of IL-12 by DCs in the T cell areas of lymphoid organs in mice.[91] This IL-12 would be expected to recruit NK cells and help polarize naïve T cells to the Th1 type. T cell area DCs make IFNγ,[92] while plasmacytoid DCs release prodigious amounts of interferon-α.[93]

Table 2. Specialized mechanisms for antigen capture and MHC-peptide formation in DCs

---

- Adsorptive endocytosis receptors
    e.g., DEC-205, FcγR, receptors for heat shock proteins
- Cross presentation/exogenous pathway to MHC class I
    e.g., dead and dying cells, immune complexes
- Regulation of the endocytic system during maturation
    e.g., distribution of MHC class II, cystatin C, endocytosis via Cdc42, proteolysis
- Macropinocytosis
    e.g., select aquaporins
- Immunologic synapse formation
    e.g., coassembly of MHC class I, II and CD86

Interferons activate NK cells and increase antigen presentation on MHC class I and II products, in addition to their more standard anti-viral functions. DCs can directly stimulate NK-T[88,94] and NK cells,[89] which can kill certain virus-infected and tumor targets. Innate NKT and NK cells themselves make protective cytokines, especially interferon-γ, to increase antigen presentation and thereby adaptive immunity. These many and powerful roles of DCs in innate responses are proving to be a major sphere of DC function.

Subsets of DCs

In addition to distinct stages of DC maturation, distinct forms of immature and mature DCs are being identified. In mice, peripheral lymphoid tissues, especially spleen, have at least two subsets: one with high expression of CD8 and the DEC-205 endocytosis receptor (below) and low expression of CD11b integrin; the other has low CD8 and DEC-205 but high CD11b integrin.[95] Following isolation, the CD8+ subset is main adjuvant for Th1 type immunity,[96] while in vivo, the CD8+ subset appears to be the main cell that presents exogenous antigens on MHC class I[97](see below). In human blood, there are separate subsets of CD11c positive and CD11c negative DC precursors that are also called DC1 and DC2.[98] The former can produce very large amounts of IL-12 and the latter very large amounts of IFNα.[93] When DCs are generated from CD34+ progenitors, at least 2 distinct types of DCs are produced: Langerhans cells, and interstitial or dermal DCs.[87] Only the latter can stimulate B cell differentiation directly,[99] while Langerhans cells may stimulate CD8+ T cells better. This is a very brief summary of DC subsets, an important emerging area of DC biology. Vaccine design has yet to exploit these subsets.

Some Mechanisms Underlying DC Function

Dozens of molecular events underlie DC function and the control of the immune response. It is helpful to divide this intricate physiology into sets of signals, though different authors have different ways of doing this. Here we refer to signal 1 as the many events involved in the formation of ligands for the TCR, i.e., antigen presentation; signal 2 as the many surface molecules that mediate T cell adhesion and activation, as well as secreted cytokines; and signal 3 as the mechanisms mediating DC function in situ.

MHC-Peptide Complex Formation—Signal 1

Immature DCs are able to take up substrates through pinocytosis and phagocytosis. A maturation stimulus subsequently regulates DC endocytic activity, proteolysis, and the formation and transport of MHC-peptide complexes. Upon maturation, uptake of fluid and particles by pinocytosis and phagocytosis decreases, through the inactivation of Cdc42, a Rho-family GTPase.[31] DCs also can downregulate the levels of cystatin C within lysosomal compartments.[100] The loss of cystatin C, an inhibitor of cathepsin S, should increase proteolysis of the invariant chain, which in turn should increase exchange of antigenic peptides with CLIP and movement of the MHC-peptide complex to the cell surface. Proteolysis of antigens through other cathepsins also can be regulated by cytokines in DCs.[101] In sum, during maturation, many DCs typically cease taking up additional substrates but efficiently convert acquired substrates into MHC-peptide complexes.

DCs have an unusual endocytic receptor termed DEC-205.[102] This receptor traffics through the endocytic system in a distinct way, being able to enter and recycle through MHC II positive late endosomes or lysosomes. The ligand recognition properties of DEC-205 are not yet known, but targeting of surrogate ligands through this receptor has the potential to increase the efficiency of antigen presentation on MHC II by 10-100 fold.[103] At this time, DEC-205 is an excellent candidate for future attempts to target vaccines better to DCs.

A striking feature of DCs is termed the exogenous pathway of presentation on MHC class I, or cross presentation. Immune complexes (Fcγ receptor) and dead or dying cells (the αvβ5 integrin is one relevant receptor) are efficiently taken up by immature DCs.[82,104] Somehow these endocytosed, nonreplicating substrates gain access to the necessary proteosomal and TAP machinery for presentation on MHC class I. DCs also present nonreplicating forms of viruses,[105] but here the viral envelope or capsid proteins likely deliver viral antigens to the cytoplasm. Normally, MHC class I molecules are charged with peptides that are newly synthesized in the cytoplasm (the endogenous pathway), but with DCs, inanimate immune complexes and dead cells are processed onto MHC I (the exogenous pathway).[81,106] DCs also present peptides from dying cells on MHC class II with very high efficiency.[107,108] A fascinating recent example is the EBNA1 molecule that DCs efficiently process from Epstein Barr Virus transformed B cells.[108] Epstein Barr Virus causes an acute lytic infection of B cells, e.g., during infectious mononucleosis. The processing of dying B cells by DCs may explain why healthy carriers of EBV typically show a Th1 polarized, CD4+ T cell response to EBNA1[109] and are able to contain this transforming virus.

The processing of dead cells at first seems to violate the beauty of MHC restricted recognition, whereby T cells focus their function on targets of replicating microbial antigens. Cross presentation of dead cells would allow T cells to attack uninfected targets that present peptides via the noninfectious, exogenous pathway. However cross presentation appears primarily to be a function of DCs, and it is proposed that its major role is to allow DCs to induce tolerance. The reasoning is as follows. Following microbe-induced cell death, phagocytic immature DCs would seem unable to distinguish microbial proteins from cellular self proteins, or proteins that are normally resident in the external environment like the airway and intestine. In other words, if "maturation", "danger", or "pattern recognition" were the only control of immunogenicity as discussed above, then during infection we would all become immunized to the many environmental and self antigens that DCs would inevitably process (during the maturation associated with microbial infection). Therefore it is argued that DCs in the steady state (e.g., while you are reading this chapter) are specialized to cross present antigens, captured as a result of normal cell turnover.[83] Without a maturation stimulus in the steady state, these self proteins are presented in a tolerogenic way, perhaps through the induction of regulatory T cells,[73,80] or through deletional and anergic tolerance.[110] Then, when the DCs are subsequently required to process cells dying during infection and associated maturation stimuli, the immune response focuses in an MHC-restricted way on the infected cells, and importantly, on the microbe rather than self and environmental proteins. As a corollary, those self and environmental peptides that are not presented by DCs in the steady state are in essence non-self. Neurons for example do not undergo extensive turnover or capture by DCs, so the exogenous pathway may not be available to elicit tolerance to many self peptides in neurons.

The capacity of DCs to present cellular antigens in a tolerogenic way is of importance in DNA vaccination. This type of vaccine stably expresses antigens in muscle or skin, and over the long term, there may not be any inflammatory stimulus at the injection site. Therefore there is a chance that the vaccine can elicit tolerance rather than immunity. As we shall discuss below, the priming of CD4[+] T helper cells during vaccination, may be critical to ensure that the stable reservoir of vaccine antigen sustains memory rather than induces tolerance.

T Cell Binding and Costimulation—Signal 2

The potency of DCs is far from explained, but several mechanisms are beginning to emerge. Again, these mechanisms can be influenced by maturation. A lot of the work has been done with what are termed myeloid, particularly DCs produced by stimulation with GM-CSF from mouse bone marrow progenitors and human monocytes.

DC-SIGN is a newly described C-type lectin that binds to ICAM-3 on resting T cells. It is proposed that DC-SIGN (DC Specific, ICAM-3 Grabbing, Non integrin) allows DCs to form loose conjugates with T cells in an antigen-independent fashion.[111] Once the loose DC-T cell conjugates form, it is further proposed that the immunologic synapse can begin to assemble and fire. More direct work is needed to examine DC-SIGN function and synapse formation in naïve T cells.

Many costimulators are expressed by mature DCs including CD54/ICAM-1, CD48 and CD58/LFA-3, several B7 family members including CD80 and CD86, and some TNF related costimulators like 4-1BB ligand and BAFF/BlyS. The powerful CD86 co-stimulator is of some interest. It is very abundant on DCs relative to other leukocytes, and it is rapidly upregulated during maturation.[112,113] At the DC surface, CD86 is clustered together with MHC-peptide in surface patches.[114] The formation of these surface aggregates suggest that DCs are fully prepared, before they contact the T cell, to co-assemble TCR and CD28 molecules and form the supramolecular immunologic synapse. If it is important for CD28 to be juxtaposed to the TCR complex for costimulation to take place, then mature DCs are designed to set up a functioning APC-T cell synapse with naïve T cells.[114] Much of the published work on immunologic synapse formation has used previously activated T cells, but the abundance and distribution of B7 molecules on DCs may be critical to set up the synapse in naïve T cells.

The B7 family is proving to have additional unusual features on DCs, particularly as it relates to Th1/Th2 polarization and the quality of the immune response. Mature DCs can lack ICOS-ligand, a key Th2 polarizing B7 family member.[115] Other DCs express the new B7-DC and B7-H3 molecules that can induce IFNγ from naïve T cells.[116] These new B7 molecules could account for the strong Th1 polarizing activity observed when DCs are used as adjuvants in humans.[73]

Upon receipt of a maturation stimulus, DCs can make several cytokines that act on other cells. IL-12 can be made in particularly large amounts. The new IL-12 relative, IL-23, is also produced,[117] as is the memory sustaining cytokine IL-15.[118] The control of IL-12 production is intricate.[119-121] Other cytokines (GM-CSF, IL-4, IFNγ) influence bioactive IL-12 synthesis following encounter of a maturation stimulus. Abundant cytokine production ceases within 12 hours of receiving a maturation stimulus,[121,122] whereupon the DCs have been termed "exhausted".[122] Nevertheless, these DCs are far from exhausted in terms of their capacity to stimulate T cell immunity, including Th1 responses in vivo.[73] IL-1β also can be made by DCs and may act back on the DCs to stimulate some of the changes in maturation. In the past, IL-1 was thought to be a major lymphocyte activating factor, but its activating role seems directed more to DCs and their development.[123,124] Plasmacytoid DCs make very large amounts of IFNα but little IL-12, while monocyte-derived DCs seem to make relatively little IFNα but very large amounts of IL-12.[93] Immature DCs can make abundant IL-10,[125,126] which would be expected to be immunosuppressive. Importantly, mature DCs resist the immunosuppressive effects of IL-10.[127] Therefore DCs can produce many costimulatory cytokines and membrane molecules, and all seem to be critically influenced by the process of maturation.

Mobilization and Movement in vivo—Signal 3

Figure 1 diagrams some of the sites where DC function can be manipulated in vivo. DCs traffic through many tissues in the steady state. For example in rat lung epithelium and mouse spleen, DCs are turning over with a half time of just 2 days.[128,129] There is a constant traffic of DCs in the lymph, since cannulation always reveals a substantial flux of several thousand DCs per hour.[52,54,55] One source of DCs in lymph could be blood monocytes. Important cues for monocyte differentiation into DCs can be provided during their reverse transmigration across endothelium.[130] MIP-3α acting on CCR6 is a candidate to explain the steady state recruitment of Langerhans cells to the epidermis,[131] and other immature DCs to mucosal associated lymphoid tissues.[132] DCs can also be mobilized in larger numbers. For example, precursor and immature forms of DCs increase 5-10 fold in blood and other tissues following systemic administration of flt-3 ligand and G-CSF.[133,134]

During two powerful immune stimuli, contact allergy and transplantation, epidermal LCs in situ begin to migrate and to mature, i.e., MHC II and costimulatory molecules are upregulated.[135] A hallmark of DC maturation is the upregulation of functional CCR7 receptors for CCL19 and CCL21, two chemokines that are made constitutively in the T cell area (the sources might be DCs, other stromal cells, and certain endothelia).[136-139] CCR7 knockout mice show poor migration of epidermal Langerhans cells in response to contact allergens.[140] Recently, it has been found that multidrug transporters (MDR-1 and MRP) can participate in DC migration, probably by pumping cysteinyl leukotrienes, which in turn improve CCR7 responsiveness.[141] CD40L, which is found on mast cells and platelets and not just activated T cells, also is critical for DC mobilization in situ.[142]

DCs produce several chemokines. The most abundant are: CCL19 (ELC or MIP-3β), which recruits naïve T cells via CCR7;[139] DC-CK1 or PARC, which also recruits naïve T cells,[143] but the chemokine receptor is not yet known; CCL18 (TARC), which recruits central memory and Th2 cells via CCR4;[144,145] CCL22 (MDC, ABCD-1), which also reacts with CCR4 on memory and Th2 cells;[146] and CX3CL1 (fractalkine), which recruits activated T cells cells via CXCR3. Interestingly, mature DCs release fractalkine,[147,148] and this could attract immature DCs such as the CXCR3 expressing plasmacytoid DCs.[149] Immature DCs also make several inflammatory chemokines, like CCL3 (MIP-1α), CCL4 (MIP-1β), and CCL5 (RANTES),[145,150] which would recruit more DCs via CCR5. The production of chemokines by DCs is so intricate that many reviews have been needed.[136,151]

A current conundrum is to identify the controls on DC migration in the steady state vs. inflammation. As mentioned above, immature DCs are migrating constitutively in lymphatics. What directs this migration? Might fractalkine produced by DCs in the T cell areas[147] control the movement of immature DCs from the periphery? What are the roles for the marked upregulation of CCR7 on maturing DCs? Does the CCR7 system have a costimulatory role for the immune response beyond a role in chemotaxis of mature DCs? Does CCR7 allow DC migration to be more rapid during inflammatory stimuli? The control of DC migration and function in the T cell area is central to understanding the consequences of DC traffic in the steady state and during mobilization with a vaccine, since immature DCs may lead to tolerance and/or T regulatory cells and mature DCs to strong Th1 and CTL immunity.

After DCs reach the lymph node, their life span is short because the cells are not found in efferent lymph. Nonetheless, lifespan can be prolonged by ligation of CD40 or TRANCE-R, thus enhancing immunogenicity.[39] Given all the mechanisms that are being identified to enhance DC numbers and function in vivo during infection and inflammation, it may be valuable to mimic these pathways to enhance the efficacy of vaccines, and also, to decrease the proposed tolerogenic roles of DCs in the steady state.

DCs as Mediators of DNA Vaccination

The implication of the above findings is that the efficacy of a DNA vaccine would be increased if the encoded antigens were presented by mature DCs. The evidence that DCs play a key role in DNA vaccination comes from several avenues of experimentation in mice. The experiments also suggest sites for further research to improve DNA vaccine efficacy.

Bone Marrow Derived Cells, not Somatic Cells, Present Antigens Encoded by DNA Vaccines

Several publications have monitored the MHC restriction of T cells that are primed by DNA vaccines in bone marrow chimeric mice. In these animals, parental marrow (MHC-A) is used to reconstitute F1 (MHC-A x B) recipients. Therefore the MHC expressed by somatic cells e.g., the muscle or skin cells that are the main sites for expression of the injected DNA, differs from the MHC of the bone marrow derived cells, e.g., the DCs. Nevertheless, the vaccinated T cells typically recognize antigen in the context of the bone marrow (MHC-A) and not the somatic cells (MHC-B as well as MHC-A).[152-154] This indicates that leukocytes, not somatic cells, prime T cells during DNA vaccination of mice, even though the vaccine is expressed primarily in nonhematopoietic cells.

DCs Are Directly Transduced During DNA Vaccination in Mice

In the above chimera experiments, the bone marrow derived cells could either be directly transduced by the DNA vaccine, or pick up proteins expressed in non bone marrow derived cells. Casares et al detected DNA in DCs from DNA vaccinated mice, but not in other cells.[155] When DCs were isolated from the vaccinated mice, these cells were the main presenters of the vaccine antigen, an influenza hemagglutinin, to MHC class II restricted CD4+ T cells.[155] Nonetheless the isolated DCs presented antigen poorly relative to that seen when antigenic peptide was added directly to the tissue culture assay for presentation to CD4+ T cells. This means that the amount of antigen presented by individual DCs from vaccinated mice was very small, or more likely, the frequency of DCs presenting antigen was small. Bot et al extended the data with an MHC I restricted epitope, the influenza nucleoprotein.[156] They visualized cells expressing native viral nucleoprotein encoded by the DNA vaccine. The protein was expressed and targeted to the nucleus in ~ 2% of the DCs from the DNA vaccination site. These DCs were sufficient to induce virus specific CTLs following injection into the spleens of naïve mice.[156] Akbari et al[157] scarified the ears of mice with a full length DNA for complement C5. They showed that DCs were the main cell expressing C5 antigen in the draining lymph node, while keratinocytes expressed the DNA in the skin. Again ~2% of the DCs were estimated to express the vaccine DNA, which could not be detected beyond 2 weeks in the node. These experiments all reveal that DCs in the draining lymph nodes have been transduced by DNA vaccines in situ, but they do not formally distinguish whether in addition, the lymph node DCs capture vaccine antigen from other cells.

Porgador et al designed an experiment to prove that directly transduced DCs were the major source of presented antigen at the early time points after vaccination.[158] They co-administered two DNA plasmids, one encoding the vaccine antigen and the other human CD4 as a marker. They again found that DCs were the main cell presenting the vaccine, but in addition, the presenting cells could be selected with an antihuman CD4 antibody. The latter would fail to react with most of the nontransduced DCs picking up antigens by the exogenous pathway.

Taken together, these results show that DCs are transduced during DNA vaccination and that these transduced DCs are the main source of presented antigen in the lymphoid tissues shortly after vaccination. Clearly the frequency of DNA transfected DCs has been very small in all four studies above, either at the DNA injection site or in the draining lymphoid tissues where the immune response is likely to be generated. Furthermore, DCs might also acquire vaccine antigen from other cells, especially later in the course of vaccination when the antigen is primarily expressed in nonhematopoietic cells at the vaccination site.

After the Priming Phase of DNA Vaccination, Are DCs Continuing to Cross Present Antigen from Other DNA Vaccinated Cells and Is This Important for Expanding Vaccine Immunity?

These questions have not yet been answered directly but may be critical unknowns in the efficacy of vaccination, whether it involves plasmid DNA vaccines or other vector-based vaccines. In the work of Akbari et al,[157] DNA bearing DCs could not be detected for >2 weeks after vaccination whereas CD4+ memory T cells persisted >40 weeks. Do small numbers of transfected DCs prime the immune system early on, but then additional DCs continue to capture antigens released or phagocytosed from transduced somatic cells (muscle, fibroblasts, keratinocytes) in vivo? Akbari et al showed that DCs could acquire antigen from keratinocytes in culture, as long as the skin cells were killed by irradiation, presumably enhancing phagocytosis and cross presentation by the DCs as discussed above. There are other studies showing that bone marrow derived cells, possibly DCs, can acquire antigens from somatic cells,[159] e.g., a muscle transplant, expressing the DNA vaccine.[154]

A critical part of immune memory is the interaction of CD4+ helpers and CD8+ cytolytic T cells. When a CD4+ T cell acts on a DC, e.g., via CD40L, the DC matures and presents antigen better to CD8+ T cells.[160-162] Antigen-specific CD4+ T cells are most likely primed by DCs acutely during DNA vaccination, but they may also continue to act chronically on DCs that capture antigens from vaccine-transduced somatic cells, especially transduced cells that die during normal cell turnover? This would allow the DNA-transduced muscle or skin to act as a source of antigen that sustains or expands the immune response. In mice, CD4+ helper T cells can sustain CD8+ killer T cell function following DNA vaccination,[163] although the time at which the T cells are acting (priming, memory) is not yet clear. CD4+ Th1 helpers appear critical for CD8+ T cell memory in other situations, especially tumors[74] and chronic viral infections.[76,77] During the priming phase of the immune response, when there is more likely to be inflammation and other stimuli (e.g., the DNA vaccine itself, next section) to mature the DCs, CD4+ T cells may not be essential. This is because mature DCs are capable of stimulating CD8+ T cells in the absence of helper cells.[16,17] In the longer term, however, there is unlikely to be a stimulus for DC maturation, except if DCs present peptides on MHC II and mature in response to CD40L[38,62] or other ligands[164,165] on CD4+ helper cells. Therefore it will be helpful in DNA vaccination to monitor more closely the presence and kinetics of different types of CD4+ and CD8+ T cells. It will probably be important to try to increase both arms of the T cell response to enhance immunogenicity in the priming and memory phases of vaccination.

Responsiveness of DCs to the Adjuvant Action of DNA Vaccines and CpG Oligodeoxynucleotides (ODNs)

Although only 1-2% of DCs express vaccine DNA in lymph nodes, these vaccines (including the DNA vector itself) globally activate most DCs in the node.[157] In other words, most DCs—not just the transfected cells—upregulate their costimulatory molecules like CD40, 54, 80 and 86 following administration of DNA. As mentioned, DCs are responsive to CpG deoxyoligonucleotides (ODN's), undergoing maturation in vitro[47-49] and in vivo.[50] Therefore a DNA vaccine has the capacity to carry out the two key features of vaccination: to express vaccine antigens in DCs and to mature even nontransfected DCs to their potent stimulatory function. While direct transduction of DCs is valuable, one could argue that as long as DCs are being matured while acquiring antigens from other sources (like DNA transfected muscle or skin), the net effect should be the same, strong immunity. It may not be straightforward to increase the number of DCs that are directly transduced by the DNA vaccine, but it may be more feasible to enhance cross presentation and maturation by many nontransduced DCs.

Conclusion

There may be many points at which DCs play a role in DNA vaccination. DCs can be directly transfected and prime Th1 CD4$^+$ helpers and CD8$^+$ killer cells that together provide strong T cell-mediated immunity. It is possible that DCs also acquire antigens from other transfected cells, to further boost the response and to provide the kind of long-term memory upon which successful vaccination depends. DNA vaccination itself can act as a distinct maturation stimulus for DCs, a vital step towards immunogenicity, whether the DC acquires antigen directly by transduction or by cross presentation. These issues will require further research, but they probably lie at the heart of improving the efficacy of DNA vaccination and eventually, the use of DNA vaccines to regulate and tolerize the immune system as well.

References

1. Seder RA, Hill AV. Vaccines against intracellular infections requiring cellular immunity. Nature 2000; 406:793-798.
2. Mackaness GB, Blanden RV. Cellular immunity. Prog Allergy 1967; 11:89-140.
3. Medawer PB. The immunobiology of transplantation. Harvey Lect 1967; 52:144-176.
4. Banchereau J, Briere F, Caux C et al. Immunobiology of dendritic cells. Annu Rev Immunol 2000; 18:767-811.
5. Thery C, Amigorena S. The cell biology of antigen presentation in dendritic cells. Curr Opin Immunol 2001; 13:145-51.
6. Steinman RM, Cohn ZA. Identification of a novel cell type in peripheral lymphoid organs of mice. II. Functional properties in vitro. J Exp Med 1974; 139:380-397.
7. Steinman RM, Kaplan G, Witmer MD et al. Identification of a novel cell type in peripheral lymphoid organs of mice. V. Purification of spleen dendritic cells, new surface markers, and maintenance in vitro. J Exp Med 1979; 149:1-16.
8. Steinman RM, Gutchinov B, Witmer MD et al. Dendritic cells are the principal stimulators of the primary mixed leukocyte reaction in mice. J Exp Med 1983; 157:613-627.
9. Van Voorhis WC, Valinsky J, Hoffman E et al. Relative efficacy of human monocytes and dendritic cells as accessory cells for T cell replication. J Exp Med 1983; 158:174-191.
10. Hart DN, McKenzie JL. Isolation and characterization of human tonsil dendritic cells. J Exp Med 1988; 168:157-170.
11. Inaba K, Steinman RM, Van Voorhis WC et al. Dendritic cells are critical accessory cells for thymus-dependent antibody responses in mouse and man. Proc Natl Acad Sci USA 1983; 80:6041-6045.
12. Inaba K, Steinman RM. Protein-specific helper T lymphocyte formation initiated by dendritic cells. Science 1985; 229:475-479.
13. Steinman RM, Witmer MD. Lymphoid dendritic cells are potent stimulators of the primary mixed leukocyte reaction in mice. Proc Natl Acad Sci USA 1978; 75:5132-5136.
14. Inaba K , Steinman RM. Resting and sensitized T lymphocytes exhibit distinct stimulatory [antigen-presenting cell] requirements for growth and lymphokine release. J Exp Med 1984; 160:1717-1735.
15. Koide S, Steinman RM. Induction of interleukin-1α mRNA during the antigen-dependent interaction of sensitized T lymphoblasts with macrophages. J Exp Med 1988; 168:409-416.
16. Inaba K, Young JW, Steinman RM. Direct activation of CD8$^+$ cytotoxic T lymphocytes by dendritic cells. J Exp Med 1987; 166:182-194.
17. Young JW, Steinman RM. Dendritic cells stimulate primary human cytolytic lymphocyte responses in the absence of CD4$^+$ helper T cells. J Exp Med 1990; 171:1315-1332.
18. Klinkert WEF, Labadie JH, Bowers WE. Accessory and stimulating properties of dendritic cells and macrophages isolated from various rat tissues. J Exp Med 1982; 156:1-19.
19. Austyn JM, Steinman RM, Weinstein DE et al. Dendritic cells initiate a two-stage mechanism for T lymphocyte proliferation. J Exp Med 1983; 157:1101-1115.
20. Bhardwaj N, Young JW, Nisanian AJ et al. Small amounts of superantigen, when presented on dendritic cells, are sufficient to initiate T cell responses. J Exp Med 1993; 178:633-642.
21. Romani N, Inaba K, Pure E et al. A small number of anti-CD3 molecules on dendritic cells stimulate DNA synthesis in mouse T lymphocytes. J Exp Med 1989; 169:1153-1168.
22. Schuler G, Steinman RM. Murine epidermal Langerhans cells mature into potent immunostimulatory dendritic cells in vitro. J Exp Med 1985; 161:526-546.

23. Holt PG, Schon-Hegrad MA, Oliver J. MHC class II antigen-bearing dendritic cells in pulmonary tissues of the rat. Regulation of antigen presentation activity by endogenous macrophage populations. J Exp Med 1987; 167:262-274.
24. Stingl G, Katz SI, Clements L et al. Immunologic functions of Ia-bearing epidermal Langerhans cells. J Immunol 1978; 121:2005-2013.
25. Stumbles PA, Thomas JA, Pimm CL et al. Resting respiratory tract dendritic cells preferentially stimulate T helper cell type 2 (Th2) responses and require obligatory cytokine signals for induction of Th1 immunity. J Exp Med 1998; 188:2019-2031.
26. Inaba K, Metlay JP, Crowley MT et al. Dendritic cells pulsed with protein antigens in vitro can prime antigen-specific, MHC-restricted T cells in situ. J Exp Med 1990; 172:631-640.
27. Romani N, Koide S, Crowley M et al. Presentation of exogenous protein antigens by dendritic cells to T cell clones: intact protein is presented best by immature, epidermal Langerhans cells. J Exp Med 1989; 169:1169-1178.
28. Inaba K, Inaba M, Naito M et al. Dendritic cell progenitors phagocytose particulates, including Bacillus Calmette-Guerin organisms, and sensitize mice to mycobacterial antigens in vivo. J Exp Med 1993; 178:479-488.
29. Reis e Sousa C, Stahl PD, Austyn JM. Phagocytosis of antigens by Langerhans cells in vitro. J Exp Med 1993; 178:509-519.
30. Sallusto F, Cella M, Danieli C et al. Dendritic cells use macropinocytosis and the mannose receptor to concentrate antigen in the major histocompatibility class II compartment. Downregulation by cytokines and bacterial products. J Exp Med 1995; 182:389-400.
31. Garrett WS, Chen LM, Kroschewski R et al. Developmental control of endocytosis in dendritic cells by Cdc42. Cell 2000;102:325-334.
32. Nijman HW, Kleijmeer MJ, Ossevoort MA et al. Antigen capture and MHC class II compartments of freshly isolated and cultured human blood dendritic cells. J Exp Med 1995;182:163-174.
33. O'Doherty U, Steinman RM, Peng M et al. Dendritic cells freshly isolated from human blood express CD4 and mature into typical immunostimulatory dendritic cells after culture in monocyte-conditioned medium. J Exp Med 1993;178:1067-1078.
34. Witmer-Pack MD, Olivier W, Valinsky J et al. Granulocyte/macrophage colony-stimulating factor is essential for the viability and function of cultured murine epidermal Langerhans cells. J ExpMed 1987;166:1484-1498.
35. O'Doherty U, Peng M, Gezelter S et al. Human blood contains two subsets of dendritic cells, one immunologically mature, and the other immature. Immunol 1994;82:487-493.
36. Feuerstein B, Berger TG, Maczek C et al. A method for the production of cryopreserved aliquots of antigen- preloaded, mature dendritic cells ready for clinical use. J Immunol Meth 2000; 245:15-29.
37. Rieser C, Bock G, Klocker H et al. Prostaglandin E2 and tumor necrosis factor a cooperate to activate human dendritic cells: synergistic activation of interleukin 12 production. J Exp Med 1997; 186:1603-1608.
38. Caux C, Massacrier C, Vanbervliet B et al. Activation of human dendritic cells through CD40 cross-linking. J Exp Med 1994; 180:1263-1272.
39. Josien R, Hi H-L, Ingulli E et al. TRANCE, a tumor necrosis family member, enhances the longevity and adjuvant properties of dendritic cells in vivo. J Exp Med 2000; 191:495-501.
40. Sauter B, Albert ML, Francisco L et al. Consequences of cell death. Exposure to necrotic tumor cells, but not primary tissue cells or apoptotic cells, induces the maturation of immunostimulatory dendritic cells. J Exp Med 2000;191:423-434.
41. Gallucci S, Lolkema M, Matzinger P. Natural adjuvants: endogenous activators of dendritic cells. Nat Med 1999; 5:1249-1255.
42. Binder RJ, Anderson KM, Basu S et al. Heat shock protein gp96 induces maturation and migration of CD11c(+) cells in vivo. J Immunol 2000; 165:6029-6035.
43. Cella M, Engering A, Pinet V et al. Inflammatory stimuli induce accumulation of MHC class II complexes on dendritic cells. Nature 1997; 388:782-787.
44. Rescigno M, Martino M, Sutherland CL et al. Dendritic cell survival and maturation are regulated by different signaling pathways. J Exp Med 1998; 188:2175-2180.
45. Cella M, Salio M, Sakakibara Y et al. Maturation, activation, and protection of dendritic cells induced by double-stranded RNA. J Exp Med 1999; 189:821-829.
46. Verdijk RM, Mutis T, Esendam B et al. Polyriboinosinic polyribocytidylic Acid (Poly(I:C)) induces stable maturation of functionally active human dendritic cells. J Immunol 1999; 163:57-61.
47. Sparwasser T, Koch E-V, Vabulas RM et al. Bacterial DNA and immunostimulatory CpG oligonucleotides trigger maturation and activation of murine dendritic cells. Eur J Immunol 1998;28:2045-2054.

48. Jakob T, Walker PS, Krieg AM et al. Bacterial DNA and CpG-containing oligodeoxynucleotides activate cutaneous dendritic cells and induce IL-12 production: implications for the augmentation of Th1 responses. Int Arch Allergy Immunol 1999; 118:457-461.
49. Sparwasser T, Vabulas RM, Villmow B et al. Bacterial CpG-DNA activates dendritic cells in vivo: T helper cell-independent cytotoxic T cell responses to soluble proteins. Eur J Immunol 2000; 30:3591-3597.
50. Kaisho T, Akira S. Dendritic-cell function in Toll-like receptor- and MyD88-knockout mice. Trends in Immunol 2001;22:78-83.
51. Witmer MD, Steinman RM. The anatomy of peripheral lymphoid organs with emphasis on accessory cells: light microscopic, immunocytochemical studies of mouse spleen, lymph node and Peyer's patch. Am J Anat 1984; 170:465-481.
52. Drexhage HA, Mullink H, de Groot J et al. A study of cells present in peripheral lymph of pigs with special reference to a type of cell resembling the Langerhans cells. Cell Tiss Res 1979; 202:407-430.
53. Knight SC, Balfour BM, O'Brien J et al. Role of veiled cells in lymphocyte activation. Eur J Immunol 1982; 12:1057-1060.
54. Pugh CW, MacPherson GG, Steer HW. Characterization of nonlymphoid cells derived from rat peripheral lymph. J Exp Med 1983; 157:1758-1779.
55. Bujdoso R, Hopkins J, Dutia BM et al. Characterization of sheep afferent lymph dendritic cells and their role in antigen carriage. J Exp Med 1989; 170:1285-1302.
56. Liu LM, MacPherson GG. Antigen acquisition by dendritic cells: intestinal dendritic cells acquire antigen administered orally and can prime naive T cells in vivo. J Exp Med 1993;177:1299-1307.
57. Freudenthal PS, Steinman RM. The distinct surface of human blood dendritic cells, as observed after an improved isolation method. Proc Natl Acad Sci USA 1990; 87:7698-7702.
58. Matsuno K, Ezaki T, Kudo S et al. A life stage of particle-laden rat dendritic cells in vivo: their terminal division, active phagocytosis and translocation from the liver to hepatic lymph. J Exp Med 1996; 183:1865-1878.
59. Ingulli E, Mondino A, Khoruts A et al. In vivo detection of dendritic cell antigen presentation to CD4+ T cells. J Exp Med 1997; 185:2133-2141.
60. Crowley M, Inaba K, Steinman RM. Dendritic cells are the principal cells in mouse spleen bearing immunogenic fragments of foreign proteins. J Exp Med 1990; 172:383-386.
61. Labeur MS, Roters B, Pers B et al. Generation of tumor immunity by bone marrow-derived dendritic cells correlates with dendritic cell maturation stage. J Immunol 1999; 162:168-175.
62. Inaba K, Turley S, Iyoda T et al. The formation of immunogenic MHC class II- peptide ligands in lysosomal compartments of dendritic cells is regulated by inflammatory stimuli. J Exp Med 2000; 191:927-936.
63. Schuurhuis DH, Laban S, Toes RE et al. Immature dendritic cells acquire CD8(+) cytotoxic T lymphocyte priming capacity upon activation by T helper cell-independent or -dependent stimuli. J Exp Med 2000; 192:145-150.
64. Ludewig B, Ehl S, Karrer U et al. Dendritic cells efficiently induce protective antiviral immunity. J Virol 1998; 272:3812-3818.
65. Mayordomo JI, Zorina T, Storkus WJ et al. Bone marrow-derived dendritic cells pulsed with synthetic tumour peptides elicit protective and therapeutic antitumour immunity. Nat Med 1995;1:1297-1302.
66. Nair SK, Snyder D, Rouse B et al. Regression of tumors in mice vaccinated with professional antigen-presenting cells pulsed with tumor extracts. Int J Canc 1997; 70:706-715.
67. Schuler-Thurner B, Dieckmann D, Keikavoussi P et al. Mage-3 and influenza-matrix peptide-specific cytotoxic T cells are inducible in terminal stage HLA-A2.1+ melanoma patients by mature monocyte-derived dendritic cells. J Immunol 2000; 165:3492-3496.
68. Banchereau J, Palucka AK, Dhodapkar M et al. Clinical and immunologic responses to CD34+ progenitor-derived dendritic cells in patients with stage IV melanoma. Submitted 2001.
69. Hsu FJ, Benike C, Fagnoni F et al. Vaccination of patients with B-cell lymphoma using autologous antigen-pulsed dendritic cells. Nat Med 1996; 2:52-58.
70. Kugler A, Stuhler G, Walden P et al. Regression of human metastatic renal cell carcinoma after vaccination with tumor cell-dendritic cell hybrids. Nat Med 2000; 6:332-336.
71. Geiger J, Hutchinson R, Hohenkirk L et al. Treatment of solid tumours in children with tumour-lysate-pulsed dendritic cells. Lancet 2000; 356:1163-1165.
72. Dhodapkar M, Steinman RM, Sapp M et al. Rapid generation of broad T-cell immunity in humans after single injection of mature dendritic cells. J Clin Invest 1999; 104:173-180.
73. Dhodapkar MV, Steinman RM, Krasovsky J et al. Antigen specific inhibition of effector T cell function in humans after injection of immature dendritic cells. J Exp Med 2001; 193:233-238.

74. Nishimura T, Iwakabe K, Sekimoto M et al. Distinct roles of antigen-specific T helper type 1 (Th1) and Th2 cells in tumor eradication in vivo. J Exp Med 1999; 190:617-628.

75. Maloy KJ, Burkhart C, Junt TM et al. CD4⁺ T cell subsets during virus infection: protective capacity depends on effector cytokine secretion and on migratory capability. J Exp Med 2000; 191:2159.

76. Cardin RD, Brooks JW, Sarawar SR et al. Progressive loss of CD8⁺ T cell-mediated control of γ-herpesvirus in the absence of CD4+ T cells. J Exp Med 1996; 184:863-871.

77. Christensen JP, Cardin RD, Branum KC et al. CD4⁺ T cell-mediated control of a γ-herpesvirus in B cell- deficient mice is mediated by IFN-γ. Proc Natl Acad Sci USA 1999; 96:5135-5140.

78. Dhodapkar MV, Krasovsky J, Steinman RM et al. Mature dendritic cells boost functionally superior T cells in humans without foreign helper epitopes. J Clin Invest 2000; 105:R9-R14.

79. Thurner B, Haendle I, Röder C et al. Vaccination with Mage-3A1 peptide-pulsed mature, monocyte-derived dendritic cells expands specific cytotoxic T cells and induces regression of some metastases in advanced stage IV melanoma. J Exp Med 1999; 190:1669-1678.

80. Jonuleit H, Schmitt E, Schuler G et al. Induction of human IL-10-producing, non-proliferating CD4⁺ T cells with regulatory properties by repetitive stimulation with allogeneic immature dendritic cells. J Exp Med 2000; 192:1213-1222.

81. Albert ML, Sauter B, Bhardwaj N. Dendritic cells acquire antigen from apoptotic cells and induce class I-restricted CTLs. Nature 1998; 392:86-89.

82. Albert ML, Pearce SFA, Francisco LM et al. Immature dendritic cells phagocytose apoptotic cells via αᵥβ₅ and CD36, and cross-present antigens to cytotoxic T lymphocytes. J Exp Med 1998; 188:1359-1368.

83. Steinman RM, Turley S, Mellman I et al. The induction of tolerance by dendritic cells that have captured apoptotic cells. J Exp Med 1999; 191:411-416.

84. Matzinger P. Tolerance, danger, and the extended family. Annu Rev Immunol 1994; 12:991-1045.

85. Janeway CA. The immune system evolved to discriminate infectious nonself from noninfectious self. Immunol Today 1992; 13:11.

86. Fayette J, Dubois B, Vandenabelle S et al. Human dendritic cells skew isotype switching of CD40-activated naive B cells towards IgA1 and IgA2. J Exp Med 1997;185:1909-1918.

87. Caux C, Vanbervliet B, Massacrier C et al. CD34+ hematopoietic progenitors from human cord blood differentiate along two independent dendritic cell pathways in response to GM-CSF+ TNF α. J Exp Med 1996; 184:695-706.

88. Kitamura H, Iwakabe K, Yahata T et al. The natural killer T (NKT) cell ligand α-galactosylceramide demonstrates its immunopotentiating effect by inducing interleukin (IL)-12 production by dendritic cells and IL-12 receptor expression on NK-T cells. J Exp Med 1999; 189:1121-1128.

89. Fernandez NC, Lozier A, Flament C et al. Dendritic cells directly trigger NK cell functions: cross-talk relevant in innate anti-tumor immune responses in vivo. Nat Med 1999; 5:405-411.

90. Carnaud C, Lee D, Donnars O et al. Cutting edge: Cross-talk between cells of the innate immune system: NKT cells rapidly activate NK cells. J Immunol 1999; 163:4647-4650.

91. Reis e Sousa C, Hieny S, Scharton-Kersten T et al. In vivo microbial stimulation induces rapid CD40L- independent production of IL-12 by dendritic cells and their re-distribution to T cell areas. J Exp Med 1997; 186:1819-1829.

92. Ohteki T, Fukao T, Suzue K et al. Interleukin 12-dependent interferon g production by CD8α+ lymphoid dendritic cells. J Exp Med 1999; 189:1981-1986.

93. Siegal FP, Kadowaki N, Shodell M et al. The nature of the principal type 1 interferon-producing cells in human blood. Science 1999; 284:1835-1837.

94. Takahashi T, Nieda M, Koezuka Y et al. Analysis of human valpha24+ CD4+ NKT cells activated by alpha-glycosylceramide-pulsed monocyte-derived dendritic cells. J Immunol 2000; 164:4458-4464.

95. Vremec D, Pooley J, Hochrein H et al. CD4 and CD8 expression by dendritic cell subtypes in mouse thymus and spleen. J Immunol 2000; 164:2978-2986.

96. Maldonado-Lopez R, De Smedt T, Michel P et al. CD8α+ and CD8α- subclasses of dendritic cells direct the development of distinct T helper cells in vivo. J Exp Med 1999; 189:587-592.

97. den Haan J, Lehar S, Bevan M. CD8+ but not CD8- dendritic cells cross-prime cytotoxic T cells in vivo. J Exp Med 2000; 192:1685-1696.

98. Grouard G, Rissoan M-C, Filgueira L et al. The enigmatic plasmacytoid T cells develop into dendritic cells with IL-3 and CD40-ligand. J Exp Med 1997; 185:1101-1111.

99. Dubois B, Vanbervliet B, Fayette J et al. Dendritic cells enhance growth and differentiation of CD40-activated B lymphocytes. J Exp Med 1997; 185:941-951.

100. Pierre P, Mellman I. Developmental regulation of invariant chain proteolysis controls MHC class II trafficking in mouse dendritic cells. Cell 1998; 93:1135-1145.

101. Fiebiger E, Meraner P, Weber E et al. Cytokines regulate proteolysis in major histocompatibility complex class II-dependent antigen presentation by dendritic cells. J Exp Med 2001; 193:881-892.

102. Jiang W, Swiggard WJ, Heufler C et al. The receptor DEC-205 expressed by dendritic cells and thymic epithelial cells is involved in antigen processing. Nature 1995; 375:151-155.
103. Mahnke K, Guo M, Lee S et al. The dendritic cell receptor for endocytosis, DEC-205, can recycle and enhance antigen presentation via MHCII+, lysosomal compartments. J Cell Biol 2000; 151:673-683.
104. Regnault A, Lankar D, Lacabanne V et al. Fcγ receptor-mediated induction of dendritic cell maturation and major histocompatibility complex class I-restricted antigen presentation after immune complex internalization. J Exp Med 1999; 189:371-380.
105. Bender A, Bui LK, Feldman MAV et al. Inactivated influenza virus, when presented on dendritic cells, elicits human CD8⁺ cytolytic T cell responses. J Exp Med 1995; 182:1663-1671.
106. Subklewe M, Paludan C, Tsang ML et al. Dendritic cells cross-present latency gene products from Epstein-Barr Virus-transformed B cells and expand tumor-reactive CD8⁺ killer T cells. J Exp Med 2001; 193:405-412.
107. Inaba K, Turley S, Yamaide F et al. Efficient presentation of phagocytosed cellular fragments on the MHC class II products of dendritic cells. J Exp Med 1998; 188:2163-2173.
108. Munz C, Bickham KL, Subklewe M et al. Human CD4⁺ T lymphocytes consistently respond to the latent Epstein-Barr Virus nuclear antigen EBNA1. J Exp Med 2000; 191:1649-1660.
109. Bickham K, Münz C, Larsson M et al. EBNA 1-specific CD4+ T cells in healthy carrierrs of Epstein- Barr virus are primarily Th1 in function. J Clin Invest 2001;107:121-130.
110. Hawiger D, Inaba K, Mahnke K et al. Peripheral tolerance induced by dendritic cells in the steady state. Submitted 2001.
111. Geijtenbeek TBH, Torensma R, van Vliet SJ et al. Identification of DC-SIGN, a novel dendritic cell-specific ICAM-3 receptor that supports primary immune responses. Cell 2000; 100:575-585.
112. Inaba K, Witmer-Pack M, Inaba M et al. The tissue distribution of the B7-2 costimulator in mice: abundant expression on dendritic cells in situ and during maturation in vitro. J Exp Med 1994; 180:1849-1860.
113. Caux C, Vanbervliet B, Massacrier C et al. B70/B7-2 is identical to CD86 and is the major functional ligand for CD28 expressed on human dendritic cells. J Exp Med 1994; 180:1841-1847.
114. Turley SJ, Inaba K, Garrett WS et al. Transport of peptide-MHC class II complexes in developing dendritic cells. Science 2000; 288:522-527.
115. Tafuri A, Shahinian A, Bladt F et al. ICOS is essential for effective T-helper-cell responses. Nature 2001; 409:105-109.
116. Tseng S-Y, Otsugi M, Gorski K et al. B7-DC, a new dendritic cell molecule with potent costimulatory properties for T cells. J Exp Med 2001; 193:839-846.
117. Oppmann B, Lesley R, Blom B et al. Novel p19 protein engages IL-12p40 to form a cytokine, IL-23, with biological activities similar as well as distinct from IL-12. Immun 2000; 13:715-725.
118. Ku CC, Murakami M, Sakamoto A et al. Control of homeostasis of CD8+ memory T cells by opposing cytokines. Science 2000; 288:675-678.
119. Hochrein H, O'Keeffe M, Luft T et al. Interleukin (IL)-4 is a major regulatory cytokine governing bioactive IL-12 production by mouse and human dendritic cells. J Exp Med 2000; 192:823-834.
120. Schulz O, Edwards AD, Schito M et al. CD40 triggering of heterodimeric IL-12 p70 production by dendritic cells in vivo requires a microbial priming signal. Immun 2000; 13:453-462.
121. Ebner S, Ratzinger G, Krosbacher B et al. Production of interleukin-12 by human monocyte-derived dendritic cells is optimal when the stimulus is given at the onset of maturation, and is further enhanced by interleukin-4. J Immunol 2001; 166:633-641.
122. Langenkamp A, Messi M, Lanzavecchia A et al. Kinetics of dendritic cell activation: impact on priming of Th1, Th2 and nonpolarized T cells. Nat Immunol 2000; 1:311-316.
123. Koide SL, Inaba K, Steinman RM. Interleukin-1 enhances T-dependent immune responses by amplifying the function of dendritic cells. J Exp Med 1987; 165:515-530.
124. Inaba K, Witmer-Pack MD, Inaba M et al. The function of Ia⁺ dendritic cells, and Ia⁻ dendritic cell precursors, in thymocyte mitogenesis to lectin and lectin plus IL-1. J Exp Med 1988; 167:149-162.
125. Buelens C, Verhasselt V, De Groote D et al. Human dendritic cell responses to LPS and CD40 ligation are differentially regulated by IL-10. Eur J Immunol 1997; 27:1848-1852.
126. Corinti S, Albanesi C, la Sala A et al. Regulatory activity of autocrine IL-10 on dendritic cell functions. J Immunol 2001; 166:4312-4318.
127. Thurner B, Röder C, Dieckmann D et al. Generation of large numbers of fully mature and stable dendritic cells from leukapheresis products for clinical application. J Immunol Meth 1999; 223:1-15.
128. Holt PG, Haining S, Nelson DJ et al. Origin and steady-state turnover of class II MHC-bearing dendritic cells in the epithelium of the conducting airways. J Immunol 1994; 153:256-261.
129. Kamath AT, Pooley J, O'Keeffe MA et al. The development, maturation, and turnover rate of mouse spleen dendritic cell populations. J Immunol 2000; 165:6762-6770.

130. Randolph GJ, Beaulieu S, Steinman RM et al. Differentiation of monocytes into dendritic cells in a model that mimics entry of cells into afferent lymph. Science 1998; 282:480-483.
131. Dieu-Nosjean MC, Massacrier C, Homey B et al. Macrophage inflammatory protein 3α is expressed at inflamed epithelial surfaces and is the most potent chemokine known in attracting Langerhans cell precursors. J Exp Med 2000; 192:705-718.
132. Iwasaki A , Kelsall BL. Localization of distinct Peyer's patch dendritic cell subsets and their recruitment by chemokines macrophage inflammatory protein (MIP)-3α, MIP-3β, and secondary lymphoid organ chemokine. J Exp Med 2000; 191:1381-1394.
133. Maraskovsky E, Brasel K, Teepe M et al. Dramatic increase in the numbers of functionally mature dendritic cells in Flt3 ligand-treated mice: Multiple dendritic cell subpopulations identified. J Exp Med 1996; 184:1953-1962.
134. Arpinati M, Green CL, Heimfeld S et al. Granulocyte-colony stimulating factor mobilizes T helper 2-inducing dendritic cells. Blood 2000; 95:2484-2490.
135. Larsen CP, Steinman RM, Witmer-Pack M et al. Migration and maturation of Langerhans cells in skin transplants and explants. J Exp Med 1990; 172:1483-1493.
136. Cyster JG. Chemokines and the homing of dendritic cells to the T cell areas of lymphoid organs. J Exp Med 1999; 189:447-450.
137. Willimann K, Legler DF, Loetscher M et al. The chemokine SLC is expressed in T cell areas of lymph nodes and mucosal lymphoid tissues and attracts activated T cells via CCR7. Eur J Immunol 1998; 28:2025-2034.
138. Kellermann SA, Hudak S, Oldham ER et al. The CC chemokine receptor-7 ligands 6Ckine and macrophage inflammatory protein-3β are potent chemoattractants for in vitro- and in vivo- derived dendritic cells. J Immunol 1999; 162:3859-3864.
139. Ngo VN, Tang HL, Cyster JG. Epstein-Barr virus-induced molecule 1 ligand chemokine is expressed by dendritic cells in lymphoid tissues and strongly attracts naive T cells and activated B cells. J Exp Med 1998; 188:181-191.
140. Forster R, Schubel A, Breitfeld D et al. CCR7 coordinates the primary immune response by establishing functional microenvironments in secondary lymphoid organs. Cell 1999; 99:23-33.
141. Robbiani DF, Finch RA, Jaeger D et al. The leukotriene C4 transporter MRP1 regulates CCL19 (MIP-3β, ELC)-dependent mobilization of dendritic cells to lymph nodes. Cell 2000; 103:757-768.
142. Moodycliffe AM, Shreedhar V, Ullrich SE et al. CD40-CD40 ligand interactions in vivo regulate migration of antigen-bearing dendritic cells from the skin to draining lymph nodes. J Exp Med 2000; 191:2011-2020.
143. Adema GJ, Hartgers F, Verstraten R et al. A dendritic-cell-derived C-C chemokine that preferentially attracts naive T cells. Nature 1997; 387:713-717.
144. Sallusto F, Lenig D, Forster R et al. Two subsets of memory T lymphocytes with distinct homing potentials and effector functions. Nature 1999; 401:708-712.
145. Lieberam I , Forster I. The murine beta-chemokine TARC is expressed by subsets of dendritic cells and attracts primed CD4⁺ T cells. Eur J Immunol 1999;29:2684-2694.
146. Tang HL, Cyster JG. Chemokine up-regulation and activated T cell attraction by maturing dendritic cells. Science 1999; 284:819-822.
147. Kanazawa N, Nakamura T, Tashiro K et al. Fractalkine and macrophage-derived chemokine: T cell-attracting chemokines expressed in T cell area dendritic cells. Eur J Immunol 1999; 29:1925-1932.
148. Papadopoulos EJ, Sassetti C, Saeki H et al. Fractalkine, a CX3C chemokine, is expressed by dendritic cells and is up-regulated upon dendritic cell maturation. Eur J Immunol 1999; 29:2551-2559.
149. Cella M, Jarrossay D, Facchetti F et al. Plasmacytoid monocytes migrate to inflamed lymph nodes and produce large amounts of type I interferon. Nat Med 1999; 5:919-923.
150. Sallusto F, Palermo B, Lenig D et al. Distinct patterns and kinetics of chemokine production regulate dendritic cell function. Eur J Immunol 1999; 29:1617-1625.
151. Sallusto F, Lanzavecchia A. Mobilizing dendritic cells for tolerance, priming, and chronic inflammation. J Exp Med 1999; 189:611-614.
152. Corr M, Lee DJ, Carson DA et al. Gene vaccination with naked plasmid DNA: mechanism of CTL priming. J Exp Med 1996; 184:1555-1560.
153. Iwasaki A, Torres CAT, Ohashi PS et al. The dominant role of bone marrow derived cells in CTL induction following plasmid DNA immunization at different sites. J Immunol 1997; 159:11-14.
154. Fu T-M, Ulmer JB, Caulfield MJ et al. Priming of cytotoxic T lymphocytes by DNA vaccines: requirement for professional antigen presenting cells and evidence for antigen transfer from myocytes. Molec Med 1997; 3:362-371.
155. Casares S, Inaba K, Brumeanu T-D et al. Antigen presentation by dendritic cells following immunization with DNA encoding a class II-restricted viral epitope. J Exp Med 1997; 186:1481-1486.

156. Bot A, Stan AC, Inaba K et al. Dendritic cells at a DNA vaccination site express the encoded influenza nucleoprotein and prime MHC class I-restricted cytolytic lymphocytes upon adoptive transfer. Int Immunol 2000; 12:825-832.

157. Akbari O, Panjwani N, Garcia S et al. DNA vaccination: transfection and activation of dendritic cells as key events for immunity. J Exp Med 1999;189:169-178.

158. Porgador A, Irvine KR, Iwasaki A et al. Predominant role for directly transfected dendritic cells in antigen presentation to CD8(+) T cells after gene gun immunization. J Exp Med 1998; 188:1075-1082.

159. Corr M, von Damm A, Lee DJ et al. In vivo priming by DNA injection occurs predominantly by antigen transfer. J Immunol 1999; 163:4721-4727.

160. Ridge JP, Di Rosa F, Matzinger P. A conditioned dendritic cell can be a temporal bridge between a CD4+ T helper and a T-killer cell. Nature 1998; 393:474-478.

161. Bennett SRM, Carbone FR, Karamalis F et al. Help for cytotoxic-T-cell responses is mediated by CD40 signalling. Nature 1998; 393:478-480.

162. Schoenberger SP, Toes REM, van der Voort EIH et al. T-cell help for cytotoxic T lymphocytes is mediated by CD40-CD40L interactions. Nature 1998; 393:480-483.

163. Chan K, Lee DJ, Schubert A et al. The roles of MHC class II, CD40, and B7 costimulation in CTL induction by plasmid DNA. J Immunol 2001; 166:3061-3066.

164. Bachmann MF, Wong BR, Josien R et al. TRANCE, a tumor necrosis factor family member critical for CD40 ligand- independent T helper cell activation. J Exp Med 1999; 189:1025-1031.

165. Lu Z, Yuan L, Zhou X et al. CD40-independent pathways of T cell help for priming of CD8+ cytotoxic T lymphocytes. J Exp Med 2000; 191:541-550.

CHAPTER 3

Activation of the Innate Immune System by DNA Vaccines

Julie Fitzgerald and Hildegund C. J. Ertl

Introduction

D NA vaccines were discovered serendipitously by gene therapists attempting to replace missing or faulty genes with bacterial expression vectors. The transgene product was found to elicit an immune response that rapidly eliminated the transduced cells[1] reducing the attractiveness of this approach for gene replacement therapy while opening the field of DNA vaccinology. DNA vaccines were initially injected intramuscularly where they were shown to transduce local muscle cells.[2] This added further to the puzzle of their immunogenicity. Why would a small piece of circular DNA injected in saline upon uptake by cells not considered professional antigen presenting cells induce both T and B cells to the encoded gene product, when even large doses of foreign protein require addition of an adjuvant to induce a robust immune response? The initial suspicion that the immunogenicity of DNA vaccines was linked to their contamination with trace amounts of lipopolysaccharides from the bacteria used to propagate the DNA vaccines was rapidly dismissed experimentally. Bacterial DNA had been known prior to the era of DNA vaccines as an activator of B and NK cells and as potential tumor therapeutics.[3] This effect could be mimicked by synthetic oligodeoxinucleotides (ODNs) with self-complemtentary palindromic sequences.[4] Eventually it was appreciated that such palindromic sequences that are unmethylated CpG motifs common in bacterial but not mammalian genome were crucial for the immunogenicity of DNA vaccines by providing inflammatory signals to the innate immune system.

The innate immune system, our first line of defense against pathogenic microorganisms, was previously thought of as a group of entirely nonspecific cells, which phagocytose foreign together with host-derived material at random. Now it is known that the cells of the innate immune system carry receptors called Toll-like receptors (TLRs), which are evolutionary highly conserved germ-line encoded receptors that recognize specific traits of microorganisms and then orchestrate an adaptive immune response most suited to combat the invading pathogen.[5,6] TLRs were named for their similarity to the Toll protein found in Drosophila where it plays a role in defense against fungal infections and dorsoventral patterning of the embryo.[6-9] According to phylogenetic analysis, the ancestral prevertebrate TLR evolved approximately 500 million years ago.[10]

Unlike the receptors of the adaptive immune system, TLRs are nonclonal receptors. They are part of a family of pattern recognition receptors (PRRs) which recognize certain unique motifs conserved among microbes called pathogen-associated molecular patterns (PAMPs).[5,11] This specificity allows the innate immune system to distinguish microbial antigens from each other and from part of the nonthreatening antigens from the host cells or other benign or even essential foreign materials such as food. An adaptive immune response is initiated by antigen displayed on professional antigen presenting cells (APCs) which are mature dendritic cells.

DNA Vaccines, edited by Hildegund C. J. Ertl. ©2003 Eurekah.com and Kluwer Academic / Plenum Publishers.

Such dendritic cells are dispersed at an immature stage throughout an organism. Immature dendritic cells take up antigen but they are ill suited to present such antigen to naïve T or B cells. Upon receiving a maturation signal, dendritic cells undergo a number of phenotypic and functional changes and migrate to the T cell rich areas of draining lymph nodes where they can now activate an adaptive immune response. Dendritic cells come in different shapes and forms and are currently roughly divided into type 1 and type 2 dendritic cells, which have distinct albeit overlapping functions. In humans, type 1 dendritic cells upon activation secrete IL-12 thus sponsoring activation of type 1 T helper cell responses.[12] Type 2 dendritic cells produce markedly lower amounts of IL-12 and favor activation of type 2 T helper cells.[12] Type 1 and 2 dendritic cells express unique patterns of TLRs as was shown for human dendritic cells: type 1 dendritic cells express high levels of TLR-1, 2, 3, low levels of TLR-5, 6, 8 and 10 and no TLR-4, 7 and 9 while type 2 dendritic cells in contrast express high levels of transcript for TLR-7 and 9, low levels of TLR-1, 6 and 10 but no TLR-2, 3, 4, 5 or 8.[13] Different types of dendritic cells thus become matured by different microbes. This further allows the innate immune system to control which type of an adaptive immune response is generated against a pathogen.

Toll-Like Receptors and Their Ligands

There are currently 10 identified TLRs, named TLR1 – TLR10, each encoded by germline genes.[6,10,14-17] The members of this family of proteins are all single transmembrane proteins with extracellular leucine-rich repeats flanked by cystein-rich regions and an intracellular domain similar to that of the interleukin-1 receptor (IL-1R).[14,18,19] Not surprisingly, signaling occurs through a pathway similar to that used by the IL-1R (Figure 1). The proteins involved in the signaling cascade include the Myeloid differentiation protein (MyD88), IL-1R-associated kinase (IRAK), TNF receptor-associated factor 6 (TRAF6) and the Toll-interacting protein Tollip. MyD88 and Tollip are proteins that link IL-1R and IRAK.[20-23] TLRs in general signal through a MyD88-dependent pathway, however, TLR4 can also signal through a MyD88-independent pathway. A third linker protein, Toll-interleukin-1 receptor domain-containing adaptor protein (TIRAP), is involved in signaling through the MyD88-independent pathway.[24] Both pathways can induce dendritic cell (DC) maturation, but the MyD88 pathway is required for cytokine production.[24]

The signaling cascade involves the phosphorylation of IRAK, which subsequently dissociates from the receptor and complexes with TRAF6. This leads to activation of two pathways: map kinase kinase activation and subsequent activation of the transcription factor AP-1, as well as disassembly of the IKK complex causing eventual NF-κB activation.[25] These transcription factors then direct processes involved in upregulating cytokines, costimulatory molecules, and markers of dendritic cell maturation.

Signaling occurs after recognition of specific PAMPs by the TLRs. Each TLR has unique specificities for these PAMPs (Table 1). TLR2 has many ligands, including peptidoglycans, lipoproteins, and phenol-soluble modulin (PSM), which allow it to recognize Gram positive bacteria, spirochetes, mycobacteria, and fungi.[26-34] TLRs 1 and 6 modulate the signaling response by TLR2.[23,35] TLR4 recognizes lipopolysaccharide (LPS),[36,37] TLR5 recognizes flagellin,[38] TLR3 recognizes double-stranded RNA, a common intermediate product of viral replication, and TLR9 recognizes CpG DNA.[16,39] Ligands for the remaining TLRs have not been characterized.

Recognition of Bacterial DNA by TLR9

DNA would seem an unlikely candidate for a PAMP. Mounting an immune response against DNA seems counterintuitive and would be expected to cause autoimmunity as DNA, being composed in all living organisms of only four components, shows limited individuality if broken down into small segments. However, bacterial and vertebrate DNA have several differences, which allow the innate immune system to recognize bacterial DNA as a PAMP. First of all, CpG suppression is a phenomenon found in vertebrate DNA: CpG dinucleotides are found

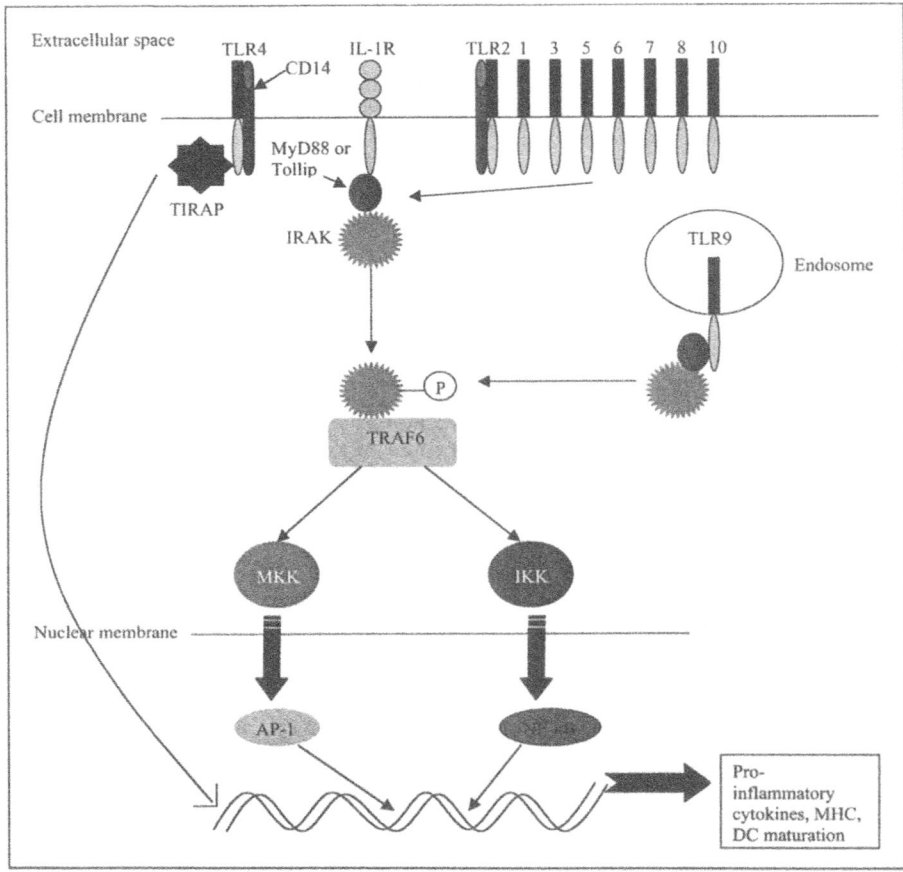

Figure 1. Signaling pathways used by PRRs. The toll-interleudin-1 receptor domain present in the intra-cellular portions of IL-1R and the TLRs connect these receptors to the MyD88-dependent signaling pathway. Activation leads to phosphorylation and dissociation of IRAK. Transcription factors AP-1 and NF-κB are activated downstream, leading to dendritic cell maturation and cytokine production. TLR4 can additionally signal through TIRAP in the MyD88-independent pathway also leading to dendritic cell maturation. The details of this pathway are not yet understood.

four times less frequently in the vertebrate genome than would be predicted based on random base usage.[40] When present these dinucleotides occur mostly in nonimmunostimulatory se-quences[41]. Finally, most CpG dinucleotides in the vertebrate genome are heavily methylated.[42] The lack of methylation, the context of the surrounding sequence of bases, and the greater frequency of CpG's in bacterial DNA enable the immune system to recognize DNA associated with a pathogen as foreign. Methylation of DNA vaccines based on the methylation-insensi-tive early promoter of Simian virus 40 was shown to completely abolish the immune response to the DNA vaccine without reducing transgene expression.[42] This provides the final proof that expression of a foreign protein by DNA vaccines is insufficient to trigger an immune response unless an inflammatory reaction is triggered concomitantly by unmethylated CpG motifs in the bacterial part of the expression vector. The CpG motifs that induce optimal activation of cells of the innate immune system are species specific. For example mice respond best to the GACGTT motif while humans cells are activated optimally by GTCGTT motifs.[39]

Table 1. TLR specificity

TLR	Specificity/Function
1	Modulates TLR2 signaling
2	Peptidoglycans, lipoproteins, phenol-soluble modulin, Gram positive bacteria, Spirochetes, mycobacteria, fungi
3	Double-stranded RNA
4	Lipopolysaccharide
5	Flagellin
6	Modulates TLR2 signaling
7	Unknown
8	Unknown
9	CpG DNA
10	Unknown

Bacterial DNA was shown to be the ligand for TLR9 by knocking out TLR9 in mice, which ablated the response by splenocytes, macrophages, and bone marrow-derived dendritic cells to CpG motifs.[16] D-galactosamine-sensitized mice die from shock when exposed to CpG motifs; however, TLR9 knock-out mice survive when challenged with CpG DNA.[16] The specificity of TLR9 for CpG DNA was additionally demonstrated by transfecting TLR9-negative cells with a plasmid encoding TLR9.[39] These cells gained responsiveness to CpG DNA. These data agree with the correlative evidence that TLR9-expressing cells (i.e., B cells and plasmacytoid dendritic cells) recognize CpG DNA whereas TLR9-negative cells (i.e., monocyte-derived dendritic cells) fail to do so.[39,43] CpG DNA causes maturation of plasmacytoid (type 2) dendritic cells, but not monoctye-derived (type 1) dendritic cells, resulting in secretion of type 1 interferon, IL-12, IL-8, TNF, and GM-CSF; increased surface levels of CD40, CD80, and MHC class I and II; increased cell survival; and acquisition of a morphology that is typical for mature dendritic cells.[43]

It is believed that most TLRs are cell surface receptors, however, there is evidence that CpG DNA is taken up into endosomes and that endosomal acidification is required for the cellular response to CpG DNA.[42,44,45] In cells transfected with TLR9 and exposed to CpG DNA, blocking endosomal acidification shuts off NF-kB induction.[39] Thus it is thought that although TLR9 has structural homology to other TLRs and signals through the MyD88-dependent pathway, it may be present in endosomes instead of at the cell surface. Indeed, Hemmi, et al, discuss unpublished confocal data showing MyD88 and DNA co-localizing in endosomes and LPS and MyD88 localizing to the cell membrane.[16]

Summary

DNA vaccines provide both signals necessary to activate the innate immune system and to stimulate an adaptive immune response: a maturation signal to immature dendritic cells by CpG motifs that signal through the TLR9 pathway and a foreign antigen encoded by the expression casette of the vector. DNA vaccines in general initiate a type 1 biased T helper cell response, yet other PAMPs can stimulate other types of immune responses. Therefore, though the signal transduction pathways upon ligation of different TLRs are similar, there are subtle differences yet to be understood. Elucidating these pathways will help in understanding how DNA vaccines work, how they could be improved especially for use in humans, which cells they are activating, and when it is appropriate to use them based on the type of immune response desired for a particular antigen, ultimately resulting in improved vaccine design.

References

1. Yang Y, Jooss KU, Su Q et al. Immune responses to viral antigens versus transgene product in the elimination of recombinant adenovirus-infected hepatocytes in vivo. Gene Ther 1996; 3:137-144.

2. Wolff JA, Malone RW, Williams P et al. Direct gene transfer into mouse muscle in vivo. Science 1990; 247:1465-168.

3. Tokunaga T, Yamamoto H, Shimada S et al. Antitumor activity of deoxyribonucleic acid fraction from Mycobacterium bovis BCG. I. Isolation, physicochemical characterization, and antitumor activity. J Natl Cancer Inst 1984; 72:955-962.

4. Yamamoto S, Yamamoto T, Kataoka T et al. Unique palindromic sequences in synthetic oligonucleotides are required to induce IFN [correction of INF] and augment IFN-mediated [correction of INF] natural killer activity. J Immunol 1992; 148:4072-4076.

5. Medzhitov R, Janeway CA Jr. Innate immunity: the virtues of a nonclonal system of recognition. Cell 1997; 91:295-298.

6. Medzithov R, Preston-Hurlburt P, Janeway CA Jr. A human homologue of the Drosophila Toll protein signals activation of adaptive immunity. Nature 1997; 388:394-397.

7. Lemaitre B, Nicolas E, Michaut L et al. The dorsoventral regulatory gene cassette spatzle/Toll/cactus controls the potent antifungal response in Drosophila adults. Cell 1996; 86:973-983.

8. Anderson KV, Jurgens G, Nusslein-Volhard C. Establishment of dorsal-ventral polarity in the Drosophila embryo: genetic studies on the role of the Toll gene product. Cell 1985; 42:779-789.

9. Anderson KV, Bokla L, Nusslein-Volhard C. Establishments of dorsal-ventral polarity in the Drosophila embryo: the induction of polarity by the Toll gene product. Cell 1985; 42:791-798.

10. Du X, Poltorak A, Wei Y et al. Three novel mammalian toll-like receptors: gene structure, expression, and evolution. Eur Cytokine Netw 2000; 11:362-371.

11. Medzhitov R, Janeway CA Jr. Innate immunity: impact on the adaptive immune response. Curr Opin Immunol 1997; 9:4-9.

12. Liu YJ, Kanzler H, Soumelis V et al. Dendritic cell lineage, plasticity and cross-regulation. Nat Immunol 2001; 2:585-589.

13. Kadowaki N, Ho S, Antonenko S et al. Subsets of human dendritic cell precursors express different toll-like receptors and respond to different microbial antigens. J Exp Med 2001; 194:863-869.

14. Rock FL, Hardiman G, Timans JC et al. A family of human receptors structurally related to Drosophila Toll. Proc Natl Acad Sci USA 1998; 95:588-593.

15. Takeuchi O, Kawai T, Sanjo H et al. TLR6: A novel member of an expanding toll-like receptor family. Gene 1999; 231:59-65.

16. Hemmi H, Takeuchi O, Kawai T et al. A Toll-like receptor recognizes bacterial DNA. Nature 2000; 408:740-745.

17. Chuang T, Ulevitch RJ. Identification of hTLR10: a novel human Toll-like receptor preferentially expressed in immune cells. Biochim Biophys Acta 2001; 1518:157-161.

18. Hoffmann JA, Kafatos FC, Janeway CA et al. Phylogenetic perspectives in innate immunity. Science 1999; 284:1313-1318.

19. O'Neill LA, Greene C. Signal transduction pathways activated by the IL-1 receptor family: ancient signaling machinery in mammals, insects, and plants. J Leukoc Biol 1998; 63:650-657.

20. Medzhitov R, Preston-Hurlburt P, Kopp E et al. MyD88 is an adaptor protein in the hToll/IL-1 receptor family signaling pathways. Mol Cell 1998; 2:253-258.

21. Muzio M, Ni J, Feng P et al. IRAK (Pelle) family member IRAK-2 and MyD88 as proximal mediators of IL-1 signaling. Science 1997; 278:1612-1615.

22. Burns K, Clatworthy J, Martin L et al. Tollip, a new component of the IL-1RI pathway, links IRAK to the IL-1 receptor. Nat Cell Biol 2000; 2:346-351.

23. Bulut Y, Faure E, Thomas L et al. Cooperation of Toll-like receptor 2 and 6 for cellular activation by soluble tuberculosis factor and Borrelia burgdorferi outer surface protein A lipoprotein: role of Toll-interacting protein and IL-1 receptor signaling molecules in Toll-like receptor 2 signaling. J Immunol 2001; 167:987-994.

24. Horng T, Barton GM, Medzhitov R. TIRAP: an adapter molecule in the Toll signaling pathway. Nat Immunol 2001; 2:835-841.

25. Akira S, Takeda K, Kaisho T. Toll-like receptors: critical proteins linking innate and acquired immunity. Nat Immunol 2001; 2:675-680.

26. Aliprantis AO, Yang RB, Mark MR et al. Cell activation and apoptosis by bacterial lipoproteins through toll-like receptor-2. Science 1999; 285:736-739.

27. Brightbill HD, Libraty DH, Krutzik SR et al. Host defense mechanisms triggered by microbial lipoproteins through toll-like receptors. Science 1999; 285:732-736.

28. Hirschfeld M, Kirschning CJ, Schwandner R et al. Cutting edge: inflammatory signaling by Borrelia burgdorferi lipoproteins is mediated by toll-like receptor 2. J Immunol 1999; 163:2382-2386.
29. Lien E, Sellati TJ, Yoshimura A et al. Toll-like receptor 2 functions as a pattern recognition receptor for diverse bacterial products. J Biol Chem 1999; 274:33419-33425.
30. Means TK, Wang S, Lien E et al. Human toll-like receptors mediate cellular activation by Mycobacterium tuberculosis. J Immunol 1999; 163:3920-3927.
31. Schwandner R, Dziarski R, Wesche H et al. Peptidoglycan- and lipoteichoic acid-induced cell activation is mediated by toll-like receptor 2. J Biol Chem 1999; 274:17406-17409.
32. Yoshimura A, Lien E, Ingalls RR et al. Cutting edge: recognition of Gram-positive bacterial cell wall components by the innate immune system occurs via Toll-like receptor 2. J Immunol 1999; 163:1-5.
33. Takeuchi O, Kaufmann A, Grote K et al. Cutting edge: preferentially the R-stereoisomer of the mycoplasmal lipopeptide macrophage-activating lipopeptide-2 activates immune cells through a toll-like receptor 2- and MyD88-dependent signaling pathway. J Immunol 2000; 164:554-557.
34. Hertz CJ, Kiertscher SM, Godowski PJ et al. Microbial lipopeptides stimulate dendritic cell maturation via Toll-like receptor 2. J Immunol 2001; 166:2444-2450.
35. Ozinsky A, Underhill DM, Fontenot JD et al. The repertoire for pattern recognition of pathogens by the innate immune system is defined by cooperation between toll-like receptors. Proc Natl Acad Sci USA 2000; 97:13766-13771.
36. Poltorak A, He X, Smirnova I et al. Defective LPS signaling in C3H/HeJ and C57BL/10ScCr mice: mutations in Tlr4 gene. Science 1998; 282:2085-2088.
37. Hoshino K, Takeuchi O, Kawai T et al. Cutting edge: Toll-like receptor 4 (TLR4)-deficient mice are hyporesponsive to lipopolysaccharide: evidence for TLR4 as the Lps gene product. J Immunol 1999; 162:3749-3752.
38. Hayashi F, Smith KD, Ozinsky A et al. The innate immune response to bacterial flagellin is mediated by Toll-like receptor 5. Nature 2001; 410:1099-1103.
39. Bauer S, Kirschning CJ, Hacker H et al. Human TLR9 confers responsiveness to bacterial DNA via species-specific CpG motif recognition. Proc Natl Acad Sci USA 2001; 98:9237-9242.
40. Krieg AM. Now I know my CpGs. Trends Microbiol 2001; 9:249-252.
41. Krieg AM, Wu T, Weeratna R et al. Sequence motifs in adenoviral DNA block immune activation by stimulatory CpG motifs. Proc Natl Acad Sci USA 1998; 95:12631-12636.
42. Krieg AM, Yi AK, Matson S et al. CpG motifs in bacterial DNA trigger direct B-cell activation. Nature 1995; 374:546-549.
43. Bauer M, Redecke V, Ellwart JW et al. Bacterial CpG-DNA triggers activation and maturation of human CD11c-, CD123+ dendritic cells. J Immunol 2001; 166:5000-5007.
44. Macfarlane DE, Manzel L. Antagonism of immunostimulatory CpG-oligodeoxynucleotides by quinacrine, chloroquine, and structurally related compounds. J Immunol 1998; 160:1122-1131.
45. Hacker H, Mischak H, Miethke T et al. CpG-DNA-specific activation of antigen-presenting cells requires stress kinase activity and is preceded by nonspecific endocytosis and endosomal maturation. EMBO Journal 1998; 17:6230-6240.

CHAPTER 4

Induction of B Cells by DNA Vaccines

K. Kilpatrick, M. Sarzotti and G. Kelsoe

Introduction

Antigen first activates T and B lymphocytes in the T cell areas of secondary lymphoid tissues where cognate- and costimulus-dependent proliferation expands the population of reactive lymphocytes. Selected T and B cell progeny from this population migrate into B cell zones to form germinal centers (GC), where intense proliferation, apoptosis, and V(D)J hypermutation takes place. Here we address the development and maturation of antigen-specific B cells responses of adult and neonatal mice following the propulsion of DNA-coated gold microparticles into the epidermal tissue using the gene gun technology. While we will not directly address the induction of humoral responses following intramuscular (i.m.) injection of plasmid, an alternative means of delivering DNA-based immunizations, we recommend a recently published review on this subject.[1]

The Initiation of Humoral Immunity

Dendritic cells (DCs) are the principal antigen-presenting cells that activate naïve T lymphocytes.[2] Antigen is processed and presented on the DC membrane as peptides in noncovalent association with MHC class II molecules. Antigen presenting DCs migrate via the lymphatics to T cell rich areas of secondary lymphoid tissues from sites of infection,[3] and initiate antigen specific T cell activation via cell-cell interaction. T cell activation requires the engagement of T cell antigen receptors (TCR) and the formation of an activating structure at the interface of the DC and T cell called an immunological synapse.[4] Imaging studies show that the immunological synapse is initially stabilized by the adhesion molecules LFA-1 and ICAM-1 to form a supramolecular activation cluster[5-7] that is the focus of TCR : MHC interaction.[4,8] The costimulatory molecule CD28 is also transported to this complex[8] where interaction between CD28 and the CD80 and CD86 expressed on activated DCs, provides costimulatory signals necessary for complete T cell activation.[9]

These activation events create a population of CD4+ T cells that proliferate and migrate to the interface of the T cell and B cell zones. This migration[10] is regulated by altered expression of chemokine receptors. The CCR7 receptor, specific for the chemokines SLC and ELC, localizes T lymphocytes within the T cell zone of secondary lymphoid tissues.[3,11] After activation by antigen-presenting DC, T cells decrease the expression of CCR7 and upregulate an alternative chemokine receptor CXCR5 specific for BLC, a chemokine that defines B cell zones in spleen and lymph nodes.[12] In this way, the migration of antigen bearing DCs, their activation of CD4+ T cells, and the expanded population of antigen-specific T cells becomes focused to provide the maximum opportunity for T lymphocytes to interact with B cells and initiate humoral immunity.

Naïve B cells collect antigen on their surface either nonspecifically or via their antigen receptors, they process and present antigen peptides on their surface as well. The interaction of immunoglobulin and antigen on the surface of B lymphocytes initiates their migration to the T-B zone interface.[13] The lymphocytes encounter antigen as it enters lymphoid tissue via the

DNA Vaccines, edited by Hildegund C. J. Ertl. ©2003 Eurekah.com and Kluwer Academic / Plenum Publishers.

blood or lymph. Internalized antigen is processed and presented at the cell surface and the context of MHC class II and at the same time, these antigen presenting B cells upregulate the costimulatory molecule CD86. Activated and antigen-pulsed B cells migrate to the T-B interface,[10,14] in response to altered CXCR5/CCR7 signals.[13] In addition, activated B cells express attractants for activated T cells, raising the probability of cognate interaction.[11,12,15,16] The concerted migration of activated T and B lymphocytes towards the interface of the T cell and B cell zones of secondary lymphoid tissue and the expression of T cell attractants by activated B cells insures the efficient interaction of antigen-activated T and B lymphocytes and results in the efficient production of antibody responses.

Activated B cells that receive T cell help progress along one of two developmental pathways: local proliferation and terminal differentiation to short-lived plasmacytes[14,17] or the initiation of the GC reaction.[18] In the spleen, a subset of activated B cells localizes to the bridging channels, regions of contact between the white and red pulps. Here, the B cells proliferate and differentiate to form foci of plasmacytes responsible for the production of early primary antibody.[14,17,19] Commitment to this pathway requires T cell help and is probably determined by OX40/OX40-ligand interaction.[20] Plasmacytic foci are detected as early as two days after immunization and by eight days, these foci reach their peak sizes.[14,17] Such mature populations are composed entirely of antigen-specific plasmacytes with few or no T cells present.[14,17] Plasmacytic foci in the spleen are relatively short-lived with half-lives of only three to five days and are virtually absent by 14 days after the administration of nonreplicating antigens.[14,21]

Initially, the antibody produced in plasmacytic foci is IgM, however these plasmacytes can undergo isotype switch to IgG within eight days post-immunization[14] without evidence of affinity maturation.[17,22] Isotype switching requires cognate interaction with T lymphocytes. However as plasmacytic foci are devoid of T lymphocytes at the time of class switch,[14,17] the required switch signals are probably provided during the initial cognate interaction at the interface of T and B cell zones.[23]

The alternative differentiation pathway that produces GCs requires subsets of antigen-activated T and B lymphocytes to migrate into the follicular dendritic cell (FDC) reticulum of follicles. FDCs produce the chemokine BLC that acts as a potent chemoattractant for CXCR5-expressing lymphocytes.[12,24,25] Migration of CD4+ T cells into B cell zones is guided by a concerted upregulation of CXCR5[26] and the lowered responsiveness to the T cell chemokines, ELC and SLC.[12]

Upon migration into the B cell follicle, antigen-activated B and T cells form a nascent GC and undergo active proliferative expansion. GCs are founded by one to three antigen-specific B cells[14,17] that persist after the initial rounds of follicular proliferation and down-regulation of membrane IgD. GC B cells proliferate rapidly with doubling times of only 6-8 hours.[14] The nascent GC becomes polarized into distinct dark and light zones marking its transition to a mature GC.[18] Dark zones of GCs contain rapidly dividing centroblasts, whereas light zones comprise noncycling centrocytes, the majority of the FDC reticulum, and activated CD4+ T cells. GCs are the primary sites of VDJ hypermutation and clonal selection that result in the generation of high affinity serum antibodies and the generation of B cell memory.[18,21,22]

DNA-Based Vaccines

Although much has been published over the past decade regarding DNA-based immunization, the evolution of this technology actually spans the last 40 years. A study published in 1962 was the first to demonstrate that antibody responses against polyoma were generated in hamsters following the injection of polyomavirus DNA.[27] Several years later, studies highlighted the potential of using DNA to transfer genes inserted into a plasmid vector as a means to genetically alter mammalian cell lines.[28] Using the reporter genes luciferase, β-galactosidase (β-gal) and chloramphenicol acetyl transferase (CAT), Wolfe et al, convincingly demonstrated in vivo transfer and expression following intramuscular injections of plasmid into mice, opening the door for further studies on DNA-based immunotherapeutic and vaccine applications.[29-33]

Ensuing studies revealed that in vivo expression of protein following DNA-based immunizations resulted in the generation of immune responses similar to those raised against live attenuated vaccines. Immune recognition is engaged even though in vivo protein production levels of the transferred gene product are only estimated to be in the picogram to nanogram range.[34] Studies have found that both arms of the immune system are activated, giving rise to MHC class I and MHC class II presentation yielding robust cellular and humoral responses.

To date, numerous routes and means of delivering plasmid have been assessed.[35] However, most studies evaluating DNA immunizations either utilize injection of plasmid into dermal or muscle tissue using a needle and syringe, or delivery of plasmid-coated gold microprojectiles using the gene gun. Alternative routes used for injecting DNA that have also been studied to date include intraperitoneal, subcutaneous and intravenous delivery. Noninjectable means of delivering DNA include solution-based ocular, intravaginal, intranasal inhalation, intranasal instillation delivery and oral feeding.[35]

Gene Gun Delivery of Microparticles

This study addresses the early, rapid induction of antibody responses following DNA-based immunizations using the gene gun. Biolistic delivery of DNA was first studied in plants[36] and later applied to live animals in order to assess in vivo transfection of somatic cells.[30] These reports were followed up by studies evaluating effects of in vivo-expressed human growth hormone (hGH) on the growth of mice.[31] While growth characteristics were not directly affected by in vivo-expressed hGH protein, analysis of serum collected from experimental mice revealed that antibodies were raised against hGH, demonstrating immune recognition of the transferred gene product. An ensuing study utilizing intramuscular delivery of DNA encoding the influenza virus nucleoprotein (NP) was the first report demonstrating the induction of protective antibodies, as well as antigen-specific cytotoxic T lymphocytes.[32,33] Gene gun delivery systems have been successfully used for delivering gold particles carrying proteins, peptides, inactivated virus particles or RNA in vivo, ex vivo or in vitro.[31,33,36] Gene gun delivery of gold particles carrying nanogram amounts of DNA stimulate both antibody and cytotoxic T cell responses, using less DNA when compared to the amount of DNA needed to generated similar responses following solution-based intramuscular immunizations.[37]

Antigen Presentation Following Gene Gun Immunizations

Cellular mechanisms involved in the induction of immune responses following gene gun delivery of DNA-gold particles into epidermal tissue likely involves two populations of bone marrow-derived DCs, Langerhans cells (LCs) and interstitial, dermal dendritic cells. LCs[38] are derived from myeloid lineage hematopoietic cells[39] and reside within the suprabasal region of the epidermis. LCs comprise approximately 5 % of the cellular makeup of the epidermis.[40,41] DNA-coated gold particles delivered into epidermal tissue directly transfect low numbers of LCs.[42,43] During the process of antigen uptake, maturation and migration to regional lymph nodes (LNs), LCs undergo morphological and phenotypic changes, that are regulated in part by cytokines including GM-CSF, TNF-α and IL-1.[44-46] Expression levels of CD80 (B7-1), CD86 (B7-2) and major histocompatibility complex (MHC) II increase following activation of LCs.[47,48] The role adhesion molecules play in LC transmigration has been recently highlighted by studies that show differential regulation of ICAM-1, E-cadherin, α6 integrins, CD11b and CD11c.[49] Following activation, LC within the epidermis rapidly disassociate from resident keratinocytes after loosening E-cadherin associated intracellular contacts,[50] then pass through the basement membrane into dermal lymphatic vessels. Upon entering lymphatic capillaries, LCs are transported by lymphatic flow to regional LNs. LCs entering LNs via the afferent lymphatics become interdigitating cells (IDCs) that pass through the subcapsular sinus and traffic to paracortical compartments, where they serve as APCs to naïve T cells.[48,50,51] Skin-derived DC transfected with a gene encoded by naked DNA delivered by the gene gun can be detected within LNs expressing the transferred gene byproduct within 24 hrs.[42] At this time, gold particles are also visualized within the cytoplasm of these migratory DCs.[42] Using a reporter gene, attempts have been made to quantitate the actual number of directly transfected DCs present in regional LNs following gene gun delivery of DNA.[43] Microscopic analysis of sectioned LNs revealed approximately 15 β-Gal-expressing DCs per inguinal LN 24 hours after immunization (0.5mg gold, 1ug plasmid, 1 abdominal shot). A linear increase in the number of cells expressing β-Gal was observed when additional nonoverlapping shots of DNA were delivered to the abdominal region.

Other cells that contribute to the magnitude of early B and T cell responses following gene gun delivery of DNA include nonmigratory epidermal keratinocytes, which are the predominant cell type transfected at the immunization site.[52] Peak production of protein by transfected keratinocytes occurs within 48-72 hours of DNA delivery, diminishing over a two-week period.[31,52-55] Most of the cells at the targeted immunization site slough off within 3 days.[53] A recent study that confirmed expression of the transferred gene product in keratinocytes following DNA delivery also reported expression in hair follicle epithelium, and low level expression in dermal macrophages, fibroblast, sebocytes and LCs as well.[56] Keratinocytes have immunoregulatory functions that influence LCs. Keratinocytes secrete GM-CSF, TNF-α and IL-1, which signal LCs to up-regulate MCH class II, adhesion molecules, and influences trafficking of LCs to LNs.[53]

Bacterial DNA used to develop DNA vaccines contains unmethylated cytosine-guanosine (CpG) dinucleotides, which are abundantly found in the backbone of prokaryotic DNA.[57] Unmethylated CpG motifs exhibit distinct molecular patterns that have been shown to activate myeloid cells. CpG motifs serve as potent adjuvants to murine LCs,[58] as well as DCs, macrophages and B lymphocytes. Immunostimulatory bacterial CpG motifs are comprised of an unmethylated cytosine-phosphate-guanosine (CpG) dinucleotide presented in the context of a defined flanking region composed of two 5' purines and two 3' pyrimidines (PuPuCPGPyPy). Studies using synthetic CpG olidodeoxyribonucleotides (ODNs) revealed that unmethylated CpG dinucleotides induce B cells to produce IL-6 and IL-12.[59] Recent studies have elucidated that a Toll-like receptor family member, TLR9, binds bacterial DNA CpG. Studies using CpG ODNs demonstrate the direct induction of APCs such as DCs to secrete IL-6, IL-12, and IFN-α.[60,61] In vitro stimulation of DC with CpG ODNs induces cellular activation, which results in the upregulation of MHC class II, and costimulatory molecules including CD40, CD80 and CD86.

Inflammation plays a role as well in early cellular events following subcutaneous injection or particle bombardment. Extensive damage to the epidermis and dermis tissue is observed in areas where the most concentrated amounts of gold particles are localized. Upon closer examination of the targeted tissue, visible erythema, low level edema, hemorrhage and full thickness epidermal necrosis is evident.[56] Damaged tissue rapidly undergoes regeneration within 24 hours, and resolves within 5 days.[56] As the result of tissue damage, there is an initial infiltration of neutrophils, followed by macrophages and lymphocytes.[56]

The effect of delivering gold to the epidermis, in the absence or presence of bacterial DNA has also been previously reported.[43] Regardless of whether or not DNA is present, gene-gun delivery of gold particles results in a large number of CD11c+ DCs infiltrating major draining LNs. Increased numbers of CD11c+ DCs are detectable for as long as 18 days after gene gun delivery of DNA.[62] Studies have shown that low levels of inflammation-induced phagocytic monocytes infiltrate the vaccination sites following subcutaneous injection of microspheres. These infiltrating cells phagocytose the microspheres, then drain to regional LNs, where they are identified by distinct DC-restricted markers.[63] Monocyte-derived DCs translocate to the paracortical T cell region within regional LNs, with peak accumulation occurring 3-4 days after immunization. These CD11c^{dim}, MHC class II^{hi}, MIDC-8$^+$, DEC-205$^{-/+}$ DCs likely become APCs after the uptake of the gene product synthesized by transduced epidermal keratinocytes at the immunization site.

Site of Induction of B Cell Responses Following Gene Gun Delivery of DNA

Previously we found that gene gun delivery of plasmid to epidermal tissue results in early antigen-specific B cell expansion, which takes place in LNs draining the immunization site, as opposed to the spleen. Our findings were based on the analysis of antigen-specific antibodies generated by hybridomas established from early somatic fusions of immune cells isolated from LNs and the spleens harvested from immunized mice. Our observation was confirmed by a report demonstrating delayed presence of APCs in spleen following gene gun immunizations.[37,62] As well, other have reported that CD11c^{dim}, MHC class II^{hi}, DEC-205$^{-/+}$ DCs are rarely observed in the spleen following subcutaneous immunization,[63] suggesting that infiltrating DCs that acquire the expressed gene product at the immunization site subsequently traffic to regional LNs.

Recent studies indicated that immune responses are initiated by small numbers of directly transfected LCs that migrate to LNs draining the immunization site within 5-12 hrs of immunization. Immediate excision of skin from the immunization site eliminates both T and B cell responses.[53,54] However, when the epidermal tissue is left intact for 3 days, maximal antibody responses develop.[53] This observations indicates that uptake of the expressed gene product by infiltrating DCs likely contributes to the magnitude of the primary B cell response over time.

T Helper Functions Involved in Early B Cell Responses

Th1 and Th2 CD4$^+$ T helper lymphocytes generate specific lymphokines that directly influence the subclass of immunoglobulin produced by immune B cells. During T-dependent B cell maturation within LNs, migrating DCs prime naïve T cells, establishing the basis for T helper co-stimulatory functions required for mediating B cell maturation, expansion and antibody production. DCs also direct Th1 or Th2 phenotypes through B7-1 or B7-2 cognative interations with CD28 on T cells,[64] influencing the subclass of immunoglobulin generated by immune B cells. Following antigenic priming, expanding Th1 and Th2 cells migrate to B cell follicles where their involvement in B cell expansion and antibody production is dependent upon CD40-CD40L interactions.[65] DCs have also been shown to play a role in initiating and regulating humoral responses by directly interacting with B lymphocytes.[66] Migrating DCs, retaining unprocessed antigen, transfer it to recirculating B cells in secondary lymphoid tissue, along with early signals involved in isotype switch.[66] Antigen-primed B cells serving as APCs also influence the Th1/Th2 pathway,[67-69] which is again dependent upon expression levels of B7-1 or B7-2.[70,71] The predominant subclass of antibody associated with this pathway of B cell priming is IgG$_1$, reflective of complement-independent Th2 responses, involving IL-4, IL-5, IL-6, IL-10. Our studies find that hybridomas producing IgG$_1$, immortalized by somatic fusion of immune lymphocytes following gene gun-based DNA immunization are the predominant type. IgG$_{2a}$ and IgG$_{2b}$ producing hybridomas are also detectable, but at a significantly lower frequency. Analyses of polyclonal antisera collected less than 2 weeks after gene gun delivery of DNA-coated gold particles, and at later time-points, confirm a Th2 bias.[72,73] Intramuscular immunizations of DNA yield complement-dependent Th1 responses effecting class-switching to IgG$_{2a}$ and IgG$_{2b}$, characteristic of IFN-γ and IL-2 secretion.[72] Polarization of T helper lymphocytes that account for differences seen between intramuscular and gene gun immunizations are thought to arise as the result of vast differences in the quantity of DNA injected. The reduced concentration of DNA likely effects the adjuvant and mitogenic effects that are mediated by unmethylated CpG dinucleotides.[57] Lymphoid cells stimulated by CpG motifs produce IFN-γ, IL-12 and IL-18, cytokines generally associated with a Th1 response.[74,75] Another likely factor effecting the Th bias results from the rapid translocation of LCs following particle bombardment, which diminishes the chance for a localized adjuvant effect that would predominately mediate Th1 reactions.[72]

Enhancement of B Cell Activation Following in vivo Expression of Human Fc-Fusion Proteins Using DNA-Based Immunization

Fc-fusion proteins also referred to as Ig-chimeras have been widely used as research reagents to aid in studying the biological function of receptor-ligand interactions. Fc fusion proteins are commonly comprised of framework sequences encoding the hinge, C_H2 and C_H3 regions from human IgG_1 combined with sequences encoding full-length or trucations of the protein of biological interest.[76,77] DNA-based immunization leading to in vivo expression in mice of gene products engineered to be expressed as secreted human Fc-chimeras (Hu IgG_1) have also been successfully used to induce antigen-specific polyclonal antibody responses.[78-81] Affinity matured antibody-producing cell lines have been successfully generated using this approach. Augmentation of B cell responses, which result from the expression of antigens as Fc chimera, is observed using either amino or carboxy terminal fusions.[78-80] Dimerization of fusion proteins that result from disulfide bonds formed by cysteine residues comprising the hinge region of Fc result in increased antibody titers, as compared to monomeric expression of secreted proteins.[80,82] Recent studies assessing the use of truncations of exon 4 and exon 3 of the hinge region of human IgG3 showed lower antibody titers likely due to the elimination of cystein residues, which reduced the numbers of inter-chain disulfide bonds thus increasing monomeric expression of the fusion proteins.[82] Dimerization of in vivo expressed antigens may increase the half-life of proteins, increasing the likelyhood of antigen uptake by APCs.[82] B cell activation may also be positively effected as the result of dimerization through the engagement of multiple immunoglobulin receptors.[82]

Rapidly maturing B cell responses are generated against both components of the fusion protein following delivery of DNA-coated gold particles encoding human Fc fusion proteins. However the predominant response is directed against human Fc. This conclusion is drawn from ELISA analysis of polyclonal antisera, as well as ELISA analysis of immunoglobulin produced by hybridomas following somatic fusion of LNs cells. In a previous study using the PowderJect gene gun with a total of only 5 μg of DNA encoding a human Fc fusion with a highly conserved gene, Phosphoprotein over Expressed in Diabetes/ Phosphoprotein Enriched in Astrocytes (PED/PEA-15), monoclonal antibodies were rapidly generated. Hybridomas resulting from somatic fusion of immune cells harvested from axillary and brachial lymph nodes 13 day after the onset of immunization exhibit binding properties indicative of affinity matured immunoglobulin to PED. The monoclonal antibodies reacted with PED/PEA-15, a protein found in patients with type 2 diabetes, in Western blotting, ELISA and immunoprecipitation. Our current study analyzes the early developing anti-human Fc response following gene gun immunization using the PED/Fc plasmid. We have confirmed that primary germinal centers are evident within LNs draining the immunization site following gene gun delivery of pPED/Fc. Analysis of polyclonal antisera, as well as immunoblobulin produced by hybridomas specific for human Fc confirm class-switching and affinity-matured immunoglobulin reactive with human Fc.

We do not know how much of a role secretion actually plays in the rapid induction of early B cell maturation. Previous reports have shown that antibody responses can be augmented using expression vectors encoding secreted immunogens.[62,83] We predict that the signal sequence more than likely enhances B cell responses by making the secreted gene product available to bystander APCs. Studies have shown antigen transfer from transfected cells to APCs following DNA-based immunizations.[42,84] Interdigitating cells (IDCs) that capture and retain unprocessed antigen, have also been shown to transfer antigen to naïve B cells providing signals that regulate isotype switch.[66,85]

Germinal Centers

GCs are formed by expanding B cells in defined cellular compartments located within LNs and the spleen in response to T helper (Th)-dependent antigenic stimulation.[86] Affinity maturation results from antigen-driven somatic mutations that arise in variable region genes during rapid cell division of centroblast cells located within the dark zone (DZ) of a germinal center. Selection of antigen-specific immunoglobulin producing B cells occurs as centrocytes, located within the light zone (LZ) of GC, interact with FDCs displaying immune complexes (iccosomes) of the antigen. These cellular events take place within newly formed GCs 7-14 day after the onset of priming.[87-91] Mutations that give rise to increased affinity can be detected as early as day 6, and continue throughout the first 2 weeks.[22,92,93] Long-lived antibody secreting cells (B220$^{+/-}$CD138$^+$) and memory B cells (B220$^+$CD138$^-$) are thought to arise from primary GC reactions.[94-97] Class switching from IgM to IgG, in the absence of somatic hypermutation has been shown to occur within B cell follicles, or in B cells undergoing extrafollicular responses.[17,89,,93,98] However, high affinity antibodies resulting from mutations in V-region genes of an antigen-driven B cell maturation depend on GC reactions.[22,90] Previous studies demonstrated class switching occurred after the selection of high affinity LZ centrocytes.[99] Questions still remain as to whether or not memory B cells can generate new germinal centers and undergo further somatic mutations and affinity maturation events following subsequent antigenic challenge.[94,100-102]

Within 3-5 days of inducing primary responses with soluble antigen, antigen primed T and B cells cognatively interact within T cell compartments of draining lymph LNs. This result in detectable Cγ1 and Cγ2a switch transcripts observed in immune B cells as early as day 3[103] with formation of GC arising by day 5. Class switching continues during B cell expansion and proliferation within GCs through day 16.[103] Plasma cells producing IgG$_1$ as early as day 5 have been reported[94] and affinity matured MAbs producing IgG$_1$ and IgG$_2$a have been generated from somatic fusions performed as early as day 7.[104] Upon antigen rechallenge, affinity matured memory B cell responders, bearing evidence of somatic mutations comparable to those detected during primary GC reactions, rapidly reemerge as antigen-binding B cells. Pooling and fusing immune LN cells, prior to any further B cell selection processes that occur within GCs or to circulating antibody,[105] may capture the height of antigen-driven immunoglobulin diversity and maturation.

Our studies use the P3XBcl-2-13 murine myeloma fusion partner (RIMMS) to immortalize B cells in LN draining the site of gene gun immunizations. We have previously found that the affinities of antibodies produced by monoclonals generated using DNA encoding human Fc fusion constructs reflect binding properties characteristic of matured immunoglobulin, as demonstrated by the isolation of immunoprecipitating monoclonal antibodies and Western reactive antibodies. Affinities of greater than 10^8 mol^{-1} are associated with highly reactive immunoprecipitating monoclonal antibodies, and strong immunoblotting signals result from antibodies with affinities of greater than 10^7 mol^{-1}. The level of maturation implied by the binding profiles seen with the anti-human Fc antibodies indicate that hyperproliferative responses giving rise to the antigen-driven somatic mutations have likely taken place prior to somatic fusions on days 11 or 12.

Previous reports show that the affinity of a monoclonal antibody-producing cell line isolated from a fusion of LN cells 7 days after the onset of immunization was 1.65 X 10^9, as determined by BIAcore analsis.[104] GC formation occurs within 1 week following primary immunization.[93,106,107] Hyperproliferating GC centroblasts have a doubling time of 6-7 hours with point mutations occurring at a rate of 10^{-3}/bp/generation. Taken together, a correlation could be made between antigen driven GC formation, centroblast hyperproliferation and somatic mutations and the diverse repertoire and affinities displayed by antibodies captured by somatic fusion of cells from immune LNs with P3xBcl-2-13 following RIMMS.

Two 8 week old C57Bl/6 female mice were immunized on day 0 and 8 with 2 overlaying shots (1.25 µg per shot) of pPED/Fc into the thoracic region with a total of 5µg of DNA. On day 12, the mice were sacrificed and axillary and brachial nodes were harvested. The total number of cells isolated for somatic fusion with P3XBcl-2-13 was 2 x 10^7. Analysis of supernatants collected from the 48 plated wells revealed that 20 of the wells were positive (41%) for

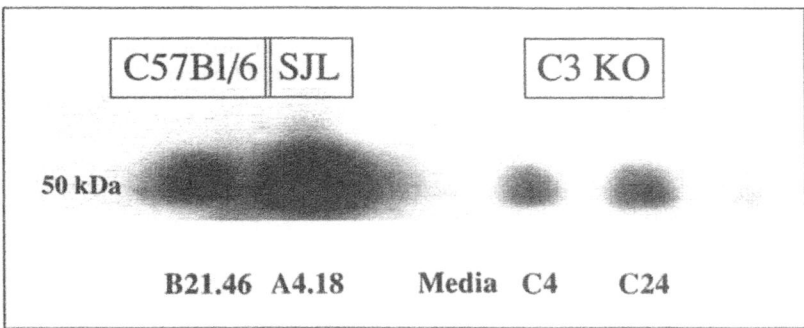

Figure 1. Western blot of anti-human IgG run under denaturing conditions probed with monoclonal antibodies generated using pAlphaPed/Fc delivered by the PowderJect gene gun. Goat anti-mouse HRP-labeled IgG and ECL were used for immunodetection.

anti-human IgG in ELISA using γ-chain specific secondary detection. Of these 20 IgG positives, 9 bound human IgG under native conditions in Western blotting. A clone isolated by limiting dilution designated A4.18, was determined to produce IgG_{2b} that bound human Fc under denaturing conditions in Western blotting as well (Fig. 1). B21 and B23 both produced IgG_1 immunoglobulin that also bound denatured human IgG as well in Western blot.

Two C3 knockout (KO) mice were immunized on days 0 and 8 using the same immunization procedure reported above. Axillary and brachial nodes and the spleen from the 2 mice were harvested for somatic fusion of immune cells on day 11.

The number of cells isolated from the LNs was 2.8×10^7. Following somatic fusion the cells were plated (144 wells). ELISA analysis of supernatants revealed that 11 wells were positive using γ-chain specific secondary detection (4 strong by ELISA bound native in western blotting; clones C4 and C24 bound native and denatured equally well) (see Table 1).

DNA immunization with pPED/Fc elicited comparable immune responses in both C3 sufficient and –deficient mice, as determined by the quality of recovered hybridoma antibodies. Antibody binding constants of hybridoma proteins generated in C3 knockouts indicated a modest, but nonsignificant, reduction in affinity maturation (Fig. 1). Nonetheless, the potent immunogenicity of pPED/Fc does not depend on local Fc/C3 interaction.

Responsiveness of Neonatal B Cells

Until about 6 months of age, human infants do not make long lasting antibody responses to several vaccines. After this time, responses become long-lived and stable.[108] In addition to quantitative differences between neonatal and adult B and T cells, other factors could contribute to the limited antibody responses in early life. Collaboration between T and B cells in the periarteriolar lymphoid sheath (PALS) requires costimulation and results in the proliferation of both cell types. During this initial proliferative time, the B cell either develops extrafollicularly into an antibody-secreting plasmacyte or re-enters into the follicle to initiate the GC reaction. It is possible that the ability to generate the GC reaction and/or long lived plasmacytes is impaired in neonates. Neonatal B cells produce less antibodies than adults after antigen-specific activation. However, B cells in the neonatal mouse respond to a primary immunization with T-dependent antigens,[108] such as hapten-carrier molecules and with conventional vaccines. In adult mice and humans, long-lived serum antibody responses depend upon the generation of bone marrow plasma cells that persist for years.[108] In turn, these memory antibody-forming cells (AFC) depend on the GC reaction for their production.[21]

Table 1. *Kinetic rate constants and calculated equilibrium dissociation constants for anti-human Fc antibodies C57A4.18 and C3K0C24 binding human IgG determined by Biacore analysis*

Antiserum	ka (1/Ms) ± S.E.	kd (1/s) ± S.E.	K_D (M)	Chi^2
C57A4.18	$3.4 \pm 0.02 \times 10^5$	$4.1 \pm 0.96 \times 10^{-5}$	1.2×10^{-10}	2.47
C3K0C24	$1.9 \pm 0.03 \times 10^5$	$1.2 \pm 0.01 \times 10^{-3}$	6.3×10^{-9}	0.73

Mouse antibody specific for human Fc were bound by immobilized rabbit anti-mouse Fc and then allowed to react with purified human IgG injected into the binding chamber. Association and dissociation data were analyzed using Biacore analysis software.

T and B cells must be in the proper location and in sufficient numbers to generate effector responses. As compared to the adult, the secondary lymphoid organs of the neonate contain few lymphocytes.[109] B220$^+$ B cells are present in the spleen by 1.5 days after birth (8% of total cells) and their relative number reaches adult levels two weeks later (68% of total cells). In the LN B cells are very infrequent at birth (0.6% of total cells) and gradually increase in number with time. In addition, the level of B220 expression in splenic and LN B cells within the first 2 weeks is low to intermediate, suggesting a more immature phenotype.[109] To determine if B cells in the neonatal mouse are phenotypically immature, we employed a monoclonal antibody (493) that recognizes an antigen (pB 130-140)[110] expressed at high levels on pro-, pre- and immature B cells and at low levels on splenic transitional B lymphocytes (493^{lo}).[110] Staining of 2 and 4-day old spleens with 493 confirmed that more than half of the splenic B220$^+$ cells in two day-old mice express the pB 130-140 antigen and low- to intermediate levels of MHC class II molecules (Fig. 2). These results indicate that the spleen of neonatal mice contains high percentages of B cells with immature phenotype.

DNA Immunization of Neonates by Gene Gun

DNA immunization is a powerful vaccine delivery system, capable of inducing cellular and humoral immune responses.[111] We[112] and others[113-115] have demonstrated the efficacy of DNA vaccines in neonatal mice. Recently, Kilpatrick et al demonstrated that gene-gun immunization of adult mice with a plasmid DNA (pAlphaPED/Fc) encoding the Fc portion of human IgG,[79] resulted in a rapid class-switched, anti-Fc antibody response as early as 13 days after immunization. This type of immunization provided a mean to test the induction of an Ab response within 2-3 weeks of immunization of neonates with a strong Ag. AlphaPED/Fc was delivered intradermally (i.d.) by gene gun in 4 day old mice. Mice were given a dose of 2.5-5 µg pDNA at 4 days of age followed by an identical dose at day 8. Mice were bled 13 days after immunization and tested for anti-HuFc Ab response. The draining LNs were prepared and stained with anti-B220 and anti-GL7 Ab to test for evidence of GC B cells (Fig. 4).[22] Serum antibody titers were 32 fold lower in neonates than in adults and the intensity of B220 staining was reduced compared to adults, suggesting an immature B cell phenotype (Figs. 3,4). Similar results were obtained in the NFS/N and in the C57BL/6 mouse strains. The antibody response was of the IgG$_1$ isotype in all of the neonatal and adult serum samples, and the staining of the LNs showed detectable GC formation in mice immunized as neonates. However the size and number of GC was lower in neonatally immunized mice (Fig. 4). These results indicate that neonatal mice are primed and not tolerized by pAlphaPED/Fc immunization, and that such immunization induces an antibody response qualitatively similar to adults but of smaller magnitude.

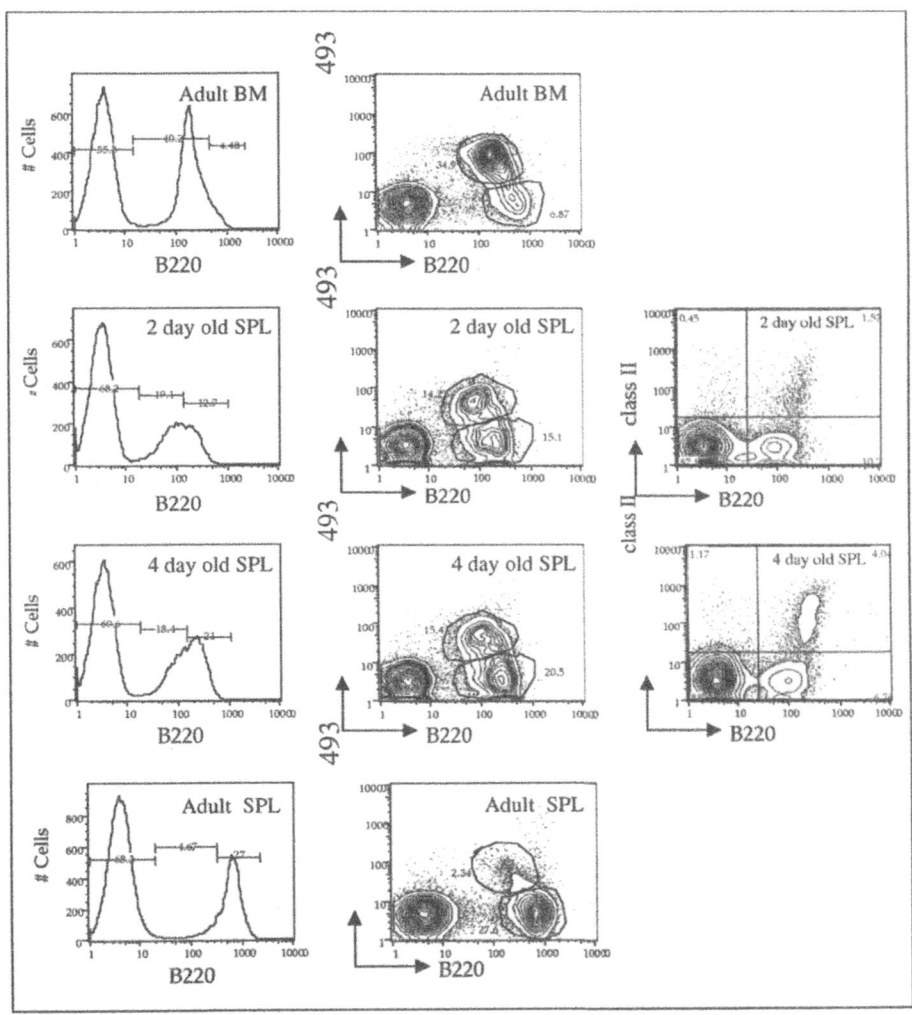

Figure 2. Staining of 2 and 4-day old spleens with anti-pB 130-140 (493) and anti-B220 monoclonal antibodies demonstrated that all splenic B cells in neonatal mice express an immature phenotype characterized by low to intermediate levels of B220 and MHC class II molecules, as well as high levels of pB 130-140.

Conclusions

DNA-based immunization strategies will undoubtedly become more common for the induction of prophylactic immunity. However, the use of these novel methods will be maximized only when we solidify our understanding of how and to what type of immunocyte this new immunization format supplies. Interestingly, virtually all classical vaccines were optimized by trial and error and without the benefit of an understanding of the cellular- and molecular processes on which humoral immunity depends. Ongoing DNA-based vaccination studies offer a new chance for contemporary immunology to focus once again on the induction of effective immunity and to understand how immunological protection arises from the interaction from the immune system's disparate components.

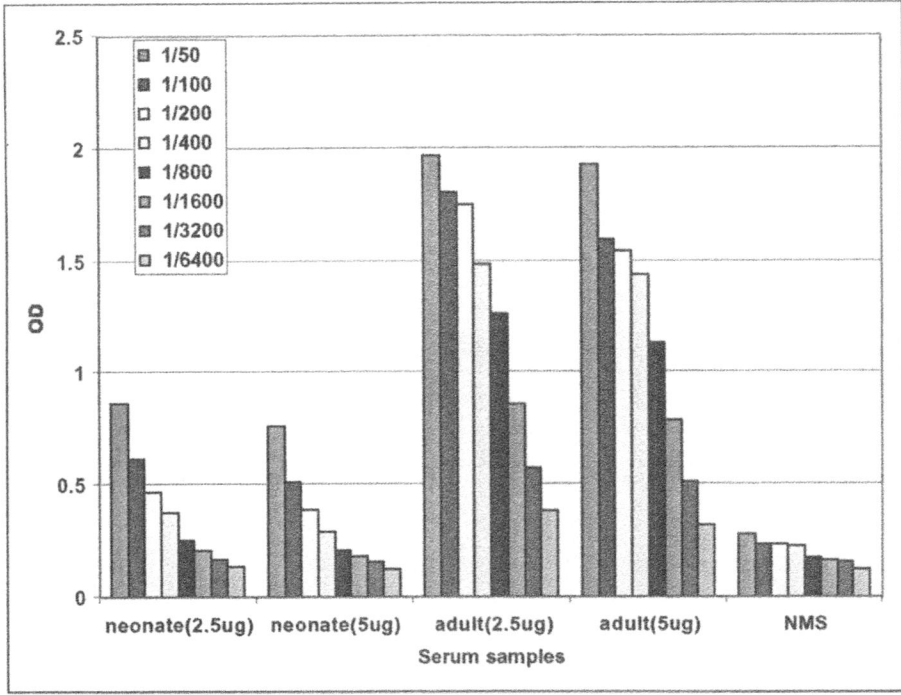

Figure 3. Serum anti-Fc antibody response of mice immunized at 4 days of age or as adults with different doses (2.5-5 µg) of pAlphaPed/Fc pDNA by gene gun, and boosted at day 8 with the same pDNA.

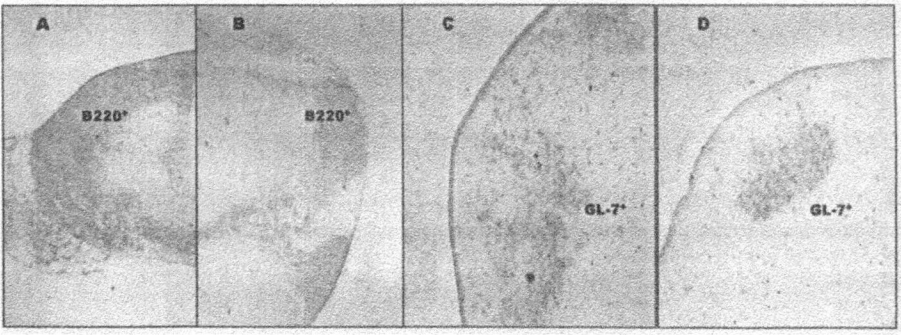

Figure 4. Histologic staining of LN sections from DNA-immunized neonatal and adult mice. LN sections from pAlphaPed/Fc DNA-immunized adult (A,C) or neonatal (B,D) mice (13 days after immunization) were stained with anti-B220 or with anti-GL7 alkaline phosphatase-labeled monoclonal antibodies. The intensity of B220 staining (A,B)(100x) is lower in neonatally immunized mice compared to adults, corresponding to the high frequency of peripheral B220^{low} B cells in neonates. GL-7$^+$ germinal centers (GC)(C,D)(200x) are detectable in LNs from neonatally immunized mice,but the size and number of LNs GCs in neonatal mice is lower than in adults.

References

1. Robinson HL. DNA vaccines: basic mechanism and immune responses (Review). Int J Mol Med 1999; 4:549.
2. Stockwin LH, McGonagle D, Martin IG et al. Dendritic cells: immunological sentinels with a central role in health and disease. Immunol Cell Biol 2000; 78:91.
3. Cyster JG. Chemokines and the homing of dendritic cells to the T cell areas of lymphoid organs. J Exp Med 1999; 189:447.
4. Dustin ML, Cooper JA. The immunological synapse and the actin cytoskeleton: molecular hardware for T cell signaling. Nat Immunol 2000; 1:23.
5. Dustin ML, Olszowy MW, Holdorf AD et al. A novel adaptor protein orchestrates receptor patterning and cytoskeletal polarity in T cell contacts. Cell 1998; 94:667.
6. Dustin ML. Making a little affinity go a long way: a topological view of LFA-1 regulation. Cell Adhes Commun 1998; 6:255.
7. Monks CR, Freiberg BA, Kupfer H et al. Three-dimensional segregation of supramolecular activation clusters in T cells. Nature 1998; 395:82.
8. Grakoui A, Bromley SK, Sumen C et al. The immunological synapse: a molecular machine controlling T cell activation. Science 1999; 285:221.
9. Lenschow DJ, Walunas TL, Bluestone JA. CD28/B7 system of T cell costimulation. Annu Rev Immunol 1996; 14:233.
10. Garside P, Ingulli E, Merica RR et al. Visualization of specific B and T lymphocyte interactions in the lymph node. Science 1998; 281:96.
11. Cyster JG. Chemokines and cell migration in secondary lymphoid organs. Science 1999; 286:2098.
12. Ansel KM, McHeyzer-Williams LJ, Ngo VN et al. In vivo-activated CD4 T cells upregulate CXC chemokine receptor 5 and reprogram their response to lymphoid chemokines. J Exp Med 1999; 190:1123.
13. Reif K, Ekland EH, Ohl L et al. Balanced responsiveness to chemoattractants from adjacent zones determines B cell position. Nature 2002; 416:94.
14. Jacob J, Kassir R, Kelsoe G. In situ studies of the primary immune response to (4-hydroxy-3-nitrophenyl)acetyl. I. The architecture and dynamics of responding cell populations. J Exp Med 1991; 173:1165.
15. Schaniel C, Pardali E, Sallusto F et al. Activated murine B lymphocytes and dendritic cells produce a novel CC chemokine which acts selectively on activated T cells. J Exp Med 1998; 188:451.
16. Tang HL, Cyster JG. Chemokine Up-regulation and activated T cell attraction by maturing dendritic cells. Science 1999; 284:819.
17. Jacob J, Kelsoe G. In situ studies of the primary immune response to (4-hydroxy-3-nitrophenyl)acetyl. II. A common clonal origin for periarteriolar lymphoid sheath-associated foci and germinal centers. J Exp Med 1992; 176:679.
18. Kelsoe G. Life and death in germinal centers (redux). Immunity 1996; 4:107.
19. Liu YJ, Zhang J, Lane PJ et al. Sites of specific B cell activation in primary and secondary responses to T cell-dependent and T cell-independent antigens. Eur J Immunol 1991; 21:2951.
20. Stuber E, Strober W. The T cell-B cell interaction via OX40-OX40L is necessary for the T cell-dependent humoral immune response. J Exp Med 1996; 183:979.
21. Takahashi Y, Cerasoli DM, Dal Porto JM et al. Relaxed negative selection in germinal centers and impaired affinity maturation in bcl-xL transgenic mice. J Exp Med 1999; 190:399.
22. Jacob J, Kelsoe G, Rajewsky K et al. Intraclonal generation of antibody mutants in germinal centres. Nature 1991; 354:389.
23. Toellner KM, Gulbranson-Judge A, Taylor DR et al. Immunoglobulin switch transcript production in vivo related to the site and time of antigen-specific B cell activation. J Exp Med 1996; 183:2303.
24. Cyster JG, Ngo VN, Ekland EH et al. Chemokines and B cell homing to follicles. Curr Top Microbiol Immunol 1999; 246:87.
25. Gunn MD, Tangemann K, Tam C et al. A chemokine expressed in lymphoid high endothelial venules promotes the adhesion and chemotaxis of naive T lymphocytes. Proc Natl Acad Sci USA 1998; 95:258.
26. Walker LS, Gulbranson-Judge A, Flynn S et al. Compromised OX40 function in CD28-deficient mice is linked with failure to develop CXC chemokine receptor 5-positive CD4 cells and germinal centers. J Exp Med 1999; 190:1115.
27. Atansiu P. Production de tumers chez le hamster par inoculation d'acide desoxribonucleique extrait de cultures des tissue unfectees par le virus de polyomie. Acad Sci 1962; 254:4228.
28. Fried M, Klein B, Murray K et al. Infectivity in mouse fibroblasts of polyoma DNA integrated into plasmid PBR322 or lambdoid phage DNA. Nature 1979; 279:811.

29. Wolff JA, Malone RW, Williams P et al. Direct gene transfer into mouse muscle in vivo. Science 1990; 247:1465.
30. Yang NS, Burkholder J, Roberts B et al. In vivo and in vitro gene transfer to mammalian somatic cells by particle bombardment. Proc Natl Acad Sci USA 1990; 87:9568.
31. Tang DC, DeVit M, Johnston SA. Genetic immunization is a simple method for eliciting an immune response. Nature 1992; 356:152.
32. Ulmer JB, Donnelly JJ, Parker SE et al. Heterologous protection against influenza by injection of DNA encoding a viral protein. Science 1993; 259:1745.
33. Wang B, Ugen KE, Srikantan V et al. Gene inoculation generates immune responses against human immunodeficiency virus type 1. Proc Natl Acad Sci USA 1993; 90:4156.
34. Gurunathan S, Wu CY, Freidag BL et al. DNA vaccines: a key for inducing long-term cellular immunity. Curr Opin Immunol 2000; 12:442.
35. McCluskie MJ, Brazolot Millan CL, Gramzinski RA et al. Route and method of delivery of DNA vaccine influence immune responses in mice and nonhuman primates. Mol Med 1999; 5:287.
36. Sanford JEA. An improved, helium driven biolistic device. Techniques 1991; 3:3.
37. Robinson HL. Nucleic acid vaccines: an overview. Vaccine 1997; 15:785.
38. Langerhans P. Uber die Nerven der menschlichen Haut. Virchows Arch 1868; 44:325.
39. Inaba K, Inaba M, Deguchi M et al. Granulocytes, macrophages, and dendritic cells arise from a common major histocompatibility complex class II-negative progenitor in mouse bone marrow. Proc Natl Acad Sci USA 1993; 90:3038.
40. Bergstresser PR, Fletcher CR, Streilein JW. Surface densities of Langerhans cells in relation to rodent epidermal sites with special immunologic properties. J Invest Dermatol 1980; 74:77.
41. Stingl G. Dendritic cells of the skin. Dermatol Clin 1990; 8:673.
42. Condon C, Watkins SC, Celluzzi CM et al. DNA-based immunization by in vivo transfection of dendritic cells. Nat Med 1996; 2:1122.
43. Porgador A, Irvine KR, Iwasaki A et al. Predominant role for directly transfected dendritic cells in antigen presentation to CD8+ T cells after gene gun immunization. J Exp Med 1998; 188:1075.
44. Witmer-Pack MD, Olivier W, Valinsky J et al. Granulocyte/macrophage colony-stimulating factor is essential for the viability and function of cultured murine epidermal Langerhans cells. J Exp Med 1987; 166:1484.
45. Heufler C, Koch F, Schuler G. Granulocyte/macrophage colony-stimulating factor and interleukin 1 mediate the maturation of murine epidermal Langerhans cells into potent immunostimulatory dendritic cells. J Exp Med 1988; 167:700.
46. Cumberbatch M, Kimber I. Dermal tumour necrosis factor-alpha induces dendritic cell migration to draining lymph nodes, and possibly provides one stimulus for Langerhans' cell migration. Immunology 1992; 75:257.
47. Aiba S, Katz SI. The ability of cultured Langerhans cells to process and present protein antigens is MHC-dependent. J Immunol 1991; 146:2479.
48. Larsen CP, Steinman RM, Witmer-Pack M et al. Migration and maturation of Langerhans cells in skin transplants and explants. J Exp Med 1990; 172:1483.
49. Price AA, Cumberbatch M, Kimber I et al. Alpha 6 integrins are required for Langerhans cell migration from the epidermis. J Exp Med 1997; 186:1725.
50. Romani N, Ratzinger G, Pfaller K et al. Migration of dendritic cells into lymphatics-the Langerhans cell example: routes, regulation, and relevance. Int Rev Cytol 2001; 207:237.
51. Fossum S. Lymph-borne dendritic leucocytes do not recirculate, but enter the lymph node paracortex to become interdigitating cells. Scand J Immunol 1988; 27:97.
52. Andree C, Swain WF, Page CP et al. In vivo transfer and expression of a human epidermal growth factor gene accelerates wound repair. Proc Natl Acad Sci USA 1994; 91:12188.
53. Torres CA, Iwasaki AA, Barber BH et al. Differential dependence on target site tissue for gene gun and intramuscular DNA immunizations. J Immunol 1997; 158:4529.
54. Klinman DM, Sechler JM, Conover J et al. Contribution of cells at the site of DNA vaccination to the generation of antigen-specific immunity and memory. J Immunol 1998; 160:2388.
55. Eisenbraun MD, Fuller DH, Haynes JR. Examination of parameters affecting the elicitation of humoral immune responses by particle bombardment-mediated genetic immunization. DNA Cell Biol 1993; 12:791.
56. Steele KE, Stabler K, VanderZanden L. Cutaneous DNA vaccination against Ebola virus by particle bombardment: histopathology and alteration of CD3-positive dendritic epidermal cells. Vet Pathol 2001; 38:203.
57. Sato Y, Roman M, Tighe H et al. Immunostimulatory DNA sequences necessary for effective intradermal gene immunization. Science 1996; 273:352.

58. Krieg AM, Yi AK, Matson S et al. CpG motifs in bacterial DNA trigger direct B cell activation. Nature 1995; 374:546.

59. Klinman DM, Yi AK, Beaucage SL et al. CpG motifs present in bacteria DNA rapidly induce lymphocytes to secrete interleukin 6, interleukin 12, and interferon gamma. Proc Natl Acad Sci USA 1996; 93:2879.

60. Jakob T, Walker PS, Krieg AM et al. Activation of cutaneous dendritic cells by CpG-containing oligodeoxynucleotides: a role for dendritic cells in the augmentation of Th1 responses by immunostimulatory DNA. J Immunol 1998; 161:3042.

61. Sparwasser T, Koch ES, Vabulas RM et al. Bacterial DNA and immunostimulatory CpG oligonucleotides trigger maturation and activation of murine dendritic cells. Eur J Immunol 1998; 28:2045.

62. Boyle JS, Koniaras C, Lew AW. Influence of cellular location of expressed antigen on the efficacy of DNA vaccination: cytotoxic T lymphocyte and antibody responses are suboptimal when antigen is cytoplasmic after intramuscular DNA immunization. Int Immunol 1997; 9:1897.

63. Randolph GJ, Inaba K, Robbiani DF et al. Differentiation of phagocytic monocytes into lymph node dendritic cells in vivo. Immunity 1999; 11:753.

64. Kuchroo VK, Das MP, Brown JA et al. B7-1 and B7-2 costimulatory molecules activate differentially the Th1/Th2 developmental pathways: application to autoimmune disease therapy. Cell 1995; 80:707.

65. Smith KM, Pottage L, Thomas ER et al. Th1 and Th2 CD4+ T cells provide help for B cell clonal expansion and antibody synthesis in a similar manner in vivo. J Immunol 2000; 165:3136.

66. Wykes M, Pombo A, Jenkins C et al. Dendritic cells interact directly with naive B lymphocytes to transfer antigen and initiate class switching in a primary T-dependent response. J Immunol 1998; 161:1313.

67. Chesnut RW, Grey HM. Studies on the capacity of B cells to serve as antigen-presenting cells. J Immunol 1981; 126:1075.

68. Janeway CA Jr, Ron J, Katz ME. The B cell is the initiating antigen-presenting cell in peripheral lymph nodes. J Immunol 1987; 138:1051.

69. Hayglass KT, Naides SJ, Scott CF Jr et al. T cell development in B cell-deficient mice. IV. The role of B cells as antigen-presenting cells in vivo. J Immunol 1986; 136:823.

70. Agrewala JN, Suvas S, Verma RK et al. Differential effect of anti-B7-1 and anti-M150 antibodies in restricting the delivery of costimulatory signals from B cells and macrophages. J Immunol 1998; 160:1067.

71. Vijayakrishnan L, Natarajan K, Manivel V et al. B cell responses to a peptide epitope. IX. The kinetics of antigen binding differentially regulates costimulatory capacity of activated B cells. J Immunol 2000; 164:5605.

72. Feltquate DM, Heaney S, Webster RG et al. Different T helper cell types and antibody isotypes generated by saline and gene gun DNA immunization. J Immunol 1997; 158:2278.

73. Tighe H, Corr M, Roman M. Gene vaccination: plasmid DNA is more than just a blueprint. Immunol Today 1998; 19:89.

74. Roman M, Martin-Orozco E, Goodman JS et al. Immunostimulatory DNA sequences function as T helper-1-promoting adjuvants. Nat Med 1997; 3:849.

75. Chu RS, Targoni OS, Krieg AM et al. CpG oligodeoxynucleotides act as adjuvants that switch on T helper 1 (Th1) immunity. J Exp Med 1997; 186:1623.

76. Fanslow WC, Anderson DM, Grabstein KH et al. Soluble forms of CD40 inhibit biologic responses of human B cells. J Immunol 1992; 149:655.

77. Kaplan JB, Sridharan L, Zaccardi JA et al. Characterization of a soluble vascular endothelial growth factor receptor-immunoglobulin chimera. Growth factors 1997; 14:243.

78. Kilpatrick KE, Cutler Y, Whitehorn E et al. Gene gun delivered DNA-based immunizations mediate rapid production of murine monoclonal antibodies to the Flt-3 receptor. Hybridoma 1998; 17:569.

79. Kilpatrick KE, Danger DP, Hull-Ryde EA et al. High-affinity monoclonal antibodies to PED/PEA-15 generated using 5 microg of DNA. Hybridoma 2000; 19:297.

80. Lew AM, Brady BJ, Boyle BJ. Site-directed immune responses in DNA vaccines encoding ligand-antigen fusions. Vaccine 2000; 18:1681.

81. Sutherland RM, McKenzie BS, Corbett AJ et al. Overcoming the poor immunogenicity of a protein by DNA immunization as a fusion construct. Immunol Cell Biol 2001; 79:49.

82. Drew DR, Boyle JS, Lew AM et al. The human IgG3 hinge mediates the formation of antigen dimers that enhance humoral immune responses to DNA immunisation. Vaccine 2001; 19:4115.

83. Inchauspe G, Vitvitski L, Major ME et al. Plasmid DNA expressing a secreted or a nonsecreted form of hepatitis C virus nucleocapsid: comparative studies of antibody and T-helper responses following genetic immunization. DNA Cell Biol 1997; 16:185.

84. Fu TM, Ulmer JB, Caulfield MJ et al. Priming of cytotoxic T lymphocytes by DNA vaccines: requirement for professional antigen presenting cells and evidence for antigen transfer from myocytes. Mol Med 1997; 3:362.

85. Gerloni M, Lo D, Zanetti M. DNA immunization in relB-deficient mice discloses a role for dendritic cells in IgM-->IgG1 switch in vivo. Eur J Immunol 1998; 28:516.

86. Kroese FG, Timens W, Nieuwenhuis P. Germinal center reaction and B lymphocytes: morphology and function. Curr Top Pathol 1990; 84:103.

87. Liu YJ, Joshua DE, Williams GT et al. Mechanism of antigen-driven selection in germinal centres. Nature 1989; 342:929.

88. Szakal AK, Kosco MH, Tew JG. Microanatomy of lymphoid tissue during humoral immune responses: structure function relationships. Annu Rev Immunol 1989; 7:91.

89. Apel M, Berek C. Somatic mutations in antibodies expressed by germinal centre B cells early after primary immunization. Int Immunol 1990; 2:813.

90. Berek C, Berger A, Apel M. Maturation of the immune response in germinal centers. Cell 1991; 67:1121.

91. Nossal GJ. The molecular and cellular basis of affinity maturation in the antibody response. Cell 1992; 68:1.

92. Weiss U, Zoebelein R, Rajewsky K. Accumulation of somatic mutants in the B cell compartment after primary immunization with a T cell-dependent antigen. Eur J Immunol 1992; 22:511.

93. McHeyzer-Williams MG, McLean MJ, Lalor PA et al. Antigen-driven B cell differentiation in vivo. J Exp Med 1993; 178:295.

94. MacLennan IC, Gray D. Antigen-driven selection of virgin and memory B cells. Immunol Rev 1986; 91:61.

95. Ahmed R, Gray D. Immunological memory and protective immunity: understanding their relation. Science 1996; 272:54.

96. Rajewsky K. Clonal selection and learning in the antibody system. Nature 1996; 381:751.

97. McHeyzer-Williams MG. Immune response decisions at the single cell level. Semin Immunol 1997; 9:219.

98. Pulendran B, Smith KG, Nossal GJ. Soluble antigen can impede affinity maturation and the germinal center reaction but enhance extrafollicular immunoglobulin production. J Immunol 1995; 155:1141.

99. Liu YJ, Malisan F, de Bouteiller O et al. Within germinal centers, isotype switching of immunoglobulin genes occurs after the onset of somatic mutation. Immunity 1996; 4:241.

100. Siekevitz M, Kocks C, Rajewsky K et al. Analysis of somatic mutation and class switching in naive and memory B cells generating adoptive primary and secondary responses. Cell 1987; 48:757.

101. Roost HP, Bachmann MF, Haag A et al. Early high-affinity neutralizing anti-viral IgG responses without further overall improvements of affinity. Proc Natl Acad Sci USA 1995; 92:1257.

102. Decker DJ, Linton PJ, Zaharevitz S et al. Defining subsets of naive and memory B cells based on the ability of their progeny to somatically mutate in vitro. Immunity 1995; 2:195.

103. Toellner KM, Luther SA, Sze DM et al. T helper 1 (Th1) and Th2 characteristics start to develop during T cell priming and are associated with an immediate ability to induce immunoglobulin class switching. J Exp Med 1998; 187:1193.

104. Bynum J, Andrews JL, Ellis B et al. Development of class-switched, affinity-matured monoclonal antibodies following a 7-day immunization schedule. Hybridoma 1999; 18:407.

105. Vora KA, Manser T. Altering the antibody repertoire via transgene homologous recombination: evidence for global and clone-autonomous regulation of antigen-driven B cell differentiation. J Exp Med 1995; 181:271.

106. Levy NS, Malipiero UV, Lebecque SG et al. Early onset of somatic mutation in immunoglobulin VH genes during the primary immune response. J Exp Med 1989; 169:2007.

107. Liu YJ, Mason DY, Johnson GD et al. Germinal center cells express bcl-2 protein after activation by signals which prevent their entry into apoptosis. Eur J Immunol 1991; 21:1905.

108. Special Issue: Immunity in Early Life. Vaccine 1998.

109. Garcia AM, Fadel SA, Cao S et al. T Cell Immunity in Neonates. Immunol Res 2000; 22.

110. Rolink G, Andersson J, Melchers F. Characterization of immature B cells by a novel monoclonal antibody, by turnover and by mitogen reactivity. Eur J Immunol 1998; 28:3738.

111. Donnelly JJ, Ulmer JB, Shiver JW et al. DNA vaccines. Annu Rev Immunol 1997; 15:617.

112. Sarzotti M, Dean TA, Remington MP et al. Induction of cytotoxic T cell responses in newborn mice by DNA immunization. Vaccine 1997; 15:795.

113. Bot A, Antohi S, Bona C. Immune response of neonates elicited by somatic transgene vaccination with naked DNA. Front Biosci 1997; 2:d173.
114. Siegrist CA, Lambert PH. Immunization with DNA vaccines in early life: advantages and limitations as compared to conventional vaccines. Springer Semin Immunopathol 1997; 19:233.
115. Pertmer TM, Robinson HL. Studies on antibody responses following neonatal immunization with influenza hemagglutinin DNA or protein. Virology 1999; 257:406.

CHAPTER 5

Immune Responses to DNA Vaccines: Induction of CD8+ T Cells

Jens A. Leifert and J. Lindsay Whitton

CD8+ T Cells Are Important in Controlling Most Virus Infections

The importance of CD8+ T cells in the control and eradication of viruses has been demonstrated in mice and men. In the mouse, they are critical in combating infection with lymphocytic choriomeningitis virus (LCMV),[1] and in humans, "experiments of nature" strongly suggest that T cells play a vital role in controlling many virus infections.[2] For example, children born with hereditary agammaglobulinemia are much more susceptible to suppurative bacterial infections,[3] and people with defects in the complement cascade show increased susceptibility to Neisserial diseases;[4-6] however, in contrast to their greatly enhanced vulnerability to bacterial infections, these individuals show only mildly elevated susceptibility to most viral diseases, with the exception of rare enteroviral meningitides, caused most often by picornaviruses such as coxsackievirus[7,8] and echovirus type 9 or 11.[9,10] For most virus infections, the incidence of disease, and disease severity, are similar in antibody-deficient and in immunocompetent individuals. These observations suggest that other factors—perhaps CD8+ T cells—are capable of resolving (most) virus infections in humans. This suggestion is supported by the finding that the frequency and severity of virus infections are markedly increased in humans with impaired T-cell responses [for example, in patients with Di George's syndrome (congenital thymic aplasia), acquired immunodeficiency syndrome (AIDS), leukemia, or recipients of immunosuppressive therapy].[11] In HIV infection, CD8+ T-cell activity correlates with clearance of initial viral load, and their absence heralds a return to high viral titers, and eventual AIDS.[12-14] The importance of T cells in controlling human virus infections is further highlighted by our responses to measles virus. In immunocompetent individuals, the infection is typified by the characteristic and diagnostic) rash, and complete recovery is the norm. In contrast, in T-cell deficiency, the disease is often fatal.[15-17] The rash itself is T-cell-mediated and does not develop in severely immunosuppressed individuals; indeed, the presence of a rash in an immunosuppressed victim (e.g., in a leukemic child with measles) is considered a positive prognostic indicator.[18] In agammaglobulinemic children, the rash develops normally, and the infection is cleared. Furthermore, these children are subsequently immune to measles,[2] suggesting that T cells can play an important role not only in controlling a primary infection, but also in preventing disease following secondary exposure; this observation was an early (and often overlooked) indication that CD8+ memory T cells might be important in vaccine-induced antiviral immunity (see below). In the next section, we shall provide a molecular explanation of why CD8+ T cells are important in controlling most virus infections; and why bacterial infections, rather than virus infections, are more severe in the absence of antibodies.

DNA Vaccines, edited by Hildegund C. J. Ertl. ©2003 Eurekah.com and Kluwer Academic / Plenum Publishers.

Antigen Presentation Pathways Determine the Type of Immune Response Mounted by the Host

In this section, we wish to make two broad points: (i) the host mounts the type of immune response best suited to eradicating the particular type of microbe which it faces; and (ii) the type of immune response mounted is dictated by the interaction of the infectious agent with the host's antigen presentation pathways. These ideas, which are relevant to our subsequent discussion of how DNA vaccination induces CD8⁺ T cells, have been reviewed elsewhere,[19] and detailed molecular, immunological, and biological perspectives are available.[20] The underlying mechanisms are cartooned in Figure 1 (the molecular details of the antigen presentation pathways are described in other chapters in this volume). Intracellular organisms (usually viruses) will synthesize proteins inside the infected cells, and these antigens therefore gain ready access to the MHC class I pathway; as a result, they often induce strong MHC class I restricted CD8⁺ T cell responses (these cells are usually cytotoxic, although they also secrete a variety of cytokines in response to antigen). Furthermore, the antigens synthesized during intracellular infections usually will be released from the infected cells into the extracellular milieu, and these soluble materials are taken up by specialized antigen presenting cells (APC) which express both classes of MHC molecule. Inside these APC, the antigens enter the MHC class II pathway, permitting them to induce MHC class II restricted CD4⁺ T cells, which usually provide "help" to B cells and, often to CD8⁺ T cells. The soluble antigens will also encounter B lymphocytes, and therefore should (in concert with CD4⁺ T cell "help") induce antibody responses. Therefore, we would expect most intracellular organisms to induce strong CD4⁺ and CD8⁺ T cell responses, and decent antibody responses; and, broadly speaking, this is the case. In contrast, the antigens of extracellular organisms (most bacteria) cannot efficiently enter the MHC class I pathway [cross-priming notwithstanding] and, therefore, most bacteria fail to induce strong CD8⁺ immunity. However, many of the proteins encoded by extracellular microbes will be taken up by APCs and introduced into the MHC class II pathway, and so extracellular organisms induce strong CD4⁺ T cell and antibody responses.

These considerations explain how the host "knows" what kind of response to produce against a given microbe—it is dictated by the intracellular or extracellular nature of the invader. Furthermore, as also shown in Figure 1, the response induced is that which is most appropriate to deal with the infection. An intracellular organism will induce CD8⁺ T cells (which recognize and eradicate infected cells) as well as antibodies (which mop up the organisms when they are in their extracellular state) and CD4⁺ T cells (which provide help). In contrast, the host will not "waste its time" producing CD8⁺ T cells against an organism whose entire life-cycle is extracellular; instead, strong CD4⁺ T cell and antibody responses predominate. Figure 1 also facilitates understanding of the biological consequences of immune dysfunction, which were described above. Individuals with agammaglobulinemia show increased susceptibility to bacterial diseases because they have no other adaptive immune response which can combat extracellular organisms; CD8⁺ T cells have little effect on our ability to combat most bacterial infections. Therefore, antibodies are absolutely critical for protection against many bacterial infections. In contrast, the absence of antibodies is often not as devastating when the microbe is intracellular, because the host can rely, to a significant extent, on CD8⁺ T cell responses; hence, agammaglobulinemic people can resist most[3] [although not all][7,10,21] virus infections. Finally, these ideas have implications for vaccine design.

Designing Vaccines Against Viruses and Bacteria

For many years, antibodies were thought to be the only important facet of vaccine-induced immunity. This misunderstanding had two causes. The first was practical; antibodies were much easier to detect than T cells. The second was conceptual; the observation that protective immunity usually coincided with strong antibody responses led many to conclude—illogically—that protection must be conferred by antibodies. As stated above, measles infection of agammaglobulinemic individuals suggested that other facets of adaptive immunity could

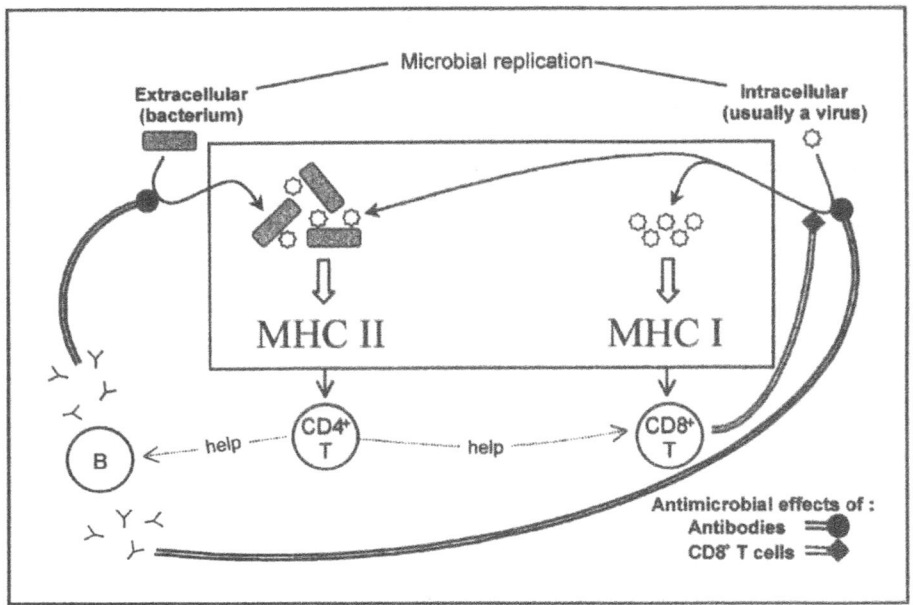

Figure 1. Antigen processing pathways ensure that the immune response which is mounted is that which is best suited to combat the invading microbe. The MHC class I and II antigen presentation pathways determine the types of adaptive immune response mounted to a microbial infection.

protect against disease following a secondary exposure to virus. More than a decade ago, we and others, working in viral model systems, asked if a vaccine which induced CD8[+] T cells alone could solidly protect the host against subsequent viral challenge. In LCMV infection[22-26] and in murine cytomegalovirus (MCMV) infection,[27,28] it proved possible to fully protect a naive animal from virus challenge by immunization with a recombinant vaccinia virus (VV) expressing a single viral internal (i.e., nonmembrane) protein. Indeed, recombinant vaccines containing "minigenes" encoding isolated CD8[+] T cell epitopes as short as 11 residues could confer protection against normally-lethal doses of challenge virus, and different epitopes could be linked in a "string of beads" to protect on several MHC backgrounds.[29,30] No virus-specific antibody responses are induced by these vaccines, proving that the outcome of infection (death or survival) is determined by vaccine-induced CD8[+] T cells. In the case of MCMV, a herpesvirus, protection is conferred by CD8[+] T cells specific for a protein expressed very early in the virus life-cycle. Presumably, it is to the host's benefit to recognize such early proteins, because it gives the host a chance to destroy infected cells before the virus can produce infectious progeny. In human herpesvirus infections, it is clear that CD8[+] T cell responses are generated against similar classes of proteins. For example, the major CTL response to human cytomegalovirus (HCMV) is to a protein expressed immediately upon infection;[31] a similar situation exists for varicella-zoster virus[32] and herpes simplex virus (HSV).[33] Of course, the immediate-early, early and late proteins, seen in herpesviruses, are not found in all virus infections; many viruses express their full complement of proteins more or less synchronously, and in these viruses, many proteins can be targets for biologically-relevant CD8[+] T cell responses. Influenza virus infection induces CTLs directed against most viral components, although a major group is specific for the virus nucleoprotein (NP). These NP-specific T cells, unlike anti-influenza antibodies, are cross-reactive; that is, they lyse HLA-matched target cells infected with a serologically distinct strain of influenza virus. However their presence fails to confer absolute immunity to infection and disease caused by a serotypically distinct influenza virus. These results are sometimes used to challenge the hypothesis that the presence of virus-specific CD8⁺ T cells confers protection against virus-induced disease. However, it is important to understand that CD8⁺ T cells cannot prevent infection; on the contrary, recognition by these cells requires that the target cell be infected. Antibodies prevent infection, and CD8⁺ T cells limit virus production and dissemination; together they protect against disease. Thus, the pre-existing anti-influenza CTLs, while failing to prevent infection, may diminish the ensuing morbidity (disease) and mortality (death).

We do not argue that induction of antibodies is unimportant for antiviral vaccines; on the contrary, a role for antiviral antibodies in vaccination is unquestionable. Passive antibody therapy can protect against or modify the course of several human virus infections. For example, infusion of antibodies specific for the Junin arenavirus is beneficial in Argentinian hemorrhagic fever,[34,35] whereas post-exposure rabies prophylaxis relies on vaccination and concurrent administration of virus-specific immunoglobulins.[36] Furthermore, in experimental models, antibodies can lower viral titers and modulate disease. For instance, recovery from ocular herpesvirus infection is hastened by administration of anti-HSV antibody.[37] Nevertheless, for the reasons discussed above, we conclude that an antiviral vaccine should not rely solely on antibodies, and instead should induce both arms of the antigen-specific response; T cells (CD8⁺ and CD4⁺) and antibodies. When viewed in the light of Figure 1, this conclusion makes good sense. What of antibacterial vaccines? For bacteria with a largely intracellular life cycle (e.g., Listeria monocytogenes), CD8⁺ T cells, as well as CD4⁺ T cells and antibodies, should be induced. The importance of CD8⁺ T cells in controlling Listeria was demonstrated years ago,[38,39] and a recombinant vaccinia vaccine encoding a single CD8⁺ T cell epitope from Listeria could confer some protection against bacterial challenge.[40] However, for a strictly extracellular organism, CD8⁺ T cells are much less likely to play a protective role. As stated above, antibodies are vital to the control of most bacterial infections, and therefore antibody induction should be an invariable goal of most bacterial vaccines.

It is appropriate, at this point, to make a general comment on the selection of vaccine target proteins. CD8⁺ T cell responses may be directed against essentially any intracellular protein, since most proteins made inside an infected cell can, potentially, contain epitopes presented by the MHC class I pathway. In many cases, the most important targets for CD8⁺ T cells are defined temporally; the most effective CD8⁺ T cells are often those specific for proteins made early in the microbial life cycle, and these proteins are appropriate vaccine candidates. Similarly, CD4⁺ T cell responses can be mounted against almost any protein, internal or external, since most proteins should be taken up by APCs, where their lysosomal degradation products become available for presentation by MHC class II molecules. In contrast to the large number of antigens open to T cell perusal, biologically-important antibody responses are most likely to be directed against "external" proteins, at the surface of the microbe (be it a virus or a bacterium), since those proteins will be most accessible. So, for most bacteria, the most promising vaccine candidates are likely to be cell surface proteins, or other cell surface structural components, which are accessible to antibodies. There is at least one additional consideration. Many bacterial diseases result not from the bacterial infection per se, but rather from the actions of bacterially-derived toxins; examples include tetanus and cholera. [In contrast, viruses generally do not produce toxins; although there are exceptions.][41] Vaccines against these diseases may, therefore, be targeted not against the bacterium, but against the toxin. In essence, such a vaccine does not necessarily allow the host to improve its control of infection; instead, it readies the host to neutralize the pathogenic capacity of the organism, and may leave the naturally-developing immune response to take care of the infection.

The Effector Functions Through Which CD8⁺ T Cells Exert Their Biological Activities

CD8⁺ T cells have two general effector functions: they can induce lysis of cells expressing the cognate antigen, or they can secrete cytokines in response to antigen contact. Lytic activity was the first criterion by which these cells were identified and, as a result, they often are termed cytotoxic T lymphocytes (CTL). However, it is becoming clear that CD8$^+$ T cells can regulate their effector functions in a subtle manner, and that some antigen-stimulated CD8$^+$ T cells can produce cytokines, but show low lytic activity.[42,43] Furthermore, as discussed below, CD8$^+$ T cells can control certain infections by cytokine release alone. Therefore, the term CTL should be used only when lytic activity is proven, and should not be used as a synonym for all antigen-responsive CD8$^+$ T cells.

CD8$^+$ T Cells Usually Contain Perforin, a Pore-Forming Protein

Most CD8$^+$ T cells contain granules, which align with the target cell upon recognition, and whose contents are released in a calcium-dependent manner onto the target cell membrane. These granules include a protein called perforin,[44] which undergoes assembly into trans-membrane pores and thus punches holes in the cytoplasmic membrane of the target cell. Perforin shares immunologic cross-reactivity with the C9 component of complement, and cloning of the murine perforin gene has allowed identification of a short stretch of amino acid homology between the two proteins. The membrane lesions caused by perforin are similar to those induced by the complement C9 complex. Thus, CTLs and complement-mediated lysis seem to share one common mechanism of action. The importance of perforin in CTL activity in vivo has been demonstrated.[45-47] Transgenic mice with a dysfunctional perforin gene are less effective at controlling infection by some (though not all) viruses; it has been argued that perforin is important for the clearance of "persistent" viruses, but is not required to counter acute virus infections which may be cleared by a combination of cytokines and antibodies.[48] This hypothesis, although intriguing, remains unproven.

CD8$^+$ T Cells Can Induce Apoptosis in Target Cells

Apoptosis, or programmed cell death, is a well-recognized phenomenon responsible for several developmental processes, including the clonal deletion of T cells in the thymus. Its most characteristic features are nuclear blebbing and disintegration, resulting in a nucleosome stepladder of fragmented DNA. The perforin channels inserted into the cell membrane by CTL permit the entry of other CTL-produced proteins (granzymes) which induce apoptosis in the target cell. Furthermore, CD8$^+$ (and sometimes CD4$^+$) T cells can induce apoptosis when the fasL protein, expressed on the T cell membrane,[49] interacts with Fas protein on the target cell, initiating a signaling cascade which ends in target cell apoptosis. Virus-specific T cells may induce this process in infected target cells.[50-52]

CD8$^+$ T Cells Release Antiviral Cytokines

Many CD8$^+$ T cells release high levels of cytokines, for example interferon-γ (IFNγ) and tumor necrosis factor-α (TNFα). Mice lacking the IFNγ receptor have increased susceptibility to several infections, despite apparently normal CTL and Th responses.[53] It has been cogently argued that a major role of the TcR/MHC/peptide interaction is to ensure that cytokine production by CD8$^+$ T cells is limited to the immediate proximity of virus-infected cells,[54,55] and convincing data from mice persistently infected with LCMV,[56,57] from hepatitis B virus (HBV) transgenic mice[58,59] and from HBV-infected primates[60] have shown that viral materials can be eradicated in vivo from neurons[56,57] and hepatocytes[58,59] in the absence of cytolysis.

Antigenic Control of CD8$^+$ T Cell Activation, and Effector Function

Antigen-specific recognition by the TcR, and signal transduction across the cell membrane, initiate a series of events in naïve CD8$^+$ T cells, including cell division and the activation of effector functions.

A Single Short Antigenic Pulse Is Sufficient to Drive Naïve CD8⁺ T Cells to Become Memory Cells

Recent data strongly suggest that the entire program of CD8⁺ T cell maturation can be initiated by a single, short-term, exposure to antigen. After antigen contact for as few as 2 hours, naïve CD8⁺ T cells bearing the appropriate TcR are irrevocably committed to divide, expand, develop their effector functions, and pass into the memory phase; it is important to note that no further antigen contact is required.[61,62] These findings have implications for DNA vaccination, indicating that long-term antigen expression should not be required and that, instead, a single strong pulse of antigen would be sufficient.

Antigen-Specific CD8⁺ T Cells Are Exquisitely Sensitive to Antigen Contact

Our lab has recently shown that, even at the peak of the antiviral immune response, when some 50% of all CD8⁺ T cells may be virus-specific,[63,64] the great majority of these cells are not actively producing cytokines; however these virus-specific cells are exquisitely sensitive to antigen contact, initiating cytokine synthesis immediately upon encountering an infected cell, and terminating production the instant the contact is broken.[65] This tight regulation is important because cytokines can be toxic to the host, and indeed systemic cytokine release is responsible for many of the symptoms of microbial infection, and for the extensive weight loss seen in certain cancers.[66]

CD8⁺ Memory T Cells Are Effector Cells

In the preceding pages, we have stressed that vaccine-induced CD8⁺ memory T cells are important components of antiviral vaccines. What are these vaccine-induced CD8⁺ memory cells, and how do they exert their biological activities? We have previously shown that the outcome of infection in immunized individuals is determined by the presence, at early times, of detectable memory CD8⁺ T cells; and we argued that the outcome of infection is decided within minutes or hours of virus challenge, even before any expansion of the memory cell population can occur.[67] One inescapable corollary of this conclusion is that the memory cells present at the time of infection must be able to display at least some of their effector functions very soon after encountering virus-infected cells. However, contrary to this contention, memory cells have, for many years, been considered a dormant population, lacking the effector functions displayed by cells during the acute phase of infection (which we shall refer to as "acute cells"). Although there are some differences—for example, memory cells are smaller than most acute cells, and usually contain less perforin—there are many similarities between the two cell populations. For example, surface markers such as CD11a, CD11b, CD44 and CD62L are present on "antigen-experienced" (acute or memory) CD8⁺ T cells,[68-73] but absent from naïve cells; this provides an inkling that memory and acute cells may share common functions. Furthermore, attempts to discriminate memory cells from acute cells using surface markers has been difficult, although a recent study showed differences in cell-surface expression of O-glycans.[74] Several years ago, Selin and Welsh showed that a small population of virus-specific memory cells was capable of immediate lytic activity, without extensive antigenic restimulation[75] and, more recently, others have shown that CD8⁺ memory T cells harvested from nonlymphoid tissues were lytic, in contrast to their counterparts harvested from the spleen.[76]

Our lab has evaluated CD8⁺ T cell effector functions during the acute and memory phases of virus infection, and has found similarities and differences.[77] As shown in Figure 2 (top row), during the acute response to primary infection, virus-specific cells comprise two populations, IFNγ⁺TNFα⁻ (single-positive) and IFNγ⁺TNFα⁺ (double-positive). After infection is cleared, and the immune response enters the memory phase, the virus-specific cells are almost uniformly double-positive. A similar pattern is observed after secondary virus challenge in immune mice (Figure 2, bottom row). Thus, the population of virus-specific T cells during active infection differs from the population present in the memory phase; but the relationship between the single-positive and double-positive cells is unknown. Importantly, we also found

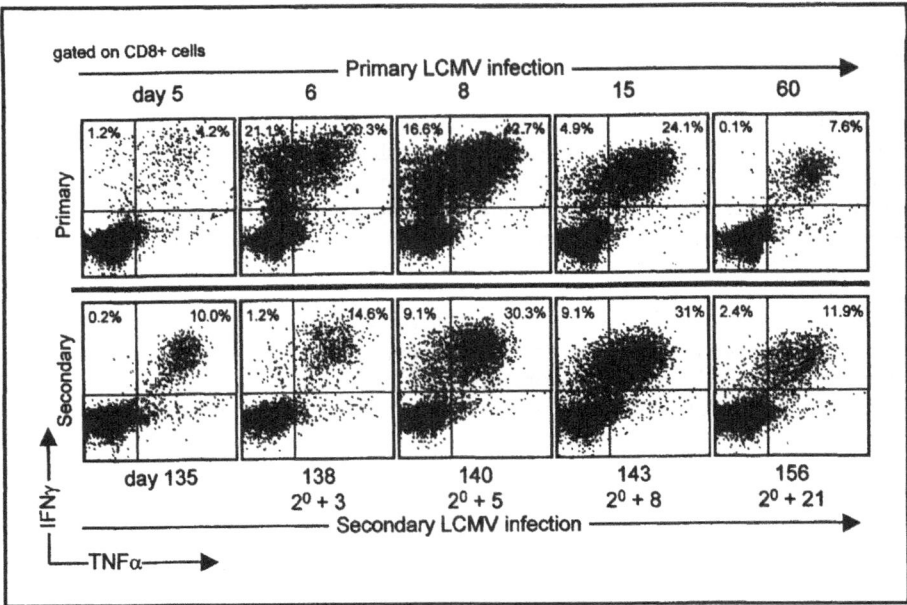

Figure 2. Cytokine production by effector and memory CD8⁺T cells. Naïve mice were infected with LCMV (top row) and at the indicated times post-infection, virus-specific CD8⁺ T cell responses were measured by ICCS. Splenocytes were gated on CD8⁺ T cells, and expression of IFNγ (y axis) and TNFα (x axis) was measured. Similar analyses were carried out in immune mice re-infected with LCMV (bottom row). The numbers indicate the number of cells in that quadrant, as a percentage of total CD8⁺ T cells. (This figure is modified from reference 77).

that, after encountering cognate antigen, memory cells initiated IFNγ production every bit as rapidly as did acute cells. During the acute phase of infection, many more virus-specific CD8⁺ T cells are present than are found during the memory phase (Figure 3A); however, the rapidity with which the virus-specific cells can respond is very similar in the two populations (Figure 3B). Thus, CD8⁺ memory cells can exert their effector functions immediately upon antigen contact. These observations are consistent with our hypothesis that the fate of an infected vaccinee is sealed very soon after infection; if the pre-existing memory cells are sufficiently numerous, they can contain the infection immediately; if not, the virus disseminates, and can cause disease.

How Are CD8⁺ T Cells Induced Following DNA Vaccination?

Surprisingly, this issue remains somewhat controversial. Doubtless, this topic will be ad-dressed in other chapters in this book, and here we give our own perspectives on this important question. Soon after the discovery of DNA immunization, two hypotheses were proposed to explain the observation that CD8⁺ T cells could be induced by DNA vaccines. First, that "nonspecialized" cells took up the DNA and expressed the encoded antigens; and these cells induced a primary CD8⁺ T cell response. A priori, this hypothesis appeared unlikely to be correct, because most somatic cells—although usually expressing class I MHC molecules—do not express the costimulatory signals needed to stimulate naïve antigen-specific T cells. Indeed, healthy muscle cells express negligible amounts of MHC molecules,[78] rendering them even less likely candidates for the role of stimulator cells. The second hypothesis was that the DNA-encoded antigen was presented to T cells by specialized antigen-presenting cells (APCs); and elegant studies showed incontrovertibly that APCs were required for successful induction of CD8⁺ T cell responses by DNA vaccines.[79,80] That APCs are important is generally

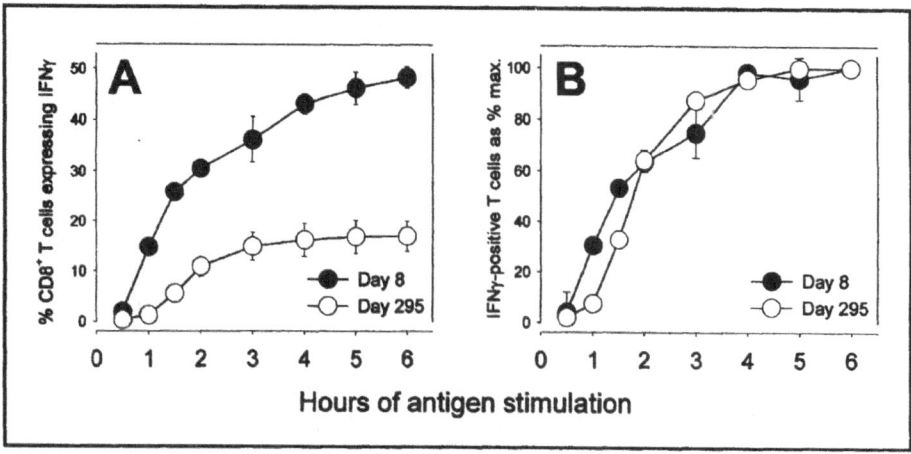

Figure 3. Acute and memory CD8+ T cells initiate IFNγ production at very similar rates. Splenocytes were harvested from mice at 8 days (l) or 295 days (●) after LCMV infection, and were incubated with antigenic peptide for the indicated number of hours. Responding cells were identified by their production of IFNγ, and were enumerated by ICCS. In panel A, the responding cells are shown as a percentage of total CD8+ T cells. In panel B, the same data are re-plotted; for both time points, the developing response is shown as a percentage of the maximum response at that time point. (This figure is modified from reference 77).

accepted; and dendritic cells (DCs) are thought likely to play a key role. DCs expressing a foreign antigen are remarkably efficient at inducing CD8+ T cells; only 100-1000 antigen-presenting DCs are needed to induce a CD8+ T cell response capable of protecting against subsequent virus challenge.[81] Here we encounter the unresolved, and controversial, issue—exactly how does the DNA-encoded antigen gain access to the MHC class I pathway in the vaccinee's APCs? Again, two hypotheses have been proposed; (a) cross-priming, in which protein is transferred from transfected cells, to APCs and (b) uptake of DNA by APCs.

Does DNA Immunization Depend on Protein Transfer from in vivo Transfected Cells to APCs?

The first hypothesis states that protein produced by transfected somatic cells (e.g., by muscle cells following intramuscular injection of plasmid) is released, and is taken up by APCs (perhaps DCs) which have the unusual ability to introduce exogenous proteins into their MHC class I pathway. The phenomenon of apparent protein transfer, first observed in studies of minor histocompatibility responses,[82] is termed "cross-priming". Some authors have suggested that cross-priming is important for[83] [and perhaps, even, central to][84] the induction of CD8+ T cell responses during virus infection; we do not believe that this stance is supported by the available evidence, but here we shall focus on CD8+ T cell induction by DNA immunization. Several arguments have been advanced to support the idea that cross-priming underlies the successful induction of CD8+ T cells by DNA vaccines.[79,80] Perhaps the strongest argument is based on the following experimental approach. Nonlymphoid cells (e.g., muscle cells) expressing a given mouse MHC haplotype (say, H-2^b) are transfected in tissue culture with a plasmid expressing a well-characterized antigen ("X") for which CD8+ T cell epitopes are know for the "self" haplotype (H-2^b) and for one other MHC haplotype (say, H-2^d). Stably-transfected cells are selected, cultured for several passages, and these cells—which, presumably, no longer contain any free plasmid DNA—are inoculated into H-2^b/H-2^d recipient mice. Induction of H-2^b-restricted CD8+ T cell responses to antigen X would suggest that some form of antigen transfer to APCs is occurring, because (as stated above) it is unlikely that the injected muscle cells could directly stimulate a primary CD8+ T cell response. An even more persuasive case can be made if H-2^d-restricted CD8$^+$ T cell responses are induced; since the inoculated muscle cells do not express H-2^d alleles, the presence of such T cells is powerful evidence that antigen X is being presented by H-2^d MHC class I molecules, presumably on the host's APCs. Several studies similar to the above have shown that such CD8$^+$ T cells can be induced by inoculation of stably-expressed cells and, as a result, many workers in the field believe that transfer of proteins from in vivo transfected cells is central to DNA immunization. Furthermore, plasmid-encoded fusion proteins in which the antigen of interest is attached to an immunoglo-bulin Ig fragment,[85] or to the extracellular domain of Flt3 ligand,[86] are more immunogenic; the authors concluded that the enhanced immunogenicity resulted from specific uptake by DCs, although other interpretations are possible. Thus, a number of findings support the con-tention that DNA vaccines work by cross-priming. However, other studies run counter to this conclusion. For example, Zinkernagel's group found that disruption of stably-transfected fi-broblasts prior to injection abrogated CD8$^+$ T cell induction, indicating that free proteins could not induce CD8$^+$ T cells via cross-priming.[87] Furthermore, excision of a DNA-injected muscle within minutes of plasmid inoculation failed to prevent the induction of immunity, suggesting that protein production by muscle cells is unlikely to underlie successful DNA immunization.[88] Finally, we[89-91] and others[92-95] have shown that targeting a DNA-encoded protein for rapid degradation within the transfected cell leads to enhanced CD8$^+$ T cell re-sponses, and pulse-chase studies showed that, at any given time point, very little intact protein could be found in the transfected cell. It is difficult to reconcile this observation with the hypothesis that transfer of intact protein is important for successful DNA immunization.

Does DNA Immunization Depend on Uptake and Expression of Injected Plasmid by APCs?

An attractive alternative to cross-priming is that the requisite antigen presentation by APCs results simply from their taking up the inoculated DNA, expressing the encoded antigen, and thereby allowing the endogenously-synthesized protein to enter the MHC class I pathway. DNA uptake and gene expression have been observed in DCs in vivo following DNA immuni-zation,[96] and adoptive transfer of the in vivo transfected cells leads to CTL induction.[97] Cotransfection studies also support the DNA uptake hypothesis.[98] Furthermore, this idea is more consistent with the result of the study in which the injected muscle was rapidly ablated; presumably, some of the injected plasmid exited the muscle (either in the bloodstream, or by "leakage") prior to its excision. The above-cited findings with DNAs encoding rapidly-degraded proteins also are more easily reconciled with this hypothesis than with cross-priming. Thus, although cross-priming probably plays some part in the induction of CD8$^+$ T cells following DNA immunization, we favor this second hypothesis. Note that the question is not merely academic; it is important that the dominant mechanism be identified, to allow us to optimize DNA immunization. If CD8$^+$ T cell induction occurs through protein transfer and cross-priming, then we should design plasmids which drive high levels of protein synthesis in the in vivo transfected cells (e.g., in muscle cells). On the other hand, if DNA immunization works via DNA uptake by APCs, we should focus on targeting the DNA to these cells, and enhancing gene expression in this cell type.

Enumeration and Characterization of CD8$^+$ T Cells Induced by DNA Immunization

Hundreds of papers have now been published which show, unequivocally, that DNA vac-cines can induce CD8$^+$ T cell responses (as well as CD4$^+$ T cells and antibodies); we shall not attempt to review all of these in this chapter. In some cases, DNA vaccines have proven more immunogenic than other recombinant delivery systems (for example, recombinant vaccinia viruses),[99] and sometimes they appear to overcome a host's previous nonresponsiveness to a particular antigen.[100] As a rule, DNA vaccines appear to induce better CD8$^+$ T cell responses than antibody responses.[101] This is true regardless of the route of immunization. CD8⁺ T cell responses have been detected in many DNA vaccine studies, but in the vast majority of cases, the responses were not measured directly ex vivo; instead, DNA-induced CD8⁺ T cells were first subjected to some form of secondary stimulation, and these restimulated cells were quantitated. The types of restimulation employed varied from study to study, but usually fell into one of two categories. (i) In vitro restimulation. Splenocytes from DNA-immunized animals were incubated for days (or sometimes weeks) in tissue culture, with stimulator cells expressing the cognate antigen, and then were assayed for cytolytic activity; or (ii) In vivo restimulation. Animals (usually mice) immunized with DNA encoding a particular antigen (often viral) were infected with the appropriate virus and, several days later, splenocytes were taken, and their cytolytic activity was measured. The use of extensive restimulation severely limited the conclusions which could be drawn from these studies; neither the number of cells which had been induced by the DNA vaccine, nor the functional attributes of the DNA-induced cells, could be confidently inferred. Therefore, it was important to develop T cell assays which would allow detection of low numbers of T cells, without extensive restimulation. For many years, in vitro cytotoxicity assays have been the "read-out" of CD8⁺ T cell function. This assay has been enormously valuable, and remains useful. However, it is not sufficiently sensitive to detect low numbers of effector cells and not all antigen-specific cells show high lytic activity. Analyses of antigen-specific CD8⁺ T cell responses has been transformed by the advent of three additional techniques: first, MHC peptide tetramer technology, which allows the detection of T cells bearing receptors of defined antigen specificity. Second, intracellular cytokine staining (ICCS), in which cytokine production by T cells (which occurs within minutes of antigen contact);[65,102] can be detected by antibody staining and subsequent flow cytometry. Third, cytokine ELIspot, in which cytokine production by immobilized antigen-specific cells can be detected.

DNA-Induced CD8⁺ T Cells Can Be Detected Directly ex vivo

Our lab has made extensive use of ICCS, and we have found that DNA vaccine-induced CD8⁺ T cells can be readily identified directly ex vivo without extensive restimulation;[67,103,104] with ICCS, the cells are exposed to antigen within minutes of their being harvested from the animal, and antigen stimulation continues for only ~6 hours, which is too short a time to allow significant expansion of CD8⁺ T cell numbers. In this way, we can be confident that we are detecting the actual CD8⁺ T cells induced by the DNA vaccine. Examples of ICCS-based detection of DNA vaccine-induced CD8⁺ memory T cells are shown in Figure 4. Neonatal or adult mice were inoculated once with DNA (either pCMV-NP, which encodes the LCMV nucleoprotein, or the "empty" plasmid pCMV, as a negative control), and a year later (without any interim boosting) the mice were sacrificed, and their splenocytes were assayed by ICCS. Antigen-specific CD8⁺ T cells (identified by IFNγ production in response to the 6 hour exposure to epitope peptide) were easily identified in the appropriate mice, and constituted ~1% of total CD8⁺ T cells. [In comparison, previous studies, in which DNA-induced memory cells were quantitated after restimulation, found much lower frequencies.][90,105] Therefore, a single DNA immunization induces very long-lived memory cells, even in mice inoculated within hours of birth. This is consistent with the recent finding, that a single pulse of antigen is enough to cause naïve CD8⁺ T cells to differentiate into memory cells. Others have shown that DNA vaccination of primates (rhesus macaques) can induce CD8⁺ T cell response which are detectable directly ex vivo, and comprise ~0.5% of CD8⁺ T cells in the animals' peripheral blood.[106] Thus, DNA vaccines induce CD8⁺ T cell responses which, although lower than those found following virus infection or vaccination using a recombinant virus,[103] are nevertheless impressive.

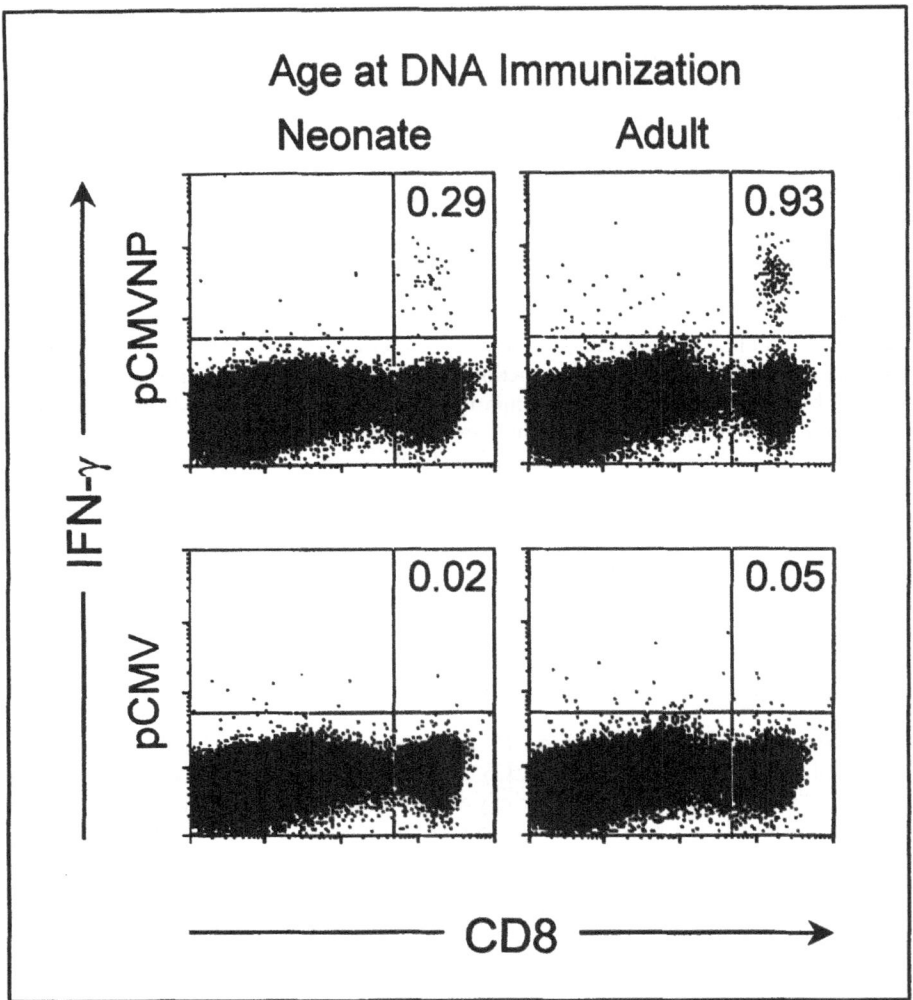

Figure 4. Typical CD8⁺ T cell responses following a single DNA immunization. Neonatal or adult mice were immunized with a single injection of plasmid DNA; either pCMV-NP (top row) or the negative control plasmid pCMV (bottom row). One year later, splenocytes were analyzed by ICCS. The y axis shows IFNγ staining; staining for CD8 is shown on the x axis. The numbers indicate the number of antigen-specific CD8⁺ T cells detected, as a percentage of total CD8⁺ T cells in the mouse. (This figure is modified from reference 104).

DNA-Induced CD8⁺ T Cells Are Qualitatively Similar to Those Induced by Virus Infection

Direct ex vivo detection of CD8⁺ T cells induced by DNA immunization enables us to ask whether or not these cells are functionally equivalent to those induced by conventional vaccines. We have evaluated such cells using five criteria; cytokine production, perforin content, lytic ability, functional avidity, and protective efficacy. By all criteria, DNA-induced cells were indistinguishable from cells induced by other means.[103,104]

CD8⁺ Cell Responses Induced by DNA Immunization in Humans

T cell responses have been identified in many animal models (see above); however, of the hundreds of papers published about DNA immunization, relatively few have reported results in humans. A review of clinical trials is included in another chapter in this volume, and here we provide only a short summary. A DNA vaccine encoding a malaria antigen was given to malaria-naïve healthy human recipients, and CD8⁺ T cell responses were detected against all 10 known epitopes, presented by at least 6 HLA alleles; promising though this may appear, the responses were not strong, and their induction required three doses of vaccine.[107] Similarly, when 15 HIV-infected patients were given escalating doses of an HIV env/rev DNA vaccine, only weak responses, of arguable significance, were noted.[108] Another study of DNA vaccination in HIV-infected individuals (3 immunizations over a 6-month period) reported MHC restricted CD8⁺ T cell responses in 8 of 9 vaccinees, although the response was transient in three patients.[109] In a phase 1 trial, an HIV-1 env/rev DNA vaccine was administered (using either 100 μg or 300 μg doses) on 4 occasions to HIV-1 seronegative individuals; antigen-specific lymphocyte proliferation and cytokine production was detected at least once in all high-dose recipients, but the response did not persist in any of the individuals.[110] In summary, so far, DNA vaccines (used alone) have not induced very strong immune responses in our species. There is, therefore, room for improvement.

Enhancing CD8⁺ T Cell Induction by DNA Vaccines

We have recently summarized some of the approaches which might be used to enhance antibody and T cell responses to DNA immunization;[111] here, we focus on improving CD8⁺ T cell induction. Several approaches have been taken, each with some success.

Route of Immunization

CD8⁺ T cell responses have been detected after DNA delivery by the intramuscular and intradermal routes but, perhaps surprisingly, the intravenous route appears less successful.[112,113] A recent study suggested that the immunogenicity of inoculated DNA could be increased 100- to 1000-fold by direct injection into lymph nodes.[114] Several laboratories have reported success using attenuated Salmonella as a delivery vehicle for plasmid DNA, and one study showed that oral Salmonella followed by antigen-expressing DC was especially effective.[115] Finally, in vivo electroporation (also called electropermeabilization)—in which an electric current is applied in vivo to potentiate cellular uptake of DNA—markedly increases in vivo transfection efficiency.[116] There is a current surge in interest in applying this technique to DNA vaccination; and the method appears to enhance both antibody and CD8⁺ T cell responses.[117-123]

Linking the Antigen to Heat-Shock Proteins

It has been reported that heat-shock proteins may play a role in cross-priming, perhaps delivering antigenic peptides to the MHC class I pathway.[124,125] Consequently, some labs have attached antigens to heat shock proteins, in an attempt to exploit this apparent pathway. For example, a fusion between *Mycobacterium bovis* Hsp65 and fragments of influenza NP led to more effective CTL induction.[126] Fusion of human papilloma virus (HPV) E7 protein to Hsp70 from *Mycobacterium tuberculosis* led to a ~30-fold increase in E7-specific CD8⁺ T cells,[127] and the same group reported similar results using Hsp linkage in a replicon-based RNA immunization system;[128] whether this enhancement was due to more efficient cross-priming, or to a nonspecific "adjuvant" effect resulting from the co-expression of a highly immunogenic bacterial protein, remains to be determined.

Improving Codon Usage

Successful translation of mRNA requires that the appropriate quantities of charged tRNAs be available to "read" the codons; since prokaryotic codon usage differs from that seen in eukaryotes, optimal translation of a bacterial gene may require codon optimization. In one study, an oligonucleotide encoding a 9-amino acid epitope from the intracellular bacterium Listeria monocytogenes was synthesized using the native bacterial sequence, or using codons optimized for murine expression; the latter construct was much more effectively translated in murine cells, and was more immunogenic when administered as a DNA vaccine.[129] Similar conclusions were reached using an optimized gene encoding tetanus toxin,[130] and in studies of Plasmodium yoelii.[131] One would not necessarily predict that codon usage would be a consideration when designing antiviral vaccines, since viruses have evolved to exploit the eukaryotic translational machinery. However, codon usage differs among eukaryote genes, presumably as a means to subtly regulate protein expression, and therefore one cannot assume that all viral genes are optimally designed for translation. Altering the codon usage of HIV-1 gp120 enhanced the antibody and CTL responses induced by a DNA vaccine,[132] and codon optimization of the HIV gag gene yielded a DNA vaccine that was effective at a 100-fold lower dose than a vaccine encoding the standard sequences.[133] In contrast, others have carried out similar experiments using HIV gp160, and have concluded that little advantage accrued from codon optimization.[134]

Proteasomal Targeting to Increase Peptide Delivery to the Endoplasmic Reticulum

A certain proportion of proteins synthesized within the cell are destined to be delivered to the proteasome, a cytosolic organelle which hydrolyses the proteins, and the resulting peptides are transported (via the TAP transporters) into the endoplasmic reticulum (ER). It is thought that mis-folded proteins, termed defective ribosomal products (or drips) may be preferentially targeted for proteasomal degradation;[135] and proteasomal targeting requires the covalent attachment of multiple copies of the cellular protein ubiquitin, each attached to the other in a head-to-tail poly-ubiquitin array.[136,137] Several laboratories have exploited the ubiquitin pathway to destabilize DNA-encoded proteins, and thereby to enhance their delivery to the proteasome. Two subtly different tacks have been taken, and both have produced similar results. One approach exploits Varshavsky's "N-end" rule,[138] which states that a protein's stability is determined, in large part, by the N-terminal amino acid residue of the mature polypeptide. Two groups have shown that, by using ubiquitin to replace a "stable" N-terminus with an "unstable" residue, one can increase protein turnover and thereby alter its immunogenicity.[92-94] Another technique places a gene encoding a modified ubiquitin in-frame with the antigen of interest in a DNA vaccine.[139] We used this approach to show that, when the fusion protein is expressed in cells, the modified ubiquitin acts as the target for the addition of a poly-ubiquitin chain, and thus the antigen of interest is very efficiently delivered to, and degraded by, the proteasome. In this way, intracellular degradation is greatly enhanced, and the protective capacity of the DNA vaccines is improved.[90,91,111] Similar success has been achieved using these ubiquitin plasmids in tumor models.[89,95] However, the rapid degradation comes at a cost; antibody responses to the destabilized proteins are much reduced, presumably because there is insufficient intact protein to induce a biologically-significant antibody response.

Direct Delivery of the Encoded Materials to the ER

Ubiquitin indirectly increases peptide delivery to the ER, but other labs have taken a more direct approach, by attaching proteins, or isolated epitope minigenes, to signal sequences; these should ensure that nascent proteins, synthesized on the rough ER, will be translocated into the ER. For example, a DNA vaccine contain an HIV epitope minigene was markedly enhanced by the addition of the E3 leader sequence from adenovirus,[140] and the immunogenicity of an SV40 CD8+ T cell epitope expressed in a recombinant vaccinia virus was greatly increased using a similar approach.[141]

Do Protein Transduction Domains Enhance DNA Vaccination?

Sequences termed protein transduction domains (PTD), which can translocate attached materials across biological membranes, have been found in proteins encoded by viruses, bacteria, insects, and other organisms.[142] For example, a short highly basic region (YGRKKRRQRRR) allows the HIV-1 nuclear transactivation protein (tat) to cross the cell membrane in a receptor-dependent manner,[143-146] and importantly, for our purposes, tat- or PTD-linked proteins are degraded inside the transduced cells, and their epitopes are presented by MHC class I molecules in a TAP-dependent manner.[147,148] We[149] and others[150] have attempted to exploit this attribute of PTD to improve antigen delivery following DNA immunization. The results are conflicting. Hung et al, using a herpesvirus PTD attached to GFP, concluded that intercellular spread of biologically-active protein was occurring.[150] However, this finding is difficult to reconcile with other data which suggested that successful in vivo transduction of full-length proteins required that the protein be denatured prior to transfer;[144,145,151] and in our study we specifically sought, but were unable to detect, evidence of extensive intercellular protein transfer following transfection.[149] However, in both studies, addition of the PTD resulted in much more effective stimulation of epitope-specific CD8+ T cells, and enhanced in vivo protective efficacy of the DNA vaccine. Attachment of Pseudomonas aeruginosa exotoxin A to the HPV E7 protein led to a ~30-fold increase in the number of CD8+ T cells primed by a DNA vaccine, with a parallel increase in biological efficacy.[152]

Prime-Boost Regimens

A number of recent studies have indicated that DNA vaccines may be valuable when used as part of a prime-boost strategy.[153] For example, studies using a recombinant poxvirus delivery system showed that CTL could be induced, but that the effect of boosting was diminished by responses to the vector itself. However, if instead the vaccinee received a DNA injection, followed some time later with a recombinant poxvirus expressing the same antigens, a much-improved outcome was noted; this was true for both HIV and malaria antigens.[154,155] Studies with hepatitis C virus antigens showed that the CD8+ T cell responses which followed recombinant poxvirus boosting were not only higher, but also were more diverse, than those seen following DNA vaccine priming;[156] and the most effective CD8+ T cell response against the papillomavirus E7 protein was induced by DNA prime / vaccinia boost.[157] The approach also works with "string of beads" constructs, which were first conceived in 1993,[22] and encode isolated epitopes. Also termed "poly-epitope" or "polytope" vaccines,[158,159] these vaccines induce responses when administered as plasmid DNA, and the responses are enhanced by subsequent poxviral boosting.[160] The results are not always quite so clear-cut, however; in one instance, poxvirus boosting after DNA priming led to increased cellular immunity, but decreased levels of antibody.[161] Effective boosting of DNA-primed responses is not limited to recombinant poxviruses. DNA prime / protein boost strategies have met with some success.[162] Boosting with additional doses of DNA also may be beneficial;[163] we have found that, after 2 boosts, we can drive epitope-specific CD8+ T cells to remarkably high levels (~20% of a mouse's total CD8+ T cells; Hassett and Whitton, unpublished).

Co-Administration of Immunostimulatory Molecules

Many studies have employed immunostimulatory molecules to skew the immune responses (for example, towards Th1 or Th2 phenotypes) or to otherwise modulate them. Since a chapter in this book focuses on this topic, here we limit ourselves to presenting a table in which the effects on CD8+ T cell responses are summarized. As can be seen, in several cases (e.g., GM-CSF) the effect of an individual immunostimulatory molecule varies from study to study. In regard to CpG motifs, several studies (not cited in Table 1) have indicated that these sequences, if present in the administered DNA, provide a "built-in" adjuvant effect. One study has evaluated the effect of co-administering CpG oligonucleotides with a DNA vaccine, and the authors found no effect; however, that study also concluded that DNA vaccines did not induce long-term antiviral immunity, a result which differs very much from the conclusions reached by most laboratories.

Table 1. Modulating CD8⁺ T cell responses by co-administration of immunostimulatory molecules

Adjuvant	CTL	Reference
AlPO₄, AlOH, Al(OH)PO₄	↑	(164)
CD40L	↑	(165, 166)
CD54 (ICAM-1)	↑	(167)
CD80 (B7-1)	=	(168-170)
CD86 (B7-2)	↑	(168-172)
CpG DNA (co-administered)	=	(173)
Erythromycin	↑	(174)
GM-CSF	=	(175-177)
GM-CSF	↓	(178)
GM-CSF	↑	(165, 168, 179-181)
GM-CSF+CD40L	↑	(165)
GM-CSF+CD80	↑	(168)
GM-CSF+CD86	↑	(168)
HSP70.1	↑	(182)
IFN-γ	↑	(179, 183, 184)
IL-1α	↑	(185)
IL-2	=	(175)
IL-2	↑	(179, 184-188)
IL-2/Ig	↑	(189)
IL-4	↓	(177, 179, 185)
IL-4	=	(175, 184)
IL-5	↓	(185)
IL-6	↑	(178)
IL-10	=	(175, 185)
IL-12	↑	(166, 168, 175-179, 183, 185, 188, 190-192)
IL-12+GM-CSF	↑	(168, 177)
IL-12+CD80	↑	(168)
IL-12+CD86	↑	(168)
IL-12+GM-CSF+ CD80	↑	(168)
IL-13	=	(184)
IL-15	↑	(185, 188)
IL-18	↑	(183, 185)
Josamycin	=	(174)
LFA-3	↑	(167)
LPS (Lipopolysaccharide)	=	(193)
MIP-1α	↑	(194, 195)
RANTES	↑	(194)
TCA-3	↑	(196)
TNF-α	↑	(185)
TNF-β	↑	(185)

Summary

Antigens encoded by DNA vaccines can induce all arms of the adaptive immune response, but to date they have proven most effective at inducing antigen-specific CD8⁺ T cells. The great majority of experiments have been carried out in small animal models, where these vaccines work quite well; in a limited number of studies in primates (including humans), their effectiveness, although demonstrable, is somewhat diminished. Therefore, to accelerate the introduction of DNA vaccines into clinical and veterinary practice, it is important that their immunogenicity be enhanced. The rational modification of DNA vaccines requires that we have a basic understanding of the mechanisms which underpin successful DNA immunization. In this chapter, we review how DNA vaccines may work, and how this information permits us to exploit biological pathways to improve the outcome of genetic immunization. In addition to reviewing "rational" vaccine modification (based on, e.g., targeting antigens to specific antigen presentation pathways; or co-administering cytokines to modulate the vaccine-induced response) we also consider "empirical" approaches, such as using different prime-boost regimens which—by a mechanism as yet unclear—appear to greatly enhance antigen-specific memory in the vaccinee. Empirical studies proved the efficacy of essentially all vaccines in current use, and this old approach may once again prove useful in launching a new technology into the clinical arena.

References

1. Whitton JL. Lymphocytic choriomeningitis virus CTL. Sem Virol 1990; 1:257-262.
2. Good RA. and Zak SJ. Disturbance in gamma-globulin synthesis as "experiments of nature". Pediatrics 1956; 18:109-149.
3. Ochs HD, Smith CI. X-linked agammaglobulinemia. A clinical and molecular analysis. Medicine 1996; 75:287-299.
4. McBride SJ, McCluskey DR, Jackson PT. Selective C7 complement deficiency causing recurrent meningococcal infection. J Infect 1991; 22:273-276.
5. McWhinney PH, Langhorne P, Love WC et al. Disseminated gonococcal infection associated with deficiency of the second component of complement. Postgrad Med J 1991; 67:297-298.
6. Orren A, Potter PC, Cooper RC et al. Deficiency of the sixth component of complement and susceptibility to Neisseria meningitidis infections: studies in 10 families and five isolated cases. Immunology 1987; 62:249-253.
7. Geller TJ, Condie D. A case of protracted coxsackie virus meningoencephalitis in a marginally immunodeficient child treated successfully with intravenous immunoglobulin. J Neurol Sci 1995; 129:131-133.
8. Hertel NT, Pedersen FK, Heilmann C. Coxsackie B3 virus encephalitis in a patient with agammaglobulinaemia. Eur J Pediatr 1989; 148:642-643.
9. McKinney REJ, Katz SL, Wilfert CM. Chronic enteroviral meningoencephalitis in agammaglobulinemic patients. Rev Infect Dis 1987; 9:334-356.
10. Misbah SA, Spickett GP, Ryba PC. et al. Chronic enteroviral meningoencephalitis in agammaglobulinemia: case report and literature review. J Clin Immunol 1992; 12:266-270.
11. Good RA. Experiments of nature in the development of modern immunology. Immunol Today 1991; 12:283-286.
12. Musey L, Hughes J, Schacker T et al. Cytotoxic-T-cell responses, viral load, and disease progression in early human immunodeficiency virus type 1 infection. N Engl J Med 1997; 337:1267-1274.
13. Rinaldo Jr CR, Gupta P, Huang XL et al. Anti-HIV type 1 memory cytotoxic T lymphocyte responses associated with changes in CD4⁺ T cell numbers in progression of HIV type 1 infection. AIDS Res Hum Retroviruses 1998; 14:1423-1433.
14. Betts MR, Krowka JF, Kepler TB et al. Human immunodeficiency virus type 1-specific cytotoxic T lymphocyte activity is inversely correlated with HIV type 1 viral load in HIV type 1-infected long-term survivors. AIDS Res Hum Retroviruses 1999; 15:1219-1228
15. Gray MM, Hann IM, Glass S et al. Mortality and morbidity caused by measles in children with malignant disease attending four major treatment centres: a retrospective review. Br Med J [Clin Res] 1987; 295:19-22.
16. Nahmias AJ, Griffith D, Salsbury C et al. Thymic aplasia with lymphopenia, plasma cells, and normal immunoglobulins. Relation to measles virus infection. JAMA 1967; 201:729-734.
17. Siegel MM, Walter TK, Ablin AR. Measles pneumonia in childhood leukemia. Pediatrics 1977; 60:38-40.

18. Kernahan J, McQuillin J, Craft AW. Measles in children who have malignant disease. Br Med J [Clin Res] 1987; 295:15-18.
19. Whitton JL, Oldstone MBA. The Immune Response to Viruses. In: Fields BN, Knipe DM, Howley PM, eds. Fields' Virology. 4th ed. Philadelphia: Lippincott Williams & Wilkins, 2001:285-320.
20. Whitton JL, ed. Curr Top Microbiol Immunol. Antigen presentation. 232. New York: Springer Verlag.
21. Kew OM, Sutter RW, Nottay BK et al. Prolonged replication of a type 1 vaccine-derived poliovirus in an immunodeficient patient. J Clin Microbiol 1998; 36:2893-2899.
22. Whitton JL, Sheng N, Oldstone MBA et al. A "string-of-beads" vaccine, comprising linked minigenes, confers protection from lethal-dose virus challenge. J Virol 1993; 67:348-352.
23. Oldstone MBA, Tishon A, Eddleston M et al. Vaccination to prevent persistent viral infection. J Virol 1993; 67:4372-4378.
24. Klavinskis LS, Whitton JL, Joly E et al. Vaccination and protection from a lethal viral infection: identification, incorporation, and use of a cytotoxic T lymphocyte glycoprotein epitope. Virol 1990; 178:393-400.
25. Klavinskis LS, Whitton JL, Oldstone MBA. Molecularly engineered vaccine which expresses an immunodominant T-cell epitope induces cytotoxic T lymphocytes that confer protection from lethal virus infection. J Virol 1989; 63:4311-4316.
26. Klavinskis LS, Oldstone MBA, Whitton JL. Designing vaccines to induce cytotoxic T lymphocytes: protection from lethal viral infection. In: Brown F, Chanock R, Ginsberg H et al, eds. Vaccines 89. Modern Approaches to New Vaccines Including Prevention of AIDS. Cold Spring Harbor: Cold Spring Harbor Laboratory, 1989:485-489.
27. Volkmer H, Bertholet C, Jonjic S et al. Cytolytic T lymphocyte recognition of the murine cytomegalovirus nonstructural immediate-early protein pp89 expressed by recombinant vaccinia virus. J Exp Med 1987; 166:668-677.
28. Jonjic S, del Val M, Keil GM et al. A nonstructural viral protein expressed by a recombinant vaccinia virus protects against lethal cytomegalovirus infection. J Virol 1988; 62:1653-1658.
29. An LL, Whitton JL. A multivalent minigene vaccine, containing B cell, CTL, and Th epitopes from several microbes, induces appropriate responses in vivo, and confers protection against more than one pathogen. J Virol 1997; 71:2292-2302.
30. An LL, Whitton JL. Multivalent minigene vaccines against infectious disease. Curr Opin Mol Ther 1999; 1:16-21.
31. Borysiewicz LK, Hickling JK, Graham S et al. Human cytomegalovirus-specific cytotoxic T cells. Relative frequency of stage-specific CTL recognizing the 72-kD immediate early protein and glycoprotein B expressed by recombinant vaccinia viruses. J Exp Med 1988; 168:919-931.
32. Arvin AM, Sharp M, Smith S et al. Equivalent recognition of a varicella-zoster virus immediate early protein (IE62) and glycoprotein I by cytotoxic T lymphocytes of either CD4+ or CD8+ phenotype. J Immunol 1991; 146:257-264.
33. Manickan E, Yu Z, Rouse RJ et al. Induction of protective immunity against herpes simplex virus with DNA encoding the immediate early protein ICP 27. Viral Immunol 1995; 8:53-61.
34. Enria DA, Fernandez NJ, Briggiler AM et al. Importance of dose of neutralizing antibodies in treatment of argentine hemorrhagic fever with immune plasma. Lancet 1984; 2:255-256.
35. Maiztegui JI, Fernandez NJ, de Damilano AJ. Efficacy of immune plasma in treatment of Argentine haemorrhagic fever and association between treatment and a late neurological syndrome. Lancet 1979; 2:1216-1217.
36. Lin FT, Chen SB, Wang YZ et al. Use of serum and vaccine in combination for prophylaxis following exposure to rabies. Rev Infect Dis 1988; 10 Suppl 4:S766-S770.
37. Lausch RN, Staats H, Metcalf JF et al. Effective antibody therapy in herpes simplex virus ocular infection. Characterization of recipient immune response. Intervirology 1990; 31:159-165.
38. Lukacs K, Kurlander R. Lyt-2+ T cell-mediated protection against listeriosis. Protection correlates with phagocyte depletion but not with IFN-γ production. J Immunol 1989; 142:2879-2886.
39. Harty JT, Bevan MJ. CD8+ T cells specific for a single nonamer epitope of Listeria monocytogenes are protective in vivo. J Exp Med 1992; 175:1531-1538.
40. An LL, Pamer EG, Whitton JL. A recombinant minigene vaccine containing a nonameric cytotoxic-T-lymphocyte epitope confers limited protection against Listeria monocytogenes infection. Infect Immun 1996; 64:1685-1693.
41. Ball JM, Tian P, Zeng CQ et al. Age-dependent diarrhea induced by a rotaviral nonstructural glycoprotein. Science 1996; 272:101-104.
42. Levy JA, Mackewicz CE, Barker E. Controlling HIV pathogenesis: the role of the noncytotoxic anti-HIV response of CD8+ T cells. Immunol Today 1996; 17:217-224.

43. Rodriguez F, Slifka MK, Harkins S et al. Two overlapping subdominant epitopes identified by DNA immunization induce protective CD8⁺ T cell populations with differing cytolytic activities. J Virol 2001; (in press).
44. Podack ER, Lowrey DM, Lichtenheld M et al. Function of granule perforin and esterases in T cell-mediated reactions. Components required for delivery of molecules to target cells. Ann N Y Acad Sci 1988; 532:292-302.
45. Kagi D, Ledermann B, Burki K et al. Cytotoxicity mediated by T cells and natural killer cells is greatly impaired in perforin-deficient mice. Nature 1994; 369:31-37.
46. Kagi D, Vignaux F, Ledermann B et al. Fas and Perforin Pathways as Major Mechanisms of T Cell-Mediated Cytotoxicity. Science 1994; 265:528-530.
47. Walsh CM, Matloubian M, Liu CC et al. Immune function in mice lacking the perforin gene. Proc Natl Acad Sci USA 1994; 91:10854-10858.
48. Kagi D, Seiler P, Pavlovic J et al. The roles of perforin- and Fas-dependent cytotoxicity in protection against cytopathic and noncytopathic viruses. Eur J Immunol 1995; 25:3256-3262.
49. Suda T, Nagata S. Purification and characterization of the Fas-ligand that induces apoptosis. J Exp Med 1994; 179:873-879.
50. Welsh RM, Nishioka WK, Antia R et al. Mechanism of killing by virus-induced cytotoxic T lymphocytes elicited in vivo. J Virol 1990; 64:3726-3733.
51. Zychlinsky A, Zheng LM, Liu CC et al. Cytolytic lymphocytes induce both apoptosis and necrosis in target cells. J Immunol 1991; 146:393-400.
52. Shresta S, Pham CT, Thomas DA et al. How do cytotoxic lymphocytes kill their targets? Curr Opin Immunol 1998; 10:581-587.
53. Huang S, Hendriks W, Althage A et al. Immune response in mice that lack the interferon-γ receptor. Science 1993; 259:1742-1745.
54. Ramsay AJ, Ruby J, Ramshaw IA. A case for cytokines as effector molecules in the resolution of virus infection. Immunol Today 1993; 14:155-157.
55. Ruby J, Ramshaw IA. The antiviral activity of immune CD8⁺ T cells is dependent on interferon-_. Lymphokine Cytokine Res 1991; 10:353-358.
56. Oldstone MBA, Blount P, Southern PJ et al. Cytoimmunotherapy for persistent virus infection reveals a unique clearance pattern from the central nervous system. Nature 1986; 321:239-243.
57. Tishon A, Lewicki H, Rall GF et al. An essential role for type 1 interferon-γ in terminating persistent viral infection. Virol 1995; 212:244-250.
58. Guidotti LG, Ishikawa T, Hobbs MV et al. Intracellular inactivation of the hepatitis B virus by cytotoxic T lymphocytes. Immunity 1996; 4:25-36.
59. Guidotti LG, Chisari FV. To kill or to cure: options in host defense against viral infection. Curr Opin Immunol 1996; 8:478-483.
60. Guidotti LG, Rochford R, Chung J et al. Viral Clearance Without Destruction of Infected Cells During Acute HBV Infection. Science 1999; 284:825-829.
61. Kaech SM, Ahmed R. Memory CD8⁺ T cell differentiation: initial antigen encounter triggers a developmental program in naive cells. Nat Immunol 2001; 2:415-422.
62. Chan K, Lee DJ, Schubert A et al. The roles of MHC class II, CD40, and B7 costimulation in CTL induction by plasmid DNA. J Immunol 2001; 166:3061-3066.
63. Murali-Krishna K, Altman JD, Suresh M et al. Counting antigen-specific CD8 T cells: a reevaluation of bystander activation during viral infection. Immunity 1998; 8:177-187.
64. Butz EA, Bevan MJ. Massive expansion of antigen-specific CD8⁺ T cells during an acute virus infection. Immunity 1998; 8:167-175.
65. Slifka MK, Rodriguez F, Whitton JL. Rapid on/off cycling of cytokine production by virus-specific CD8⁺ T cells. Nature 1999; 401:76-79.
66. Slifka MK, Whitton JL. Clinical implications of dysregulated cytokine production. J Mol Med 2000; 78:74-80.
67. An LL, Rodriguez F, Harkins S et al. Quantitative and qualitative analyses of the immune responses induced by a multivalent minigene DNA vaccine. Vaccine 2000; 18:2132-2141.
68. Lau LL, Jamieson BD, Somasundaram T et al. Cytotoxic T-cell memory without antigen. Nature 1994; 369:648-652.
69. Andersson EC, Christensen JP, Scheynius A et al. Lymphocytic choriomeningitis virus infection is associated with long- standing perturbation of LFA-1 expression on CD8⁺ T cells. Scand.J Immunol 1995; 42:110-118.
70. McFarland HI, Nahill SR, Maciaszek JW et al. CD11b (Mac-1): a marker for CD8⁺ cytotoxic T cell activation and memory in virus infection. J Immunol 1992; 149:1326-1333.

71. Budd RC, Cerottini JC, Horvath C et al. Distinction of virgin and memory T lymphocytes. Stable acquisition of the Pgp-1 glycoprotein concomitant with antigenic stimulation. J Immunol 1987; 138:3120-3129.
72. Tripp RA, Hou S, Doherty PC. Temporal loss of the activated L-selectin-low phenotype for virus-specific CD8⁺ memory T cells. J Immunol 1995; 154:5870-5875.
73. Galvan M, Murali-Krishna K, Ming LL et al. Alterations in cell surface carbohydrates on T cells from virally infected mice can distinguish effector/memory CD8⁺ T cells from naive cells. J Immunol 1998; 161:641-648.
74. Harrington LE, Galvan M, Baum LG et al. Differentiating between memory and effector CD8 T cells by altered expression of cell surface O-glycans. J Exp Med 2000; 191:1241-1246.
75. Selin LK, Welsh RM. Cytolytically active memory CTL present in lymphocytic choriomeningitis virus-immune mice after clearance of virus infection. J Immunol 1997; 158:5366-5373.
76. Masopust D, Vezys V, Marzo AL et al. Preferential localization of effector memory cells in nonlymphoid tissue. Science 2001; 291:2413-2417.
77. Slifka MK, Whitton JL. Activated and memory CD8⁺ T cells can be distinguished by their cytokine profiles and phenotypic markers. J Immunol 2000; 164:208-216.
78. Karpati G, Pouliot Y, Carpenter S. Expression of immunoreactive major histocompatibility complex products in human skeletal muscles. Ann Neurol 1988; 23:64-72.
79. Doe B, Selby M, Barnett S et al. Induction of cytotoxic T lymphocytes by intramuscular immunization with plasmid DNA is facilitated by bone marrow- derived cells. Proc Natl Acad Sci USA 1996; 93:8578-8583.
80. Corr M, Lee DJ, Carson DA et al. Gene vaccination with naked plasmid DNA: mechanism of CTL priming. J Exp Med 1996; 184:1555-1560.
81. Ludewig B, Ehl S, Karrer U et al. Dendritic cells efficiently induce protective antiviral immunity. J Virol 1998; 72:3812-3818.
82. Bevan MJ. Minor H antigens introduced on H-2 different stimulating cells cross- react at the cytotoxic T cell level during in vivo priming. J Immunol 1976; 117:2233-2238.
83. Sigal LJ, Crotty S, Andino R et al. Cytotoxic T-cell immunity to virus-infected non-haematopoietic cells requires presentation of exogenous antigen. Nature 1999; 398:77-80.
84. Carbone FR, Kurts C, Bennett SR et al. Cross-presentation: a general mechanism for CTL immunity and tolerance. Immunol Today 1998; 19:368-373.
85. You Z, Huang X, Hester J et al. Targeting dendritic cells to enhance DNA vaccine potency. Cancer Res 2001; 61:3704-3711.
86. Hung CF, Hsu KF, Cheng WF et al. Enhancement of DNA vaccine potency by linkage of antigen gene to a gene encoding the extracellular domain of Fms-like tyrosine kinase 3-ligand. Cancer Res 2001; 61:1080-1088.
87. Kundig TM, Bachmann MF, DePaolo C et al. Fibroblasts as efficient antigen-presenting cells in lymphoid organs. Science 1995; 268:1343-1347.
88. Torres CA, Iwasaki A, Barber BH et al. Differential dependence on target site tissue for gene gun and intramuscular DNA immunizations. J Immunol 1997; 158:4529-4532.
89. Xiang R, Lode HN, Chao TH et al. An autologous oral DNA vaccine protects against murine melanoma. Proc Natl Acad Sci USA 2000; 97:5492-5497.
90. Rodriguez F, An LL, Harkins S et al. DNA immunization with minigenes: low frequency of memory CTL and inefficient antiviral protection are rectified by ubiquitination. J Virol 1998; 72:5174-5181.
91. Rodriguez F, Zhang J, Whitton JL. DNA immunization: ubiquitination of a viral protein enhances CTL induction, and antiviral protection, but abrogates antibody induction. J Virol 1997; 71:8497-8503.
92. Tobery T, Siliciano RF. Induction of enhanced CTL-dependent protective immunity in vivo by N-end rule targeting of a model tumor antigen. J Immunol 1999; 162:639-642.
93. Tobery TW, Siliciano RF. Targeting of HIV-1 antigens for rapid intracellular degradation enhances cytotoxic T lymphocyte (CTL) recognition and the induction of de novo CTL responses in vivo after immunization. J Exp Med 1997; 185:909-920.
94. Wu Y, Kipps TJ. Deoxyribonucleic acid vaccines encoding antigens with rapid proteasome-dependent degradation are highly efficient inducers of cytolytic T lymphocytes. J Immunol 1997; 159:6037-6043.
95. Velders MP, Weijzen S, Eiben GL et al. Defined flanking spacers and enhanced proteolysis is essential for eradication of established tumors by an epitope string DNA vaccine. J Immunol 2001; 166:5366-5373.
96. Condon C, Watkins SC, Celluzzi CM et al. DNA-based immunization by in vivo transfection of dendritic cells. Nat Med 1996; 2:1122-1128.

97. Bot A, Stan AC, Inaba K et al. Dendritic cells at a DNA vaccination site express the encoded influenza nucleoprotein and prime MHC class I-restricted cytolytic lymphocytes upon adoptive transfer. Int Immunol 2000; 12:825-832.

98. Porgador A, Irvine KR, Iwasaki A et al. Predominant role for directly transfected dendritic cells in antigen presentation to CD8⁺ T cells after gene gun immunization. J Exp Med 1998; 188:1075-1082.

99. Chen CH, Wang TL, Ji H et al. Recombinant DNA vaccines protect against tumors that are resistant to recombinant vaccinia vaccines containing the same gene. Gene Ther 2001; 8:128-138.

100. Schirmbeck R, Bohm W, Ando K et al. Nucleic acid vaccination primes hepatitis B virus surface antigen-specific cytotoxic T lymphocytes in nonresponder mice. J Virol 1995; 69:5929-5934.

101. Otten GR, Doe B, Schaefer M et al. Relative potency of cellular and humoral immune responses induced by dna vaccination. Intervirology 2000; 43:227-232.

102. Slifka MK, Whitton JL. Antigen-specific regulation of T cell-mediated cytokine production. Immunity 2000; 12:451-457.

103. Hassett DE, Slifka MK, Zhang J et al. Direct ex vivo kinetic and phenotypic analyses of CD8⁺ T cell responses induced by DNA immunization. J Virol 2000; 74:8286-8291.

104. Hassett DE, Zhang J, Slifka MK et al. Immune responses following neonatal DNA vaccination are long-lived, abundant, and qualitatively similar to those induced by conventional immunization. J Virol 2000; 74:2620-2627.

105. Chen Y, Webster RG, Woodland DL. Induction of CD8⁺ T cell responses to dominant and subdominant epitopes and protective immunity to Sendai virus infection by DNA vaccination. J Immunol 1998; 160:2425-2432.

106. Allen TM, Vogel TU, Fuller DH et al. Induction of AIDS virus-specific CTL activity in fresh, unstimulated peripheral blood lymphocytes from rhesus macaques vaccinated with a DNA prime/modified vaccinia virus Ankara boost regimen. J Immunol 2000; 164:4968-4978.

107. Wang R, Doolan DL, Le TP et al. Induction of antigen-specific cytotoxic T lymphocytes in humans by a malaria DNA vaccine. Science 1998; 282:476-480.

108. MacGregor RR, Boyer JD, Ugen KE et al. First human trial of a DNA-based vaccine for treatment of human immunodeficiency virus type 1 infection: safety and host response. J Infect Dis 1998; 178:92-100.

109. Calarota S, Bratt G, Nordlund S et al. Cellular cytotoxic response induced by DNA vaccination in HIV-1- infected patients. Lancet 1998; 351:1320-1325.

110. Boyer JD, Cohen AD, Vogt S et al. Vaccination of seronegative volunteers with a human immunodeficiency virus type 1 env/rev DNA vaccine induces antigen-specific proliferation and lymphocyte production of β-chemokines. J Infect Dis 2000; 181:476-483.

111. Rodriguez F, Whitton JL. Enhancing DNA immunization. Virol 2000; 268:233-238.

112. Yokoyama M, Zhang J, Whitton JL. DNA immunization: effects of vehicle and route of administration on the induction of protective antiviral immunity. FEMS Immunol Med Microbiol 1996; 14:221-230.

113. Bohm W, Mertens T, Schirmbeck R et al. Routes of plasmid DNA vaccination that prime murine humoral and cellular immune responses. Vaccine 1998; 16:949-954.

114. Maloy KJ, Erdmann I, Basch V et al. Intralymphatic immunization enhances DNA vaccination. Proc Natl Acad Sci USA 2001; 98:3299-3303.

115. Zoller M, Christ O. Prophylactic tumor vaccination: comparison of effector mechanisms initiated by protein versus DNA vaccination. J Immunol 2001; 166:3440-3450.

116. Mathiesen I. Electropermeabilization of skeletal muscle enhances gene transfer in vivo. Gene Ther 1999; 6:508-514.

117. Smith LC, Nordstrom JL. Advances in plasmid gene delivery and expression in skeletal muscle. Curr Opin Mol Ther 2000; 2:150-154.

118. Mir LM. Therapeutic perspectives of in vivo cell electropermeabilization. Bioelectrochemistry 2001; 53:1-10.

119. Bachy M, Boudet F, Bureau M et al. Electric pulses increase the immunogenicity of an influenza DNA vaccine injected intramuscularly in the mouse. Vaccine 2001; 19:1688-1693.

120. Zucchelli S, Capone S, Fattori E et al. Enhancing B- and T-cell immune response to a hepatitis C virus E2 DNA vaccine by intramuscular electrical gene transfer. J Virol 2000; 74:11598-11607.

121. Selby M, Goldbeck C, Pertile T et al. Enhancement of DNA vaccine potency by electroporation in vivo. J Biotechnol 2000; 83:147-152.

122. Kadowaki S, Chen Z, Asanuma H et al. Protection against influenza virus infection in mice immunized by administration of hemagglutinin-expressing DNAs with electroporation. Vaccine 2000; 18:2779-2788.

123. Widera G, Austin M, Rabussay D et al. Increased DNA vaccine delivery and immunogenicity by electroporation In vivo. J Immunol 2000; 164:4635-4640.

124. Blachere NE, Li Z, Chandawarkar RY et al. Heat shock protein-peptide complexes, reconstituted in vitro, elicit peptide-specific cytotoxic T lymphocyte response and tumor immunity. J Exp Med 1997; 186:1315-1322.

125. Ciupitu A-MT, Petersson M, O'Donnell CL et al. Immunization with a lymphocytic choriomeningitis virus peptide mixed with heat shock protein 70 results in protective antiviral immunity and specific cytotoxic T lymphocytes. J Exp Med 1998; 187:685-691.

126. Anthony LS, Wu H, Sweet H et al. Priming of CD8⁺ CTL effector cells in mice by immunization with a stress protein-influenza virus nucleoprotein fusion molecule. Vaccine 1999; 17:373-383.

127. Chen CH, Wang TL, Hung CF et al. Enhancement of DNA vaccine potency by linkage of antigen gene to an HSP70 gene. Cancer Res 2000; 60:1035-1042.

128. Cheng WF, Hung CF, Chai CY et al. Enhancement of Sindbis virus self-replicating RNA vaccine potency by linkage of Mycobacterium tuberculosis heat shock protein 70 gene to an antigen gene. J Immunol 2001; 166:6218-6226.

129. Uchijima M, Yoshida A, Nagata T et al. Optimization of codon usage of plasmid DNA vaccine is required for the effective MHC class I-restricted T cell responses against an intracellular bacterium. J Immunol 1998; 161:5594-5599.

130. Stratford R, Douce G, Zhang-Barber L et al. Influence of codon usage on the immunogenicity of a DNA vaccine against tetanus. Vaccine 2000; 19:810-815.

131. Nagata T, Uchijima M, Yoshida A et al. Codon optimization effect on translational efficiency of DNA vaccine in mammalian cells: analysis of plasmid DNA encoding a CTL epitope derived from microorganisms. Biochem Biophys Res Commun 1999; 261:445-451.

132. Andre S, Seed B, Eberle J et al. Increased immune response elicited by DNA vaccination with a synthetic gp120 sequence with optimized codon usage. J Virol 1998; 72:1497-1503.

133. zur Megede J, Chen MC, Doe B et al. Increased expression and immunogenicity of sequence-modified human immunodeficiency virus type 1 gag gene. J Virol 2000; 74:2628-2635.

134. Vinner L, Nielsen HV, Bryder K et al. Gene gun DNA vaccination with Rev-independent synthetic HIV-1 gp160 envelope gene using mammalian codons. Vaccine 1999; 17:2166-2175.

135. Schubert U, Anton LC, Gibbs J et al. Rapid degradation of a large fraction of newly synthesized proteins by proteasomes. Nature 2000; 404:770-774.

136. Johnson ES, Bartel B, Seufert W et al. Ubiquitin as a degradation signal. EMBO J 1992; 11:497-505.

137. Chau V, Tobias JW, Bachmair A et al. A multiubiquitin chain is confined to specific lysine in a targeted short-lived protein. Science 1989; 243:1576-1583.

138. Varshavsky A. The N-end rule: functions, mysteries, uses. Proc Natl Acad Sci USA 1996; 93:12142-12149.

139. Barry MA, Lai WC, Johnston SA. Protection against mycoplasma infection using expression-library immunization. Nature 1995; 377:632-635.

140. Ciernik IF, Berzofsky JA, Carbone DP. Induction of cytotoxic T lymphocytes and antitumor immunity with DNA vaccines expressing single T cell epitopes. J Immunol 1996; 156:2369-2375.

141. Fu TM, Mylin LM, Schell TD et al. An endoplasmic reticulum-targeting signal sequence enhances the immunogenicity of an immunorecessive simian virus 40 large T antigen cytotoxic T-Lymphocyte epitope. J Virol 1998; 72:1469-1481.

142. Lindgren M, Hallbrink M, Prochiantz A et al. Cell-penetrating peptides. Trends Pharmacol Sci 2000; 21:99-103.

143. Tyagi M, Rusnati M, Presta M et al. Internalization of HIV-1 TAT requires cell surface heparan sulfate proteoglycans. J Biol Chem 2000; 276:3254-4261.

144. Schwarze SR, Dowdy SF. In vivo protein transduction: intracellular delivery of biologically active proteins, compounds and DNA. Trends Pharmacol Sci 2000; 21:45-48.

145. Schwarze SR, Ho A, Vocero-Akbani A et al. In vivo protein transduction: delivery of a biologically active protein into the mouse. Science 1999; 285:1569-1572.

146. Vives E, Brodin P, Lebleu B. A truncated HIV-1 Tat protein basic domain rapidly translocates through the plasma membrane and accumulates in the cell nucleus. J Biol Chem 1997; 272:16010-16017.

147. Moy P, Daikh Y, Pepinsky B et al. Tat-mediated protein delivery can facilitate MHC class I presentation of antigens. Mol Biotechnol 1996; 6:105-113.

148. Kim DT, Mitchell DJ, Brockstedt DG et al. Introduction of soluble proteins into the MHC class I pathway by conjugation to an HIV tat peptide. J Immunol 1997; 159:1666-1668.

149. Leifert JA, Lindencrona JA, Charo J et al. Enhancing T cell activation and antiviral protection by introducing the HIV-1 protein transduction domain into a DNA vaccine. Hum Gene Ther 2001; 12:1881-1892.

150. Hung CF, Cheng WF, Chai CY et al. Improving vaccine potency through intercellular spreading and enhanced MHC class I presentation of antigen. J Immunol 2001; 166:5733-5740.

151. Nagahara H, Vocero-Akbani AM, Snyder EL et al. Transduction of full-length TAT fusion proteins into mammalian cells: TAT-p27Kip1 induces cell migration. Nat Med 1998; 4:1449-1452.
152. Hung CF, Cheng WF, Hsu KF et al. Cancer immunotherapy using a DNA vaccine encoding the translocation domain of a bacterial toxin linked to a tumor antigen. Cancer Res 2001; 61:3698-3703.
153. Ramshaw IA, Ramsay AJ. The prime-boost strategy: exciting prospects for improved vaccination. Immunol Today 2000; 21:163-165.
154. Schneider J, Gilbert SC, Blanchard TJ et al. Enhanced immunogenicity for CD8+ T cell induction and complete protective efficacy of malaria DNA vaccination by boosting with modified vaccinia virus Ankara. Nat Med 1998; 4:397-402.
155. Hanke T, Blanchard TJ, Schneider J et al. Enhancement of MHC class I-restricted peptide-specific T cell induction by a DNA prime/MVA boost vaccination regime. Vaccine 1998; 16:439-445.
156. Pancholi P, Liu Q, Tricoche N et al. DNA prime-canarypox boost with polycistronic hepatitis C virus (HCV) genes generates potent immune responses to HCV structural and nonstructural proteins. J Infect Dis 2000; 182:18-27.
157. Chen C, Wang T, Hung C et al. Boosting with recombinant vaccinia increases HPV-16 E7-specific T cell precursor frequencies of HPV-16 E7-expressing DNA vaccines. Vaccine 2000; 18:2015-2022.
158. Thomson SA, Khanna R, Gardner J et al. Minimal epitopes expressed in a recombinant polyepitope protein are processed and presented to CD8+ cytotoxic T cells: implications for vaccine design. Proc Natl Acad Sci USA 1995; 92:5845-5849.
159. Suhrbier A. Multi-epitope DNA vaccines. Immunol Cell Biol 1997; 75:402-408.
160. Hanke T, McMichael A. Pre-clinical development of a multi-CTL epitope-based DNA prime MVA boost vaccine for AIDS. Immunol Lett 1999; 66:177-181.
161. Kent SJ, Zhao A, Best SJ et al. Enhanced T-cell immunogenicity and protective efficacy of a human immunodeficiency virus type 1 vaccine regimen consisting of consecutive priming with DNA and boosting with recombinant Fowlpox Virus. J Virol 1998; 72:10180-10188.
162. Tanghe A, D'Souza S, Rosseels V et al. Improved immunogenicity and protective efficacy of a tuberculosis DNA vaccine encoding Ag85 by protein boosting. Infect Immun 2001; 69:3041-3047.
163. Fuller DH, Simpson L, Cole KS et al. Gene gun-based nucleic acid immunization alone or in combination with recombinant vaccinia vectors suppresses virus burden in rhesus macaques challenged with a heterologous SIV. Immunol Cell Biol 1997; 75:389-396.
164. Wang S, Liu X, Fisher K et al. Enhanced type I immune response to a hepatitis B DNA vaccine by formulation with calcium- or aluminum phosphate. Vaccine 2000; 18:1227-1235.
165. Burger JA, Mendoza RB, Kipps TJ. Plasmids encoding granulocyte-macrophage colony-stimulating factor and CD154 enhance the immune response to genetic vaccines. Vaccine 2001; 19:2181-2189.
166. Gurunathan S, Irvine KR, Wu CY et al. CD40 ligand/trimer DNA enhances both humoral and cellular immune responses and induces protective immunity to infectious and tumor challenge. J Immunol 1998; 161:4563-4571.
167. Kim JJ, Tsai A, Nottingham LK et al. Intracellular adhesion molecule-1 modulates β-chemokines and directly stimulates T cells in vivo. J Clin Invest 1999; 103:869-877.
168. Iwasaki A, Stiernholm BJ, Chan AK et al. Enhanced CTL responses mediated by plasmid DNA immunogens encoding costimulatory molecules and cytokines. J Immunol 1997; 158:4591-4601.
169. Kim JJ, Bagarazzi ML, Trivedi N et al. Engineering of in vivo immune responses to DNA immunization via codelivery of costimulatory molecule genes. Nat Biotechnol 1997; 15:641-646.
170. Tsuji T, Hamajima K, Ishii N et al. Immunomodulatory effects of a plasmid expressing B7-2 on human immunodeficiency virus-1-specific cell-mediated immunity induced by a plasmid encoding the viral antigen. Eur J Immunol 1997; 27:782-787.
171. Agadjanyan MG, Kim JJ, Trivedi N et al. CD86 (B7-2) can function to drive MHC-restricted antigen-specific CTL responses in vivo. J Immunol 1999; 162:3417-3427.
172. Kim JJ, Nottingham LK, Wilson DM et al. Engineering DNA vaccines via co-delivery of co-stimulatory molecule genes. Vaccine 1998; 16:1828-1835.
173. Oehen S, Junt T, Lopez-Macias C et al. Antiviral protection after DNA vaccination is short lived and not enhanced by CpG DNA. Immunology 2000; 99:163-169.
174. Sato Y, Shishido H, Kobayashi H et al. Adjuvant effect of a 14-member macrolide antibiotic on DNA vaccine. Cell Immunol 1999; 197:145-150.
175. Kim JJ, Simbiri KA, Sin JI et al. Cytokine molecular adjuvants modulate immune responses induced by DNA vaccine constructs for HIV-1 and SIV. J Interferon Cytokine Res 1999; 19:77-84.
176. Kim JJ, Ayyavoo V, Bagarazzi ML et al. In vivo engineering of a cellular immune response by coadministration of IL-12 expression vector with a DNA immunogen. J Immunol 1997; 158:816-826.

177. Okada E, Sasaki S, Ishii N et al. Intranasal immunization of a DNA vaccine with IL-12- and granulocyte- macrophage colony-stimulating factor (GM-CSF)-expressing plasmids in liposomes induces strong mucosal and cell-mediated immune responses against HIV-1 antigens. J Immunol 1997; 159:3638-3647.

178. Lee SW, Youn JW, Seong BL et al. IL-6 induces long-term protective immunity against a lethal challenge of influenza virus. Vaccine 1999; 17:490-496.

179. Chow YH, Chiang BL, Lee YL et al. Development of Th1 and Th2 populations and the nature of immune responses to hepatitis B virus DNA vaccines can be modulated by codelivery of various cytokine genes. J Immunol 1998; 160:1320-1329.

180. Sedegah M, Weiss W, Sacci Jr JB et al. Improving protective immunity induced by DNA-based immunization: priming with antigen and GM-CSF-encoding plasmid DNA and boosting with antigen-expressing recombinant poxvirus. J Immunol 2000; 164:5905-5912.

181. Weiss WR, Ishii KJ, Hedstrom RC et al. A plasmid encoding murine granulocyte-macrophage colony-stimulating factor increases protection conferred by a malaria DNA vaccine. J Immunol 1998; 161:2325-2332.

182. Chen W, Lin Y, Liao C et al. Modulatory effects of the human heat shock protein 70 on DNA vaccination. J Biomed Sci 2000; 7:412-419.

183. Kim JJ, Nottingham LK, Tsai A et al. Antigen-specific humoral and cellular immune responses can be modulated in rhesus macaques through the use of IFN-γ, IL-12, or IL-18 gene adjuvants. J Med Primatol 1999; 28:214-223.

184. Kim JJ, Yang JS, Montaner L et al. Coimmunization with IFN-γ or IL-2, but not IL-13 or IL-4 cDNA can enhance Th1-type DNA vaccine-induced immune responses in vivo. J Interferon Cytokine Res 2000; 20:311-319.

185. Kim JJ, Trivedi NN, Nottingham LK et al. Modulation of amplitude and direction of in vivo immune responses by co- administration of cytokine gene expression cassettes with DNA immuno-gens. Eur J Immunol 1998; 28:1089-1103.

186. Geissler M, Gesien A, Wands JR. Inhibitory effects of chronic ethanol consumption on cellular immune responses to hepatitis C virus core protein are reversed by genetic immunizations augmented with cytokine-expressing plasmids. J Immunol 1997; 159:5107-5113.

187. Xin KQ, Hamajima K, Sasaki S et al. Intranasal administration of human immunodeficiency virus type-1 (HIV- 1) DNA vaccine with interleukin-2 expression plasmid enhances cell- mediated immunity against HIV-1. Immunology 1998; 94:438-444.

188. Xin KQ, Hamajima K, Sasaki S et al. IL-15 expression plasmid enhances cell-mediated immunity induced by an HIV-1 DNA vaccine. Vaccine 1999; 17:858-866.

189. Barouch DH, Craiu A, Kuroda MJ et al. Augmentation of immune responses to HIV-1 and sim-ian immunodeficiency virus DNA vaccines by IL-2/Ig plasmid administration in rhesus monkeys. Proc Natl Acad Sci USA 2000; 97:4192-4197.

190. Gherardi MM, Ramirez JC, Esteban M. Interleukin-12 (IL-12) enhancement of the cellular im-mune response against human immunodeficiency virus type 1 env antigen in a DNA prime/vac-cinia virus boost vaccine regimen is time and dose dependent: suppressive effects of IL-12 boost are mediated by nitric oxide. J Virol 2000; 74:6278-6286.

191. Kim JJ, Maguire Jr HC, Nottingham LK et al. Coadministration of IL-12 or IL-10 expression cas-settes drives immune responses toward a Th1 phenotype. J Interferon Cytokine Res 1998; 18:537-547.

192. Tsuji T, Hamajima K, Fukushima J et al. Enhancement of cell-mediated immunity against HIV-1 induced by coinnoculation of plasmid-encoded HIV-1 antigen with plasmid expressing IL-12. J Immunol 1997; 158:4008-4013.

193. Boyle JS, Brady JL, Koniaras C. Inhibitory effect of lipopolysaccharide on immune response after DNA immunization is route dependent. DNA Cell Biol 1998; 17:343-348.

194. Boyer JD, Kim J, Ugen K et al. HIV-1 DNA vaccines and chemokines. Vaccine 1999; 17 Suppl 2:S53-S64.

195. Lu Y, Xin KQ, Hamajima K et al. Macrophage inflammatory protein-1 alpha (MIP-1α) expression plasmid enhances DNA vaccine-induced immune response against HIV-1. Clin Exp Immunol 1999; 115:335-341.

196. Tsuji T, Fukushima J, Hamajima K et al. HIV-1-specific cell-mediated immunity is enhanced by co-inoculation of TCA3 expression plasmid with DNA vaccine. Immunology 1997; 90:1-6.

CHAPTER 6

Minigene-Based Vaccines for Eliciting CD8⁺ T Cell Responses

Jonathan W. Yewdell

Building a Better Mouse Trap

There is increasing interest in developing vaccines that specifically elicit CD8⁺ T cell responses. In part this has been driven simply by advances in understanding that make such a thing possible. The past 15 years has witnessed an explosion in our knowledge of the molecular basis for CD8⁺ T cell recognition of target cells, the cell biological processes involved in generating the structures recognized by CD8⁺ T cells, and how CD8⁺ T cells are activated and become memory cells. In at least equal measure, however, is a genuine need to induce effector arms of the immune system that are poorly elicited by traditional vaccines. Two diseases stand out as candidates for CD8⁺ T cell-vaccines. The first is infection with human immunodeficiency virus, where traditional vaccines based on induction of antibody responses have fared poorly and where there is evidence that CD8⁺ T cells can play a beneficial role.[1-3] The second is cancer, where it has been demonstrated both in experimental animals and humans that tumor-specific CD8⁺ T cells can reduce tumor burdens and occasionally effect eradication.[4]

Given a suitable target antigen for vaccination (discussed below), what is the best method for inducing CD8⁺ T cells? In this Chapter, I will review recent (and not so recent) progress in using minigenes to elicit CD8⁺ T cell responses. Before I get too far, however, it is necessary to review briefly some basic principles of CD8⁺ T cell recognition of foreign antigens.

Back to Basics

MHC class I molecules are integral membrane glycoproteins that are expressed on the surface of many cell types in vertebrate organisms. Class I molecules function to present oligopeptides to the immune system. Class I genes are highly polymorphic in most vertebrate species, and most of the allelism alters the peptide specificity of class I molecules. Peptide binding to class I molecules is largely governed by a few simple rules regarding peptide length (8 to 11 residues), and the presence of certain residues in two to three locations of the peptide. Using increasingly sophisticated algorithms based on these rules (visit http://134.2.96.221/scripts/MHCServer.dll/home.htm), it is possible to predict a reasonably high percentage of peptides from a given target molecule that will bind to class I molecules with a biologically significant affinity (this usually means a K_d less than 1 μM).

Class I-peptide complexes on the cell surface are perused by CD8⁺ T cells, which circulate through central and peripheral lymphoid organs, ever alert to the presence of class I molecules bearing foreign peptides. CD8⁺ T cells express clonally restricted receptors (T cell receptor [TCR]) that are selected in the thymus for their capacity to interact weakly with self-MHC class I molecules. Inherent to the design of the TCR is the chance that it will bind to a class I molecule with a novel peptide with a higher affinity. Upon such an encounter, the TCR

DNA Vaccines, edited by Hildegund C. J. Ertl. ©2003 Eurekah.com and Kluwer Academic / Plenum Publishers.

initiates the signaling cascade resulting in CD8$^+$ T cell activation. This induces the rapid expansion of naïve CD8$^+$ T cells and results in the generation of armed effector cells as well as long-lived memory cells. The ratio of effector to memory cells is not constant, and deciphering rules that govern this split, as well as the functional diversity of activated CD8$^+$ T cells remains a central quest of cellular immunology.

The stable cell surface expression of class I molecules requires the presence of a peptide ligand. Most class I molecules are loaded with high affinity peptides and have half-lives in the range of tens of hours. A significant fraction (perhaps a quarter) possess low affinity ligands [5]. After dissociation of peptides, these molecules rapidly denature on the cell surface but during their brief lifespan (half life of ~10 min) they are capable of binding peptides present in extracellular fluids. Such molecules account for the ability of synthetic peptides to sensitize antigen presenting cells (APCs) to CD8$^+$ T cell recognition.[6]

Most peptides presented by class I molecules are derived from proteins biosynthesized by the APC. Recent evidence suggests that many peptides are derived from a cohort of newly synthesized proteins that are targeted for rapid destruction, presumably because they are defective in some way.[7] Most antigenic peptides are initially liberated from their polypeptides by the proteasome, an abundant highly intricate ATP-dependent protease.[8] Proteasomes are present in all eukaryotic cells (and in Archaea and even some Eubacteria) where they play an essential role in degrading damaged, denatured, or unwanted proteins.[9] In eukaryotic cells, proteasomes are located in both the cytosol and the nucleus.[10] They are capable of generating the peptides of correct size and sequence for class I binding, but most often, they probably generate peptides with the proper COOH-terminus for class I binding, with additional trimming provided by aminopeptidases in the cytosol and endoplasmic reticulum (ER).[11] Cytosolic peptides of less than 18 residues and greater than 7 residues in length are transported into the ER by TAP (transporter associated with antigen processing).[12] TAP serves as a major assembly site of class I molecules, which consist of an allelically variable integral membrane heavy chain and the invariant β_2-microglobulin (β_2m). Newly synthesized heavy chains with loosely tethered β_2m are delivered to TAP in association with general purpose- (calreticulin, ERp56) and dedicated- (tapasin) molecular chaperones. Binding of a high affinity peptide increases the affinity of β_2m for heavy chain and results in the release of newly synthesized class I molecules from the ER. Class I molecules then reach the cell surface via the Golgi complex by the standard secretory pathway.

Numbers Games

Antigen processing is inherently inefficient. Even though a virus infected cell may produce millions of copies of a given gene product, the number of class I molecules that bear a peptide derived from the molecule is low, usually in the range of hundreds to thousands of copies per cell.[13-16] Deliberately targeting proteins for rapid degradation increases the number of class I peptide complexes generated, but the gains are relatively unimpressive. For example, cells infected with a recombinant vaccinia virus (rVV) producing influenza virus nucleoprotein (NP) express ~30 copies of a K^d-restricted peptide and 1800 copies of a K^k-restricted peptide.[14] Expression of a rapidly degraded fragment of the protein (the first 168 residues) resulted in only a roughly 4-fold increase in recovery of each peptide. Even this modest increase in presentation, however, can be reflected in enhanced immunogenicity relative to a rVV encoding full length NP, as defined by the primary lytic activity induced by infection with a rVV.[17] "Can" is used advisedly, since for some full-length proteins expressed by rVVs (such as chicken ovalbumin) increasing the number of complexes generated by manipulating the antigen does not enhance immunogenicity.[17] The correlation between enhanced generation of MHC-peptide complexes with immunogenicity was first described by Towsend et al.[18]

These findings are consistent with the idea that proteolytic liberation of peptides from their precursors is the rate-limiting step in generating peptide-MHC complexes. This conclusion is borne out by findings made using rVVs expressing oligopeptides encoded by "minigenes", a term first used in this context by Whitton and Oldstone,[19] who showed that a rVVs expressing a 22 amino residue peptide containing a naturally processed nonameric peptide could sensitize cells for lysis by antigen-specific CD8$^+$ T cells. Gould et al[20] simultaneously described similar findings using a rVV expressing a 15 residue peptide. These observations extended previous findings of Sweetser et al who demonstrated recognition of cells infected with a rVV expressing a 41-residue peptide. Ultimately, Eisenlohr et al[21] demonstrated that a rVV expressing a minimal peptide with a single additional NH$_2$-terminal Met (required for efficient translation initiation) could sensitize cells for CD8$^+$ T cell lysis. Bacik et al[22] then showed that such minimal peptides expressed by rVV are antigenic and, in general, are presented in a TAP-dependent manner. Such TAP-dependence is relative, rather than absolute as first noted by Zweerink et al.[23] Though this can also occur with full length gene products [24], minigene products in general exhibit more leakiness.

There were early hints that presentation of rVV-encoded minigene products is more efficient than presentation of the same determinants from their source proteins.[25] The extent of this effect was not fully appreciated, however, until direct quantitation was performed by titration of acid eluted peptides recovered by HPLC[14] or by use of a monoclonal antibody specific for a mouse class I molecule complexed with SIINFEKL, a peptide derived from ovalbumin.[15] Amazingly, the number of peptide class I complexes generated from minigenes was up to 2,000-fold higher than from expression of the source protein (e.g., from 30 to 55,000 for a determinant from influenza virus nucleoprotein). Since minigene products are degraded extremely rapidly by cells (unless they are protected by binding to a class I molecule), it has not been possible to measure their translation rate. Inasmuch as they utilize the same VV promoter as full-length gene products, it is predicted that they are translated at the same rate. While it is not difficult to imagine that the small size of the message could alter this in a positive or negative manner, the likely difference in the potential amounts of peptide synthesized (millions to tens of millions) versus the amount of complexes generated (tens of thousands) suggests that minigene products are produced in saturating amounts. It would be of interest to examine whether this is indeed the case and to determine to what extent the various steps of the antigen-processing pathway are saturated.

The enhanced immunogenicity of rVV-encoded minigene products relative to longer gene products was first demonstrated by Restifo et al[17] who measured the lytic activity of secondary in vitro cultures to gauge immunogenicity. This effect varied, however, depending on the determinant studied. For some determinants, expression as a minigene product resulted in no apparent enhancement of immunogenicity, even compared to immunization with a rVV encoding a full length gene product. A number of studies have confirmed the concept, however, that in general, expression of determinants as minigenes enhances their immunogenicity in the context of rVV immunization. The relationship between class I-peptide complex number and immunogenicity of rVVs was elegantly demonstrated by Eisenlohr and colleagues[26] who controlled gene expression by creative manipulation of the translation rate.

Take Me to Your Leader

Consistency is the hobgoblin of biology, where exceptions make life interesting. Not all minigene products are presented at high efficiency by APCs, in fact, there are several examples where minigenes are presented at lower efficiency (and are less immunogenic) than their counterparts expressed as full length proteins. This is due at least in part, to destruction by proteasomes,[27] which both giveth and taketh away.[28] This can be mitigated by addition of a flanking sequences,[27] but only on a strictly empirical basis, limiting its use as a general-purpose vaccine strategy.

A better tactic for avoiding the destructive potential of proteasomes and other cytosolic-proteases is to target peptides to the ER. This approach was pioneered by Anderson et al,[29] who appended to a 14 residue peptide the NH$_2$-terminal ER-targeting sequence of the E19K glycoprotein of adenovirus. The minigene encoding this peptide was used to transfect TAP-deficient cells. This provided the initial demonstration that ER-targeted peptides can bypass the requirement for TAP. These findings were extended Bacik et al[22] who demonstrated that:

1. this phenomenon extends to rVV-expressed peptides targeted by the E19-leader sequence, or another leader sequence.
2. high efficiency TAP-independent presentation requires the leader sequence to be at the NH_2-terminus of the peptide, suggesting that it functions as a true leader sequence and not simply by facilitating peptide diffusion into the ER by virtue of increased hydrophobicity.
3. very low amounts of the ER-targeted peptides are secreted (as detected by a lack of sensitization of bystander cells), with the implication that peptides associate with class I molecules in the ER.
4. each of five peptides tested was able to bypass TAP by addition of the NH_2-terminal leader sequence.

Later studies revealed that like their cytosolic counterparts, rVV expressed ER-targeted minigene products are much more efficient sources of peptides for class I molecules[14] in TAP-expressing cells, and can be much more immunogenic than rVVs expressing full length proteins.[17,30]

It is a working assumption that ER-targeted peptides are translocated into the ER via the translocon and that the leader sequence is cleaved by signal peptidase. This really is an assumption however, as the mechanism has never been carefully examined. Also, it should not be forgotten, that in TAP-expressing cells leader peptides can also be processed by cytosolic enzymes with the resulting peptides being transported by TAP. Snyder et al[31] showed that it is also possible to directly target peptides to the ER using Jaw1, a prototypic protein that is inserted into the ER post-translationally with a type II membrane orientation.[32] This resulted in the generation of a similar number of peptide class I complexes as the corresponding ER-targeted peptide, but offers the advantage of having a measurable protein (Jaw1) to monitor levels of synthesis.

Polydeterminant Minigenes

It is clearly an advantage to include as many antigenically relevant determinants in a vaccine as feasible. Everything comes with a price, however. Although rVVs and other vectors encoding multiple determinants strung together with or without intervening sequences have been shown to induce CD8+ T cell responses to each of the constituent determinants,[33-35] a number of caveats are in order. First, there have been few (any?) attempts to quantitate from such constructs relative to individual cytosolic or ER-targeted minigene products either the immunogenicity or the efficiency of antigen processing. Although the cytosol has a considerable capacity for liberating peptides from polypeptides, any increase in the size of polypeptides is sure to be accompanied by a decrease in the efficiency of liberation of the determinant except, as discussed above in the relatively infrequent cases where determinants are destroyed by cytosolic proteases, in which case addition of a few flanking residues may be beneficial.[27] This limitation is even more severe for ER-targeted peptides, since the ER has a very limited capacity for generating short peptides from longer peptides.[21,36-40]

There are two potential solutions to this problem. The first is to express multiple determinants from multiple promoters. The second is to express multiple determinants in a context from which they are efficiently liberated by proteases. Unlike the first approach, which is technically feasible, the world awaits an expression system that enables liberating peptides in a precise manner. Until this time, the best solution may be to target multi-determinant polypeptides to the proteasome by genetic fusion with NH_2-terminal ubiquitin rendered uncleavable by alteration of COOH-terminal Gly_{76} to Val or Ala.[41]

The strategy of generating vaccines that express large quantities of multiple determinants using multiple promoters, while clearly feasible from a technical standpoint, would still face the formidable problem of immunodominance. This is the inherent tendency of the immune system to respond to determinants in a hierarchical fashion.[42] Of particular relevance to poly-determinant minigene vaccines is immunodomination, where CD8+ T cell specific for the highest determinant in a hierarchy suppress responses to lower ranked determinants. The extent of this problem will probably vary in a highly unpredictable way in a manner dependent upon the determinants expressed, and even more troublesome, on the genotype of vaccine recipients (HLA of course, but possibly other genes as well). The extent of this problem has not been systematically examined in experimental animals, let alone man, but it could spell trouble for polydeterminant vaccines.

Since immunodominance appears to reflect competition of some sort at the level of the afferent APC,[43,44] it may be possible to bypass the problem by vaccination with several vectors at multiple sites, with each vector expressing a limited number of determinants.[45]

Minigenes in Other Vectors

The use of minigenes as vaccines has been pioneered using rVVs. While poxviruses have been invaluable as research tools, they are ill-suited as vaccine vectors due to their antigenic complexity. Poxviruses are large viruses that express a considerable number of proteins. These proteins interfere with the immunogenicity of even minigene determinants.[46] Moreover, it is pointless at best, and harmful at worst (potential autoimmunity, "scarring" of the repertoire, i.e., loss of repertoire due to expansion of some clones[47]) to generate CD8+ T cells against irrelevant antigens.

The ultimate solution to this problem is to use vectors that provide only essential information to the immune system. This is, of course, one of the major attractions of DNA vaccines. There are a number of reports demonstrating enhanced immunogenicity associated with the use of DNA vaccines expressing cytosolic, and particularly, ER targeted minigene products.[35,48,49] Immunogenicity was enhanced by including a peptide capable of activating CD4+ T cells,[35,49,50] or by coexpressing a cytokine.[51] No doubt other improvements are possible by mimicking the signals elicited by agents that induce vigorous CD8+ T cell responses.

An Immunological Oxymoron: Rational CD8+ T Cell-Vaccine Design

So far, I have avoided mentioning the proverbial 2-ton elephant in the room: our ignorance of the manner in which antigens are presented to naïve CD8+ T cells in vivo. After all, how can we rationally design vaccines when we can only conjecture about the nature of the APC (probably a dendritic cell, but never conclusiveley demonstrated in vivo for any form of antigen administered to the animal), the site of antigen presentation (probably a lymph node draining the site of immunization, but again, never demonstrated directly), and the nature of the antigen presented (i.e., contribution of endogenous antigens versus exogenous antigens). It is likely that these parameters will vary depending on the:

1. nature of the vaccine (naked versus virus-packaged DNA and then between different viral vectors)
2. route of administration
3. nature of the gene product expressed.

Until now, vaccine development has relied nearly exclusively on empirical approaches. While empiricism will probably always play a role in vaccine development, it is important to continue to increase the knowledge base to enable rational design of vaccines for eliciting CD8+ T cell responses.

References

1. Jin X, Bauer DE, Tuttleton SE et al. Dramatic rise in plasma viremia after CD8(+) T cell depletion in simian immunodeficiency virus-infected macaques. J Exp Med 1999; 189:991-998.
2. Schmitz JE, Kuroda MJ, Santra S et al. Control of viremia in simian immunodeficiency virus infection by CD8+ lymphocytes. Science 1999; 283:857-860.
3. Ogg GS, Jin X, Bonhoeffer S et al. Quantitation of HIV-1-specific cytotoxic T lymphocytes and plasma load of viral RNA. Science 1998; 279:2103-2106.
4. Wang RF,Rosenberg SA. Human tumor antigens for cancer vaccine development. Immunol Rev 1999; 170:85-100.

5. Day PM, Esquivel F, Lukszo J et al. Effect of TAP on the generation and intracellular trafficking of peptide-receptive major histocompatibility complex class I molecules. i 1995; 2:137-147.

6. Rock KL, Gamble S, Rothstein L et al. Dissociation of β₂-microgluobulin leads to the accumulation of a substantial pool of inactive class I MHC heavy chains on the cell surface. Cell 1991; 65:611-620.

7. Schubert U, Anton LC, Gibbs J et al. Rapid degradation of a large fraction of newly synthesized proteins by proteasomes. Nature 2000; 404:770-774.

8. Rock KL, Goldberg AL. Degradation of cell proteins and the generation of MHC class I-presented peptides. Annu Rev Immunol 1999; 17:739-779.

9. Voges D, Zwickl P, Baumeister W. The 26S proteasome: A molecular machine designed for controlled proteolysis. Annu Rev Biochem 1999; 68:1015-1068.

10. Reits EAJ, Benham AM, Plougastel B et al. Dynamics of proteasome distribution in living cells. EMBO J 1997; 16:6087-6094.

11. Yewdell J, Anton LC, Bacik I et al. Generating MHC class I ligands from viral gene products. Immunol Rev 1999; 172:97-108.

12. Elliott T. Transporter associated with antigen processing. Adv Immunol 1997; 65:47-109, 47-109.

13. Falk K, Rötzschke O, Deres K et al. Identification of naturally processed viral nonapeptides allows their quantification in infected cells and suggests an allele-specific T cell epitope forecast. J Exp Med 1991; 174:425-434.

14. Antón LC, Yewdell JW, Bennink JR. MHC class I-associated peptides produced from endogenous gene products with vastly different efficiencies. J Immunol 1997; 158:2535-2542.

15. Porgador A, Yewdell JW, Deng Y et al. Localization, quantitation, and in situ detection of specfic peptide-MHC class I complexes using a monoclonal antibody. Immunitiy 1997; 6:715-726.

16. Tsomides TJ, Aldovini A, Johnson RP et al. Naturally processed viral peptides recognized by cytotoxic T lymphocytes on cells chronically infected by human immunodeficiency virus type 1. J Exp Med 1994; 180:1283-1293.

17. Restifo NP, Bacík I, Irvine KR et al. Antigen processing in vivo and the elicitation of primary CTL responses. J Immunol 1995; 154:4414-4422.

18. Townsend A, Bastin J, Gould K et al. Defective presentation to class I-restricted cytotoxic T lymphocytes in vaccinia-infected cells is overcome by enhanced degradation of antigen. J Exp Med 1988; 168:1211-1224.

19. Whitton JL, Oldstone MBA. Class I MHC can present an endogenous peptide to cytotoxic T lymphocytes. J Exp Med 1989; 170:1033-1038.

20. Gould K, Cossins J, Bastin J et al. A 15 amino acid fragment of influenza nucleoprotein synthesized in the cytoplasm is presented to class I-restricted cytotoxic T lymphocytes. J Exp Med 1989; 170:1051-1056.

21. Eisenlohr LC, Yewdell JW, Bennink JR. Flanking sequences influence the presentation of an endogenously synthesized peptide to cytotoxic T lymphocytes. J Exp Med 1992; 175:481-487.

22. Bacik I, Cox JH, Anderson R et al. TAP-independent presentation of endogenously synthesized peptides is enhanced by endoplasmic reticulum insertion sequences located at the amino but not carboxy terminus of the peptide. J Immunol 1994; 152:381-387.

23. Zweerink HJ, Gammon MC, Utz U et al. Presentation of endogenous peptides to MHC class I-restricted cytotoxic T lymphocytes in transport deletion mutant T2 cells. J Immunol 1993; 150:1763-1771.

24. Esquivel F, Yewdell JW, Bennink JR. RMA/S cells present endogenously synthesized cytosolic proteins to class I-restricted cytotoxic T lymphocytes. J Exp Med 1992; 175:163-168.

25. Yewdell JW, Lapham CK, Bacik I et al. MHC-encoded proteasome subunits LMP2 and LMP7 are not required for efficient MHC class I restricted-presentation of viral antigens. J Immunol 1994; 152:1163-1170.

26. Wherry EJ, Puorro KA, Porgador A et al. The induction of virus-specific CTL as a function of increasing epitope expression: responses rise steadily until excessively high levels of epitope are attained. J. Immunol. 1999; 163:3735-3745.

27. Fu TM, Mylin LM, Schell TD et al. An endoplasmic reticulum-targeting signal sequence enhances the immunogenicity of an immunorecessive simian virus 40 large T antigen cytotoxic T-lymphocyte epitope. J Virol 1998; 72:1469-1481.

28. Luckey CJ, King GM, Marto JA et al. Proteasomes can either generate or destroy MHC class I epitopes: Evidence for nonproteasomal epitope generation in the cytosol. J Immunol 1998; 161:112-121.

29. Anderson K, Cresswell P, Gammon M et al. Endogenously synthesized peptide with an endoplasmic reticulum signal sequence sensitizes antigen processing mutant cells to class I-restricted cell-mediated lysis. J Exp Med 1991; 174:489-492.

30. Lawson CM, Bennink JR, Restifo NP et al. Primary pulmonary cytotoxic T lymphocytes induced by immunization with a vaccinia virus recombinant expressing influenza A virus nucleoprotein peptide do not proctect mice against challenge. J Virol 1994; 68:3505-3511.
31. Snyder HL, Bacik I, Bennink JR et al. Two novel routes of transporter associated with antigen processing (TAP)-independent major histocompatibility complex class I antigen processing. J Exp Med 1997; 186:1087-1098.
32. Behrens TW, Kearns GM, Rivard JJ et al. Carboxyl-terminal targeting and novel post-translational processing of JAW1, a lymphoid protein of the endoplasmic reticulum. J Biol Chem 1996; 271:23528-23534.
33. Whitton JL, Sheng N, Oldstone MBA et al. A "string of beads" vaccine, comprising linked minigenes, confers protection from lethal-dose virus challenge. J Virol 1993; 67:348-352.
34. Thomson SA, Khanna R, Gardner J et al. Minimal epitopes expressed in a recombinant polyepitope protein are processed and presented to CD8$^+$ cytotoxic T cells: Implications for vaccine design. Proc Natl Acad Sci USA 1995; 92:5845-5849.
35. Ishioka GY, Fikes J, Hermanson G et al. Utilization of MHC class I transgenic mice for development of minigene DNA vaccines encoding multiple HLA-restricted CTL epitopes. J Immunol 1999; 162:3915-3925.
36. Snyder HL, Bacik I, Yewdell JW et al. Promiscuous liberation of MHC-class I-binding peptides from the C termini of membrane and soluble proteins in the secretory pathway. Eur J Immunol 1998; 28:1339-1346.
37. Snyder HL, Yewdell JW, Bennink JR. Trimming of antigenic peptides in an early secretory compartment. J Exp Med 1994; 180:2389-2394.
38. Hammond SA, Bollinger RC, Tobery TW et al. Transporter-independent processing of HIV-1 envelope protein for recognition by CD8$^+$ T cells. Nature 1993; 364:158-161.
39. Elliott T, Willis A, Cerundolo V et al. Processing of major histocompatibility class I-restricted antigens in the endoplasmic reticulum. J Exp Med 1995; 181:1481-1491.
40. Lobigs M, Chelvanayagam G, Mullbacher A. Proteolytic processing of peptides in the lumen of the endoplasmic reticulum for antigen presentation by major histocompatibility class I. Eur J Immunol 2000; 30:1496-1506.
41. Rodriguez F, An LL, Harkins S et al. DNA immunization with minigenes: Low frequency of memory cytotoxic T lymphocytes and inefficient antiviral protection are rectified by ubiquitination. J Virol 1998; 72:5174-5181.
42. Yewdell JW, Bennink JR. Immunodominance in major histocompatibility complex class I-restricted T lymphocyte responses. Ann Rev Immunol 1999; 17:51-88.
43. Sandberg JK, Grufman P, Wolpert EZ et al. Superdominance among immunodominant H-2Kb-restricted epitopes and reversal by dendritic cell-mediated antigen delivery. J Immunol 1998; 160:3163-3169.
44. Kedl RM, Rees WA, Hildeman DA et al. T cells compete for access to antigen-bearing antigen-presenting cells. J Exp Med 2000; 192:1105-1113.
45. Chen W, Anton LC, Bennink JR et al. Dissecting the multifactorial causes of immunodominance in class I-restricted T cell responses to viruses. Immunity 2000; 12:83-93.
46. Sourdive DJ, Murali-Krishna K, Altman JD et al. Conserved T cell receptor repertoire in primary and memory CD8 T cell responses to an acute viral infection. J Exp Med 1998; 188:71-82.
47. Ciernik IF, Berzofsky JA, Carbone DP. Induction of cytotoxic T lymphocytes and antitumor immunity with DNA vaccines expressing single T cell epitopes. J Immunol 1996; 156:2369-2375.
48. Fomsgaard A, Nielsen HV, Kirkby N et al. Induction of cytotoxic T-cell responses by gene gun DNA vaccination with minigenes encoding influenza A virus HA and NP CTL-epitopes. Vacccine 1999; 18:681-691.
49. An LL, Whitton JL. Multivalent minigene vaccines against infectious disease. Curr Opin Mol Ther 1999; 1:16-21.
50. Thomson SA, Sherritt MA, Medveczky J et al. Delivery of multiple CD8 cytotoxic T cell epitopes by DNA vaccination. J Immunol 1998; 160:1717-1723.

CHAPTER 7

DNA Vaccines Against RNA Viruses

Jeffrey B. Ulmer

Introduction

DNA vaccines have been used successfully in many animal models of infectious and non-infectious diseases. The former has included viruses, bacteria, parasites, and other pathogens. However, because virus proteins are expressed by infected cells of the host, DNA vaccines may, in principle, be particularly useful for induction of anti-viral immunity. Perhaps because of this, and the ready availability of animal models of many viral diseases, most papers published on "DNA vaccines" have utilized virus models. Of these papers, slightly more than half have dealt with RNA viruses. Most of the work on DNA vaccines against RNA viruses has focused on a handful of human pathogens (see Table 1), and by far, most of the effort on RNA viruses has been with HIV. DNA vaccines against HIV are covered in a different chapter. However, many other human and animal pathogens also have been investigated as targets for DNA vaccines (see Table 2). This review summarizes the published work on DNA vaccines against several of the most commonly studied RNA virus targets.

DNA Vaccines Against Specific RNA Viruses

Influenza Virus

Influenza viruses comprise the Orthomyxoviridae and contain a negative-sense, segmented, ssRNA genome. Influenza A is a pathogen of humans and several animal species (most commonly pigs, horses and birds), while influenza B is thought to be strictly a human pathogen. Disease is spread via upper respiratory mucosa and results in an acute respiratory illness that can be lethal in susceptible populations. Eight individual influenza genes encode ten proteins, including the surface proteins hemagglutinin (HA) and neuraminidase (NA), and internal proteins nucleoprotein (NP) and matrix (M1). The primary antigenic determinants of influenza viruses are HA and NA, against which virus neutralizing antibodies are directed. Cell-mediated immune (CMI) responses, including CD4+ helper T cells and CD8+ cytotoxic T lymphocytes (CTL) are directed toward NP and M1. Immunity in humans is believed to be mediated primarily by anti-HA neutralizing antibodies, which provide durable subtype-specific protection. However, because mutation in the amino acid sequence of HA readily occurs (termed antigenic drift), heterosubtypic immunity is low in humans. Furthermore, reassortment of RNA segments can result in influenza strains with novel HA molecules (termed antigenic shift). In these situations, pandemics can occur due to a complete lack of cross-reactive anti-HA antibodies from prior influenza exposure. Influenza vaccines have been available for some time, but are based on inactivated whole virus or subvirion preparations. As such, they primarily induce humoral immune responses and, hence, are not well suited to provide protection against antigenic drift or antigenic shift. Thus, annual reformulation of influenza virus vaccines is required, and novel vaccine approaches are needed to overcome this problem. To this end, a live, attenuated virus has recently been approved for human use and many investigators have evaluated DNA vaccines in animal models leading to human clinical trials.

DNA Vaccines, edited by Hildegund C. J. Ertl. ©2003 Eurekah.com and Kluwer Academic / Plenum Publishers.

Table 1. Top ten list of RNA virus targets for DNA vaccines

No.	RNA Virus	Selected References
1	Human immunodeficiency virus	See Chapter 9
2	Influenza virus	1-46
3	Hepatitis C virus	47-66
4	Rabies virus	67-82
5	Measles virus	83-90
6	Respiratory syncitial virus	91-95
7	Rotavirus	96-101
8	Dengue virus	102-107
9	Japanese encephalitis virus	108-113
10	Tick-borne encephalitis virus	114-117

Most commonly studied animal model systems for DNA vaccines against RNA viruses.

The availability of well characterized animal models of influenza and reagents for generation of DNA vaccine constructs and measurement of anti-influenza immune responses provided an ideal situation for investigating the efficacy of the DNA vaccine technology. As a consequence, proof of principle for induction of CTL and protective immunity was demonstrated using the influenza model.[1] Specifically, anti-NP CTL were primed in mice by i.m. injection of naked NP DNA, and mice were protected from subsequent challenge with lethal doses of influenza virus. Importantly, protection was conferred against both homologous and heterosubtypic virus challenge, suggesting that targeting conserved internal viral proteins may be a strategy for broader protection in humans against antigenically drifted influenza virus strains. DNA vaccines were subsequently shown effective at priming humoral immune responses and protection mediated by anti-HA antibodies-.[2-5] This was demonstrated in chickens,[2-4] mice,[3, 5-7] and ferrets,[8,9] and protection was durable.[5] The level of protection correlated well with the presence of pre-existing hemagglutination-inhibiting (HI) antibodies,[7] which provide a surrogate measure of virus neutralizing antibodies. In some cases, though, protection was afforded without detectable pre-challenge anti-HA antibodies,[3, 4] but the rapid increase in anti-HA antibodies immediately after challenge suggests that B cell priming had mediated protection. Unlike protection afforded by NP DNA, in many cases only homologous protection was achieved by HA DNA, likely due to subtype-specific neutralizing antibodies induced against HA. In chickens, though, HA DNA was protective against challenge with an antigenically variant strain.[4] In ferrets, increased breadth of protection against antigenically drifted influenza strains was attained by inclusion of DNA vaccines encoding the internal viral proteins M1 and NP, likely due to contribution of CMI responses.[8, 9] DNA vaccines encoding other influenza antigens have been tested, including neuraminidase (NA) and non-structural protein 1 (NS1). NA DNA induced anti-NA antibodies in mice and protection from lethal virus challenge,[10-12] and showed an additive protective effect when mixed with HA DNA.[11] In contrast, NS1 DNA induced neither antibodies nor protection.[10] Immunogenicity and/or efficacy of influenza DNA vaccines have been demonstrated in neonatal,[13-16] as well as aged animals.[17, 18] With respect to neonates, DNA vaccines were able to mostly overcome the inhibiting effect of circulatingmaternal antibodies on vaccine-induced immune responses, as antibody and cellular immune responses to NP were not affected but antibodies to HA were reduced.[16] With respect to aged mice, HA and NP DNA were able to prime HI antibodies and CTL, respectively; and protection from homologous and heterosubtypic virus challenge, respectively. Influenza DNA vaccines also

Table 2. Other pathogenic RNA viruses as targets for DNA vaccines

Human Pathogens	Animal Pathogens
Coxsackie virus	Bovine leukemia virus
Ebola virus	Bovine parainfluenzavirus
Hantavirus	Bovine viral disease virus
Human T-lymphotrophic virus	Canine distemper virus
LaCrosse virus	Caprine arthritis-encephalitis virus
Murray Valley encephalitis virus	Foot-and-mouth disease virus
Polio virus	Infectious bursal disease virus
Rubella virus	Infectious hematopoietic necrosis virus
	Murine hepatitis virus

Other human and animal RNA viruses for which DNA vaccine technologies have been evaluated.

have been shown to be effective in large animal species, as well as in the small animal species described above. For example, influenza DNA vaccines have been immunogenic in baboons [15] and monkeys,[8] and have conferred protective efficacy in pigs [19] and horses.[20] Taken together, these data demonstrate the potential of the DNA technology for vaccine development against animal and human influenza viruses.

The immune mechanisms involved in influenza DNA vaccine-induced protection have been investigated. HA DNA primes antibody secreting B cells found initially in local lymph nodes, then long term in the bone marrow.[21] These cells secrete circulating virus neutralizing antibodies that provide substantial homologous protection, which can achieve apparent sterilizing levels of immunity with sufficient antibody titers.[7] In addition, HA DNA primes Th1-type helper T cells, which may contribute directly to protection through secretion of anti-viral cytokines and indirectly by enhancement of antibody secretion by B cells.[22] Likewise, NP DNA induces both humoral and cellular immune responses. However, anti-NP antibodies do not play a role in protection.[1] Rather, both Th1-type CD4$^+$ T cells and CTL appear to be involved, as demonstrated by studies using adoptive transfer of lymphocytes, depletion of T cell subsets in vivo, and knockout mice.[23-26]

Animal models of influenza have been used extensively to determine the mode of action of CTL priming by DNA vaccines. Early studies with DNA vaccines demonstrated that vaccine-encoded protein was produced primarily, if not exclusively, by muscle cells. Thus, three means of CTL priming were possible: 1) direct CTL priming by muscle cells, 2) cross-priming, and 3) direct transfection of antigen presenting cells (APCs). Implantation of NP-transfected myoblasts demonstrated that expression of an antigen in myocytes was sufficient to prime CTL.[27] However, the use of bone marrow chimeric mice demonstrated unequivocally that muscle cells were not directly priming CTL, rather cells derived from the bone marrow were required.[28-30] This was true even when expression of the antigen was restricted to muscle cells,[29] suggesting that cross-priming was involved. Evidence for direct transfection of APCs also was generated using the influenza system. Torres et al [31] showed that expression of antigen in muscle was not required for CTL priming after i.m. injection of DNA, indicating that some other cell (possibly an APC) must have been transfected. Other evidence includes observations that APCs isolated from DNA-injected tissues contain antigen and are capable of presenting antigen to antigen-specific T cells.[32, 33] Subsequent work in other animal models has confirmed that APCs are directly transfected by DNA vaccines, particularly when the DNA was delivered by the gene gun into the skin. The relative contribution of cross-priming and direct transfection to

CTL priming by DNA vaccines appears to depend on the means of administration of the vaccine, since cross-priming predominates after i.m. injection [34] while direct transfection is most important after gene gun administration.[35] This is perhaps not surprising, since APCs are much more readily transfected by the gene gun than by i.m. injection.

The robust immunogenicity of DNA vaccines in small animals, yet modest effectiveness in primates, triggered much activity on increasing the potency of influenza DNA vaccines. A direct approach was modification of the plasmid DNA to encode an antigen targeted to APCs. Specifically, HA was expressed as a fusion protein with CTLA4, which resulted in targeting of the antigen to APCs via B7 and enhanced vaccine potency in mice.[36] A second approach was to directly stimulate the immune system through the use of adjuvants. Enhanced DNA vaccine potency was accomplished by administration of a simple mixture of HA DNA with the conventional adjuvant aluminum phosphate,[37] or with DNA plasmids encoding GM-CSF,[38, 39] IL-12 [39] or costimulatory molecules.[40] A third approach was to increase DNA delivery using physical methods. Needle-free injection into the skin with a jet injector [41] and the Powderject system [42] both resulted in induction of potent immune responses in mice. In addition, DNA delivery can be substantially facilitated into muscle cells by electroporation in vivo.[43] This technique has been used to enhance the immunogenicity [44] and protective efficacy [45] of influenza DNA vaccines. Finally, increased vaccine effectiveness has been achieved in many animal models with a prime-boost regimen, involving a DNA vaccine prime and a recombinant viral vector boost. This also was true for influenza, as a DNA vaccine prime followed by a modified vaccinia virus Ankara boost resulted in superior immunogenicity and protection.[46]

Hepatitis C Virus

Hepatitis C viruses (HCV) belong to the *Flaviviridae* family and consist of a positive-sense, non-segmented, ssRNA genome. Humans are the only natural hosts of HCV and pathology is manifest as liver damage due to cytopathogenicity of infection and host immune responses. Most exposed individuals have an acute course and resolve infection, indicating that some form of immunity can control disease. However, approximately 40% of infections result in chronicity leading to hepatitis, cirrhosis and hepatocellular carcinoma. Transmission is predominantly percutaneous (e.g., blood transfusion, i.v. drug use) and the development of effective and sensitive blood screening methods for HCV has substantially reduced the incidence of disease. However, it is estimated that there are 170 million infected people worldwide and treatment is costly, effective in only a minority of patients and has considerable side effects. HCV vaccine research and development have been hampered by the lack of 1) a cell culture system for studying the virus and 2) a practical animal model of infection and disease (only chimpanzees can be infected). Nevertheless, much is now known about HCV. The primary translation product of the virus is a polyprotein that is cleaved by viral and host proteases to yield individual proteins. E1 and E2 encode glycoproteins that are the likely targets of virus neutralizing antibodies. But, there is considerable diversity among virus isolates, due mostly to the hypervariable regions of E1 and E2. The more conserved non-structural and core proteins are targets of CMI responses. The critical need for preventive and therapeutic HCV vaccines has prompted much effort on HCV DNA vaccines.

Several HCV antigens are potential vaccine candidates, including the surface E1 and E2 glycoproteins, and the internal core (C) and non-structural proteins (NS3-4-5). As such, each of these proteins has been investigated in the context of DNA vaccines, but most of the work has focused on E1 and E2 for the induction of both humoral and cellular immune responses. With regard to antibody responses, several challenges are posed by HCV. First, because of the inability to grow virus in cell culture there is no virus neutralizing assay (although a receptor binding inhibition assay—NOB—serves as a surrogate). Second, E1 and E2 are thought to exist as a heterodimer on the virus surface. Third, HCV is thought to bud from the endoplasmic reticulum, hence the glycoproteins possess only immature oligosaccharides. The latter two points raise potential problems in expressing relevant E1 and E2 structures by DNA vaccines. Antibody responses have been elicited in mice and macaques by DNA encoding E1 and/or E2, either as separate DNA vaccines or as part of a polyprotein containing other HCV antigens.[47-56] However, naked DNA alone was not as effective as a recombinant E2 protein plus adjuvant for both quantity (ELISA) and quality (NOB) of antibody responses.[54] Several approaches have been taken to increase the potency of anti-E2 antibody responses. These include, E2 expression as a fusion protein with the hepatitis B surface antigen;[48] co-expression of E2 with GM-CSF in a bicistronic plasmid vector;[49] expression of truncated forms of E2 rather than E1E2 heterodimers;[51, 54] prime-boost regimens involving E2 DNA vaccine priming followed by recombinant E2 protein [52] or canary pox vector [53] boosting; DNA vectors encoding Semliki Forest virus RNA replicons;[55] and electroporation in vivo.[56] Many of these approaches were effective at increasing anti-E2 antibodies, but without a good correlate of protection for HCV and until some of these technologies are tested in the chimpanzee challenge model, it is unclear if any of these represent enabling technologies for HCV.

In addition to neutralizing antibodies, CMI responses are also thought to be important for resolution of HCV infection in humans. CMI responses, including helper T cells and CTL, have been induced in mice by DNA vaccines encoding E2,[49, 52, 53, 55-58] core,[55, 57-63] and NS proteins.[64-66] CD4[+] helper T cell responses have been, as with other types of DNA vaccines, of the Th1 phenotype, with measurable production of IL-2 and interferon-γ.[53, 59, 65] Several different means have been used to detect HCV-specific CTL. These include standard lytic activity after restimulation in vitro, flow cytometry to measure Th1 cytokine production in CD8[+] T cells after restimulation with a MHC class I-restricted peptide,[65] and CTL-mediated protection from challenge with tumor cells [52] or recombinant vaccinia virus [62] expressing HCV antigens. The latter study also provided a means of transition from mouse to human HCV CTL responses by using HLA –A2.1-transgenic mice. Different routes of immunization also have been used to prime CTL by HCV DNA vaccines, including standard i.m. injection and gene gun administration, but also means to deliver DNA directly to the liver, such as intraportal [66] or intrahepatic [63] injection. The ability to express HCV proteins in the liver with subsequent induction of CTL provides a method of mimicking immune priming during an HCV infection.

Taken together, these data demonstrate that humoral and CMI responses directed against relevant HCV antigens can be primed in animals by DNA vaccines, and that improvements can be achieved through the used of new DNA vaccine technologies. The next logical steps in the development of these vaccines are to demonstrate protective efficacy in the chimpanzee model, then safety and potency in human clinical trials.

Rabies Virus

Rabies viruses belong to the *Rhabdoviridae* family and comprise the lyssavirus genus. They contain a negative-sense, non-segmented, ssRNA genome encoding only five gene products. Rabies viruses can infect all warm-blooded animals and are usually passed to humans through bites from infected dogs, cats and bats. Infection is initially confined to the site of virus entry (i.e., bitten skin), but then progresses to the peripheral and central nervous systems and brain. The onset of pathology can be from weeks to months and appears to depend on the proximity of the bitten skin to the brain. The course of disease involves neurological impairment, coma and death. Rabies vaccines have been used for over a hundred years, with modern versions consisting of cell culture-derived, inactivated virus preparations. Neutralizing antibodies directed toward the surface glycoprotein (G) confer immunity to rabies. Anti-G CTL and anti-nucleoprotein (N protein) CD4[+] helper T cell responses also are induced but their role in protection is unclear.

The well-characterized nature of rabies virus proteins and availability of good animal models of disease have facilitated the study of rabies virus DNA vaccines. Consequently, some of the early DNA vaccine work was conducted in this model system. Ertl and colleagues demonstrated that i.m. injection of DNA encoding the surface glycoprotein G induced virus neutralizing antibodies, Th1-type helper T cells, CTL and long-term protective efficacy in mice.[67, 68]

Immunogenicity and long-term protection also were shown in mice after gene gun administration of G DNA.[69] In addition, G DNA was shown to be effective at priming immune responses and conferring protection in dogs,[70, 71] cats,[70] and non-human primates.[72] Importantly, G DNA induces antibody responses that are capable of neutralizing a spectrum of rabies virus variants [72, 73] and potency is comparable to the licensed human diploid cell vaccine.[72, 74] Attempts to increase the breadth and potency of G DNA vaccine-induced immune responses have included the adjuvant MPL,[73, 75] prime-boost regimens involving recombinant vaccinia virus [74] or replication-defective adenovirus,[76] coexpression of GM-CSF,[77] and expression of chimeric lyssavirus glycoproteins.[78-80] The ability of G DNA vaccines to elicit protective immunity in large animal species, including primates, suggests that this vaccine may be useful for prevention of disease in humans, as well as blocking transmission from animal to humans.

The rabies mouse model has been useful for assessing the utility of DNA vaccines in neonatal animals. Some problems that arise in vaccination of newborns are 1) immature immune systems that do not respond adequately to vaccines, and 2) the presence of maternal antibodies that can interfere with priming of immune responses against vaccine antigens. With regard to the former, it was demonstrated that G DNA induced immune responses of similar magnitude in neonatal and adult mice,[81] suggesting that this may not be an issue with DNA vaccines. With respect to maternal antibodies, in principle, antigen-specific antibodies should not prevent the delivery of DNA vaccines. However, it was shown that immune responses induced by G DNA were not as strong in suckling newborns of rabies-immune mothers or in mice with passively transferred anti-G antibodies, although the inhibition was not as pronounced as was seen for an inactivated virus vaccine,[82] Therefore, even though it is unlikely that anti-G antibodies affected DNA vaccine delivery and uptake in situ, they may have interacted with newly synthesized G protein thereby interfering with antibody induction by B cells.

Measles Virus

Measles virus belongs to the *Paramyxoviridae* family and has a negative-sense, non-segmented ssRNA genome that encodes eight proteins. The hemagglutinin (H) protein shows variability among virus isolates, but overall measles viruses are antigenically stable, as recovery from infection confers life-long immunity. Measles viruses infect only humans and non-human primates and are spread via the respiratory route. Infection proceeds from respiratory epithelial cells to lymphoid organs to monocytes circulating in the blood, and disease symptoms include fever, malaise, respiratory disease and gastrointestinal effects. Mortality is seen primarily in the young and aged, particularly in poor countries. During infection, humoral and cellular immune responses, including CD4$^+$ helper T cells and CTL, are induced against several of the viral proteins. However, the responses most important for protection have not been delineated. Measles vaccine development was first attempted more than 250 years ago, but it wasn't until the 1960s before licensed vaccines were available. The first of these were inactivated virus preparations, which were protective if exposure to virus occurred shortly after immunization. In addition, immunization with these vaccines predisposed individuals to a more severe form of disease, termed atypical measles. These inactivated vaccines were replaced by live attenuated vaccines, which are highly effective and remain in use today. However, morbidity and mortality from measles continues to be high in developing countries, despite the fact that measles elimination programs have been in place for more than 20 years. Thus, alternative measles vaccine technologies are under investigation.

DNA vaccines encoding the hemagglutinin (H), fusion (F), and nucleoprotein (N) proteins of measles virus all have been shown to be immunogenic and, in some cases, protective against virus challenge in animal models. Antibody responses were generated in mice or non-human primates against all three proteins,[83-88] but it is likely that only anti-H and anti-F antibodies play a major role in protection. This is supported by protection in animals in which anti-H and/or anti-F antibodies were raised [85, 87] and lack of protection in mice immunized with N DNA alone.[83, 85, 87] Anti-H antibody responses were found to be best primed by expression of a plasma membrane-bound form of the protein, compared to a secreted version.[88]

CTL responses against N and H proteins also were generated.[84-87, 89, 90] In one study, immunization of neonatal mice was investigated, where Th1 helper T cell and CTL responses were primed,[84] thereby demonstrating the potential of DNA vaccines to overcome some of the problems inherent in vaccination of newborns. CTL responses against measles virus proteins have been induced by i.m. injection of DNA encoding the entire protein, a minigene DNA vaccine encoding a single CTL epitope,[90] or a *Shigella flexneri* DNA vaccine delivery system.[86] CTL were also primed after mucosal immunization of H DNA either by direct injection into the buccal mucosa or by intranasal administration.[89] In this case, increased CTL responses were achieved by inclusion of the mucosal adjuvant cholera toxin or by DNA formulation in cationic lipids. Protection mediated by anti-N CTL against measles virus-induced encephalitis was achieved,[90] but overall CTL responses do not appear to be as important for protection in mouse or monkey models as do neutralizing antibodies directed toward H and F proteins.[85, 87] As discussed above, one caveat of measles vaccine development is the possibility of predisposing individuals for atypical measles due to inappropriate or suboptimal immune priming. Importantly, immunization of rhesus macaques with DNA encoding measles virus H and F proteins did not predispose the animals to developing atypical measles.[87] Therefore, the DNA vaccine technology seems to be both a safe and effective measles vaccine strategy.

Respiratory Syncitial Virus

Respiratory syncitial viruses (RSV), like measles viruses, belong to *Paramyxoviridae* and have a negative-sense, non-segmented ssRNA genome. Ten proteins are present in the virus, with the surface glycoprotein (G) and fusion protein (F) being the major antigenic determinants. The G proteins show a high degree of diversity among strains, except in the receptor-binding domain. The F proteins, on the other hand, do not appear to undergo antigenic drift, thus are more structurally conserved. RSV disease is a major cause of lower respiratory tract infection worldwide and is commonly manifest as cold-like symptoms. However, with underlying illness or immunosuppression severe forms of disease and death can occur. Shortly after exposure to virus, infection is limited to the upper respiratory tract, but progresses to the lower respiratory tract. Symptoms result from cyopathic effects of the virus as well as a host immune-mediated component. The latter is suggested by the potentiation of disease in individuals exposed to virus after vaccination with an experimental, inactivated virus vaccine in the 1960s. Protection from disease appears to be mediated by humoral, including both local IgA and serum IgG, and CMI responses, including both CD4+ helper T cells and CTL, directed against the F protein. Thus, it is likely that a fine balance exists between the protective and toxic effects of immune responses to RSV antigens. Following the failure of the inactivated virus vaccine other types of vaccines have been investigated, including subunit proteins, live-attenuated viruses, viral vectors (e.g., vaccinia, adenovirus) and DNA vaccines.

RSV F DNA vaccines have been very effective in animal models. Li et al demonstrated that F DNA primed neutralizing antibodies, CTL, Th1-type helper T cells, and protection against virus challenge.[91] F DNA was as immunogenic and efficacious as live RSV infection. In addition, F DNA was able to modulate a preexisting Th2 response to a Th1 response. Similarly, other studies also have shown the efficacy of F DNA.[92-94] In contrast, RSV G DNA also was immunogenic, but was only partially protective.[92] The failure of the inactivated RSV vaccine was associated with eosinophilic lung pathology, possibly as a consequence of a helper T cell response of the Th2-type. RSV DNA vaccines, when administered by i.m. injection, have induced a Th1 or balanced Th1/Th2 response [91, 93, 94] and, importantly, have not induced lung pathology. Interestingly, even when a Th2-type response was primed by G DNA, either by gene gun administration [92] or by co-administration of IL-4 DNA,[93] no significant lung pathology was seen. This is in contrast to Th2 responses and lung pathology observed after vaccination of mice with inactivated virus vaccines or recombinant vaccinia virus expressing G protein.[93] Therefore, the type of vaccine-induced helper T cell response may be only one of several factors associated with the pathological lung response.

DNA vaccines encoding RSV genes also have been used as tools for the evaluation of neonatal immune responses and minigene constructs. First, as has been demonstrated in several other models,[81, 82, 84] RSV F DNA was effective at priming immune responses in neonates.[95] In this case, antibodies, CTL and Th1-type helper T cells were induced, despite the presence of maternal antibodies. Moreover, it was shown that priming with F DNA, followed by boosting with F protein was particularly effective. Second, minigene constructs encoding a discrete CTL epitope from RSV M2 protein were able to induce CTL responses and protection from RSV challenge in mice, as measured by a reduction in viral load.[90]

Rotavirus

Rotaviruses comprise the *Reoviridae* family and contain a positive-sense, segmented, dsRNA genome. Rotaviruses are pathogens of humans and many other animal species, causing diarrheal disease through cytocidal effects on intestinal epithelial cells. Diarrheal diseases caused by rotaviruses result in substantial mortality and morbidity in children world-wide. Therefore, an effective vaccine would have an enormous impact on childhood health globally.

Rotaviruses can be classified based on several distinct groups, subgroups and serotypes, with varying cross-reactivity. The 11 RNA segments of the rotavirus genome code for six structural and five non-structural proteins, with the prime immunological determinants being the outer capsid proteins VP4 and VP7. Immunity to rotaviruses is not well understood, but appears to be mediated, at least in part, by antibodies directed against VP4 and VP7. In addition, local intestinal immunity seems to be particularly important for protection. Hence, vaccine development efforts have focused on live attenuated vaccines that can be given orally. These types of vaccines have shown good efficacy in clinical trials and one vaccine was licensed for human use in 1998. However, serious gastrointestinal side effects resulted in the withdrawal of the product from the market. Thus, there is strong impetus for development of alternative rotavirus vaccines.

Rotavirus DNA vaccines have been investigated for the VP4, VP6 and VP7 genes. Herrmann et al demonstrated that each of the three DNA vaccines was immunogenic in mice, as seen by serum antibodies and CTL.[96] However, only mice vaccinated with VP4 or VP7 DNA had measurable virus neutralizing antibodies and were protected from virus challenge, as measured by a reduction in virus excretion. Two subsequent studies using VP6 DNA similarly showed no protection.[97, 98] In contrast, two other studies demonstrated that VP6 DNA could induce protective immunity in mice.[99, 100] The main differences in these sets of studies were the means of DNA administration and the type of immune responses engendered. Parenteral administration of VP6 DNA, by the gene gun, did not prime measurable local IgA antibodies and was not protective.[97, 98] Whereas, mucosal administration of VP6 DNA, either by oral delivery [99] or gene gun administration to the anorectal epithelium,[100] did induce local IgA antibodies and protection. Similarly, oral delivery of VP4 or VP7 DNA primed local IgA antibodies and was protective.[101] Therefore, mucosal antibody responses appear to be very important for protection in the mouse model. These studies also demonstrate that effective oral delivery of DNA can be achieved. This was accomplished by encapsulation of DNA in poly-L-lactide co-glycolide (PLG) microspheres, which have been widely used as a means of delivering drugs and proteins in a slow release depot. The effectiveness observed for oral delivery of DNA likely was due, at least in part, to a protective effect against the harsh conditions of the gut. Oral vaccination with DNA offers two potentially significant advantages over parenteral administration: 1) priming of mucosal immune responses, and 2) a more facile means of vaccination than needle injection or gene gun administration.

Dengue Virus

Dengue viruses, like HCV, are in the *Flaviviridae* family and contain a positive-sense, non-segmented, ssRNA genome encoding three structural and five non-structural proteins. Disease associated with Dengue virus infection is a worldwide public health problem. Dengue fever is an acute disease with fever, headache and myalgia, and affects more than 100 million people each year. A more severe form of disease, Dengue hemorrhagic fever, results in hemorrhage, shock and death in approximately 250,000 people per year. Transmission of Dengue viruses proceeds between mosquitoes, where high titers in various tissues are observed, and humans, where a rather self-limiting infection is restricted primarily to cells of the reticuloendothelial lineage. Dengue viruses also infect animals, such as mice and monkeys, but disease is achieved only with adapted strains. Dengue viruses are comprised by four distinct serotypes, characterized primarily by antigenic determinants in the envelope glycoprotein (E). Homotypic protection is strong and durable, but cross-protection is short-lived. Thus, it is possible to be infected sequentially by different Dengue virus serotypes, and an effective vaccine must protect against each serotype. Immune responses directed toward the E protein, and perhaps also NS1, seem to be required for protection, and these responses include both neutralizing antibodies and CTL. No vaccines are currently available for human use, but experimental vaccines under investigation include subunit proteins, recombinant vaccinia and adenovirus vectors, live-attenuated Dengue viruses and DNA Vaccines.

Dengue virus DNA vaccines have been investigated for immunogenicity and protective efficacy in mouse and non-human primate models. Work has primarily focused on the E glycoprotein and induction of virus neutralizing antibodies directed against it. Kochel et al [102] and Konishi [103] used DNA constructs encoding fusion proteins consisting of the pre-M and E proteins to demonstrate immunogenicity in mice, as measured by neutralizing antibodies. These DNA vaccines were subsequently shown to induce protection in mice from a lethal intracerebral virus challenge.[104] Expressing the dengue virus proteins in this manner produced virus-like particles (VLP) and was more effective than expressing a truncated from of E, which did not produce VLP.[105] The VLP-producing DNA vaccine also was tested in two different non-human primate models, namely aotus monkeys and rhesus macaques. In both models, pre-M/E DNA primed virus neutralizing antibodies and conferred protection from virus challenge, as seen by either complete protection or significant reduction in viremia.[106, 107] Interestingly, though, i.d. administration was superior to i.m. injection in aotus monkeys,[106] but the converse was true in rhesus macaques.[107] The vagaries of these primate models notwithstanding, these results demonstrate the potential utility of dengue virus DNA vaccines.

Summary

This review of DNA vaccines against selected RNA viruses provides ample evidence of the applicability of the technology for viral diseases. There are many other examples that have not been discussed here, including RNA viruses that cause substantial morbidity and mortality in animals, as well as RNA viruses that are emerging as human pathogens. This large and growing database of information, together with new potentially enabling technologies for DNA vaccines, provide a solid foundation for the development of effective DNA vaccines for humans and animals, including vaccines against other RNA viruses of high current significance (e.g., West Nile virus).

References

1. Ulmer JB, Donnelly JJ, Parker SE et al. Heterologous protection against influenza by injection of DNA encoding a viral protein. Science 1993; 259(5102):1745-9.
2. Robinson HL, Hunt LA, Webster RG. Protection against a lethal influenza virus challenge by immunization with a haemagglutinin-expressing plasmid DNA. Vaccine 1993; 11(9):957-60.
3. Fynan EF, Webster RG, Fuller DH et al. DNA vaccines: protective immunizations by parenteral, mucosal, and gene-gun inoculations. Proc Natl Acad Sci USA 1993; 90(24):11478-82.
4. Kodihalli S, Haynes S, Robinson HL et al. Cross-protection among lethal H5N2 influenza viruses induced by DNA vaccine to the hemagglutinin. J Virol 1997; 71(5):3391-6.
5. Justewicz DM, Webster RG. Long-term maintenance of B cell immunity to influenza virus hemagglutinin in mice following DNA-based immunization. Virol 1996; 224(1):10-7.
6. Ulmer JB, Deck RR, De Witt CM et al. Protective immunity by intramuscular injection of low doses of influenza virus DNA vaccines. Vaccine 1994; 12(16):1541-4.

7. Deck RR, DeWitt CM, Donnelly JJ et al. Characterization of humoral immune responses induced by an influenza hemagglutinin DNA vaccine. Vaccine 1997; 15:71-8.

8. Donnelly JJ, Friedman A, Martinez D et al. Preclinical efficacy of a prototype DNA vaccine: Enhanced protection against antigenic drift in influenza virus. Nat Med 1995; 1(6):583-7.

9. Donnelly JJ, Friedman A, Ulmer JB et al. Further protection against antigenic drift of influenza virus in a ferret model by DNA vaccination. Vaccine 1997; 15(8):865-8.

10. Chen Z, Sahashi Y, Matsuo K et al. Comparison of the ability of viral protein-expressing plasmid DNAs to protect against influenza. Vaccine 1998; 16(16):1544-9.

11. Chen Z, Matsuo K, Asanuma H et al. Enhanced protection against a lethal influenza virus challenge by immunization with both hemagglutinin- and neuraminidase-expressing DNAs. Vaccine 1999; 17(7-8):653-9.

12. Chen Z, Yoshikawa T, Kadowaki S et al. Protection and antibody responses in different strains of mouse immunized with plasmid DNAs encoding influenza virus haemagglutinin, neuraminidase and nucleoprotein. J Gen Virol 1999; 80(Pt 10):2559-64.

13. Pertmer TM, Robinson HL. Studies on antibody responses following neonatal immunization with influenza hemagglutinin DNA or protein. Virol 1999; 257(2):406-14.

14. Bot A, Bot S, Bona C. Enhanced protection against influenza virus of mice immunized as newborns with a mixture of plasmids expressing hemagglutinin and nucleoprotein. Vaccine 1998; 16(17):675-82.

15. Bot A, Shearer M, Bot S et al. Induction of antibody response by DNA immunization of newborn baboons against influenza virus. Viral Immunol 1999; 12(2):91-6.

16. Pertmer TM, Oran AE, Moser JM et al. DNA vaccines for influenza virus: differential effcts ofmaternal antibody on immune responses to hemagglutinin and nucleoprotein. J Virol 2000; 74(17):787-93.

17. Bender BS, Ulmer JB, De Witt CM et al. Immunogenicity and efficacy of DNA vaccines encoding influenza A proteins in aged mice. Vaccine 1998; 16(18):1748-55.

18. Radu DL, Antohi S, Bot A et al. Plasmid expressing the influenza HA gene protects old mice from lethal challenge with influenza virus. Viral Immunol 1999; 12(3):217-26.

19. Macklin MD, McCabe D, McGregor MW et al. Immunization of pigs with a particle-mediated DNA vaccine to influenza A virus protects against challenge with homologous virus. J Virol 1998; 72(2):1491-6.

20. Lunn DP, Soboll G, Schram BR et al. Antibody responses to DNA vaccination of horses using the influenza virus hemagglutinin gene. Vaccine 1999; 17(18):2245-58.

21. Boyle CM, Morin M, Webster RG et al. Role of different lymphoid tissues in the initiation and maintenance of DNA-raised antibody responses to the influenza virus H1 glycoprotein. J Virol 1996; 70(12):9074-8.

22. Johnson PA, Conway MA, Daly J et al. Plasmid DNA encoding influenza virus haemagglutinin induces Th1 cells and protection against respiratory infection despite its limited ability to generate antibody responses. J Gen Virol 2000; 81 Pt 7(17):1737-45.

23. Fu TM, Friedman A, Ulmer JB et al. Protective cellular immunity cytotoxic T-lymphocyte responses against dominant and recessive epitopes of influenza virus nucleoprotein induced by DNA immunization. J Virol 1997; 71:2715-2721.

24. Ulmer JB, Fu TM, Deck RR et al. Protective CD4$^+$ and CD8$^+$ T cells against influenza virus induced by vaccination with nucleoprotein DNA. J Virol 1998; 72(7):5648-53.

25. Fu TM, Guan L, Friedman A et al. Dose dependence of CTL precursor frequency induced by a DNA vaccine and correlation with protective immunity against influenza virus challenge. J Immunol 1999; 162(7):4163-4170.

26. Epstein SL, Stack A, Misplon JA et al. Vaccination with DNA encoding internal proteins of influenza virus does not require CD8$^+$ cytotoxic T lymphocytes: either CD4$^+$ or CD8$^+$ T cells can promote survival and recovery after challenge. Int Immunol 2000; 12(1):91-101.

27. Ulmer J, Deck R, Dewitt C et al. Generation of MHC class I-restricted cytotoxic T lymphocytes by expression of a viral protein in muscle cells: antigen presentation by non-muscle cells. Immunol 1996; 89:59-67.

28. Corr M, Lee DJ, Carson DA et al. Gene vaccination with naked plasmid DNA: Mechanism of CTL priming. J Exp Med 1996; 184(4):1555-60.

29. Fu TM, Ulmer JB, Caulfield MJ et al. Priming of cytotoxic T lymphocytes by DNA vaccines: requirement for professional antigen presenting cells and evidence for antigen transfer from myocytes. Mol Med 1997; 3:362-371.

30. Iwasaki A, Torres CA, Ohashi PS et al. The dominant role of bone marrow-derived cells in CTL induction following plasmid DNA immunization at different sites. J Immunol 1997; 159(1):11-4.

31. Torres CA, Iwasaki A, Barber BH et al. Differential dependence on target site tissue for gene gun and intramuscular DNA immunizations. J Immunol 1997; 158(10):4529-32.

32. Casares S, Inaba K, Brumeanu TD et al. Antigen presentation by dendritic cells after immunization with DNA encoding a major histocompatibility complex class II-restricted viral epitope. J Exp Med 1997; 186(9):1481-1486.

33. Bot A, Stan AC, Inaba K et al. Dendritic cells at a DNA vaccination site express the encoded influenza nucleoprotein and prime MHC class I-restricted cytolytic lymphocytes upon adoptive transfer. Int Immunol 2000; 12(6):825-32.

34. Corr M, von Damm A, Lee DJ et al. In vivo priming by DNA injection occurs predominantly by antigen transfer. J Immunol 1999; 163(9):4721-7.

35. Porgador A, Irvine KR, Iwasaki A et al. Predominant role for directly transfected dendritic cells in antigen presentation to CD8⁺ T cells after gene gun immunization. J Exp Med 1998; 188(6):1075-82.

36. Deliyannis G, Boyle JS, BradyJL et al. A fusion DNA vaccine that targets antigen-presenting cells increases protection from viral challenge. Proc Natl Acad Sci USA 2000; 97(12):6676-80.

37. Ulmer JB, DeWitt CM, Chastain M et al. Enhancement of DNA vaccine potency using conventional aluminum adjuvants. Vaccine 1999; 18(1-2):18-28.

38. Lee SW, Youn JW, Seong BL et al. IL-6 induces long-term protective immunity against a lethal challenge of influenza virus. Vaccine 1999; 17(5):490-6.

39. Operschall E, Schuh T, Heinzerling L et al. Enhanced protection against viral infection by co-administration of plasmid DNA coding for viral antigen and cytokines in mice. J Clin Virol 1999; 13(1-2):17-27.

40. Iwasaki A, Stiernholm BJ, Chan AK et al. Enhanced CTL responses mediated by plasmid DNA immunogens encoding costimulatory molecules and cytokines. J Immunol 1997; 158(10):4591-601.

41. Haensler J, Verdelet C, Sanchez V et al. Intradermal DNA immunization by using jet-injectors in mice and monkeys. Vaccine 1999; 17(7-8):628-38.

42. Degano P, Sarphie DF, Bangham CR. Intradermal DNA immunization of mice against influenza A virus using the novel PowderJect system. Vaccine 1998; 16(4):394-8.

43. Widera G, Austin M, Rabussay D et al. Increased DNA vaccine delivery and immunogenicity by electroporation in vivo. J Immunol 2000.

44. Bachy M, Boudet F, Bureau M et al. Electric pulses increase the immunogenicity of an influenza DNA vaccine injected intramuscularly in the mouse. Vaccine 2001; 19(13-14):1688-1693.

45. Kadowaki S, Chen Z, Asanuma H et al. Protection against influenza virus infection in mice immunized by administration of hemagglutinin-expressing DNAs with electroporation. Vaccine 2000; 18(25):2779-88.

46. Degano P, Schneider J, Hannan CM et al. Gene gun intradermal DNA immunization followed by boosting with modified vaccinia virus Ankara: enhanced CD8⁺ T cell immunogenicity and protective efficacy in the influenza and malaria models. Vaccine 1999; 18(7-8):623-32.

47. Tedeschi V, Akatsuka T, Shih JW et al. A specific antibody response to HCV E2 elicited in mice by intramuscular inoculation of plasmid DNA containing coding sequences for E2. Hepatol 1997; 25(2):459-62.

48. Nakano I, Maertens G, Major ME et al. Immunization with plasmid DNA encoding hepatitis C virus envelope E2 antigenic domains induces antibodies whose immune reactivity is linked to the injection mode. J Virol 1997; 71(9):7101-9.

49. Lee SW, Cho JH, Sung YC, Optimal induction of hepatitis C virus envelope-specific immunity by bicistronic plasmid DNA inoculation with the granulocyte-macrophage colony-stimulating factor gene. J Virol 1998; 72(10):8430-6.

50. Forns X, Emerson SU, Tobin GJ et al. DNA immunization of mice and macaques with plasmids encoding hepatitis C virus envelope E2 protein expressed intracellularly and on the cell surface. Vaccine 1999; 17(15-16):1992-2002.

51. Fournillier A., Depla E, Karayiannis P et al. Expression of noncovalent hepatitis C virus envelope E1-E2 complexes is not required for the induction of antibodies with neutralizing properties following DNA immunization. J Virol 1999; 73(9):7497-504.

52. Song MK, Lee SW, Suh YS et al. Enhancement of immunoglobulin G2a and cytotoxic T-lymphocyte responses by a booster immunization with recombinant hepatitis C virus E2 protein in E2 DNA-primed mice. J Virol 2000; 74(6):2920-5.

53. Pancholi P, Liu Q, Tricoche N et al. DNA prime-canarypox boost with polycistronic hepatitis C virus (HCV) genes generates potent immune responses to HCV structural and nonstructural proteins. J Infect Dis 2000; 182(1):18-27.

54. Heile JM, Fong YL, Rosa D et al. Evaluation of hepatitis C virus glycoprotein E2 for vaccine design: an endoplasmic reticulum-retained recombinant protein is superior to secreted recombinant protein and DNA-based vaccine candidates [In Process Citation]. J Virol 2000; 74(15):6885-6892.
55. Vidalin O, Fournillier A, Renard N et al. Use of conventional or replicating nucleic acid-based vaccines and recombinant Semliki forest virus-derived particles for the induction of immune responses against hepatitis C virus core and E2 antigens. Virol 2000; 276(2):259-70.
56. Zucchelli S, Capone S, Fattori E et al. Enhancing B- and T-cell immune response to a hepatitis C virus E2 DNA vaccine by intramuscular electrical gene transfer. J Virol 2000; 74(24):11598-607.
57. Gordon EJ, Bhat R, Liu Q et al. Immune responses to hepatitis C virus structural and nonstructural proteins induced by plasmid DNA immunizations. J Infect Dis 2000; 181(1):42-50.
58. Nishimura Y, Kamei A, Uno-Furuta S et al. A single immunization with a plasmid encoding hepatitis C virus (HCV) structural proteins under the elongation factor 1-alpha promoter elicits HCV-specific cytotoxic T-lymphocytes (CTL). Vaccine 1999; 18(7-8):675-80.
59. Inchauspe G, Vitvitski L, Major ME et al. Plasmid DNA expressing a secreted or a nonsecreted form of hepatitis C virus nucleocapsid: comparative studies of antibody and T-helper responses following genetic immunization. DNA Cell Biol 1997; 16(2):185-95.
60. Hu GJ, Wang RY, Han DS et al. Characterization of the humoral and cellular immune responses against hepatitis C virus core induced by DNA-based immunization. Vaccine 1999; 17(23-24):3160-70.
61. Vidalin O, Tanaka E, Spengler U et al. Targeting of hepatitis C virus core protein for MHC I or MHC II presentation does not enhance induction of immune responses to DNA vaccination. DNA Cell Biol 1999; 18(8):611-621.
62. Arichi T, Saito T, Major ME et al. Prophylactic DNA vaccine for hepatitis C virus (HCV) infection: HCV- specific cytotoxic T lymphocyte induction and protection from HCV- recombinant vaccinia infection in an HLA-A2.1 transgenic mouse model. Proc Natl Acad Sci USA 2000; 97(1):297-302.
63. Kamei A, Tamaki S, Taniyama H et al. Induction of hepatitis C virus-specific cytotoxic T lymphocytes in mice by an intrahepatic inoculation with an expression plasmid. Virol 2000; 273(1):120-6.
64. Encke J, zu Putlitz J, Geissler M et al. Genetic immunization generates cellular and humoral immune responses against the nonstructural proteins of the hepatitis C virus in a murine model. J Immunol 1998; 161(9):4917-23.
65. Lee AY, Polako NKs, Otten GR et al. Quantification of the number of cytotoxic T cells specific for an immunodominant HCV-specific CTL epitope primed by DNA immunization. Vaccine 2000; 18(18):1962-1968.
66. Lee AY, Manning WC, Arian CL et al. Priming of Hepatitis C virus-Specific Cytotoxic T lymphocytes in Mice Following Portal Vein Injection of a Liver-Specific Plasmid DNA. Hepatol 2000; 31(6):1327-1333.
67. Xiang ZQ, Spitalnik S, Tran M et al. Vaccination with a plasmid vector carrying the rabies virus glycoprotein gene induces protective immunity against rabies virus. Virol 1994; 199(1):132-140.
68. Xiang ZQ, Spitalnik SL, Cheng J et al. Immune-responses to nucleic-acid vaccines to rabies virus. Virol 1995; 209(2):569-579.
69. Lodmell DL, Ray NB, Ewalt LC. Gene gun particle-mediated vaccination with plasmid DNA confers protective immunity against rabies virus infection. Vaccine 1998; 16(2-3):115-8.
70. Osorio JE, Tomlinson CC, Frank RS et al. Immunization of dogs and cats with a DNA vaccine against rabies virus. Vaccine 1999; 17(9-10):1109-16.
71. Perrin P, Jacob Y, Aguilar-Setien A et al. Immunization of dogs with a DNA vaccine induces protection against rabies virus. Vaccine 1999; 18(5-6):479-86.
72. Lodmell DL, Ray NB, Parnell MJ et al. DNA immunization protects nonhuman primates against rabies virus. Nat Med 1998; 4(8):949-52.
73. Ray NB, Ewalt LC, Lodmell DL. Nanogram quantities of plasmid DNA encoding the rabies virus glycoprotein protect mice against lethal rabies virus infection. Vaccine 1997; 15(8):892-5.
74. Lodmell DL, Ewalt LC. Rabies vaccination: comparison of neutralizing antibody responses after priming and boosting with different combinations of DNA, inactivated virus, or recombinant vaccinia virus vaccines. Vaccine 2000; 18(22):2394-8.
75. Lodmell DL, Ray NB, Ulrich JT et al. DNA vaccination of mice against rabies virus: effects of the route of vaccination and the adjuvant monophosphoryl lipid A (MPL). Vaccine 2000; 18(11-12):1059-66.
76. Xiang ZQ, Pasquini S, Ertl HC. Induction of genital immunity by DNA priming and intranasal booster immunization with a replication-defective adenoviral recombinant. J Immunol 1999; 162(11):6716-23.

77. Xiang Z, Ertl HC. Manipulation of the immune response to a plasmid-encoded viral antigen by coinoculation with plasmids expressing cytokines. Immunity 1995; 2(2):129-35.
78. Bahloul C, Jacob Y, Tordo N et al. DNA-based immunization for exploring the enlargement of immunological cross-reactivity against the lyssaviruses. Vaccine 1998; 16(4):417-25.
79. Desmezieres E, Jacob Y, Saron MF et al. Lyssavirus glycoproteins expressing immunologically potent foreign B cell and cytotoxic T lymphocyte epitopes as prototypes for multivalent vaccines. J Gen Virol 1999; 80(Pt 9):2343-51.
80. Jallet C, Jacob Y, Bahloul C et al. Chimeric lyssavirus glycoproteins with increased immunological potential. J Virol 1999; 73(1):225-33.
81. Wang Y, Xiang Z, Pasquini S et al. Immune response to neonatal genetic immunization. Virol 1997; 228(2):278-84.
82. Wang Y, Xiang Z, Pasquini S et al. Effect of passive immunization or maternally transferred immunity on the antibody response to a genetic vaccine to rabies virus. J Virol 1998; 72(3):1790-6.
83. Fooks AR, Jeevarajah D, Warnes A et al. Immunization of mice with plasmid DNA expressing the measles virus nucleoprotein gene. Viral Immunol 1996; 9(2):65-71.
84. Martinez X, Brandt C, Saddallah F et al. DNA immunization circumvents deficient induction of T helper type 1 and cytotoxic T lymphocyte responses in neonates and during early life. Proc.s Natl. Acad. Sci. USA 1997; 94(16):8726-8731.
85. Schlereth B, Germann PG, ter Meulen V et al. DNA vaccination with both the haemagglutinin and fusion proteins but not the nucleocapsid protein protects against experimental measles virus infection. J Gen Virol 2000; 81 Pt 5(10):1321-5.
86. Fennelly GJ, Khan SA, Abadi MA et al. Mucosal DNA vaccine immunization against measles with a highly attenuated Shigella flexneri vector. J Immunol 1999; 162(3):1603-10.
87. Polack FP, Lee SH, Permar S et al. Successful DNA immunization against measles: Neutralizing antibody against either the hemagglutinin or fusion glycoprotein protects rhesus macaques without evidence of atypical measles. Nat Med 2000; 6(7):776-81.
88. Torres CA, Yang K, Mustafa F et al. DNA immunization: effect of secretion of DNA-expressed hemagglutinins on antibody responses. Vaccine 1999; 18(9-10):805-14.
89. Etchart N, Buckland R, Liu MA et al. Class I-restricted CTL induction by mucosal immunization with naked DNA encoding measles virus haemagglutinin. J Gen Virol 1997; 78(Pt 7):1577-80.
90. Hsu SC, Obeid OE, Collins M et al. Protective cytotoxic T lymphocyte responses against paramyxoviruses induced by epitope-based DNA vaccines: involvement of IFN-gamma. Int Immunol 1998; 10(10):1441-7.
91. Li X, Sambhara S, Li CX et al. Protection against respiratory syncytial virus infection by DNA immunization. J Exp Med 1998; 188(4):681-8.
92. Bembridge GP, Rodriguez N, Garcia-Beato R et al. Respiratory syncytial virus infection of gene gun vaccinated mice induces Th2-driven pulmonary eosinophilia even in the absence of sensitisation to the fusion (F) or attachment (G) protein. Vaccine 2000; 19(9-10):1038-46.
93. Bembridge GP, Rodriguez N, Garcia-Beato R et al. DNA encoding the attachment (G) or fusion (F) protein of respiratory syncytial virus induces protection in the absence of pulmonary inflammation. J Gen Virol 2000; 81(Pt 10):2519-23.
94. Li X, Sambhara S, Li CX et al. Plasmid DNA encoding the respiratory syncytial virus G protein is a promising vaccine candidate. Virology 2000; 269(1):54-65.
95. Martinez X, Li X, Kovarik J et al. Combining DNA and protein vaccines for early life immunization against respiratory syncytial virus in mice. Eur J Immunol 1999; 29(10):3390-400.
96. Herrmann JE, Chen SC, Fynan EF et al. Protection against rotavirus infections by DNA vaccination. J Infect Dis 1996; 174(1):93-7.
97. Choi AH, Knowlton DR, McNeal MM et al. Particle bombardment-mediated DNA vaccination with rotavirus VP6 induces high levels of serum rotavirus IgG but fails to protect mice against challenge. Virol 1997; 232(1):129-38.
98. Choi AH, Basu M, Rae MN et al. Particle-bombardment-mediated DNA vaccination with rotavirus VP4 or VP7 induces high levels of serum rotavirus IgG but fails to protect mice against challenge. Virol 1998; 250(1):230-40.
99. Chen SC, Jones DH, Fynan EF et al. Protective immunity induced by oral immunization with a rotavirus DNA vaccine encapsulated in microparticles. J Virol 1998; 72(7):5757-5761.
100. Chen SC, Fynan EF, Greenberg HB et al. Immunity obtained by gene-gun inoculation of a rotavirus DNA vaccine to the abdominal epidermis or anorectal epithelium. Vaccine 1999; 17(23-24):3171-6.
101. Herrmann JE, Chen SC, Jones DH et al. Immune responses and protection obtained by oral immunization with rotavirus VP4 and VP7 DNA vaccines encapsulated in microparticles. Virol 1999; 259(1):148-53.

102. Kochel T, Wu SJ, Raviprakash K et al. Inoculation of plasmids expressing the dengue-2 envelope gene elicit neutralizing antibodies in mice. Vaccine 1997; 15(5):547-552.

103. Konishi E, Yamaoka M, Kurane I et al. A DNA vaccine expressing dengue type 2 virus premembrane and envelope genes induces neutralizing antibody and memory B cells in mice. Vaccine 2000; 18(11-12):1133-9.

104. Porter KR, Kochel TJ, Wu SJ et al. Protective efficacy of a dengue 2 DNA vaccine in mice and the effect of CpG immuno-stimulatory motifs on antibody responses. Arch Virol 1998; 143(5):997-1003.

105. Raviprakash K, Kochel TJ, Ewing D et al. Immunogenicity of dengue virus type 1 DNA vaccines expressing truncated and full length envelope protein. Vaccine 2000; 18(22):2426-34.

106. Kochel TJ, Raviprakash K, Hayes CG et al. A dengue virus serotype-1 DNA vaccine induces virus neutralizing antibodies and provides protection from viral challenge in Aotus monkeys. Vaccine 2000; 18(27):3166-73.

107. Raviprakash K, Porter KR, Kochel TJ et al. Dengue virus type 1 DNA vaccine induces protective immune responses in rhesus macaques. J Gen Virol 2000; 81(Pt 7):1659-67.

108. Lin YL, Chen LK, Liao CL et al. DNA immunization with Japanese encephalitis virus nonstructural protein NS1 elicits protective immunity in mice. J Virol 1998; 72(1):191-200.

109. Konishi E, Yamaoka M, Khin Sane W et al. The anamnestic neutralizing antibody response is critical for protection of mice from challenge following vaccination with a plasmid encoding the Japanese encephalitis virus premembrane and envelope genes. J Virol 1999; 73(7):5527-34.

110. Ashok MS, Rangarajan PN. Immunization with plasmid DNA encoding the envelope glycoprotein of Japanese encephalitis virus confers significant protection against intracerebral viral challenge without inducing detectable antiviral antibodies. Vaccine 1999; 18(1-2):68-75.

111. Chen HW, Pan CH, Liau MY et al. Screening of protective antigens of Japanese encephalitis virus by DNA immunization: a comparative study with conventional viral vaccines. J Virol 1999; 73(12):10137-45.

112. Konishi E, Yamaoka M, Kurane I et al. Japanese Encephalitis DNA vaccine candidates expressing premembrane and envelope genes induce virus-specific memory B cells and long-lasting antibodies in swine. Virol 2000; 268(1):49-55.

113. Chang GJ, Hunt AR, Davis B. A single intramuscular injection of recombinant plasmid DNA induces protective immunity and prevents Japanese encephalitis in mice. J Virol 2000; 74(9):4244-52.

114. Schmaljohn C, Vanderzanden L, Bray M et al. Naked DNA vaccines expressing the prM and E genes of Russian spring summer encephalitis virus and Central European encephalitis virus protect mice from homologous and heterologous challenge. J Virol 1997; 71(12):9563-9.

115. Schmaljohn C, Custer D, VanderZanden L et al. Evaluation of tick-borne encephalitis DNA vaccines in monkeys. Virol 1999; 263(1):166-74.

116. Aberle JH, Aberle SW, Allison SL et al. A DNA immunization model study with constructs expressing the tick- borne encephalitis virus envelope protein E in different physical forms. J Immunol 1999; 163(12):6756-61.

117. Morozova OV, Maksimova TG, Bakhvalova VN. Tick-borne encephalitis virus NS3 gene expression does not protect mice from homologous viral challenge. Viral Immunol 1999; 12(4):277-80.

CHAPTER 8

DNA Vaccines Against Herpesviruses

Christopher Pack and Barry T. Rouse

Abstract

Herpesviruses are significant pathogens of mankind, and vaccines of proven efficacy remain unavailable. This review briefly examines the current state of vaccinology against herpesviruses and then discusses experiments in animal model systems that have evaluated DNA vaccines as a means of prevention or control of lesions. The DNA vaccine approach appears promising as a viable strategy to provide a needed prophylactic vaccine. Moreover, since plasmid DNA encoding immunomodulatory molecules may be useful to control lesion severity, the approach may prove beneficial to contain lesions observed during disease recurrence.

Introduction

Vaccines have successfully controlled few DNA virus pathogens, other than smallpox virus. This situation holds true for all save one of the human herpesviruses. Thus, currently effective vaccines are unavailable for herpes simplex viruses (HSV) 1 and 2, Epstein-Barr Virus, human cytomegalovirus (HCMV) and human herpesvirus 6, 7, and 8. The exception is Varicella Zoster virus (VZV) for which an attenuated vaccine for use in children is claimed to be 85% or more effective.[1] In fact, the success of the vaccine, which unfortunately is not given to all children, has reduced the outbreaks of chicken pox. Consequently, many humans reach adulthood without VZV exposure and hence immunity. Primary infection of adults often results in severe symptoms and the attenuated vaccine fails to protect them. Furthermore, the vaccine also fails to boost immunity in adults, which would be required to protect the elderly against an outbreak of shingles.[2]

Vaccines have qualified success against several animal herpes viruses. Short-term protection results following vaccination against feline and bovine rhinotracheitis virus, as well as against infectious laryngotracheitis of poultry.[3-5] An effective vaccine protects mares against abortion, although the same vaccine is ineffective against respiratory infection caused by equine herpesvirus.[6] There is also an effective vaccine against Marek's disease of chickens. Unfortunately, this vaccine protects only against neoplasia and not against infection.[7] The conclusion is obvious, current vaccines are far from ideal and mediate only brief and partial protection. Improvement is clearly warranted.

New vaccine approaches are certainly required to control both human and animal herpesvirus infections, and DNA vaccines represent appealing candidates that require evaluation. In this review, we discuss experiments performed by our group and others aimed to assess the value of DNA vaccines against herpesviruses and other DNA viruses. Almost all studies with human herpesviruses have been directed at HSV 1 or 2, and test systems to measure in vivo immunity were performed in laboratory animals. Such model systems can provide an optimistic scenario and must be interpreted with caution. Thus many previous vaccines that provided robust immunity in mice or guinea pigs proved inadequate when subsequently tested in humans.[8] Moreover, at least with HSV, there is a more pressing need to develop therapeutic

DNA Vaccines, edited by Hildegund C. J. Ertl. ©2003 Eurekah.com and Kluwer Academic / Plenum Publishers.

vaccines that protect against recrudescent lesions resulting from reactivation of latent infection. Animal models to test such immunity are far from ideal.

Animal models assume crucial importance, however, to understand how to manipulate the quality and extent of immunity and to discover which components of immune defense function best to contain infection. These issues remain unresolved for herpesvirus infections, especially in humans. Consensus opinion nevertheless favors a crucial role for T cell-mediated immunity, but whether CD8+ or CD4+ T cells assume the principal role is still a subject of debate.[9] In vitro systems reveal that multiple mechanisms may mediate defense, but the herpesviruses, especially HCMV, appear endowed with numerous tactics that confer a level of immune evasion. Indeed, studies on both human and murine CMV infections have revealed an abundance of viral encoded proteins that interfere in a myriad of ways with the induction or expression of various defense mechanisms.[10] Such evasive maneuvers are assumed to provide the virus with an "edge" over potential defenses, thereby prolonging the opportunity for replication and spread to new hosts. Of course, the most effective strategy possessed by all herpesviruses is that of latency. In this situation, the virus adopts an alternative, non-productive interaction with some populations of cells in the body, most often cells of the nervous system. Few or no viral proteins are expressed, so the virus is in effect hidden from the immune system.[11] In consequence, latent infections remain refractory to immune elimination. This would not be an issue if latency was maintained permanently, but such is usually not the case especially with HSV and VZV. Reactivation from latency can occur, dramatically in the case of VZV, involving potentially all latently infected cells. Shingles is the resultant lesion. With HSV, reactivation is invariably subtotal and involves a minority of latently infected cells. Lesions may or may not occur and the episode is contained by the immune system. Productive episodes can occur frequently since the supply of latently infected cells appears endless. In the case of HSV, therapeutic vaccines are required to effectively limit virus expression following reactivation so that lesions fail to manifest.[12] At present, suitable therapeutic vaccines are unavailable. Conceivably, DNA vaccines could subserve this need.

Use of DNA Vaccines Against Herpesviruses

A number of groups have reported success using expression plasmid DNA encoding various herpes virus proteins to immunize against herpesviruses. Such studies revealed that DNA vaccines could induce both virus specific cell-mediated as well as humoral responses, conferring protection to these vaccinated animals against lethal challenge with the virus (Table 1). The Babiuk group was the first to apply a DNA vaccine approach to vaccinate against a herpes virus, bovine herpes virus. They demonstrated that inoculation of mice with a plasmid expressing the gIV protein of bovine herpes virus I resulted in a specific antibody response.[13] In addition, cattle injected with this plasmid also generated a protective antibody response and decreased viral titers after challenge. Our group and others also explored whether DNA vaccine approaches could be utilized in a mouse model of HSV, demonstrating that protection could be granted to mice receiving plasmid DNA encoding proteins from HSV.[14-21]

Our early studies revealed that DNA vaccination of mice with a plasmid encoding glycoprotein B (gB) of HSV-1 induced a protective immune response in a zosteriform model.[14] Adoptive transfer studies indicated that CD4+ T cells were primarily responsible for protection, not cytolytic T lymphocytes (CTL), as is often the case when DNA vaccines induce protective immunity against viral infections.[14] In these studies the level of neutralizing antibodies remained low, and the response was dominated by Th1 cytokines. Other early work by Kriesal et al[18] and Ghiasi et al [19] demonstrated that a nucleic acid vaccine encoding gD2 was able to induce a specific antibody response and mediate protection from HSV-2. Nass et al investigated the protective capacity of plasmids encoding gC, gD, or gE of HSV-1, with the most effective protection being granted with gD.[20] The Bernstein group investigated the protective capacity of immunization with plasmids encoding both gB and gC of HSV-2 and found that such an approach generated neutralizing antibodies, but failed to dramatically lower viral

Table 1. DNA Vaccine approaches against herpesviruses

Virus	Antigen	Species	Immunity	References
Bovine herpesvirus	gIV	Mice, cattle	Humoral	13
Herpes simplex	HSV-1			
virus (1 and 2)	gB, ICP27	Mice	CTL (in vitro), Humoral, Th cell	14,15,17
	gC, gD, gE	Mice	Humoral	19,20,22,23
	HSV-2			
	gB, gD	Mice, guinea pigs	Humoral	16,18,24
	gB, gC, gD	Mice	Humoral, T cell	21, 35
Pseudorabies	gD	Pigs	Humoral	30
Varicella-Zoster	gB, gE	Mice	Humoral	25
Equine herpes virus	gD	Mice	Humoral, Th1 cell	28
Cytomegalovirus	Human CMV			
	pp65	Mice	Humoral	26
	Murine CMV			
	pp89	Mice	Humoral CTL	27

clearance.[21] Higgins et al demonstrated that the expression pattern (secreted or cell-associated) also impacts the level of immune induction.[22] Such investigations revealed that plasmids encoding a secreted form of gD2 stimulated humoral responses, while non-secreted forms induced cell-mediated responses.[22] Unfortunately, the level of immunity achieved by DNA vaccine administration in these studies never equaled that obtained from live viral infection. In our hands immunization with a DNA vaccine encoding gB imparted a level of immunity 10 to 100 fold less than can be achieved by immunization with a recombinant vaccinia vector or UV inactivated HSV.[14] Of particular interest, the Weir group recently compared the strength of protection against lethal challenge between plasmid DNA vaccination (encoding gD) and natural infection.[23] They concluded that immunity to HSV using the plasmid DNA vaccine is mediated by B cells and CD8+ T cells and approaches the strength of immunity following sub-lethal infection. Such observations have not been the experience of others, and, clearly in primates, DNA vaccines are usually found to be inferior immunogens.[23]

The DNA vaccine approach has also been shown to be effective in a therapeutic sense, a far more challenging scenario. Therapeutic vaccines would be of great benefit to those individuals with established latent infections, but few groups have addressed this issue. Bourne et al and Mester et al demonstrated that DNA vaccines could be used in a therapeutic environment, reducing the recurrence of lesions in guinea pigs.[16,21] In addition, the Liu group has shown that glycoprotein-encoding (gB and gD) plasmids have been effective in suppressing the frequency and severity of recurrences in a guinea pig model of HSV.[24] However, DNA vaccines did not perform as well in this model as did other forms of vaccines previously tested. [24] This model for vaginal HSV-2 infection represents the most accurate model available for the study of genital herpes, since these animals exhibit spontaneous recurrent lesions.

DNA vaccination approaches have also been investigated for a number of other herpesviruses. Concerning VZV, improvements of the current vaccine are required, and animal studies have been conducted using DNA vaccines encoding glycoproteins (gB and gE) from the virus.[25] Such studies demonstrated strong humoral responses to these glycoproteins, an encouraging finding. Reports concerning genetic vaccination against CMV have been scarce, but one study demonstrated that plasmids expressing phosphorylated protein (pp) 65 of human CMV, a tegument protein, induced antibody responses in mice.[26] Another study involving murine CMV included vaccination with plasmid DNA encoding pp89, which carries a known protective epitope. Despite low levels of antibody, a significant CTL response was observed.[27] The Whalley group has investigated the potential of DNA-mediated vaccination for equine herpes virus (EHV) in a murine model. EHV-1 gD plasmid immunized mice exhibited neutralizing antibodies of the IgG2a isotype.[28] Unfortunately, a common recurrence from these early studies was that DNA vaccines failed to elicit the strong immune responses observed with classical vaccine approaches or natural infection. Levels of antibody production were modest, and animals receiving DNA immunizations could only resist a low dose challenge with virus.

Several groups have also addressed whether plasmid DNA vaccines can elicit a protective immune response against herpesviruses in the neonate. The neonatal period is a highly susceptible period for infection. Therefore, generation of a protective immune response at this early time would be advantageous.maternal antibodies, which exert a transient protective effect, may also inhibit the induction of immunity in the neonate. We have demonstrated that immunization of neonates with a plasmid encoding gB of HSV-1 resulted in protection, mediated by both humoral and cell-mediated responses, when such mice were challenged as adults.[29] Such protection was evident even when the neonatal regimen was administered in the presence of interferingmaternal antibodies. This pattern was also observed in studies involving pseudorabies virus, a common pathogen of pigs. Neonatal plasmid DNA encoding gD of pseudorabies, when injected into neonatal pigs, resulted in generation of antibodies.[30] However, such a response was not protective against infectious challenge. In contrast, an independent study revealed that DNA vaccination of adult pigs against pseudorabies conferred protection against disease.[28] Gerdts et al addressed an even earlier timepoint, delivering plasmid DNA vaccine to the fetus by way of the oral cavity. They demonstrated that injection into the amniotic fluid of a lamb fetus with a DNA vaccine encoding a truncated version of gD from bovine herpesvirus induced high humoral and cell-mediated responses, but the vaccinated animals were not subjected to viral challenge.[31]

Early studies using DNA vaccines for herpes viruses utilized DNA that encoded only single proteins from the virus, with such plasmids being administered intramuscularly. Later, improvement was noted when multiple proteins were included.[21,24] Recently, several groups have investigated multiple protein expression approaches through the use of bacterial artificial chromosomes (BAC-VAC). This unique strategy allows one the opportunity to include multiple genes of interest from the HSV genome.[33] The BAC-VAC system appears promising, especially when used prophylactically, since it is capable of inducing multiple immune parameters that confer protection to animals from a high dose, intracerebral viral challenge. However, other HSV vaccine approaches were not directly compared to the BAC-VAC system, so the relative superiority of the approach is difficult to judge at this point.

Minigene cassettes incorporating defined epitopes from HSV have also been examined. Yu et al designed a plasmid encoding two small CTL epitopes from gB and infected cell protein (ICP) 27, along with a larger peptide from gD, which is known to induce an antibody and CD4⁺ T cell response. Immunization with this construct yielded significant CTL production, but Th and antibody responses were modest.[34]

Enhancing DNA Vaccines Against Herpesviruses

Other approaches to boost the immune response following DNA immunization have focused on the elevation of protein production from the administered plasmid DNA. Various approaches have been investigated to attain this goal. One such approach utilized the anaesthetic bupivacaine to increase protein production in the muscle.[35] Another attempted to target expression of the plasmid to muscle cells by placing viral glycoprotein expression under the control of the creatine kinase promoter.[36] Other strategies incorporated the use of liposomes to facilitate plasmid DNA delivery across the membrane.[37] Cochleates, rigid calcium-induced spiral layers of anioinic phospholipids, have a unique structure unlike liposomes. In addition, they are resistant to harsh environments, such as acidic pH, and are therefore prime candidates for enteric administration approaches. Cochleates were shown to dramatically enhance the delivery of plasmid DNA when given orally, as measured by increased immune responses.[38] Our own group explored the use of cochleates to deliver gB DNA intragastrically, but only modest improvements were noticed in comparison to naked DNA (unpublished data). In addition, when cochleates were compared to attenuated viruses in their ability to induce immunity via the same route, the response to cochleate-shuttled plasmid DNA was diminutive in comparison.

Another interesting approach to enhancing plasmid DNA uptake and resulting protein production involves what is referred to as in vivo electroporation. In this method the plasmid DNA is given intramuscularly as before, but a small current is then passed through the muscle tissue. This treatment claims to increase protein production approximately 100 fold, resulting in enhanced cell-mediated and humoral responses. This method has been utilized in studies investigating the influenza virus protein hemagglutinin[39] and the hepatitis B virus surface antigen.[40] No attempts have been made for HSV proteins, but studies are underway.

Optimization of a DNA vaccine is not solely dependent on the level of protein produced. Other factors, such as production in a professional antigen presenting cells (APC), can prove critical. In order to prime naïve T cells, it is believed that dendritic cells must present the antigen under the appropriate conditions. Therefore, gene-gun approaches, allowing plasmid DNA to be specifically delivered to the cytosol of dendritic cells of the skin (Langerhans cells), appear attractive. The Falo group demonstrated that gene-gun mediated plasmid delivery resulted in significant delivery to such dendritic cells.[41] In addition, we have further demonstrated that enhanced immunity can be achieved if dendritic cells are transfected with HSV gB DNA in vitro and then used to generate immune responses. Although this strategy yielded significant cell-mediated responses, it was still not superior to an attenuated HSV strain.[42] Moreover, such an artificial approach is neither economical nor practical.

Although direct transfection of dendritic cells is not a viable option, bacterial or viral vectors carrying the DNA of interest might be used to target these cells.[43] Several groups have examined the use of *S. typhimurium* to achieve this means. Such a system was used to deliver DNA encoding gD and ICP27 proteins from HSV and resulted in substantial increases in T cell-mediated immunity.[44] Another favorable attribute of the approach stems from the fact that the bacterial vector is given orally, making it extremely practical and easy to administer. Another approach aiming to enhance deliver is the use of alpha-viruses, such as Sindbis virus, as well as adeno-associated vectors. The Banks group has demonstrated that mice immunized with Sindbis virus encoding the gB protein from HSV induced a broad spectrum of immune responses, including virus specific antibodies, specific CTLs, as well as protection from lethal challenge.[45] Of particular interest, such a system required 100 to 1000 fold less nucleic acid to induce the same response as naked plasmid DNA approaches. In addition, Manning et al revealed that delivery of gB and gD antigens from HSV-2 via adeno-associated vectors led to increased humoral responses in comparison to plasmid-based approaches.[46]

HSV, like many viral pathogens, infects new hosts via the mucous membrane. Therefore, one would assume that mucosal immunity would prove critical in prevention of both initial infection and clearance of an ongoing recurrent lesion. Several groups have suggested that the factors controlling the induction and maintenance of mucosal immunity are separate from those governing systemic immune responses. Therefore, we investigated if plasmid DNA encoding gB DNA might generate a protective immune response if administered mucosally, since systemic immunization generally produces little HSV-specific IgA. Mucosal application of plasmid DNA encoding gB protein of HSV resulted in enhanced specific antibody, especially IgA.[47] In addition, such specific humoral responses were observed at mucosal sites distal from the site of administration. Such mucosal immunization also generated systemic humoral and T cell-mediated gB-specific responses. However, these responses were decreased in comparison to those achieved by traditional intramuscular administration, even when the mucosal adjuvant cholera toxin B was included.[47] Although immunity against HSV proteins and overall reduction of clinical disease were demonstrated at the mucosal site, such immunity was incapable of preventing viral invasion.[47] Mucosal immunity is also critical in the control of ocular infection, which, if not effectively controlled, can lead to blindness. DNA vaccines may also prove useful in attempts to control such infections of the eye. Daheshia et al determined that plasmids encoding gB of HSV-1 could be administered to the eye, expressed, and induce immunity (both local and distal).[48]

Clearly, there exists a need to improve the overall immunogenicity of mucosally administered DNA vaccines. Recent advances such as bacterial vector systems (*S. typhimurium*) or mucosal adjuvants (cholera toxin B) might circumvent these shortcomings. Other mucosal adjuvants, such as heat-labile enterotoxin B from *E. coli*, might prove effective in augmenting the generation of protective mucosal responses induced by DNA vaccines, as results involving coadministration with HSV-1 glycoproteins appear promising. Richards et al demonstrated that coadministration of an HSV-1 glycoprotein and *E. coli* enterotoxin B induced protection from herpetic stromal keratitis as well as from zosteriform spread and encephalitis.[49]

Immunomodulation of DNA Vaccine Induced Responses

Since multiple cellular and mediator events influence the magnitude and nature of immune responses, it is possible that ineffective responses induced by DNA vaccines could be explained by their failure to engage some of the crucial events necessary for the generation of a protective immune response. Accordingly, administering DNA vaccines along with the appropriate immunomodulators may serve both to enhance immunity and shift the balance of parameters induced. This notion was first assessed by the Ertl group, who demonstrated that coinoculation of a plasmid encoding a viral protein from rabies virus along with a plasmid expressing GM-CSF yielded enhanced B and T cell activity to rabies virus.[50] In contrast, coinoculation with a plasmid encoding IFN-γ instead of GM-CSF resulted in a diminished immune response. Subsequently, other groups, including our own, have tested the immunomodulatory effects of several molecules given before or in conjunction with DNA vaccines (see Table 2). In some studies protein modulators were used, but, more recently, immunomodulators administered in the form of DNA expression plasmids have been utilized.

Comprehensive studies by the Weiner group analyzed the potential modulatory effects of multiple cytokines, both in protein as well as plasmid DNA format, on the response of mice to immunization with DNA encoding HSV proteins, such as gD and gB. The group showed enhanced immune responses following administration of certain cytokines. A wide variety of cytokines were investigated, including IL-2, IL-4, IL-10, IL-12, IL-15, IL-18, IFN-α1, IFN-γ, and GM-CSF.[51-55] Moreover, cytokine plasmid coadministration could push responses to either Th1 or Th2, depending on the cytokine or combination of cytokines used. For example, Weiner's studies demonstrated that coadministration of Th1 cytokines, especially IL-12 dramatically enhanced both cell-mediated and T-helper responses to HSV antigens (gD) and conferred increased levels of protection to such mice from a lethal challenge with HSV.[51,53] The Carr group demonstrated that administration of a plasmid encoding IFN-α1 enhanced the survival rate of HSV-2 infected mice.[54] They attributed the observed reduction of viral replication to the ability of IFN-α1 to antagonize viral replication.

Other studies have evaluated the effects of the coadministration of plasmids encoding various immunomodulators, chemokines, and costimulatory molecules, such as CD80, CD86, LFA-3, and ICAM-1 on the magnitude and quality of the immune response (See Table 2).[56,57] Weiner's group also examined the effects of various chemokines, including IL-8, MCP-1, MIP1-α, and RANTES.[58] It was observed that IL-8 and RANTES plasmid DNA, when given in conjunction with plasmid DNA encoding gD, significantly enhanced cellular immune responses and increased the protective efficacy of the DNA vaccine.[58] Our own group has also examined the effects of the coadministration of plasmids encoding chemokines along with a plasmid encoding an HSV antigen (gB). Eo et al showed that plasmids encoding MIP-1α or MIP-2,

Table 2. Immunomodulation of DNA vaccines against herpesviruses

Immunomodulator	Antigen	IgG Level	Th Response	Mortality	Reference
Th1 Type Cytokines					
IL-2 (Th0)	gD	+	+	-	52
IL-12	gB, gD	-	+Th1	-	52,53
IL-15	gD	+	ND[a]	-	52
IL-18	gB, gD	+	+Th1	-	52, UP[b]
Th2 Type Cytokines					
IL-4	gD	ND	+Th2	+	52
IL-10	gD	ND	ND	+	52,70,72
Others					
TGF-β	HSV	ND	-	+	71
GM-CSF	gB, gD	+	+Th1 (gD)	-	51,55
Chemokines					
MIP-1α	gB	+	+Th1	-	59
MIP-1β	gB	ND	+Th2	ND	59
MIP-2	gB	ND	+Th1	-	59
MCP-1	gB	+	+Th2	ND	59
RANTES	gD	ND	+Th1	-	58
IL-8	gD	ND	+Th1	-	58
SLC	gB	+	+Th1	ND	60
ELC	gB	+	+Th1	ND	60
Costimulators					
CD80	gD	ND	+Th1	-	56
CD86	gD	ND	+Th2	ND	56
Adhesion Molecules					
LFA-3	gD	+	+Th1	-	57
ICAM-1	gD	ND	ND	ND	57

[a]ND, none detected
[b] UP, unpublished

along with DNA encoding gB, augmented the gB-specific immune response and fostered a Th1 profile, leading to increased resistance to infection with HSV. Conversely, plasmids encoding the chemokines MIP1-α and MCP-1 pushed the immune response to a Th2 profile, increasing susceptibility and mortality.[59] Additional studies by Eo et al addressed the effects of the chemokines SLC and ELC, which are both CCR7 ligands. Coadministration of either along with DNA encoding an HSV protein resulted in an enhanced Th1 response, with the augmentation of both cell-mediated responses as well as Th1 cytokine production.[60] An increase in the concentration of dendritic cells in the secondary lymphoid tissues was also observed.

DNA vaccines, due to their prokaryotic origins, possess immunostimulatory sequence motifs that play a critical role in their ability to generate immune responses. Such sequences are referred to as CpG motifs and are characterized by two 5'purines and two 3' pyrimidines surrounding the central CG sequence, such as AACGTT.[61] These motifs are 20 fold more common in prokaryotic than eukaryotic DNA, due to differences in usage and methylation patterns.[62] CpG motifs have been shown to directly activate a variety of immune cells including B cells, macrophages, dendritic cells, and T cells.[63-65] Their ability to induce cytokine production in professional antigen presenting cells (APCs) has proved particularly interesting. Such CpG motifs bind to Toll-like receptor (TLR) 9 and induce the production of proinflammatory cytokines such as IL-1β, TNF-α, IL-6, and IL-12.[61,62,66] Addition of CpG motifs (up to a point) to a plasmid results in increased immunogenicity, and removal of all CpG residues significantly reduces immune reactivity.[67] Therefore, the relative amount of CpG sequence motifs might determine the overall immunogenicity of the plasmid DNA, since such CpG sequences appear to trigger an innate immune response that polarizes the adaptive response towards a Th1 phenotype.[68] Accordingly, the Rosenthal group investigated the ability of CpG sequences to augment an HSV DNA vaccination program by intranasally administering gB protein with CpG oligonucleotides.[69] They then found elevated levels of gB-specific IgA in vaginal washes and protection against HSV infection was demonstrated. However, elevated CTL activity was not demonstrated. Our own laboratory has undertaken preliminary experiments that suggest that immunization with a peptide from HSV (gB498-505) in conjunction with CpG motifs results in the generation of a protective CTL response (unpublished data). Unfortunately, this approach is limited, since a defined epitope must be used. In addition, relatively large amounts, in comparison to live virus or plasmid DNA, must be administered to achieve significant CTL responses.

Clearly, plasmid DNA encoding immunomodulators represents a powerful tool to manipulate the immune response. The current state of the field is at the stage of documenting how and which individual molecules can achieve the various desired effects. Notable virtues of the DNA vaccine approach include the fact that it is apparently unaffected by pre-existing immunity, and it is technically simple to administer multiple expression plasmids at the same time. However, it is more challenging to achieve the desired expression levels in a defined tissue site. Improvements in the systems of delivery and control of expression could help to address these concerns. This will likely be critical before the use of plasmid DNA encoding cytokines or chemokines is suggested for use in man, as expression at unwanted sites could yield detrimental side-effects.

Immunomodulation: Using DNA Vaccines to Ameliorate Immunopathology

In certain instances, immune responses against HSV-1 in the eye result in infiltration of immune cells into the stroma, causing opacity and edema of the cornea. The cornea may become highly vascularized and thickened, resulting in a blinding condition known as herpetic stromal keratitis. In contrast to Th1 cytokines, coinjection of Th2 type cytokine plasmids, such as IL-4 and IL-10, results in an increased susceptibility to challenge with HSV. In our own early studies Chun et al examined the modulatory effects of expression plasmids encoding various cytokines on the subsequent pathological immune response resulting from ocular infection with HSV.[70]

Our previous work indicated that the modulatory effects of a cytokine, such as IL-10, could be achieved if treatments were given via the mucosal as well as systemic routes.[70] Indeed, as shown later by Kuklin et al, mucosal administration of an immunomodulatory cytokine exerted more profound effects on the mucosal immune response than a systemic approach. In such studies the mucosal administration of plasmid DNA encoding TGF-β resulted in increased susceptibility to vaginal challenge with HSV. However, the prophylactic administration of TGF-β plasmid DNA resulted in a clinical reduction of ocular inflammation following infection of the eye with HSV.[71] In some instances administration of DNA encoding cytokines can yield a desired effect on an ongoing disease process, even after it has been initiated. Thus, Daheshia et al showed that an ongoing ocular inflammatory reaction could be suppressed by the topical administration of plasmid DNA encoding IL-10.[72] We believe that this long-term suppression of virus-induced inflammation is mediated by the silencing of inflammatory CD4+ T cells. Such silenced T cells appear to be in an anergic state, and T cells exhibiting a regulatory phenotype (IL-10 producing) may mediate this phenomenon. However, bystander suppression mediated by IL-10 producing regulatory T cells (Tr1) appeared to not be responsible, since neither an anti-IL-10 monoclonal antibody (mAb) nor an anti-IL-10 receptor mAb could block the suppression. The nature of the silencing appears to be of a reversible nature, since addition of exogenous IL-2 restores T cell reactivity (unpublished data).

The Carr group has investigated an alternative means for controlling ocular HSV lesions. They noticed that topical ocular administration of plasmid DNA encoding IFN-α1 granted resistance to ocular infection with HSV-1. This protection was apparently mediated by CD4+ and CD8+ T cells, but IFN-α1 may also be exerting direct anti-viral properties. [73]

Prime-Boost Strategies

Although DNA vaccines offer a convenient approach for the expression of a myriad of viral antigens, they generally fail to induce the robust immune response and subsequent memory correlating to natural infection. However, several groups have recently linked the flexibility of a DNA approach for priming and the immunostimulatory characteristics associated with viral infections for boosting of the desired response. For instance, Ramshaw and colleagues demonstrated that if plasmid DNA encoding hemagglutinin (HA) protein from influenza was administered as a priming agent, followed by boosting with an attenuated poxvirus expressing the same protein, an effective immune response could be generated.[74,75] Strangely, reversal of this order of prime-boost abrogated protection. Such approaches were soon applied to HSV DNA vaccination programs. The Weiner group investigated if priming with either plasmid DNA encoding gD or gD protein, followed by boosting with the opposite, generated both antibody as well as Th1 type cytokine responses. Such studies revealed that DNA priming-protein boosting resulted in production of IgG2a and Th1 cytokines, such as IFN-γ.[76] Conversely, protein priming-DNA boosting yielded primarily an IgG1 isotype (Th2 indicator).

In addition, our own laboratory assessed the potential of the prime-boost approach for the generation of mucosal immunity (see Table 3). Mice were primed intranasally with recombinant vaccinia virus expressing the gB protein, followed by boosting with plasmid DNA encoding gB and vice versa.[77] In such a mucosal prime-boost approach priming with virus and boosting with plasmid DNA, although opposite in order from prior systemic administration regimens, elicited strong systemic as well as mucosal immune responses and granted resistance to immunized mice against genital HSV infection. Unexpectedly, the observed protective mucosal immunity was vastly more pronounced when DNA vaccines were utilized in the boosting step, not the priming one. More importantly, the demonstration of a strong mucosal response, which was undetectable following systemic prime-boost strategies, may significantly enhance the protective efficacy of such an approach. Since HSV infects mucosal tissues first, a strong mucosal immune response will most likely prove critical in efforts to prevent or contain infection. Therefore, the mucosal route offers an effective, attractive avenue for a prime-boost approach.

The prime-boost approach has also been applied to a number of other viruses. The Whalley group employed the prime–boost strategy in a novel manner, using DNA to prime and recombinant baculovirus-expressed gD protein of equine herpesvirus for booster immunization.[78] Results indicated that mice inoculated with the combination of gD DNA and protein had enhanced antibody titers and more efficient viral clearance from the lungs. Therefore, the prime-boost approach may prove to be a valuable tool in efforts to induce protective antibody and cell-mediated responses against HSV and other pathogens.

DNA Vaccines and the Role of Cross-Priming

The original observation that injection of plasmid DNA into myocytes primed immune responses was rather perplexing to most immunologists, since myocytes are not professional APCs. However, it is now well known that once plasmid DNA is injected, no matter the route, it is widely disseminated throughout the tissues of the body.[70] More relevant, it has been demonstrated that professional APCs, such as dendritic cells (DCs), take up the plasmid DNA

Table 3. Comparison of mucosal and systemic prime-boost strategies [a]

Priming	Boosting	Systemic Immune Response	Mucosal Immune Response
Mucosal			
gBDNA	gBDNA	+	+/-
gBDNA	rvvgB	++	++
rvvgB	gBDNA	++++	++++
rvvgB	rvvgB	++	++
Systemic			
gBDNA	gBDNA	++	+/-
gBDNA	rvvgB	+++	+/-
rvvgB	gBDNA	++	+/-
rvvgB	rvvgB	+	+/-

[a]BALB/c mice (6 per group) were immunized mucosally (i.n.) or systemically (i.m.) with plasmid DNA encoding gB protein from HSV-1 (gBDNA) or a recombinant vaccinia virus expressing gB (rvvgB) and boosted 10 days later via the same route used for priming. Eight weeks after boosting the mice were challenged vaginally with HSV-1 McKrae strain. After 4 days the iliac lymph node (systemic response) and the vaginal tract (mucosal response) were removed and analyzed for gB-specific IgA and IgG antibodies and IFN-γ producing cells as determined by ELISPOT.

and express the protein, eliciting specific immune responses. The Falo group upon gene gun administration of plasmid initially observed this.[38] Later, Bona and colleagues demonstrated that DCs take up DNA following injection in the skin or muscle, travel to the draining lymph node, and prime T cell responses.[79] Although mature DCs are known to play a primary role in the generation of primary T cell responses, they are present in low numbers in muscle tissue and are poorly phagocytic.[80] Therefore, other cells, such as macrophages might play a crucial role in initial uptake, expression, and transport of the processed protein antigen.

The evidence for cross-priming or cross-presentation as a significant factor has been gaining more acceptance over the past few years. This suggests that the plasmid DNA is initially ingested and expressed in one cell (such as a myocyte, endothelial cell, or macrophage) and then transferred by some means to another cell (such as a DCs). The receiving cell is then able to present the exogenous antigen via the endogenous pathway, leading to priming of CTL.[81] Cross-priming as a potential factor in genetic vaccination approaches was originally implicated following a series of bone marrow radiation chimera experiments carried out by the Walker group.[82] These studies revealed that if splenocytes from immunocompetent F1H-2^{bxd} mice were transferred to either H-2^d or H-2^b hosts (both immunoincompetent due to radiation), the resulting CTL response was restricted to the hosts' haplotype. This suggests that the immune cell responsible for inducing the response likely acquired the antigen from the donor cell. Further support for this cross-presentation event was provided by the Liu group, who demonstrated that cross-presentation occurs in vivo and involves APCs from bone marrow.[83] In addition, the magnitude of the immune response may depend on the expression of antigen by non-lymphoid cells and subsequent transfer to APCs.[84] Bypassing the need for transfer, the Germain group demonstrated that CTL priming could be enhanced by direct augmentation of transgene product expression in DCs when plasmid was administered via gene gun.[85] Another mechanism that might be responsible for cross-priming was elucidated by studies by the Bhardwaj group, who showed that apoptotic bodies could be taken up by DCs. Such DCs were then able to prime CTL responses against antigen acquired from these apoptotic bodies.[86] This

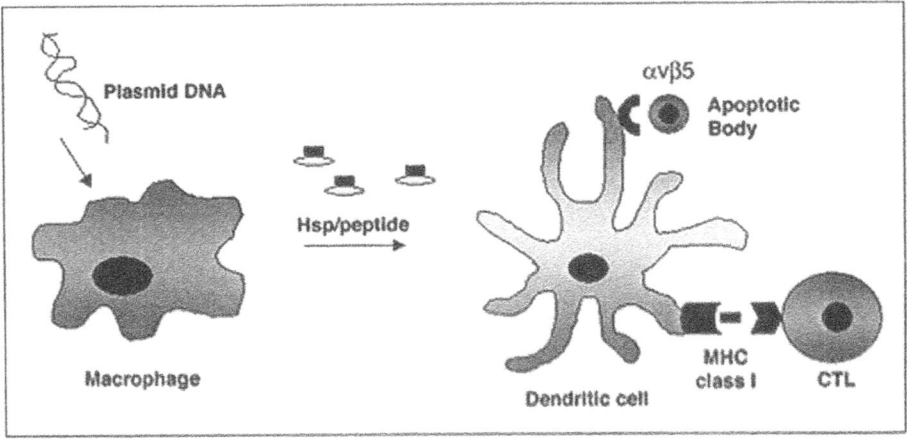

Figure 1. Possible cross-priming mechanisms in genetic vaccination.

mechanism cannot be ruled out in efforts to explain the cross-priming events that take place after DNA injection.

Our own group has also uncovered a possible cross-priming event, at least in vitro. It involves macrophages, which take up and express the plasmid DNA, and DCs, which acquire the protein antigen from the macrophages (Fig. 1). Such studies demonstrated that when macrophages were exposed to plasmid DNA encoding an HSV protein, gB, they released a factor that allowed DCs to prime naïve T cells and to generate CTLs. Further investigation revealed that this material, termed macrophage-released factor (MRF), was neither DNA nor RNA in composition. In fact, it most likely is a chaperone bound peptide conjugate, and Hsp70 was shown to be the dominant chaperone involved.[87] Moreover, rhsp70, when coupled to an HSV peptide CTL epitope, has been shown to induce CTL responses (unpublished data). In fact, CpG sequences may stimulate the production of MRF.

Conclusions and Speculations

DNA vaccines have been investigated and shown to confer immunity against herpesvirus infections. The majority of observations were made with vaccines designed to protect against HSV infection in laboratory animals, with the bulk of such studies emphasizing the type of immune response induced and whether or not immunity to viral challenge resulted. Concerning efforts involving plasmid DNA encoding single HSV proteins, levels of immunity, when appropriate comparisons were made, were lower than those achieved by vaccination with UV inactivated virus or attenuated vaccinia viruses encoding HSV proteins. Reasons for the apparent minimal immunogenicity remain to be defined, particularly if the DNA vaccine approach is to become the sought after practical vaccination strategy for herpesvirus infections.

Improvements in immunogenicity have arisen when various immunostimulatory regimens were administered along with the DNA vaccine. However, it is unclear if immune stimulators, such as cytokines and chemokines, will ever be acceptable for incorporation into vaccines given to healthy individuals. The most promising approach for effective vaccination using DNA vaccines appears to be the prime-boost approach. When priming and boosting vaccines are administered systemically, it is critical that the DNA is used in the priming step. Reversing the sequence, for unknown reasons, yields inferior responses. A thorough exploration of the various routes of priming or different types of boosting has yet to be made. However, it was observed that if prime-boost immunization was carried out by mucosal administration, immunity was generated at mucosal as well as systemic sites. Indeed, the systemic route failed to induce significant mucosal immunity and preferential barrier protection, a likely shortcoming with agents such as those of the herpes family, which generally infect via the mucosal route. The most promising results from the mucosal prime-boost strategy were obtained from reversing the order of vaccine administration previously demonstrated to be optimal for systemic priming. Thus, in this instance, DNA vaccines were most effective when used to boost the response rather than prime it. An explanation for this observation is not forthcoming, but it is assumed to relate to the fact that DNA vaccines are unaffected by preexisting immunity and that optimal memory responses usually require a vigorous priming step. At present, mucosal prime-boost represents an encouraging approach that might be further improved by the incorporation of extra effects such as the use of physiological adjuvants during priming or boosting. The final issue will be to establish if results in the mouse translate to use in human and veterinary medicine. DNA vaccines could prove to be the long sought after solution to control herpesviruses. Only the results of controlled clinical trials with therapeutic as well as prophylactic DNA vaccines will evaluate if the promise can become reality.

References

1. Kamiya H, Ito M. Update on varicella vaccine. Curr Opin Pediatr 1999; 1:3-8.
2. Raeder CK, Hayney MS. Immunology of varicella immunization in the elderly. Ann Pharmacother 2000; 34:228-234.
3. Bittle JL, Rubic WJ. Studies of feline viral rhinotracheitis virus. Vet Med Small Anim Clin 1974; 69:1503-1505.
4. Straub OC. Infectious bovine rhinotracheitis virus: History and recent developments. Dev Biol Stand 1975; 28:530-533.
5. Seimenis A, Menasse I. Studies on vaccination against infectious laryngotracheitis by the drinking water. Dev Biol Stand 1976; 33:328-331.
6. Studdert MJ. Vaccines for equine herpesvirus type 1. Vet Rec 1983; 112:334.
7. Kreager KS. Chicken industry strategies for control of tumor virus infections. Poult Sci 1998; 77:1213-1216.
8. Bernstein DI, Stanberry LR. Herpes simplex virus vaccines. Vaccine 1999; 17:1681-1689.
9. Milligan GN, Bernstein DI, Bourne N. T-lymphocytes are required for protection of the vaginal mucosa and sensory ganglia of immune mice against reinfection with herpes simplex virus type 2. J Immunol 1998; 160:6093-6100.
10. Alcami A, Koszinowski JH. Viral mechanisms of immune evasion. Immunol Today 2000; 21:447-455.
11. Daheshia M, Feldman LT, Rouse BT. Herpes simplex virus latency and the immune response. Curr Opin Microbiol 1998; 1:430-435.
12. Deshpande S, Kumaraguru U, Rouse BT. Why do we lack an effective vaccine against herpes simplex virus? Microbes Infect 2000; 2:973-978.
13. Cox GJM, Zamb TJ, Babiuk LA. Bovine herpes virus I. Immune responses in mice and cattle injected with plasmid DNA. J Virol 1993; 67:5664-5667.
14. Manickan E, Rouse RJD, Yu Z et al. Genetic immunization against herpes simplex virus: protection is mediated by CD4+ T-lymphocytes. J Immunol 1995; 155:259-265.
15. Manickan E, Yu Z, Rouse RJD et al. Induction of protective immunity against herpes simplex virus with DNA encoding the immediate early protein ICP27. Viral Immunol 1995; 8:53-61.
16. Bourne N, Stanberry LR, Bernstein DI et al. DNA immunization against experimental genital herpes simplex virus infection. J Inf Dis 1996; 173:800-807.
17. Manickan E, Francottle M, Kuklin NA et al. Vaccination with recombinant vaccinia virus expressing ICP27 induces protective immunity against herpes simplex virus through CD4+ Th1 T-cells. J Virol 1995; 69:4711-4716.
18. Kriesal JD, Spruance SL, Daynes RA et al. Nucleic acid vaccine encoding gD2 protects mice from herpes simplex virus type 2 disease. J Inf Dis 1996; 173:536-541.
19. Ghiasi H, Cai S, Slanina S et al. Vaccination of mice with HSV type 1 glycoprotein D DNA produces low levels of protection against lethal HSV-1 challenge. Antiviral Res 1995; 28:147-157.
20. Nass PH, Elkins KL, Weir JP. Antibody response and protective capacity of plasmid vaccines expressing three different herpes simplex virus glycoproteins. J Inf Dis 1998; 178:611-617.
21. Mester JC, Twomey TA, Tepe ET et al. Immunity induced by DNA immunization with herpes simplex virus type 2 glycoproteins B and C. Vaccine 1999; 18:875-883.

22. Higgins TJ, Herold KM, Arnold RL et al. Plasmid DNA-expressed secreted and nonsecreted forms of herpes simplex virus glycoprotein D2 induce different types of immune responses. J Inf Dis 2000; 182:1311-1320.
23. Nass PH, Elkins KL, Weir JP. Protective immunity against herpes simplex virus generated by DNA vaccination compared to natural infection. Vaccine 2001; 19:1538-1546.
24. McClements WL, Armstrong ME, Keys RD et al. Immunzation with DNA vaccine encoding glycoprotein D or glycoprotein B, alone or in combination, induces protective immunity in animal models of herpes simplex virus-2 disease. Proc Natl Acad Sci USA 1996; 93:11414-11420.
25. Masser M, Haumont M, Garcia L et al. Differential neutralizing antibody responses to varicella-zoster virus glycoproteins B and E following naked DNA immunization. Viral Immunol 1999; 12:227-236.
26. Pande H, Campo K, Tanamachi B et al. Direct DNA immunization of mice with plasmid DNA encoding the tegument protein pp65 (ppUL83) of human cytomegalovirus induces high levels of circulating antibody to the encoded protein. Scand J Infect Dis Suppl 1995; 99:117-120.
27. Gonzales Armas JC, Morello CS, Cranmer LD et al. DNA immunization confers protection against murine cytomegalovirus infection. J Virol 1996; 70:7921-7928.
28. Ruitenberg KM, Walker C, Wellington JE et al. Potential of DNA-mediated vaccination for equine herpesvirus 1. Vet Microbiol 1999; 68:35-48.
29. Manickan E, Yu Z, Rouse BT. DNA immunization of neonates induces immunity despite the presence of maternal antibody. J Clin Invest 1997; 100:2371-2375.
30. Monteil M, LePotier MF, Guillotin J et al. Genetic immunization of seronegative one day old piglets against pseudorabies induces neutralizing antibodies but not protection and is ineffective in piglets from immune dams. Vet Res 1996; 27:443-452.
31. GerdtsV, Jons A, Macoschey B et al. Protection of pigs against Aujeszky's disease by DNA immunization. J Gen Virol 1997; 78:2139-2146.
32. Gerdts V, Babiuk LA, van Drunen Little-van den Hurk S et al. Fetal immunization by a DNA vaccine delivered into the oral cavity. Nat Med 2000; 6:929-932.
33. Suter M, Lew AM, Grob P et al.BAC-VAC, a novel generation of (DNA) vaccines: A bacterial artificial chromosome (BAC) containing a replication-competent, packaging-defective virus genome induces protective immunity against herpes simplex virus 1. Proc Natl Acad Sci USA 1999; 96:12697-12702.
34. Yu Z, Karem KL, Kanangat S et al. Protection by minigenes: a novel approach of DNA vaccines. Vaccine 1998; 16:1660-1667.
35. Pachuk CJ, Arnold R, Herold K et al. Humoral and cellular immune responses to herpes simplex virus-2 glycoprotein D generated by facilitated DNA immunization of mice. Curr Top Microbiol Immunol 1998; 226:79-89.
36. Gebhard JR, Zhu J, Cao X et al.DNA immunization utilizing a herpes simplex virus type 2 myogenic DNA vaccine protects mice from mortality and prevents genital herpes. Vaccine 2000; 18:1837-1846.
37. Gould-Fogerite S, Mannino RJ. Mucosal and systemic immunization using cochleate and liposome vaccines. J Liposome Res 1996; 2:357-379.
38. Zarif L, Mannino RJ. Cochleates. Lipid-based vehicles for gene-delivery: Concept, achievements, and future directions. Adv Exp Med Biol 2000; 465:83-93.
39. Kadowski SE, Chen Z, Asanuma H et al. Protection against influenza virus infection in mice immunized by administration of hemagglutinin-expressing DNAs with electroporation. Vaccine 2000; 18:2779-2788.
40. Widera G, Austin M, Rabussay D et al. Increased DNA vaccine delivery and immunogenicity by electroporation in vivo. J Immunol 2000; 164:4635-4640.
41. Condon C, Watkins SC, Celluzzi CM et al. DNA based immunization by in vivo transfection of dendritic cells. Nat Med 1996; 2:1122-1128.
42. Manickan E, Kanangat S, Rouse RJ et al. Enhancement of immune response to naked DNA vaccine by immunization with transfected dendritic cells. J Leukoc Biol 1997; 61:125-132.
43. Darji A, Guzman CA, Gerstel B et al. Oral somatic transgene vaccination using attenuated S. typhimurium. Cell 1997; 91:765-775.
44. Karem K, Bowen J, Kuklin N et al. Protective immunity against herpes simplex virus (HSV) type 1 following oral administration of recombinant Salmonella typhimurium vaccine strains expressing HSV antigens. J Gen Virol 1997; 78:427-434.
45. Hariharen MJ, Driver DA, Townsend K et al. DNA immunization against herpes simplex virus: Enhanced efficacy using a sindbis virus-based vector. J Virol 1998; 72:950-958.
46. Manning WC, Paliard X, Zhou S et al. Genetic immunization with adeno-associated virus vectors expressing herpes simplex virus type 2 glycoproteins B and D. J Virol 1997; 71:7960-7962.

47. Kuklin NA, Daheshia M, Karem K et al. Induction of mucosal immunity against herpes simplex virus by DNA immunization. J Virol 1997; 71:3138-3145.
48. Daheshia M, Kuklin N, Manickan E et al. Immune induction and modulation by topical ocular administration of plasmid DNA encoding antigens and cytokines. Vaccine 1998; 16:1103-1110.
49. Richards CM, Aman AT, Hirst TR et al. Protective mucosal immunity to ocular herpes simplex virus type 1 infection in mice by using *Escherichia coli* heat-labile enterotoxin B subunit as an adjuvant. J Virol 2001; 75:1664-1671.
50. Xiang Z, Ertl HC. Manipulation of the immune response to a plasmid-encoded viral antigen by coinoculation with plasmid expressing cytokines. Immunity 1995; 2:129-135.
51. Sin JI, Kim JI, Ugen KE et al. Enhancement of protective humoral (Th2) and cell-mediated (Th1) immune responses against herpes simplex virus-2 through co-delivery of granulocyte-macrophage colony stimulating factor expression cassettes. Eur J Immunol 1998; 28:3530-3540.
52. Sin JI, Kim JI, Boyer JD et al. In vivo modulation of vaccine-induced immune responses toward a Th1 phenotype increase potency and vaccine effectiveness in a herpes simplex virus type 2 mouse model. J Virol 1999; 73:501-509.
53. Sin JI, Kim JI, Arnold RL et al. IL-12 gene as a DNA vaccine adjuvant in a herpes mouse model: IL-12 enhances Th1-type CD4+ T cell-mediated protective immunity against herpes simplex virus-2 challenge. J Immunol 1999; 162:2912-2921.
54. Harle P, Noisakran S, Carr DJ. The application of a plasmid DNA encoding IFN-alpha1 postinfection enhances cumulative survival of herpes simplex virus type 2 vaginally infected mice. J Immunol 2001; 166:1803-1812.
55. Flo J, Beatriz Perez A, Tisminetzky S et al. Superiority of intramuscular route and full length glycoprotein D for DNA vaccination against herpes simplex 2. Enhancement of protection by the co-delivery of the GM-CSF gene. Vaccine 2000; 18:3242-3253.
56. Flo J, Tisminetzky S, Baralle F. Modulation of the immune response to DNA vaccine by co-delivery of costimulatory molecules. Immunology 2000; 100:259-269.
57. Sin JI, Kim J, Dang K et al. LFA-3 plasmid DNA enhances antigen specific humoral-and-cellular-mediated protective immunity against herpes simplex virus-2 in vivo: involvement of CD4+ T cells in protection. Cell Immunol 2000; 203:19-28.
58. Sin JI, Kim JI, Pachuk C et al. DNA vaccines encoding interleukin-8 and RANTES enhance specific Th1-type CD4 (+) T-cell-mediated protective immunity against herpes simplex virus type 2 in vivo. J Virol 2000; 74:11173-11180.
59. Eo SK, Lee S, Chun S et al. Modulation of immunity against herpes simplex virus infection via mucosal genetic transfer of plasmid DNA encoding chemokines. J Virol 2001; 75:569-578.
60. Eo SK, Lee S, Kumaraguru U et al. Immunopotentiating of DNA vaccine against herpes simplex virus via co-delivery of plasmid DNA expressing CCR7 ligands. (submitted).
61. Krieg AM. The role of CpG motifs in innate immunity. Curr Opin Immunol 2000; 12:35-43.
62. Krieg AM. Immune effects and mechanisms of action of CpG motifs. Vaccine 2001; 19:618-622.
63. Krieg AM, A-K Yi, Matson S et al. CpG motifs in bacterial DNA trigger direct B cell differentiation. Nature 1995; 374:546-549.
64. Jacob T, Walker DS, Krieg AM et al. Activation of cutaneous dendritic cells by CpG-containing oligodeoxynucleotides: a role for dendritic cells in the augmentation of Th1 responses by immunostimulating DNA. J Immunol 1998; 161:3042-3049.
65. Klinman DM, Yi AK, Beaucage SL et al. CpG motifs present in bacterial DNA rapidly induced lymphocytes to secrete interleukin-6, interleukin-12, and interferon gamma. Proc Natl Acad Sci USA 1996; 93:2879-2883.
66. Hemmi H, Takeuchi O, Kawai T et al. A toll-like receptor recognizes bacterial DNA. Nature 2000; 408:740-744.
67. Klinman DM, Yamshchikov G, Ishigatsubo Y. Contribution of CpG motifs to the immunogenicity of DNA vaccines. J Immunol 1997; 158:3635-3639.
68. Chu RS, Targoni OS, Krieg AM et al. CpG oligodeoxynucleotides act as adjuvants that switch on T helper 1 (Th1) immunity. J Exp Med 1997; 186:1623-1631.
69. Gallichan WS, Woolstencroft RN, Gaurasci T et al. Mucosal immunization with CpG ODN induces significant IgA and protection against HSV-2 in the genital tract. Proceedings of the 25th International Herpesvirus Workshop 2000; 5:59.
70. Chun S, Daheshia M, Lee S et al. Distribution fate and mechanism of immune modulation following mucosal delivery of plasmid DNA encoding IL-10. J Immunol 1999; 163:2393-2402.
71. Kuklin NA, Daheshia M, Chun S et al. Immunomodulation by mucosal gene transfer using TGF-β DNA. J Clin Invest 1998; 102:438-444.
72. Daheshia M, Kuklin N, Kanangat S et al. Suppression of ongoing ocular inflammatory disease by topical administration of plasmid DNA encoding IL-10. J Immunol 1997; 159:1945-1952.

73. Noisakran S, Carr DJ. Plasmid DNA encoding IFN-alpha1 antagonizes herpes simplex virus type 1 ocular infection through CD4+ and CD8+ T lymphocytes. J Immunol 2000; 164:6435-6443.
74. Ramsay AJ, Leong KH, Ramshaw IA. DNA vaccination against virus infection and enhancement of antiviral immunity following consecutive immunization with DNA and viral vectors. Immunol Cell Biol 1997; 75:382-388.
75. Ramshaw IA, Ramsay AJ. The prime-boost strategy: Exciting prospects for improved vaccination. Immunol Today 2000; 21:163-165.
76. Sin JI, Bagarazzi M, Pachuk C et al. DNA-priming-protein boosting enhances both antigen-specific antibody and Th1-type cellular immune responses in a murine herpes simplex virus-2 gD vaccine model. DNA Cell Biol 1999; 18:771-779.
77. Eo SK, Gierynska M, Kamar AA et al. Prime-boost immunization mith DNA vaccine: Mucosal route of administration changes the rules. J Immunol 2001; In press.
78. Ruitenberg KM, Walker C, Lore DN et al. A prime-boost immunization strategy with DNA and recombinat baculovirus-expressed protein enhances protective immunogenicity of glycoprotein D of equine herpesvirus 1 in naive and infection-primed mice. Vaccine 2000; 18:1367-1373.
79. Casares S, Inaba K, Brumeanu TD et al. Antigen presentation by dendritic cells after immunization with DNA encoding a major histocompatibility complex class II-restricted viral epitope. J Exp Med 1997; 186:1481-1486.
80. Witmer-Pack MD, Croley MT, Inaba K et al. Macrophage, but not dendritic cells, accumulate colloidal carbon following administration in situ. J Cell Sci 1993; 105:965-973.
81. Bevan MJ. Antigen presentation to cytotoxic T lymphocytes in vivo. J Exp Med 1995; 182:639-641.
82. Doe B, Selby M, Barnett S et al. Induction of cytotoxic T lymphocytes by intramuscular immunization with plasmid DNA is facilitated by bone-marrow derived cells. Proc Natl Acad Sci USA 1996; 93:8578-8583.
83. Fu TM, Ulmer JB, Caulfield MJ et al. Priming of cytotoxic T lymphocytes by DNA vaccines: Requirement for professional antigen presenting cells and evidence for antigen transfer from myocytes. Mol Med 1997; 3:362-371.
84. Corr M, von-Damm A, Lee DJ et al. In vivo priming by DNA injection occurs predominantly by antigen transfer. J Immunol 1999; 163:4721-4727.
85. Porgador A, Irvine KR, Iwasaki A et al. Predominant role for directly transfected dendritic cells in antigen presentation to CD8+ T cells after gene gun immunization. J Exp Med 1998; 188:1075-1082.
86. Albert MC, Sauter B, Bhardwaj N. Dendritic cells acquire antigen from apoptotic cells and induce class-I-restricted CTLs. Nature 1998; 392:86-89.
87. Kumaraguru U, Rouse RJ, Nair SK et al. Involvement of an ATP-dependent peptide chaperone in cross-presentation after DNA immunization. J Immunol 2000; 165:750-759.

CHAPTER 9

Genetic Immunization Against HIV

Britta Wahren, Karl Ljungberg, Anne Kjerrström Zuber and Bartek Zuber

Worldwide Spread of HIV

The HIV virus and its subtypes are spread worldwide. AIDS has become the world's most threatening infectious disease. The picture of the epidemic has changed: several subtypes and recombinants of subtypes are becoming frequent in all countries. Treatment with sophisticated drugs improves the HIV-infected patient's condition. It reduces viral replication for a time but does not eliminate the virus. The drugs have side effects, are expensive and associated with the emergence of resistant virus strains. In some regions where antiviral drugs have been used frequently, the resistant strains are more common than the original wild-type strains.

Safe and effective vaccines for prophylaxis and therapy are essential to control the global epidemic.[1] Unique features make DNA vaccines particularly interesting for HIV immunization. The ease with which a new DNA sequence can be added is important because of the high mutation rate of HIV virus. Several genes can be combined in a single vaccine to be expressed in the same cell. Alternatively, plasmids with single genes can be mixed in the vaccine and expressed in different cells of the patient. The antigen is presented by both MHC class I and II molecules, stimulating all types of immune responses.[2-5] The DNA also induces a relatively long-lived immunological memory, particularly of the T-cell mediated type.

Desirable Immune Responses

No single immune mechanism has been found to be prognostic for either protection against HIV infection or disease progression. Theoretically, a good first-line defence should include local IgA and IgG antibodies, together with a cytotoxic T-cell (CTL) response. Strongly reacting cross-neutralizing antibody together with CTL should confer at least partial protection. DNA genetic immunization appears second only to attenuated viruses in the added capacities to induce both humoral and cellular memory immune responses.[6]

In therapeutic immunization, the most effective way of improving the clinical course appears to be to counter the deterioration of CD4+ cell levels and reduce the viral load.[7,8] There are indications that IgA antibodies against the HIV envelope gp160 together with CTLs are protective at the mucosal lining.[9,10] Antibodies are active against the free virus particles and may also serve to inhibit toxic substances such as secreted HIV nef and tat molecules.

CTL responses are highly desirable since they can eliminate or reduce virus production by killing virus-producing cells. Many experiments have shown that the initial DNA immunization is particularly effective in inducing cytotoxic CD8+ T cells, a characteristic it shares with viral recombinants. Vectors such as a canarypox vectors containing HIV genes of subtype B were shown to elicit broad CTL responses that recognized genetically diverse HIV-1 strains.[11] The effects of depleting CD8+ cytotoxic cells were demonstrated in primates,[12] where the lack of CD8+ cells led to an increased viral load in infected macaques.[12,13]

DNA Vaccines, edited by Hildegund C. J. Ertl. ©2003 Eurekah.com and Kluwer Academic / Plenum Publishers.

Different CTL epitopes appear to be recognized during different phases of infection, irrespective of the HLA phenotype of the host. Exposed noninfected women showed lower and more varied CTL epitope responses than chronically infected women. The longer the exposure to potentially infectious sexual encounters, the more frequent were the responses.[14] Indeed, the primary immune CTL response of patients with a new infection differs from the later reponses. Thus, acutely infected patients of the A.2 phenotype recognized diverse epitopes, while chronically infected persons mainly recognized CTL epitopes of Gag.[15] Primary responses may accordingly be more diverse, indicating that the virus exposes different peptides during the initial infection. This gives the development of vaccines yet another challenge.

One approach to accommodating as many epitopes from as many HIV strains as possible has been to devise miniepitopic sequences. Such constructs can induce strong CTL and even antibody responses[16-18] and have been prepared for human use with multiple CTL epitopes.[19] Mini-CTL epitope sites appear to be more efficient for CTL induction than whole DNA.[20] The difficulty in covering all polymorphic sites in humans and the ease with which a virus escapes the CTL responses prompt a cautious use of this approach for vaccination in humans. It seems that vaccine failure may sometimes result if the vaccine sequence varies from the virus strain by as little as a few amino acids.[21,22]

Another effective way of destroying virally infected cells would be the destruction of changed cell surfaces by NK cells.[23,24] The aim of increasing NK activity may be feasible, although several HIV-1 proteins down-regulate NK cell ligands on the HIV-infected cell.

Vaccines Based on the HIV-1 Envelope Glycoprotein Gene

gp160 Characteristics

Env is the most important inducer of both specific and cross-reacting anti HIV antibodies. It contains a number of CTL and T helper epitopes, and is an important component of all vaccine strategies against HIV-1. The *env* gene is expressed during the late phase of viral transcription as the gp160 precursor protein. The translation of the precursor is dependent on the viral Rev protein. It binds to the rev responsive element (RRE) in the *env* mRNA and mediates its nuclear export.[25] During the maturation of the virus, the gp160 is proteolytically processed to the membrane spanning gp 41 and the extracellular gp120, which is noncovalently attached to gp 41. *Env* proteins become heavily glycosylated during their passage through the Golgi apparatus. On the viral surface, the gp120 proteins form trimeric spikes that can bind to the CD4 molecule, which is the primary receptor of HIV-1. Several parts of the gp120 envelope form a structure to bind the CD4 molecule. The gp120-binding site of CD4 is similar to the binding site of CD4 molecules to MHC class II molecules. This interaction reduces the effectiveness of helper T cell responses. After primary attachment, the gp120 protein interacts with CCR5 or CXCR4, which are the most common cellular co-receptors for HIV-1. This induces conformational changes in the viral envelope proteins gp120 and gp 41, and the fusogenic region of gp 41 is displayed. This in turn leads to fusion of the viral and cellular membranes and release of the viral core particle into the cytoplasm of the cell.

Comparison of the sequences of the *env* genes reveals five hypervariable regions in the gp120 subunit. The sequence variation in the hypervariable loops is due to nucleotide changes and subsequent accumulation of point mutations, which results in amino acid substitutions. Deletions and insertions occur during the replication process. The mutation-induced changes in the envelope tertiary structure are selected by the immune response of the host and provide a means of viral escape. One way of counteracting the escape would be to provide many conserved epitopes in a vaccine, preferably from several subtypes or recombinant strains.

Immune Responses to gp160

HIV production in the infected patient is massive. Therefore the gp160 is a major target for immune responses to HIV-1, which includes both cytotoxic T cells, helper T cells and high antibody production in infected individuals. The gp120 molecule appears to suppress antibody and particularly cellular immune responses both to itself and to other antigens.[26,27] This is one reason for the poor antigenicity of HIV-1, the rapid loss of Th1 effector cells and the prolonged and profound immunosuppression following HIV infection.

The gp 41 is a strongly antigenic transmembrane glycoprotein. It acts as a helper molecule for immune responses to gp120. Antigen from this protein can serve to detect the vast majority of persons infected by HIV-1. It also possesses conserved epitopes that induce potent and broadly neutralizing antibodies.[28] Due to its conformation during infection,[29] it is unlikely that these antibodies can interfere in the infection process.

Neutralizing antibody titers are usually low, compared to such titers found in other virus diseases. The trimeric Env complex is the target. Neutralizing antibodies to the Env could, at least in theory, block virus attachment and fusion upon binding to the CD4 binding site, the co-receptor binding site or the fusion domain. However, these sites are poorly immunogenic since they are masked by the hypervariable loops and sugar moieties, and are only exposed temporarily. Most antibodies are instead directed to the hypervariable loops, especially the V3 loop. These antibodies may indeed be neutralizing, but over time will lose their neutralizing ability in the patient, due to the evolution of the virus.[30,31]

Cellular immune responses to the *env* gene product are readily elicited and a number of CTL epitopes have been defined (Los Alamos HIV sequence database, http://hiv-web.lanl.gov). There is a window of approximately 12 hours from onset of viral protein synthesis to budding of virus and it is during this time that the CTL may act. Escape from immune responses occurs also in the Env CTL epitopes. For instance, the HLA B44 molecule binds the Env derived peptide AENLWVTVY but not the mutated epitope AXNLWVTVY.[2]

Vaccine Studies

For vaccination it would be interesting to be able to induce a broadly cross-reactive antibody response capable of neutralizing various HIV-1 isolates in combination with strong CTLs to conserved Env epitopes. This could be done by immunizing with *env* DNA from several different subtypes. New genetic recombinants from several subtypes are presently being produced in several laboratories. The *env* gene or gene product has been used for vaccination in a number of studies.[7] In two large therapeutic projects, asymptomatic seropositive patients were given recombinant gp160 or placebo control every three months over several years. No effective antiviral treatment was given. Significantly fewer of the vaccinated patients had reduced CD4 counts as compared to the placebo control group.[8] A transient improvement in survival was noted in the vaccinated group in one of these studies.[8]

In two other ongoing large-scale immunization studies, recombinant gp160, gp160 subtypes B and the circulating recombinant forms of subtype A - E are being used in Thailand and subtype B in the US. In a preclinical study, a bivalent vaccine with two Env antigens was able to induce broadly neutralizing antibodies.[32] DNA vaccination with plasmids encoding HIV-1 *env* was first shown to elicit Env-specific humoral and cellular responses in mice.[5,33-35] *Env* genes from several subtypes of HIV-1 have been prepared. A comparison of immune responses against the cloned genes showed that subtype C appears to be superior in inducing cross-reactivity to other HIV subtypes, both with DNA and with protein immunization.[36]

In a multi-subtype approach to HIV-1 vaccination, mice were immunized with HIV-1 *env* gp160 genes from subtypes A, B, and C.[37] Experimental challenge after immunization of mice was performed with a subtype B HIV-1/MuLV pseudovirus.[38] Specific immune responses, in vitro neutralization and protection from challenge were better in the group of mice immunized with the gp160 gene derived from a combination of subtypes A, B and C than from single genes. The protection appeared to be type specific and was enhanced by cross-reactive *env* immunization to give a more potent immunity.

It has also proven possible to protect or partially protect challenged primates after vaccination with *env* DNA constructs.[39,40] Partial resistance to SHIV challenge has been seen in primate studies after immunization with viral vectors such as modified Vaccinia Ankara (MVA) or recombinant adenovirus carrying the *env* gene.[41,42] The partial protection is defined as lowered virus titers in infected animals and a prolonged period to development of AIDS. Using plasmids containing *env* and *rev* DNA, protective immunity was evaluated in macaques and chimpanzees. Retardation of virus replication in chimpanzees has been reported after DNA immunization with the *env* and *rev* genes.[43]

Nucleocapsid Genes

The Gag antigen is probably the most suitable for induction of CTL responses and is less prone to variation than the *env* gene. Improved expression and augmented induction of immune responses to the proteins produced from this gene have been achieved by mutation of inhibitory sequences or by altering the codons by selecting those optimal for animal cell expression in the p24 sequence.[44,45] The *gag* gene *p37* encodes p24 and p17, the best cytotoxicity inducing proteins.[46] Even here, however, there is a natural variation[47] and differences in immune responses between separate MHC haplotyoes have been described, as well as variations leading to immune escape.[48,49]

In infected patients, Gag proteins are better recognized by CTL than for example the polymerase (Pol) protein. One explanation for this may be that the ratio between the number of Gag and Pol molecules made in a cell during infection is unequal. In one study the CTL response to two well-defined HLA-A*020 restricted CTL epitopes: SLYNTVATL (p17, gag) and ILKEPVHGV (reverse transcriptase, RT) of 24 HIV-infected HLA A.2 individuals was investigated.[50] Binding studies showed that the RT peptide bound significantly better than the Gag peptide for MHC presentation. Despite this, the majority of donors generated dominant responses to the Gag peptide. An explanation may be that over 30 times more of the Gag peptides than the RT peptides were presented on the cell surface.[51]

The Polymerase Gene

Reverse transcriptase (RT) of HIV transcribes the viral genome into double stranded DNA, which is then ligated by integrase into the chromosomal DNA as a linear cDNA. These two enzymes, RT and integrase, together with the viral protease, therefore play key roles in the replication and maintenance of the infection in a latent form. The HIV genome may reside for a long time in metabolically inactive cells. A potent primary MHC class I-restricted CTL response was reported against functional domains of RT.[52] CTL against the HIV proteins produced early in the viral life cycle have been suggested to be effective in controlling viral load. Such responses have however not been linked to long-term survival in HIV patients.[53] Recent data suggest that CTL can lyse acutely HIV-infected cells only after production of progeny virus has started.[54] Although RT is an intracellular protein, antibodies to RT might have HIV-1 inhibitory effects. Antibodies to certain epitopes of RT inhibit its enzymatic activity.[55,56] A high prevalence of these antibodies was associated with asymptomatic infection.[57]

Immune Responses to RT in Humans

RT-specific CTLs in HIV infected patients were first reported by Walker et al.[58] A large number of CTL epitopes in RT have been described for different HLA types. Epitopes positioned in the "fingers" and "palm" subdomains of RT are the most conserved and among the most often detected by the immune responses of humans. Epitopes have also been found in the "thumb" and "connection" subdomains and in the RNAse H domain. In one study of the French IMMUNOCO cohort, CTL recognition of HIV-1 Pol products was found in 78% of all patients. RT was recognized by 81% of these patients and integrase and protease were recognized by 51% and 24%, respectively.[59] It was demonstrated that both progressing patients and long-term survivors recognized both conserved and nonconserved epitopes of RT. One of the progressors also recognized an epitope in the YMDD motif, which is conserved and essential for virus survival.[53]

RT Drugs and CTL

Epitopes with drug escape mutations are attractive new targets for HIV-1 specific vaccines.[60] Three classes of drugs directed to HIV enzymatic actitivies are currently on the market: nucleoside RT inhibitors (NRTI), nonnucleoside RT inhibitors (NNRTI) and protease inhibitors (PI). All three classes of drugs inhibit viral growth but resistant virus evolves rapidly if only one drug is used, or if adherence to medication is less than stringent. RT-specific drug treatment of HIV-1 induces a number of escape mutations in the RT enzyme. It was demonstrated that mutations with NRTI, M41L, L74V, M184V, and T215Y/F induced by mono- or bi-therapy did not impair CTL recognition of the RT sites of drug-induced mutations in treated patients. On the contrary, the drug-induced mutations were associated with twofold more frequent recognition of the unmutated HIV-1 RT regions.[61] The RT M184V mutation within an HLA-A2 restricted HIV-1 CTL epitope (RT 179-187) mediates drug escape against the NRTI lamivudine (3TC) and abolishes recognition by CTL in HIV-1 infected patients. It has been shown that an HIV-1 patient receiving lamivudine treatment could generate RT-specific CTL that targeted lamivudine-resistant viruses.[62]

An interesting question is whether CTL targeted to domains of RT containing drug-induced mutations could be used to put additional pressure on the virus, acting in synergy with the drugs. Synthetic genes have been constructed, containing known CTL epitopes which include amino acids that mutate as a result of drug treatment. The immunogenicity of these constructs appears to be related to expression capacity.

RT Immunization

Vaccination with the *RT* gene has been shown to induce antibody and cellular responses in mice.[38,63] Immunizations with RT have shown protection against HIV in an experimental model. Plasmid DNAs containing either the enzymatically active *RT* (YMDD) or a mutated inactive *RT* (YMLL) were used. Both constructs induced similar antibody and T-cell responses, with a tendency towards antibodies directed to peptides representing the active or the mutated sites. Immunized mice were challenged with pseudotype HIV-1 infected spleen cells. Interestingly, homologous protection against the wild-type HIV was seen after immunization with wild-type DNA, but not after vaccination with the mutated gene.[64]

DNA vaccination with the *RT* gene can give both antibody and cellular responses in rabbits.[38] Plasmids with *RT* genes[60,64] express RT, which interferes with the gene products of the other viral genes. Pairwise immunizations with RT and other genes have shown that RT interferes with transcription or translation and immunogenicity of several HIV-1 genes. Immunization with the RT gene induced weak antibody and cellular responses in Cynomolgus macaques. The responses increased substantially after a boost with the RT protein. The DNA priming was important however; monkeys primed with water and boosted with protein and oligodeoxinucleotide (ODN) adjuvants did not respond as well. A good challenge virus in such an experiment is the SHIV-RT virus, which contains the SIV backbone with HIV-1 *RT*.[65] Although the *RT* genes are theoretically ideal for human immunization, this has not yet been done.

HIV-1 Early and Accessory Proteins

Natural Immune Response

HIV-1 encodes six accessory proteins in addition to the structural and enzymatic proteins. By inducing immune responses to the regulatory proteins of HIV-1, it should be feasible to prevent the burst of newly synthesized virions from infected cells. This should limit primary and chronic infection. In HIV-1-infected patients it has been shown that antigen-specific CTL and T helper cells are directed to regulatory proteins. They appear crucial in controlling viral replication.[66] The Tat protein is the viral transactivator that enhances the elongation step in transcription of proviral DNA. The Rev protein regulates the gene expression at the post-transcriptional level by allowing transport from the nucleus to the cytoplasm. It thus regulates translation of the mRNA that encodes the structural and enzymatic HIV proteins. Nef downregulates the MHC class I and CD4 receptors from the surface of infected cells and also alters normal T cell signaling in the infected cells.

The three proteins Tat, Rev and Nef are synthesized early in the infected cell and are the first to be exposed to the immune system of the host. Several papers have been published on Tat immunogenicity.[67-71] Monoclonal antibodies against Tat have been shown to completely inhibit Tat transactivation[72,73] and to delay HIV-1 replication in peripheral blood mononuclear cells (PBMC).[74] CTL and CTL precursor cells against Tat and Rev are related to less rapid disease progression to AIDS.[54,75,76] Tat and Rev proteins are important targets for CTL; up to 37% of infected individuals develop CTLs specific for Tat and Rev.[77] Tat-specific CTL induced in rhesus macaques infected with simian immunodeficiency virus (SIV) can control primary infection. Immune escape occurs however, which prevents CTL killing of Tat-expressing targets.[78] The *rev* gene has been used in many immunization studies since Rev is necessary for production of the gp160 protein.

Following primary infection, antibodies and CTL directed against Nef have been found in approximately 50% of infected individuals.[79-82] Studies of a group of Gambian women who remained seronegative for HIV-1 infection despite exposure have revealed high levels of Nef-specific CTL, which is considered to contribute to resistance to infection.[9]

The small accessory proteins Vpu, Vif and Vpr appear essential for viral replication, with important functions such as viral assembly and disassembly, nuclear transport of the preintegration complex, and down-regulation of CD4 from the surface of an infected cell. Altfeld et al characterized CTL responses directed against Vpu, Vif and Vpr and found that the Vpr and Vif proteins are important targets of cellular host defences, whereas Vpu is infrequently targeted by CTL. Vpr seemed to be one of the most frequently used CTL-targeted antigens when occurrence of cytotoxicity was related to the length of the protein.[83] No studies have been made to date of the protection given by CTLs against Vpr, Vpu or Vif targets.

Accessory Gene Immunization

Several groups have evaluated DNA vaccines against the early accessory proteins Tat, Rev and Nef of HIV-1 and SIV in mice.[20,35,46,84-97] In an attempt to characterize many Nef epitopes by DNA immunization, it was noted that epitopes presented by a DNA vaccine induce a few immunodominant CTL populations. On the other hand, epitopes presented as 9-10 amino acid long peptides could all induce epitope-specific CTL.[20] In a nonhuman primate animal model, the products of the three genes *nef, rev* and *tat* were immunogenic and primarily induced an immune response of the Th 1 type.[40]

In HIV-1 infected individuals, all three early gene products are potent in inducing cellular responses.[98-100] The immunized patients could respond to either single or combined early genes with new cellular responses to HIV-1.

Not many studies have been conducted with the *vif, vpu* and *vpr* genes. Using the *vif, vpu* and *nef* genes to encode immunogens, Ayyavoo et al[101,102] showed that attenuated *vif* clones from HIV-1 patients' isolates can effectively induce both humoral and cellular immune responses against Vif in mice and that these responses were also capable of destroying human targets in vitro. Further work was concentrated on using a multicomponent cassette encoding the Vif, Vpu and Nef as a single immunogen.[101,102] Both humoral and cellular responses were detected in mice. Taken together, all these studies show the potential of including the accessory genes to produce immunogens against HIV-1 infection.

Receptor Genes

Certain cellular proteins may be of interest as autoantigens for protection from HIV infection. Inhibition at the receptor level of an infection will be an important step towards the identification of new targets for immunization.[103-105] The HIV virion enters the cell by CD4 and either of the two major co-receptors, CCR5 or CXCR4. These molecules are expressed on cells of lymphocytic and monocytoid/dendritic cell origin. CCR5 is the co-receptor preferentially used by HIV-1 in primary infection of human macrophages and lymphocytes.

In humans, a 32 bp deletion in the *CCR5* gene prevents the CCR5 protein from being expressed on the cell surface. This deletion has no obvious effects on the general immunocompetence of the affected individuals. Persons homozygous for the 32 bp deletion in *CCR5* are protected against primary HIV-1 infection, and individuals heterozygous for the deletion show a protected early course of infection.[106] These findings indicate that blockage of the CCR5 receptor by drugs or antibodies may confer in vivo protection against primary HIV-1 infection.

Chemokines can block HIV infection by occupying the HIV co-receptors. Monoclonal antibodies directed to different parts of CCR5 can also block HIV-1 infection of cells in vitro. The ability of DNA vaccination to induce HIV-1 blocking antibodies directed at CCR5 structures was investigated by using plasmids expressing the human CCR5 protein alone or in-frame with a promiscuous tetanus epitope.[104] Epidermal but not intramuscular delivery of the *CCR5* gene to mice elicited CCR5 specific T cell proliferative responses. It also gave strong IgG antibody responses towards a linear peptide representing the N-terminus of CCR5.

When Cynomolgus macaques were immunized intrarectally with a vector expressing human CCR5, strong T cell proliferative responses were detected against several CCR5 epitopes. The most prominent IgG and IgA antibody responses found in plasma and vaginal washings were those against the N-terminus of CCR5. Plasma from the immunized macaques inhibited primary CCR5 HIV-1 from infecting human peripheral blood lymphocytes. After boosting with peptides representing different parts of the CCR5 protein, enhanced IgG and IgA titers against CCR5 peptides were seen. The macaques were not protected, but developed a viral setpoint below that of control macaques.[104] Immunization of macaques with human CD4+ cells or CCR5 protein/peptides induced antibodies that inhibited simian immunodeficiency virus replication in vitro, possibly by downregulation of CD4+ molecules.[103,105]

Combination of Genes or Multigene Constructs

The ease with which the DNA constructs can be made will permit mixing of plasmids with different genes or introduction of many epitopes of different genes in one plasmid construct. Expression library strategies have been used to identify antigenic and immunogenic sites.[107] DNA libraries can be constructed and the whole genetic make-up of a microbe screened for sequences that give rise to immune responses or even protective immunity. Minigenes may be designed to directly represent small selected epitopes of an immunogen. This can be of particular interest for enhancing immunity or inducing reactivities to a tolerated protein. Such minigenes have been constructed.[17,19] Fragments of nucleic acid sequences or epitopes are being placed together in combinations, so that T-cell epitopes can be identified by immunologically competent hosts.

Small Animal Models for Measurement of Challenge

Several rodent models have been developed for challenge with HIV-1 virus or infected cells.[33,64,108-111] A murine model for acute infection with HIV-1 was established.[110] HIV-1/MuLV pseudovirus infected cells were used for challenge. This permits a detailed study of primary HIV-1 immune responses by identifying precursor T cells that become activated to respond to HIV after infection. The model may be suitable also for investigation of immune responses to and protection from subtypes of HIV-1.

By using mice with severe combined immunodeficiency (SCID),[112,113] models were established for testing human immunoreactivity in vivo. SCID mice were repopulated with human peripheral lymphoid cells, which responded with antibodies of the IgM class and T cell responses to DNA vaccination with HIV genes.[85]

Primate Models

In primates, the DNA immunogens have been less effective in protecting against challenges with homologous or heterologous viruses of simian immunodeficiency virus (SIV) or simian human recombinant immunodeficiency virus (SHIV). Only few studies show total protection from challenge after genetic immunization (or any other type of immunization) against SIV, SHIV or HIV-1 or -2 challenges.[114] Protection in a fraction of challenged animals has been obtained for HIV/SIV.[115-118] Immunization with the full, mutated proviral HIV-1 DNA led to lower viral load after challenge, but not to complete protection.[119]

Vaccines that stimulate CD8$^+$ T cell responses in monkeys have been shown to provide complete or partial protection from challenge with SIV. Most often these responses result in partial protection with a reduced viral load setpoint. DNA combined with the interleukin-12 (IL-12) gene gave protection from AIDS but not from infection by chimeric SHIV.[120] The combination induces strong CTL responses in rhesus macaques, which controlled viremia and prolonged the time to clinical AIDS, suggesting that CTL response initially decreases viral load.[117]

To induce broad immunity, some groups have used the *env* gene together with either the *gag* and *pol* genes, regulatory genes or most of the viral genome[40,118,119,121] or multi-gene DNA constructs.[122] In the SHIV primate studies, monkeys are usually immunized with a mixture of genes, for example HIV *env/rev* and SIV *gag/pol* and challenged with a SIV or SHIV virus. These strategies have been successful in reducing viral loads and preventing AIDS. To date however, there has been no vaccine or vaccine combination that has completely protected every immunized animal from virus challenge.

DNA Vaccine Studies

A study of DNA immunization in monkeys which stimulated the cellular immune responses, resulted in 1000-fold lower virus peak values after SHIV challenge, compared to nonvaccinated controls.[123]

Prime-boost approaches, such as prime with DNA and boost with protein or recombinant virus, have been used with encouraging results. The prime-boost shedule of a DNA vaccine followed by boosting with the same viral sequence in poxvirus vectors stimulates strong CD8$^+$ T-cell responses.[78,115,124-126] Prime boost with DNA followed by live avipox viral vector induced better immune responses than either modality alone, and protected against challenge with the virus.[115] A most promising containment of challenge in rhesus macaques was intradermal DNA priming followed by live vector boosts. The effect was independent of neutralizing antibody.[127] Macaques primed with plasmid DNA encoding multiple genes of SIV and HIV and boosted with proteins or MVA carrying structural genes showed good immunological responses and a reduced viral SHIV load following a mucosal challenge[118] or intravenous challenge.[128]

Mucosal administration of the DNA immunogens has proven effective also in macaques.[129] Two different methods of administration were compared for a combination of DNA and MVA-based multicomponent HIV-1/SIV vaccines. A combination of mucosal and intramuscular immunization gave the best protection from SHIV challenge.[128] Stronger cellular immune responses and more effective control of challenge virus replication were induced when the vaccine was administered intramuscularly and mucosally than when it was given intramuscularly only.

The most appropriate challenge virus and the best suited route of infection remains to be identified for HIV vaccine efficacy studies in nonhuman primates. Both SIV, SHIV with low pathogenicity, SHIV with high pathogenicity and HIV-2 have been used as intravenous or mucosal challenges in macaques. In chimpanzees, both humoral and cellular responses to HIV-1 were induced without toxic side effects.[43,130] Challenge of these animals showed reduced viral load compared to control animals. The latter species study is most closely related to human

HIV-1 vaccination. The chimpanzees were protected from the induction of HIV-1 infection in that virus could be recovered only for a very short period after challenge.

The most interesting results in view of what may be expected from human immunization were performed by Kim et al.[121] A first priming by *env, gag* and *pol* genes protected three of eight monkeys from SHIV challenge. Continued challenges with more pathogenic strains of SIV and SHIV showed that these individuals resisted also these challenges. Perhaps there will have to be repeated vaccination against HIV over time, to obtain enduring protective immunity in man.

Experimental Immunotherapy

A clinical benefit from therapeutic vaccination is suggested by a limited number of studies. Data from He et al indicate that acutely SIV-infected macaques might benefit from therapeutic immunization.[131] During antiretroviral treatment, animals were immunized with a vaccinia construct (NYVAC) encoding SIV Gag, Pol, and Env proteins. Vaccination elicited anti-SIV specific CD4$^+$ T cell responses in animals with a low viral load. Vaccine-induced CD8$^+$ T cell responses were elicited only in vaccinated animals receiving anti-retroviral treatment. After structured therapy interruption (STI), two out of four animals in the vaccinated group had transient viremia that was quickly suppressed. The results point to a role for therapeutic immunization in protecting against viral rebound upon withdrawal of anti-retroviral treatment.

Improvement of Immunization Related to Protection Against HIV

Codon Usage

An optimized codon usage can be prepared by generating synthetic viral gene sequences according to codons known to be highly expressed in human genes. The synthetic gp120/gp160 *env* gene acquired a rev-independent expression. The in vitro expressions of a synthetic gp120 *env* gene and a p24 *gag* gene were considerably increased in comparison with the wild type sequences.[45,132] It was also shown that both the humoral and cellular cytotoxic responses increased after vaccination with the humanized sequences, suggesting a direct correlation between expression levels and immune responses. Recently the HIV-1 *RT* gene was modified with codons most frequently used in humans. This synthetic gene elicited much stronger cellular immunity in rodents than the virus derived gene.[63] Immunization of rhesus macaques with the humanized *RT* gene coupled to a human tissue plasminogen activator leader sequence led to strong in vitro CTL killing activities and enhanced levels of RT-specific T-cells.

Adjuvants for DNA Immunization

Approaches to circumvent low antigenicity or defective T cell help are being studied by several investigators. One potential solution to this problem is the use of adjuvants that induce immune responses in the absence of T cell help. Cytokines delivered simultaneously with plasmid vaccines or proteins have proven useful in directing the immune response following systemic immunization with plasmid DNA carrying several different cytokine genes.[133-136] IL-12 DNA administered together with HIV DNA plasmids has been shown to enhance the T helper type 1 immunity and to decrease the Th2 reaction. An IL-2 plasmid enhanced delayed type hypersensitivity and CTL responses in mice. Concomitant administration of cytokines IL-2 and GM-CSF augmented both B and T cell responses in rats.[135] In rhesus macaques, antibody and CD8$^+$ T cells increased upon co-administration of vectors encoding HIV antigen and an IL-2/Ig chimeric protein.[117,120] GM-CSF has been used sucessfully in several studies to amplify the primary responses to HIV DNA.[134,135]

The bacterial DNA sequences of plasmids by themselves confer immune stimulation.[137,138] They have the ability to induce immunity-collaborating products such as interferon gamma and tumor necrosis factors. The contribution to immune activation by the CpG motif of the plasmid sequence itself has raised the possibility that endogenous effects can be achieved through careful selection of the plasmid sequences. Another possibility is to deliver these sequences separately as stable phosphotiorate ODN.[139] ODN DNA has an adjuvant effect when delivered with peptides encoding CD8+ T cell epitopes to mice deficient in T helper cell activity.[140] Immunostimulatory CpG motifs, which are species specific, activate innate immune responses through binding to the Toll-like receptor 9.[141] This in turn facilitates induction of an antigen-specific immune response.

HIV DNA transfer by apoptotic cells appears to induce strong immune responses. This uptake is receptor independent. Apoptotic cell bodies containing replication-defective HIV were sucessful in transferring infectious HIV genes. The DNA present in the apoptotic bodies can be transferred into antigen-presenting cells. This results in presentation of viral epitopes by MHC class I molecules.[142] In a challenge system in mice, it was shown that this method induced strong immune responses and protection from HIV/MuLV challenge. Induction of protection was more prominent than upon immunization with single HIV genes and similar to immunization with multiple HIV genes.[143]

Route of Inoculation

The classic pathway for immunization is the intramuscular route and this is also a route of preference in DNA immunization. Skeletal muscle appears to have an affinity for uptake and presentation of nucleic acids. HIV often enters the body through the mucosa as a sexually transmitted microbe. Direct administration of DNA vaccines at these surfaces may therefore be capable of conferring local protection. The immunization of mucosal sites yields locally protective IgA. Delivery of DNA to mucosal surfaces has been shown to be effective in inducing strong immunity.[89,129,144,145] In order to obtain mucosal immune responses, immunizations with HIV DNA could also be given intranasally.[146] C57Bl/6 mice responded well to plasmids carrying HIV-1 regulatory genes *rev, tat, nef* and structural genes p37 *gag* and gp160 *env* with or without cytokine IL-12 genes.

Antigen can be deposited intradermally with the gene gun.[147] This mode of application is more effective than intradermal needle injection, probably because the DNA-loaded particles are delivered directly into cells. The DNA is precipitated onto 1-3 μm gold particles which are injected at high speed. The method requires less DNA than intramuscular injections, perhaps because cells such as dendritic cells needed for antigen presentation[148] are transduced efficiently. HIV DNA formulated on particles given directly to skin or mucosa induces optimal responses at 100-1000 times lower doses than those required for optimal immunization following injection.[89,149]

A dental jet gun is another efficient mode of delivery which was used in humans at the oral mucosa covering the mandibular bone structure. An accumulation of cells representing activated Th 1 helper cell populations was found at the local site of jet-gun inoculation.[145]

The use of electric current to enhance cell penetration is a potent way of transducing plasmid DNA into muscle or skin cells. Electroporation creates holes in cellular membranes, facilitating entry and increasing the uptake of plasmid DNA.[150] Electroporation appears active in permitting access of DNA to intracellular locations. Strong immune responses were obtained in several species.[151]

Improvements by Viral Vectors or Attenuation

Attenuated live HIV is considered a good immunogen, since all genes and all antigens would be expressed, and since attenuated virus vaccines such as measles, rubella and mumps have been both immunologically effective and safe. The first studies of attenuated SIV showed a good immune response and long-term protection from a virulent challenge virus in macaques.[152] However, the defective live viruses could replicate in newborn monkeys and rapidly became virulent.[153] This has dampened the enthusiasm for attenuated live retroviral vaccines. Unlike the childhood vaccines, the retroviruses such as HIV establish latency and integrate into the host DNA. Over time, the incorporated HIV genomes replicate and evolve into virulent viruses.

An HIV strain that spread through transfusion was found to contain a defective *nef* gene. This natural attenuation initially appeared to inhibit initial progress of HIV disease.[154-155] Large viral and bacterial vectors can harbor many foreign genes.[156-159] These constructs have the properties of the vector, which includes induction of immunostimulating cytokines but also induction of immunity against antigens of the vector. Several boosting strategies have become very effective in combining DNA priming with live vector genes.[6] Most probably, for human use one would select one of the larger vectors such as vaccinia or modified vaccinia because of their competence in carrying large foreign genomic portions. It is of interest however, to note that small replication-incompetent vectors such as adenovirus may induce good CTL immunity.[160]

Human HIV Vaccination with DNA

A few DNA vaccines have been tested for safety and immunogenicity in humans and several clinical trials are ongoing. In a phase I study, 300 μg of HIV *env/rev* DNA given three times induced antigen-specific T-cell responses and production of cytokines.[43] Plasmid immunizations have been performed in healthy individuals with plasmids encoding HIV-1 Env/Rev and Gag/Pol constructs with up to four injections of 1000 μg each. Plasmids encoding malarial proteins have been given in doses up to three 2500 μg injections.[161] Hepatitis B virus DNA representing the HBs surface protein has been given by gene gun.[149] These limited trials have corroborated the low toxicity of DNA plasmids found in animal studies.

The canary pox virus vector has been used to transfer HIV viral DNA into noninfected humans.[162-165] Clade B-based canarypox vaccines could elicit broad CTL reactivities in HIV negative humans. Neutralizing antibodies and CTL were each induced in 33% of the vaccinated individuals.[165] The canarypox vaccine primed for antibody induction by gp120.[163] Attenuated vaccinia strains carrying HIV gene segments have entered clicinal trials.[166]

Immunization of HIV-Infected Immunosuppressed Individuals

Patients infected by HIV are immunosuppressed, even during the period just after the primary infection. Therapeutic vaccines face the additional hurdle that the recipients, the HIV-infected individuals, have functional deficits in their immune systems, particularly in the T helper cell responses. A variety of approaches are being considered to circumvent the effects of defects in the $CD4^+$ T cell population that might otherwise prevent the development of strong immune responses following immunization.

Therapeutic Immunization

The escape that occurs in prophylactically vaccinated macaques when challenged raises the question of whether prophylactic vaccination will encounter the same problems as the therapeutical trials. In therapy as well as in natural infection, replication of the infecting virus will permit virus variation and perhaps sooner or later a breakthrough.[22,167] Therapeutic studies have been performed with DNA vaccines carrying gp160 or regulatory genes. The human trials have shown that immune responses are obtained without evident side effects.[168] In a dose escalation study, *env* and *rev* genes appeared safe in HIV-1 infected individuals.[168,169] These are small studies, focused on immunogenicity and possible side effects of the genes. Data on seropositive patients immunized intramuscularly with DNA plasmids encoding a variety of HIV proteins indicate that reproducible, sustained T cell responses are rare. The responses have not been strictly dependent on the plasmid dose within the ranges explored.

Three viral regulatory genes, *tat*, *rev* and *nef*, were used for the composition of a genetic vaccine. CTL responses to these antigens have been described to be related to a better prognosis in HIV-infected individuals. The composition induced MHC class I restricted new CTL reactivities.[98] The genes could be identified in mucosal biopsies 48 hours after immunization, indicating that expression of the proteins continues locally for at least a few days. Increases in CTL precursor cells to target cells infected with the HIV-1 virus showed that new memory cells could be induced. However, the immunizations did not reduce the viral load. The oral mucosal route induced specific systemic T-cell proliferative responses.[145] Immunohistochemical analysis of oral biopsies after immunization revealed increased levels of granulocytes and T cells as well as expression of HLA-DR. Cells expressing the T-cell markers CD3, CD4 and CD8 were significantly increased in the vaccinated mucosa. Vaccine-specific antibodies were present during the immunization, but did not increase further.

Therapeutic vaccination in humans would be expected to modify ongoing disease.[170] This will give an indication of the effectiveness of defined vaccine components when planning prophylactic immunization.

Immunization and Anti-Retroviral Treatment

With the widespread introduction of highly active anti-retroviral therapy (HAART), the treatment of HIV infections has undergone dramatic changes in the last few years. The profound reduction of viral load in most patients has been related to improved immunity to opportunistic infections. The immunity to HIV has however not been improved in chronically infected patients. In contrast to the overall improvements in the immune system that are associated with HAART, specific immune responses to HIV appear to remain low or even diminish.[170-172] DNA immunization did, however, induce anti-HIV CTL also during anti-retroviral therapy.[99] Continued high levels of lymphocyte proliferation to Nef, Rev and Tat were observed when immunization with HIV DNA followed the initiation of HAART. There was no indication that HIV DNA immunization decreased viral load or that HAART induced immunity. This suggests that the combination of DNA vaccination with HAART might result both in improved immune responses and in decreased viral load.

Therapeutic vaccination studies of chronically infected patients on HAART were performed also with Env-depleted killed HIV. Vaccination with this whole-killed, Env-depleted vaccine induced CD4$^+$ cell responses to the vaccine antigens.[173,174] An ALVAC construct with structural HIV-1 genes was used for immunization early in HIV infection during anti-retroviral treatment.[175] Both antibody and cellular immune responses were seen to the expressed HIV proteins, arguing for a possibility to induce additional immune responses in infected individuals.[100,175] Increases in antigen-specific CTLs were reported in all these studies but the frequency was lower in patients who already demonstrated CTL responses.[100,169]

Structured Therapy Interruption (STI)

In patients given HAART very early after infection, protective responses can be induced or maintained that are able to control viral replication after HAART withdrawal. Increases were noted in the breadth and magnitude of HIV-specific T helper and CTL responses during STI.[176-178] Thus, even without further immune interventions, it is clear that some patients may be able to hold viral replication at levels that may not require therapy. Patients on HAART who had short therapy interruptions were more likely to exhibit strong HIV p24-specific CD4$^+$ T cell responses than patients with no interruptions.[176] Therefore it seems justified to immunize HIV-infected individuals who have been successfully treated with HAART and then halt all therapy when the CD4$^+$ T cell count is in a range with negligible risk of opportunistic infections. Such attempts to immunize during HAART treatment and study the virological and immunological effects following STI are being initiated.

Benefits from an experimental therapeutic vaccine with HAART followed by an STI have been reported. Jin et al[175] used an HIV-1 subtype B construct in HIV-infected persons on HAART. Six patients were identified who initiated HAART during acute infection. All patients were immunized four times with ALVAC 1452 and recombinant gp160. ALVAC 1452 is a recombinant canarypox virus carrying sequences for HIV *gag, pol, env* and *nef.* The vaccines appeared to be immunogenic since increases in anti-Env Ab responses of 2-3 logs were noted in all subjects and new CTL in four. One week after their last immunization, four patients underwent STI. Two experienced a rapid rebound in viral load but two had a slower rebound.

Many of the critical systems and cell types required for the development of immune responses improve with HAART, but specific immunity to HIV wanes. Results to date suggest that when anti-retroviral treatment is discontinued, T-cells may be necessary to control viral replication in cases where HAART is discontinued. Continued attempts to control HIV viral burden by immunotherapeutic vaccination will give leads also to decisive immune responses important for prophylactic immunization.

Summary

HIV's genetic variation faces vaccine production with unprecedented challenges. The virus comprises numerous subtypes, recombinants, immune escape variants and drug-resistant virus populations. The DNA vaccine technique is the most flexible in producing continuously new immunogens. Understanding the genetic variation of HIV and adapting viral vaccines to the prevalent strains may turn out to be the only way by which the spread of HIV can be predicted and countered. At the site where HIV DNA is injected, the patient's cells are made to express a number of antigens typical for HIV. The genes produce proteins that resemble those of a live infection but the immunogens are not infectious. Experimental studies have shown induction of potent cellular immune responses and partial or complete protection from human or simian immunodeficiency virus challenge. Clinical studies have shown induction of new cytotoxic T cells. However, currently used DNA vectors will probably need adjuvant formulations for effective prophylaxis. For therapy, assistance from anti-retroviral drugs will be needed. A successful HIV-1 vaccine should confer protection from as many viral subtypes and variants as possible.

References

1. Nabel G. Challenges and opportunities for development of an AIDS vaccine. Nature 2001; 410:1002-1007.
2. McMichael AJ, Rowland-Jones SL. Cellular immune responses to HIV. Nature 2001; 410:980-987.
3. Robinson HL, Hunt LA, Webster RG. Protection against a lethal influenza virus challenge by immunization with a hemagglutinin-expressing plasmid DNA. Vaccine 1993; 11:957-960.
4. Ulmer JB, Donnelly JJ, Parker SE et al. Heterologous protection against influenza by injection of DNA encoding a viral protein. Science 1993; 259:1745-1749.
5. Wang B, Boyer J, Srikantan V et al. DNA inoculation induces neutralizing immune responses against human immunodeficiency virus type 1 in mice and nonhuman primates. DNA Cell Biol 1993; 12:799-805.
6. Seder RA, Hill AV. Vaccines against intracellular infections requiring cellular immunity. Nature 2000; 406:793-798.
7. Redfield R, Birx D, Ketter N et al. A phase I evaluation of the safety and immunogenicity of vaccination with recombinant gp160 in patients with early human immunodeficiency virus infection. New Engl J Med 1991; 324:1677-1684.
8. Sandstrom E, Wahren B. Therapeutic immunisation with recombinant gp160 in HIV-1 infection: a randomised double-blind placebo-controlled trial. Nordic VAC-04 Study Group. Lancet 1999; 353:1735-1742.
9. Rowland-Jones S, Sutton J, Ariyoshi K et al. HIV-specific cytotoxic T-cells in HIV-exposed but uninfected Gambian women. Nat Med 1995; 1:59-64.
10. Kaul R, Plummer FA, Kimani J et al. HIV-1-specific mucosal CD8+ lymphocyte responses in the cervix of HIV-1- resistant prostitutes in Nairobi. J Immunol 2000; 164:1602-1611.
11. Ferrari G, Berend C, Ottinger J et al. Replication-defective canarypox (ALVAC) vectors effectively activate anti-human immunodeficiency virus-1 cytotoxic T lymphocytes present in infected patients: implications for antigen-specific immunotherapy. Blood 1997; 90:2406-2416.
12. Jin X, Bauer DE, Tuttleton SE et al. Dramatic increases in plasma viremia after CD8+ T cell depletion in simian immunodeficiency virus-infected macaques. J Exp Med 1999; 189:991-998.
13. Schmitz JE, Kuroda MJ, Santra S et al. Control of viremia in simian immunodeficiency virus infection by CD8+ lymphocytes. Science 1999; 283:857-860.
14. Kaul R, Dong T, Plummer F et al. CD8+ lymphocytes respond to different HIV epitopes in seronegative and infected subjects. J Clin Invest 2001; 107:1303-1310.
15. Goulder P, Altfeld M, Rosenberg E et al. Substantial differences in specificity of HIV-specific cytotoxic T cells in acute and chronic HIV infection. J Exp Med 2001; 193:181-193.

16. Hanke T, Neumann VC, Blanchard TJ et al. Effective induction of HIV-specific CTL by multi-epitope using gene gun in a combined vaccination regime. Vaccine 1999; 17:589-596.
17. Ishioka GY, Fikes J, Hermanson G et al. Utilization of MHC class I transgenic mice for development of minigene DNA vaccines encoding multiple HLA-restricted CTL epitopes. J Immunol 1999; 162:3915-3925.
18. Kieber-Emmons T, Monzavi-Karbassi B, Wang B et al. Cutting edga: DNA immunization with minigenes of carbohydrate mimotopes induce functional anti-carbohydrate antibody response. J Immunol 2000; 165:623-627.
19. Hanke T, McMichael A. Pre-clinical development of a multi-CTL epitope-based DNA prime MVA boost vaccine for AIDS. Immunol Lett 1999; 66:177-181.
20. Sandberg JK, Leandersson AC, Devito C et al. Human immunodeficiency virus type 1 Nef epitopes recognized in HLA-A2 transgenic mice in response to DNA and peptide immunization. Virology 2000; 273:112-129.
21. Allen T, Mal E. Induction of AIDS virus-specific CTL activity in fresh, unstimulated peripheral blood lymphocytes from rhesus macaques vaccinated with a DNA prime/modified vaccinia virus Ankara boost regimen. J Immunol 2000; 164:4968-4978.
22. Barouch DH, Kunstman J, Kuroda MJ et al. Eventual AIDS vaccine failure in a rhesus monkey by viral escape from cytotoxic T lymphocytes. Nature 2002; 415:335-339.
23. Cohen GB, Gandhi RT, Davis DM et al. The selective downregulationof class I major histocompatibility complex proteins by HIV-1 protects HIV-infected cells from NK cells. Immunity 1999; 10:661-671.
24. Hultström AL, Hejdeman B, Leandersson AC et al. Human natural killer cells in asymptomatic human immunodeficiency virus-1 infection. Intervirology 2000; 43:294-301.
25. Richard N, Iacampo SCochrane A. HIV-1 Rev is capable of shuttling between the nucleus and cytoplasm. Virology 1994; 204:123-131.
26. Wahren B, Morfeldt-Månsson L, Biberfeld G et al. Impaired specific cellular response to HTLV-III before other immune defects in patients with HTLV-III infection. New Engl J Med 1986; 315:393-394.
27. Hioe CE, Jones GJ, Rees AD et al. Anti-CD4-binding domain antibodies complexed with HIV type 1 glycoprotein 120 inhibit CD4+ T cell-proliferative responses to glycoprotein 120. AIDS Res Hum Retroviruses 2000; 16:893-905.
28. Purtscher M, Trkola A, Grassauer PM et al. Restricted antigenic variability of the epitope recognized by the neutralizing gp41 antibody 2F5. AIDS 1996; 10:587-593.
29. Chan DC, Fass D, Berger JM et al. Core structure of gp41 from the HIV envelope glycoprotein. Cell 1997; 889:263-277.
30. Albert J, Abrahamsson B, Nagy K et al. Rapid development of isolate-specific neutralizing antibodies after primary HIV-1 infection and consequent emergence of virus variants which resist neutralization by autologous sera. AIDS 1990; 4:107-112.
31. Moore JP, Parren PW, Burton DR. Genetic subtypes, humoral immunity, and human immunodeficiency virus type 1 vaccine development. J Virol 2001; 75:5721-5729.
32. Berman PW, Huang W, Riddle L et al. Development of bivalent (B/E) vaccines able to neutralize CCR5- dependent viruses from the United States and Thailand. Virology 1999; 265:1-9.
33. Wang B, Ugen KE, Srikantan V et al. Gene inoculation generates immune responses against human immunodeficiency virus type 1. Proc Natl Acad Sci USA 1993; 90:4156-4160.
34. Lu S, Santoro JC, Fuller DH et al. Use of DNAs expressing HIV-1 Env and noninfectious HIV-1 particles to raise antibody responses in mice. Virology 1995; 209:147-154.
35. Hinkula J, Lundholm PWahren B. Nucleic acid vaccination with HIV regulatory genes: a combination of HIV-1 genes in separate plasmids induces strong immune responses. Vaccine 1997; 15:874-8.
36. Gilljam G, Svensson A, Ekström A et al. Immunological responses to envelope glycoprotein 120 from subtypes of human immunodeficiency virus type 1. AIDS Res Hum Retroviruses 1999; 15:899-907.
37. Wahren B, Ljungberg K, Rollman E et al. HIV subtypes and recombination strains—strategies for induction of immune responses in man. Vaccine 2002; 20:1988-1993.
38. Isaguliants MG, Gudima SO, Ivanova OV et al. Immunogenic properties of reverse transcriptase of HIV type 1 assessed by DNA and protein immunization of rabbits. AIDS Res Hum Retroviruses 2000; 16:1269-1280.
39. Wang B, Boyer J, Srikantan V et al. Induction of humoral and cellular immune responses to the human immunodeficiency type 1 virus in nonhuman primates by in vivo DNA inoculation. Virology 1995; 211:102-112.

40. Putkonen P, Quesada-Rolander M, Leandersson AC et al. Immune responses but no protection against SHIV by gene-gun delivery of HIV-1 DNA followed by recombinant subunit protein boosts. Virology 1998; 250:293-301.
41. Buge SL, Murty L, Arora K et al. Factors associated with slow disease progression in macaques immunized with an adenovirus-simian immunodeficiency virus (SIV) envelope priming- gp120 boosting regimen and challenged vaginally with SIVmac251. J Virol 1999; 73:7430-7440.
42. Ourmanov I, Brown CR, Moss B et al. Comparative efficacy of recombinant modified vaccinia virus Ankara expressing simian immunodeficiency virus (SIV) Gag-Pol and/or Env in macaques challenged with pathogenic SIV. J Virol 2000; 74:2740-51.
43. Boyer JD, Ugen KE, Wang B et al. Protection of chimpanzees from high-dose heterologous HIV-1 challenge by DNA vaccination. Nat Med 1997; 3:526-32.
44. Schwartz S, Campbell M, Nasioulas G et al. Mutational inactivation of an inhibitory sequence in human immunodeficiency virus type 1 results in Rev-independent gag expression. J Virol 1992; 66:7176-7182.
45. Zur Megede J, Chen MC, Doe B et al. Increased expression and immunogenicity of sequence-modified human immunodeficiency virus type 1 gag gene. J Virol 2000; 74:5997-6005.
46. Hinkula J, Svanholm C, Schwartz S et al. Recognition of prominent viral epitopes induced by immunization with human immunodeficiency virus type 1 regulatory genes. J Virol 1997; 71:5528-5539.
47. Yoshimura FK, Diem FK, Learn GH et al. Intrapatient sequence variation of the gag gene of human immunodeficiency virus type 1 plasma virions. J Virol 1996; 70:8879-8887.
48. Phillips RE, Rowland-Jones S, Nixon D et al. Human immunodeficiency virus genetic variation that can escape cytotoxic T cell recognition. Nature 1991; 354:453-459.
49. Iroegbu J, Birk M, Lazdina U et al. Variability and immunogenicity of human immunodeficiency virus type 1 p24 gene quasispecies. Clin Diagn Lab Imm 2000; 7:377-383.
50. Goulder PJ, Sewell AK, Lalloo DG et al. Patterns of immunodominance in HIV-1-specific cytotoxic T lymphocyte responses in two human histocompatibility leukocyte antigens (HLA)- identical siblings with HLA-A*0201 are influenced by epitope mutation. J Exp Med 1997; 185:1423-1433.
51. Tsomides TJ, Aldovini A, Johnson RP et al. Naturally processed viral peptides recognized by cytotoxic T lymphocytes on cells chronically infected by human immunodeficiency virus type 1. J Exp Med 1994; 180:1283-1293.
52. van der Burg SH, Klein MR, van de Velde CJ et al. Induction of a primary human cytotoxic T-lymphocyte response against a novel conserved epitope in a functional sequence of HIV-1 reverse transcriptase. Aids 1995; 9:121-127.
53. van der Burg SH, Klein MR, Pontesilli O et al. HIV-1 reverse transcriptase-specific CTL against conserved epitopes do not protect against progression to AIDS. J Immunol 1997; 159:3648-3654.
54. Van Baalen CA, Schutten M, Huisman RC et al. Kinetics of antiviral activity by human immunodeficiency virus type 1- specific cytotoxic T lymphocytes (CTL) and rapid selection of CTL escape virus in vitro. J Virol 1998; 72:6851-6857.
55. Örvell C, Unge T, Bhikhabhai R et al. Immunological characterization of the human immunodeficiency virus type 1 reverse transcriptase protein by the use of monoclonal antibodies. J Gen Virol 1991; 72:1913-1918.
56. Restle T, Pawlita M, Sczakiel G et al. Structure-function relationships of HIV-1 reverse transcriptase determined using monoclonal antibodies. J Biol Chem 1992; 267:14654-14661.
57. Laurence J, Saunders AKulkosky J. Characterization and clinical association of antibody inhibitory to HIV reverse transcriptase activity. Science 1987; 235:1501-1504.
58. Walker BD, Flexner C, Paradis TJ et al. HIV-1 reverse transcriptase is a target for cytotoxic T lymphocytes in infected individuals. Science 1988; 240:64-66.
59. Haas G, Samri A, Gomard E et al. Cytotoxic T-cell responses to HIV-1 reverse transcriptase, integrase and protease. Aids 1998; 12:1427-36.
60. Zuber B, Bottiger D, Benthin R et al. An in vivo model for HIV resistance development. AIDS Res Hum Retroviruses 2001; 17:631-635.
61. Samri A, Haas G, Duntze J et al. Immunogenicity of mutations induced by nucleoside reverse transcriptase inhibitors for human immunodeficiency virus type 1-specific cytotoxic T cells. J Virol 2000; 74:9306-9312.
62. Schmitt M, Harrer E, Goldwich A et al. Specific recognition of lamivudine-resistant HIV-1 by cytotoxic T lymphocytes. Aids 2000; 14:653-658.
63. Casimiro DR, Tang A, Perry HC et al. Vaccine-Induced Immune Responses in Rodents and Non-human Primates by use of a Humanized Human Immunodeficiency Virus Type 1 pol gene. J Virol 2002; 76:185-194.

64. Isaguliants MG, Petrakova NN, Zuber B et al. DNA-encoding enzymatically active HIV-1 reverse transcriptase, but not the inactive mutant, confers resistance to experimental HIV-1 challenge. Intervirology 2000; 43:288-293.

65. Zuber B, Mäkitalo B, Kjerrström Zuber A et al. A novel potent strategy for induction of immunity to HIV-1 reverse transcriptase in primates. AIDS 2002; 16:1839-1840.

66. Brander C, Walker BD. T lymphocyte responses in HIV-1 infection: implications for vaccine development. Curr Opin Immunol 1999; 11:451-9.

67. Reiss P, Lange JM, de Ronde A et al. Speed of progression to AIDS and degree of antibody response to accessory gene products of HIV-1. J Med Virol 1990; 30:163-8.

68. Re MC, Furlini G, Vignoli M et al. Effect of antibody to HIV-1 Tat protein on viral replication in vitro and progression of HIV-1 disease in vivo. J Acquir Immune Defic Syndr Hum Retrovirol 1995; 10:408-416.

69. Tähtinen M, Ranki A, Valle SL et al. B-cell epitopes in HIV-1 Tat and Rev proteins colocalize with T-cell epitopes and with functional domains. Biomed Pharmacother 1997; 51:480-487.

70. Zagury JF, Sill A, Blattner W et al. Antibodies to the HIV-1 Tat protein correlated with nonprogression to AIDS: a rationale for the use of Tat toxoid as an HIV-1 vaccine. J Hum Virol 1998; 1:282-292.

71. Cafaro A, Caputo A, Fracasso C et al. Control of SHIV-89.6P-infection of cynomolgus monkeys by HIV-1 Tat protein vaccine. Nature Medicine 1999; 5:643-650.

72. Demirhan I, Chandra A, Hasselmayer O et al. Detection of distinct patterns of anti-tat antibodies in HIV-infected individuals with or without Kaposi's sarcoma. J Acquir Immune Defic Syndr 1999; 22:364-368.

73. Demirhan I, Chandra A, Hasselmayer O et al. Intercellular traffic of human immunodeficiency virus type 1 transactivator protein defined by monoclonal antibodies. FEBS Lett 1999; 445:53-56.

74. Zauli G, La Placa M, Vignoli M et al. An autocrine loop of HIV type-1 Tat protein responsible for the improved survival/proliferation capacity of permanently Tat-transfected cells and required for optimal HIV-1 LTR transactivating activity. J Acquir Immune Defic Syndr Hum Retrovirol 1995; 10:306-316.

75. Froebel KS, Aldhous MC, Mok JY et al. Cytotoxic T lymphocyte activity in children infected with HIV. AIDS Res Hum Retroviruses 1994; 10:S83-88.

76. Van Baalen CA, Pontesilli O, Huisman RC et al. Human immunodeficiency virus type 1 Rev- and Tat-specific cytotoxic T lymphocyte frequencies inversely correlate with rapid progression to AIDS. J Gen Virol 1997; 78:1913-1918.

77. Addo MM, Altfeld M, Rosenberg ES et al. The HIV-1 regulatory proteins Tat and Rev are frequently targeted by cytotoxic T lymphocytes derived from HIV-1-infected individuals. Proc Natl Acad Sci U S A 2001; 98:1781-1786.

78. Allen TM, O'Connor DH, Jing P et al. Tat-specific cytotoxic T lymphocytes select for SIV escape variants during resolution of primary viraemia. Nature 2000; 407:386-390.

79. Ameisen JC, Guy B, Chamaret S et al. Persistent antibody response to the HIV-1-negative regulatory factor in HIV-1-infected seronegative persons. N Engl J Med 1989; 320:251-2.

80. Chenciner N, Michel F, Dadaglio G et al. Multiple subsets of HIV-specific cytotoxic T lymphocytes in humans and in mice. Eur J Immunol 1989; 19:1537-1544.

81. Culmann B, Gomard E, Kieny MP et al. An antigenic peptide of the HIV-1 NEF protein recognized by cytotoxic T lymphocytes of seropositive individuals in association with different HLA-B molecules. Eur J Immunol 1989; 19:2383-2386.

82. Koenig S, Fuerst TR, Wood LV et al. Mapping the fine specificity of a cytolytic T cell response to HIV-1 nef protein. J Immunol 1990; 145:127-35.

83. Altfeld M, Addo MM, Eldridge RL et al. Vpr Is Preferentially Targeted by CTL During HIV-1 Infection. J Immunol 2001; 167:2743-2752.

84. Shiver JW, Perry HC, Davies ME et al. Cytotoxic T lymphocyte and helper T cell responses following HIV polynucleotide vaccination. Ann N Y Acad Sci 1995; 772:198-208.

85. Wahren B, Hinkula J, Stähle EL et al. Nucleic acid vaccination with HIV regulatory genes. Ann N Y Acad Sci 1995; 772:278-281.

86. Asakura Y, Hamajima K, Fukushima J et al. Induction of HIV-1 Nef-specific cytotoxic T lymphocytes by Nef- expressing DNA vaccine. Am J Hematol 1996; 53:116-117.

87. Svanholm C, Löwenadler BWigzell H. Amplification of T-cell and antibody responses in DNA-based immunization with HIV-1 Nef by co-injection with a GM-CSF expression vector. Scand J Immunol 1997; 46:298-303.

88. Moynier M, Kavsan V, Gales C et al. Characterization of humoral immune responses induced by immunization with plasmid DNA expressing HIV-1 Nef accessory protein. Vaccine 1998; 16:1523-1530.

89. Asakura Y, Lundholm P, Kjerrström A et al. DNA-plasmids of HIV-1 induce systemic and mucosal immune responses. Biol Chem 1999; 380:375-379.
90. Caselli E, Betti M, Grossi MP et al. DNA immunization with HIV-1 tat mutated in the trans activation domain induces humoral and cellular immune responses against wild-type Tat. J Immunol 1999; 162:5631-5638.
91. Collings A, Pitkanen J, Strengell M et al. Humoral and cellular immune responses to HIV-1 nef in mice DNA- immunised with nonreplicating or self-replicating expression vectors. Vaccine 1999; 18:460-7.
92. Kjerrström A, Wahren B. Expression of HIV regulatory DNA vaccine constructs. Biogenic Amines 1999; 15:93-112.
93. Moureau C, Moynier M, Kavsan V et al. Specificity of anti-Nef antibodies produced in mice immunized with DNA encoding the HIV-1 nef gene product. Vaccine 1999; 18:333-341.
94. Osterhaus AD, Van Baalen CA, Gruters RA et al. Vaccination with Rev and Tat against AIDS. Vaccine 1999; 17:2713-2714.
95. Svanholm C, Bandholtz L, Lobell A et al. Enhancement of antibody responses by DNA immunization using expression vectors mediating efficient antigen secretion. J Immunol Methods 1999; 228:121-130.
96. Kjerrström A, Hinkula J, Engström G et al. Interactions of single and combined human immunodeficiency virus type 1 (HIV-1) DNA vaccines. Virology 2001; 284:46-61.
97. Tähtinen M, Strengell M, Collings A et al. DNA vaccination in mice using HIV-1 nef, rev and tat genes in self- replicating pBN-vector. Vaccine 2001; 19:2039-2047.
98. Calarota S, Bratt G, Nordlund S et al. Cellular cytotoxic response induced by DNA vaccination in HIV-1- infected patients. Lancet 1998; 351:1320-1325.
99. Calarota SA, Leandersson AC, Bratt G et al. Immune responses in asymptomatic HIV-1-infected patients after HIV-DNA immunization followed by highly active antiretroviral treatment. J Immunol 1999; 163:2330-2338.
100. Calarota SA, Kjerrström A, Islam KB et al. Gene combination raises broad human immunodeficiency virus-specific cytotoxicity. Human Gene Therapy 2001; 12:1623-1637.
101. Ayyavoo V, Nagashunmugam T, Phung MT et al. Construction of attenuated HIV-1 accessory gene immunization cassettes. Vaccine 1998; 16:1872-1879.
102. Ayyavoo V, Kudchodkar S, Ramanathan MP et al. Immunogenicity of a novel DNA vaccine cassette expressing multiple human immunodeficiency virus (HIV-1) accessory genes. Aids 2000; 14:1-9.
103. Lehner T, Wang Y, Doyle C et al. Induction of inhibitory antibodies to the CCR5 chemokine receptor and their complementary role in preventing SIV infection in macaques. Eur J Immunol 1999; 29:2427-2435.
104. Zuber B, Hinkula J, Levi M et al. Induction of immune responses and break of tolerance by DNA against the HIV-1 co-receptor CCR5 but no protections from SIV SM challenge. Virology 2000; 633:1-10.
105. Lehner T, Doyle C, Wang Y et al. Immunogenicity of the extracellular domains of C-C chemokine receptor 5 and the in vitro effects on Simian immunodeficiency virus or HIV infectivity. J Immunol 2001; 166:7446-7455.
106. Dean M, Carrington M, Winkler C et al. Genetic restriction of HIV-1 infection and progression to AIDS by a deletion allele of the CCR5 structural gene. Hemophilia growth and development study, multicenter AIDS cohort study, multicenter hemophilia cohort study, San Fransisco city cohort, ALIVE study. Science 1996; 273:1856-1862.
107. Barry MA, Lai WCJohnston SA. Protection against mycoplasma infection using expression-library immunization. Nature 1995; 377:632-635.
108. Leonard JM, Abramczuk JW, Pezen DS et al. Development of disease and virus recovery in transgenic mice containing HIV proviral DNA. Science 1988; 242:1665-1670.
109. Wang B, Merva M, Dang K et al. DNA inoculation induces protective in vivo immune responses against cellular challenge with HIV-1 antigen-expressing cells. AIDS Res Hum Retroviruses 1994; 10 Suppl. 2:35-41.
110. Andäng M, Hinkula J, Hotchkiss G et al. Dose-response resistance to HIV-1/MuLV pseudotype virus ex vivo in a hairpin ribozyme transgenic mouse model. Proc Natl Acad Sci 1999; 96:12749-12753.
111. van der Ende ME, Guillon C, Boers PHM et al. Broadening of coreceptor usage by human immunodeficiency virus type 2 does not correlate with increasing pathogenicity in an in vivo model. J Gen Virol 2000; 81:507-513.
112. McCune J, "The SCID-hu mouse as a model for human immunodeficiency virus (HIV) infection.," Academic Press, Inc., 1991.

113. Mosier DE, Gulizia RJ, MacIsaac PD et al. Resistance to human immunodeficiency virus 1 infection of SCID mice reconstituted with peripheral blood leukocytes from donors vaccinated with vaccinia gp160 and recombinant gp160. Proc Natl Acad Sci USA 1993; 90:2443-2447.

114. Walther-Jallow L, Nilsson C, Söderlund J et al. Cross-protection against mucosal simian immunodeficiency virus (SIVsm) challenge in huamn immunodeficiency virus type 2-vaccinated cynomolgus monkeys. J Gen Virol 2001; 82:1601-1612.

115. Kent SJ, Zhao A, Best SJ et al. Enhanced T-cell immunogenicity and protective efficacy of a human immunodeficiency virus type 1 vaccine regimen consisting of consecutive priming with DNA and boosting with recombinant fowlpox virus. J Virol 1998; 72:10180-10188.

116. Haigwood NL, Pierce CC, Robertson MN et al. Protection from pathogenic SIV challenge using multigenic DNA vaccines. Immunol Lett 1999; 66:183-188.

117. Barouch D, Hal E. Augmentation of immune responses to HIV-1 and simian immunodeficiency virus DNA vaccines by IL-2/Ig plasmid administration in rhesus monkeys. Proc Natl Acad Sci USA 2000; 97:4192-4197.

118. Amara RR, Villinger F, Altman JD et al. Control of a mucosal challenge and prevention of AIDS by a multiprotein DNA/MVA vaccine. Science 2001; 292:69-74.

119. Akahata W, Ido E, Shimada T et al. DNA vaccination of macaques by a full genome HIV-1 plasmid which produces noninfectious virus particles. Virology 2000; 275:116-124.

120. Barouch DH, Santra S, Schmitz JE et al. Control of viremia and prevention of clinical AIDS in rhesus monkeys by cytokine-augmented DNA vaccination. Science 2000; 290:486-492.

121. Kim JJ, Yang JS, Nottingham LK et al. Protection from immunodeficiency virus challenges in rhesus macaques by multicomponent DNA immunization. Virology 2001; 285:204-217.

122. Wee EG, Patel S, McMichael AJ et al. A DNA/MVA-based candidate human immunodeficiency virus vaccine for Kenya induces multi-specific T cell repsonses in macaques. J Gen Virol 2002; 83:75-80.

123. Barouch DH, Santra S, Kuroda MJ et al. Reduction of simian-human immunodeficiency virus 89.6P viremia in rhesus monkeys by recombinant modified vaccinia virus Ankara vaccination. J Virol 2001; 75:5151-5158.

124. Hanke T, Samuel RV, Blanchard TJ et al. Effective induction of simian immunodeficiency virus-specific cytotoxic T lymphocytes in macaques by using a multiepitope gene and DNA prime-modified vaccinia virus Ankara boost vaccination regimen. J Virol 1999; 73:7524-7532.

125. Barouch DH, Craiu A, Santra S, et al. Elicitation of high-frequency cytotoxic T-lymphocyte responses against both dominant and subdominant simian-human immunodeficiency virus epitopes by DNA vaccination of rhesus monkeys. J Virol 2001; 75:2462-2472.

126. Hel Z, Tsai WP, Thornton A et al. Potentiation of simian immunodeficiency virus (SIV)-specific CD4+ and CD8+ T cell responses by a DNA-SIV and NYVAC-SIV prime/boost regimen. J Immunol 2001; 167:7180-7191.

127. Robinson HL, Montefiori DC, Johnson RP et al. DNA priming and recombinant pox virus boosters for an AIDS vaccine. Dev Biol 2000; 104:93-100.

128. Mäkitalo B, Hinkula J, Nilsson C et al., Cellular immune responses in relation to route of immunization of cynomolgus monkeys vaccinated with a DNA prime/modified vaccinia Ankara (MVA) boost regimen., Keystone Meeting "AIDS vaccines in the new millenium", Keystone, CO., 2001, p. Abstract.

129. Belyakov IM, Hel Z, Kelsall B et al. Mucosal AIDS vaccine reduces disease and viral load in gut reservoir and blood after mucosal infection of macaques. Nature Medicine 2001; 12:1320-1326.

130. Bagarazzi ML, Boyer JD, Ugen KE et al. Safety and immunogenicity of HIV-1 DNA constructs in chimpanzees. Vaccine 1998; 16:1836-1841.

131. Hel Z, Venzon D, Poudyal M et al. Viremia control following antiretroviral treatment and therapeutic immunization during primary SIV251 infection of macaques. Nature Medicine 2000; 6:1140-1146.

132. André S, Seed B, Eberle J et al. Increased immune response elicited by DNA vaccination with a synthetic gp120 sequence with optimized codon usage. J Virol 1998; 72:1497-1503.

133. Kim JJ, Ayyavoo V, Bagarazzi ML et al. In vivo engineering of a cellular immune response by coadministration of IL-12 expression vector with a DNA immunogen. J Immunol 1997; 158:816-826.

134. Okada E, Sasaki S, Ishii N et al. Intranasal immunization of a DNA vaccine with IL-12- and granulocyte-macrophage colony-stimulating factor (GM-CSF)-expressing plasmids in liposomes induces strong mucosal and cell-mediated immune responses against HIV-1 antigens. J Immunol 1997; 159:3638-3647.

135. Lee AH, Suh YSSung YC. DNA inoculations with HIV-1 recombinant genomes that express cytokine genes enhance HIV-1 specific immune responses. Vaccine 1999; 17:473-439.

136. Kim JJ, Yang JS, Manson KH et al. Modulation of antigen-specific cellular immune responses to DNA vaccination through the use of IL-2, IFN-gamma, or IL-4 gene adjuvants. Vaccine 2001; 19:2496-2505.

137. Sato Y, Roman M, Tighe H et al. Immunostimulatory DNA sequences necessary for effective intrademal gene immunization. Science 1996; 273:299-302.

138. Klinman D, Yamshchikov GIshigatsubo Y. Contribution of CpG motifs to the immunogenicity of DNA vaccines. J Immunol 1997; 158:3635-3639.

139. Krieg AM, Yi AK, Schorr J et al. The role of CpG dinucleotides in DNA vaccines. Trends Microbiol 1998; 6:23-27.

140. Vabulas R, Pircher H, Lipford G et al. CpG-DNA activates in vivo T-cell epitope presenting dendritic cells to trigger protective antiviral cytotoxic T-cell responses. J Immunol 2000; 164:2372-2378.

141. Hemmi H, Takeuchi O, Kawai T et al. A Toll-like receptor recognizes bacterial DNA. Nature 2000; 408:740-745.

142. Spetz AL, Sörensen AS, Wahren B et al. Induction of HIV-1 specific immunity after vaccination with apoptotic bodies derived from HIV-1/murine leukemia virus infected cells. J Immunol 2002; in press.

143. Spetz A-L, Rolen U, Andersson J et al., Induction of HIV-1 specific immunity in mice after vaccination with apoptotic HIV-1/MuLV cells., Proc of ICAC, Chicago, USA, 2001, p. Abstract.

144. Xin KQ, Hamajima K, Sasaki S et al. Intranasal administration of human immunodeficiency virus type-1 (HIV-1) DNA vaccine with interleukin-2 expression plasmid enhances cell-mediated immunity against HIV-1. challenges by DNA. Immunology 1998; 94:438-444.

145. Lundholm P, Leandersson A-C, Christensson B et al. DNA mucosal HIV vaccine in humans. Virus Research 2002; 82:141-145.

146. Lundholm P, Asakura Y, Hinkula J et al. Induction of mucosal IgA by a novel jet delivery technique for HIV-1 DNA. Vaccine 1999; 17:2036-2042.

147. Haynes JR, Fuller DH, McCabe D et al. Induction and characterization of humoral and cellular immune responses elicited via gene gun-mediated nucleic acid immunization. Adv Drug Delivery Rev 1996; 2:3-18.

148. Fuller DH, Murphey-Corb M, Clements J et al. Induction of immunodeficiency virus-specific immune responses in rhesus monkeys following gene gun-mediated DNA vaccination. J Med Primatol 1996; 25:236-241.

149. Roy MJ, Wu MS, Barr LJ et al. Induction of antigen-specific CD8+ T cells, T helper cells, and protective levels of antibody in humans by particle-mediated administration of a hepatitis B virus DNA vaccine. Vaccine 2001; 19:764-778.

150. Mathiesen J. Electropermeabilization of sceletal muscle enhances gene transfer in vivo. Gene Ther 1999; 6:508-514.

151. Widera G, Austin M, Rabussay D et al. Increased DNA vaccine delivery and immunogenicity by electroporation in vivo. J Immunol 2000; 164:4635-4640.

152. Desrosier RC, Wyand MS, Kodama T et al. Vaccine protection against simian immunodeficiency virus infection. Proc Natl Acad Sci USA 1989; 86:6353-6357.

153. Baba TW, Liska V, Khimani AH et al. Live attenuated, multiply deleted simian immunodeficiency virus causes AIDS in infant and adult macaques. Nature Medicine 1999; 5:194-203.

154. Deacon NJ, Tsykin A, Solomon A et al. Genomic structure of an attenuated quasi species of HIV-1 from a blood transfusion donor and recipients. Science 1995; 270:988-991.

155. Learmont JC, Gaczy AF, Mils J et al. Immunologic and virologic status after 14 to 18 years of infection with an attenuated strain of HIV-1. New Engl J Med 1999; 340:1715-1722.

156. Sizemore DR, Branstrom AASadoff JC. Attenuated Shigella as a DNA delivery vehicle for DNA-mediated immunization. Science 1995; 270:299-302.

157. Berglund P, Smerdou C, Fleeton MN et al. Enhancing immune responses using suicidal DNA vaccines. Nature Biotechnology 1998; 16:562-565.

158. Tartaglia J, Al E. Canarypox virus-based vaccines: prime-boost strategies to induce cell-mediated and humoral immunity against HIV. AIDS Res Hum Retrovir 1998; 14 (Suppl 3):S291-S298.

159. Sutter G, Haas J. Novel vaccine delivery systems: solutions to HIV vaccine dilemmas? AIDS 2001; 15:S139-S145.

160. Shiver JW, Fu TM, Chen L et al. Replication-incompetent adenoviral vaccine vector elicits effective anti-immunodeficiency-virus immunity. Nature 2002; 415:331-335.

161. Wang R, Doolan DL, Le TP et al. Induction of antigen-specific cytotoxid T lymphocytes in humans by a malaria DNA vaccine. Science 1998; 282:476-480.

162. Egan MA, Pavlat WA, Tartaglia J et al. Induction of human immunodeficiency virus type 1 (HIV-1)-specific cytolytic T lymphocyte responses in seronegative adults by a nonreplicating, host range restricted canarypox vector (ALVAC) carrying the HIV-1 MN gene. J Infect Dis 1995; 171:1623-1627.

163. Belshe RB, Gorse GJ, Mulligan MJ et al. Induction of immune reponses to HIV-1 by canarypox virus (ALVAC) HIV-1 and gp120 SF-2 recombinant vaccines in healthy volunteers. AIDS 1998; 12:2407-2415.

164. Evans TG, Keefer MC, Weinhold KJ et al. A canarypox vaccine expressing multiple human immunodeficiency virus type 1 genes given alone or with rgp120 elicits broad and durable CD8+ cytotoxic T lymphocyte responses in seronegative volunteers. J Infect Dis 1999; 180:290-298.

165. Salmon-Ceron D, Excler JL, Finkielsztein L et al. Safety and immunogenicity of a live recombinant canarypox virus expressing HIV type 1 gp120 MN tm/gag/protease LAI(ALVAC-HIV,vCP2050 followed by a p24E-V3 MN synthetic peptide (CLTB-36) administered in healthy volunteers at low risk for HIV infection. AIDS Res Human Retroviruses 1999; 15:633-645.

166. Hanke T, McMichael AJ. Design and construction of an experimental HIV-1 vaccine for a year-2000 clinical trial in Kenya. Nature Medicine 2000; 6:951-955.

167. Rowland-Jones S, Tan R, McMichael A. Role of cellular immunity in protection against HIV infection. Adv Immunol 1997; 65:277-346.

168. MacGregor RR, Boyer JD, Ugen KE et al. First human trial of a DNA-based vaccine for treatment of human immunodeficiency virus type 1 infection: safety and host response. J Infect Dis 1998; 178:92-100.

169. Boyer JD, Chattergoon MA, Ugen KE et al. Enhancement of cellular immune response in HIV-1 seropositive individuals: a DNA-based trial. Clin Immunol 1999; 90:100-107.

170. Leandersson A-C, Bratt G, Hinkula J et al. Induction of specific T-cell responses in HIV infection. AIDS 1998; 12:157-166.

171. Autran B, Carcelain G, Li T et al. Positive effects of combined antiretroviral therapy on CD4$^+$ T cell homeostasis and function in advanced HIV disease. Science 1997; 277:112-116.

172. Pitcher CJ, Quittner C, Peterson DM et al. HIV-1-specific CD4+ T cells are detectable in most individuals with active HIV-1 infection, but decline with prolonged viral suppression. Nature Medicine 1999; 5:518-525.

173. Maino VC, Suni MA, Wormsley SB et al. Enhancement of HIV type 1 antigen-specific CD4+ T cell memory in subjects with chronic HIV type 1 infection receiving an HIV type 1 immunogen. AIDS Res Hum Retroviruses 2000; 16:2065-2066.

174. Moss RB, Giermakowska W, Wallace MR et al. T-helper cell proliferative responses to whole-killed human immunodeficiency virus type 1 (HIV-1) and p24 antigens of different clades in HIV-1-infected sujects vaccinated with HIV-1 immunogen (Remune). Clin Diagn Lab Immunol 2000; 7:724-727.

175. Jin X, Ramanathan JM, Barsoum S et al. Safety and immunogenicity of ALVAC vCP1452 and recombinant gp160 in newly human immunodeficiency virus type 1-infected patients treated with prolonged highly active antiretroviral therapy. J Virol 2002; 76:2206-2216.

176. Haslett PA, Nixon DF, Shen Z et al. Strong human immunodeficiency virus (HIV)-specific CD4+ T cell responses in a cohort of chronically infected patients are associated with interruptions in anti-HIV chemotherapy. J Infect Dis 2000; 181:1264-1272.

177. Carcelain G, Tubiana R, Samri A et al. Transient mobilization of human immunodeficiency virus (HIV)-specific CD4 T-helper cells fails to control virus rebounds during intermittent antiretroviral therapy in chronic HIV type 1 infection. J Virol 2001; 75:234-241.

178. Leandersson L. "The central role of the CD4 T-helper cell in HIV infection.," Karolinska Institute, Stockholm, 2001.

CHAPTER 10

DNA Vaccines Against Bacterial Pathogens

M. A. Chambers, H. M. Vordermeier, R. G. Hewinson and D. B. Lowrie

Bacterial pathogens against which DNA vaccines are being developed encompass both intracellular and extracellular pathogens as well as vaccines against bacterial toxins. DNA vaccination has an inherent bias towards generating cellular immunity by virtue of the intracellular origin of the antigen, resulting in particular efficacy against intracellular pathogens. However, by manipulating the formulation and delivery, effective antibody responses can also be obtained. The overwhelming majority of publications describe efforts to produce DNA vaccines against tuberculosis, and therefore the first part of this chapter will be dedicated to this organism and other pathogens of the mycobacterial genus like *M. leprae* and *M. avium*. In the second part we will give a detailed overview of the endeavors to develop DNA vaccines against other bacterial pathogens.

DNA Vaccines Against Mycobacterial Infection

Pathogenic Mycobacteria

Mycobacterium is a bacterial genus containing more than 30 different species including *M. tuberculosis* and *M. bovis* causing human and bovine tuberculosis, respectively, and *M. leprae* the causative agent of leprosy in humans. Other mycobacterial human pathogens comprise *M. avium*, a pathogen causing opportunistic infections in immuno-compromised individuals like AIDS patients, and *M. ulcerans*, which causes buruli ulcer, an emerging health problem in developing countries.

The Challenges Presented by Tuberculosis

Human tuberculosis, caused by *M. tuberculosis*, is a major global human health problem that claims more than 2 millions lives every year.[1] The advent of the HIV pandemic has tragically increased the incidence of tuberculosis in developing countries, e.g., sub-Saharan Africa, where AIDS-related tuberculosis is becoming the main cause of mortality in young adults. Indeed, it has been estimated that *M. tuberculosis* is responsible for the deaths of more youths and adults than any other infectious agent. The WHO has therefore declared tuberculosis a global health emergency.[2]

BCG (*M. bovis* Bacille Calmette Guerin), the TB vaccine, is a major component of many national vaccination programs. About 90 million doses are supplied by the World Health Organization (WHO) every year. However, trials performed in a number of countries have revealed that BCG imparts a very heterogeneous degree of protection, ranging from 0 to about 80 %.[3-5] Although there is usually quite good protection imparted against the severe childhood forms of tuberculosis, such as tuberculous meningitis, there can be poor protection against the infectious adult form.[6] Hence a better vaccine is needed, or one that can be given in addition to BCG and results in better protection against both childhood and adult disease. Critically, vaccination must prevent the establishment of the persistent latent state of infection that can follow primary infection and provides a major source of adult disease many years later.

DNA Vaccines, edited by Hildegund C. J. Ertl. ©2003 Eurekah.com
and Kluwer Academic / Plenum Publishers.

A vaccine against tuberculosis in cattle is also needed. Bovine tuberculosis, caused by the closely related organism *M. bovis*, can be a significant source of disease in man. For example, around years 1930-1940 about 40% of cows in the UK were infected and 6 – 7 % of total human deaths due to tuberculosis were caused by *M. bovis*.[7] BCG can be effective in cattle, but as in man, the efficacy varies widely and the vaccine response interferes with the hypersensitivity tests deployed in control programs. The strategy of detecting (by skin test hypersensitivity) and slaughtering infected animals is being used to reduce and contain the problem (in conjunction with pasteurization of raw milk) but is not universally applicable especially in the developing world and cannot eliminate the disease in the face of reservoirs of infection in wildlife.[8] A recent review of the situation of bovine TB in cattle and wildlife in England and Wales for example has concluded that the development of an effective cattle vaccine against bovine TB has the best prospect of TB control in the National herd.[9]

The Nature of Protective Immunity Against Tuberculosis

Most constituents of the cellular and humoral immune system have been documented to be involved following infection with *M. tuberculosis* and several comprehensive review papers have been published recently.[10-13] The view that antibody has no role to play in protection may be changing slightly,[14,15] but as yet the paradigm of protection in tuberculosis remains cellular immunity. Unfortunately, the disease is also entirely a consequence of the cellular immune response. Accordingly, vaccination is required to enhance the protective and not the harmful aspects of cellular immunity. Numerous studies have shown that DNA vaccination is able to achieve this, although we do not fully understand how, and most remarkably DNA vaccines are even able to exert a therapeutic effect in infected animals.

The general principles characterizing protective immune responses are represented graphically in Figure 1. Tubercle bacilli reside and multiply within infected macrophages. Antigen-specific T lymphocytes produce lymphokines including macrophage activating factors such as IFN-γ that enable the macrophages to at least restrict growth of the bacteria and maybe even to kill them. In 90% of people who become infected the organisms are eliminated without ever causing disease, but it has been hard to establish that activated macrophages actually kill the bacteria. However, it may be significant that killing of tubercle bacteria can occur when infected macrophages are themselves killed by the cytotoxic lymphocytes either through induction of apoptosis or by granule-mediated lysis.

TB Vaccines for Prophylactic Vaccination and Identification of Protective Antigens

The era of DNA vaccination against mycobacterial diseases began in 1994 when Silva and Lowrie demonstrated protection against *M. tuberculosis* after vaccination of mice with the macrophage cell line J774 that had been stably transfected with the heat-shock protein HSP65 from *M. leprae*.[16] This provided proof of principle that a mycobacterial antigen expressed by a eukaryotic expression vector could induce protective immunity and was quickly followed by the successful use of a naked DNA vaccine expressing the same antigen.[17] Similar protection could be obtained with DNA vaccines expressing either the secreted fibronectin binding protein antigen 85A (Ag85A)[18] of *M. tuberculosis* or the *M. leprae* HSP65.[19] This was followed soon after by a report indicating that another secreted TB antigen, the phosphate transporter PstS1 (38 kDa antigen) also induced protective immunity.[20] Since then a large number of antigens have been expressed as DNA vaccines and tested in mouse, guinea pig, or cattle models of tuberculosis. Antigens, which in at least one of these animal models have been proven protective, are listed in Table 1. The overwhelming majority of antigens are secreted antigens. This is consistent with (and in part a consequence of) the long-held belief that such antigens will be the earliest targets seen by the immune system upon infection and therefore likely to be protective.[21] Nevertheless, a number of other antigens that are probably not secreted also impart protection when given as DNA vaccines. Examples are the heat-shock protein HSP65, or

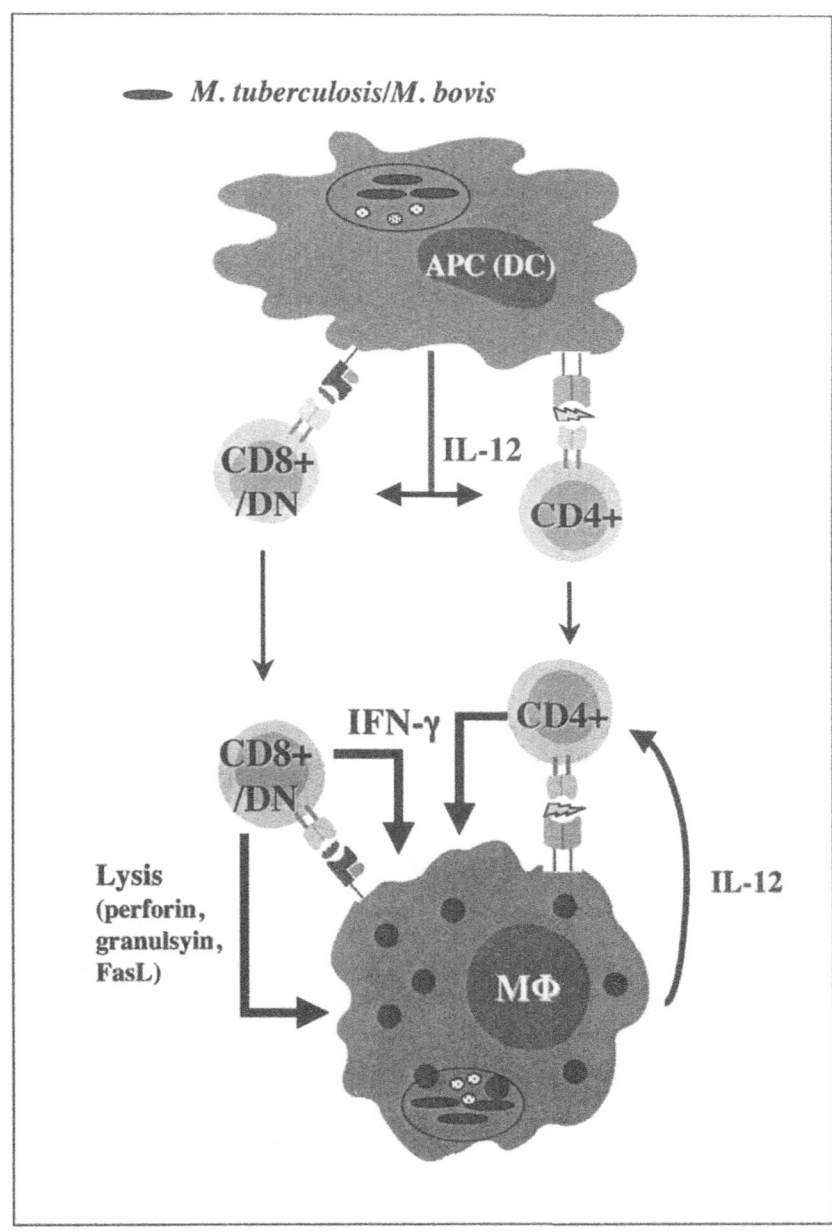

Figure 1. Immune responses against tuberculosis. The upper panel represents the induction phase of the immune response involving infection of dendritic cells by tubercle bacilli and presentation of mycobacterial antigens to naive T cells. The lower panel depicts the immune effector phase with T cells producing cytokines like IFN-γ that activate infected macrophages, or lyse infected macrophages through CTL activity (reviewed in refs. 10, 11, 138). APC, antigen presenting cell; DC, dendritic cell; MΦ, macrophage; CD4⁺, CD4-positive T cell; CD8⁺, CD8-positive T cell; DN, CD4-CD8- ('double negative') T cell; IFN-γ, interferon-gamma; IL-12, interleukin-12.

Table 1. Mycobacterial antigens protecting against tuberculosis when applied as prophylactic vaccine

Antigen	ORF Designation[a]	Source[b]	Function/ Characteristic	Animal Model	Immune Responses[c]	Challenge Organism / Route	Protection[e] (relative protection values)	Ref
A. Secreted proteins:								
Ag85A	Rv3804c	MT[f]	Fibronectin-binding, mycocyl transferase	Mouse	CD4/CTL/IFN	MT / air/iv/in	0.55 to 1.0	[18, 28-30, 139]
				Guinea pig	NT	MT/air	Prolonged survival: 181 d. vs. 119 d (control)	[31]
				Cattle	No responses	NT	NT	[32]
Ag85B	Rv1886c	MT	as above	Mouse	CD4/CTL/IFN	BCG/iv MT/air	BCG: 1.0 MT: 0.63	[29] [40]
ESAT-6	Rv3875	MT	unknown	Mouse	CD4/CTL/IFN	MT/air/iv/ip	No protection to 0.72	[22-24, 40, 140]
MPB83/ MPT83	Rv2873	MT/MB	lipoprotein/ glycoprotein	Mouse	NT	MT/air MB/iv	MT: 0.36 MB: 1.30	[22] [53]
				Guinea pig Cattle	NT CD4, ^{3}H (IFN)	MT/air MB/it	Reduction in pathology No protection[g]	[54] [32]
PstS-1/ 38 kDa	Rv0934	MT	phosphate transporter	Mouse	CD4,CTL,IFN	MT/ ip,iv,air	No protection to 0.64	[20, 30]
PstS-3	Rv0928	MT	phosphate transporter	Mouse	IFN	MT/iv	0.63 to 0.74	[30]
MPT64	Rv1980c	MT	unknown	Mouse	CD4,CTL,IFN	MT/air	0.50 to 0.55	[22, 23, 40]

continued on next page

Table 1. Mycobacterial antigens protecting against tuberculosis when applied as prophylactic vaccine (Cont'd.)

Antigen	ORF Designation[a]	Source[b]	Function/ Characteristic	Animal Model	Immune Responses[c]	Challenge Organism / Route	Protection[e] (relative protection values)	Ref
MPT32/ Apa	Rv1860	MT	glycoprotein	Mouse		BCG/iv	0.47 (spleen)	[43]
						MT/air	None	[22]
				Cattle		MB/it	No protection[g]	
MPT63	Rv1926c	MT	unknown	Mouse		MT/air	0.40	[22]
Mtb8.4	Rv1174c	MT	unknown	Mouse	CTL, IFN	MT/iv	1.12	[141]
B. Heat-shock proteins:								
HSP65	ML0381	ML	chaperone	Mouse	CD4/CTL/IFN	MT/ip	0.67 to 1.0	[19, 24]
				Cattle	IFN	MB/it	No protection[g]	
HSP65	Rv3417c	MT	chaperone	Mouse	IFN	MT/in	No protection	[25]
				Guinea pig	NT	MT/air	No protection	[25]
HSP70	Rv0350	MT	chaperone	Mouse		MT/ip	0.40	[24]
				Cattle		MB/it	No protection[g]	
C. Other proteins:								
KatG	Rv1908c	MT	cytoplasmic enzyme	Mouse	IFN	MT/air	0.61	[22,23]
MTB39	Rv1196	MT	PE/PPE	Mouse		MT/air	0.61	[142]
MTB41	Rv0915c	MT	PE/PPE	Mouse Mouse	CD4/IFN	MT/air	0.80	[143]
	Rv1818c	MT	PE/PPE		IFN	MT/air	0.58	[144]

a ORF designations of M. tuberculosis genes according to[62] M. leprae genes according to.[63] b Mycobacterial species from which vaccine antigens have been derived. c Prominent immune response described. CD4: Responses of CD4+ T cells; CTL: cytotoxic T lymphocyte responses; IFN: IFN-γ responses; [3]H: proliferative responses; H: IFN-γ responses. d Routes: air: aerosol; iv: intravenous; ip: intraperitoneal; in: intranasal, it: intratracheal. e If not specifically mentioned, lung responses are given as relative protection values compared to BCG (vaccine-induced lung protection/BCG-induced protection, i.e., relative protection for BCG = 1.0). Lowest and highest protective efficacies are shown if several references were consulted. f Mycobacterial species: MT: M. tuberculosis; MB: M. bovis; ML: M. leprae. g Vordermeier, Hewinson and Buddle, unpublished data. CFU, colony forming units; NT, not tested.

the cytoplasmic enzyme KatG (Table 1).[19,22,23] Although immune responses to the antigens listed in Table 1 conferred protection against *M. tuberculosis* to some degree, only very rarely did the protection surpass that observed after BCG vaccination (Table 1). We will now briefly discuss the results obtained so far for several of the more intimately studied antigens, namely HSP65, Ag85A, ESAT-6, and phosphate transporters.

HSP65

As mentioned above, vaccination of mice with *M. leprae* derived HSP65 imparted a good degree of protection when vaccinated mice were challenged with *M. tuberculosis*.[19,24] In initial studies, several inbred strains of mice could be protected and subsequently good protection was also observed in outbred mice. The immune responses described following vaccination included strong IFN-γ production by both CD4[+] and CD8[+] T cells as well as the induction of cytotoxic T cells. We have recently found that cattle vaccinated with *M. leprae* HSP65 generated strong IFN-γ and T cell proliferative responses, but did not respond subsequently to a tuberculin skin test positively (Vordermeier, and Hewinson, unpublished data). This is a significant observation because it suggests that vaccination strategies can be developed which would allow the continuation of tuberculin-based test and slaughter control strategies of bovine tuberculosis alongside vaccination. If this were also shown to be the case for humans, it would allow, by tuberculin skin testing, the identification of those vaccinated but infected individuals that required therapy. In subsequent studies cattle were not protected when vaccinated with DNA expressing a cocktail of antigens containing *M. leprae* HSP65 and the *M. tuberculosis* HSP70 and Apa proteins (Vordermeier, Hewinson and Buddle, unpublished results, see Table 1). However, this cocktail significantly improved the efficacy of BCG when applied in a prime-boost protocol (see below). In contrast to the results obtained after vaccination with *M. leprae* HSP65, the group of Orme vaccinated mice and guinea pigs with DNA expressing the *M. tuberculosis* homologue of this protein and were unable to demonstrate protection.[25] However, the vaccine was designed to give enhanced secretion of the antigen by including a secretion signal from tissue plasminogen activator. This approach has been reported to enhance the antibody response[23,26] and may have contributed to the severe lung damage seen in guinea pigs that was characterized by necrotizing bronchointerstitial pneumonia and bronchiolitis.[25] The authors voiced grave safety concerns against vaccination with heat-shock proteins that have close homologues in eukaryotic organisms as T cells recognizing cross-reactive determinants could give rise to autoimmune responses. Interestingly, naked DNA vaccination with *M. leprae* HSP65, rather than exacerbating adjuvant arthritis, an experimental autoimmune disease, protected rats against disease development.[27] The striking differences observed between *M. leprae* and *M. tuberculosis* HSP65 DNA vaccines await adequate explanation.

Ag85A

Antigen 85A is part of a complex of secreted antigens (consisting of Ag85A, Ag85B, Ag85C) that constitute a major fraction of the secreted proteins in culture filtrates. These fibronectin-binding proteins are also mycolyl transferases, which play an important role in the synthesis of mycobacterial cell wall components. Huygen and colleagues demonstrated that DNA vaccination with Ag85A induced substantial cellular and humoral immune responses (CTL, IFN-γ etc) and conferred a significant degree of protection to mice against aerosol challenge with virulent *M. tuberculosis*.[18] This group also provided evidence that the inclusion of the secretion signal peptide from tissue plasminogen activator (tPA) resulted not only in stronger immune responses but also in sustained protection when the resting period between vaccination and challenge was increased from 30 to 90 days. The nonmodified form of Ag85A was not protective after 90 days.[28] DNA vaccination with Ag85B has also been shown to be protective in mice, whereas vaccination with Ag85C is not.[29] Although Ag85A DNA vaccinated mice could control the numbers of viable bacilli in lungs and spleens for prolonged periods (up to 10 weeks post-infection), this protection waned rapidly and bacterial loads 12 weeks following infection were indistinguishable from those of vector-vaccinated control animals.[30]

Ag85A has also been tested in a guinea pig aerosol challenge model of tuberculosis. DNA vaccination did not lead to a reduction of bacterial burdens in the lungs of infected guinea pigs but did reduce dissemination to the spleen and resulted in prolonged survival. The mean survival time for DNA vaccinated guinea pigs was still shorter than for BCG vaccinated animals.[31] Surviving animals had extensive granulomatous pneumonia in which a high percentage of lymphocytes was present (this was less prominent in dying animals), albeit they lacked pulmonary necrosis and caseation, which is the hallmark of tuberculosis in the guinea pig.[31] Significantly, DNA vaccination of guinea pigs with Ag85A did not result in positive tuberculin skin responses[31] suggesting that this vaccination strategy could be implemented alongside tuberculin-based surveillance mechanisms. Cattle vaccinated with this plasmid did not produce any demonstrable immune responses after vaccination and therefore the vaccine's ability to protect cattle against bovine tuberculosis has not been tested.[32] It should be noted that members of the Ag85 complex also impart protection against mycobacterial diseases other than tuberculosis (see below).

Phosphate Transport Receptors (PstS-1, PstS-2, PstS-3)

Analysis of the *M. tuberculosis* genome identified three putative phosphate binding protein homologues to the periplasmic ATP-cassette (ABC) phosphate-binding receptor PstS of *Escherichia coli*. These *M. tuberculosis* proteins were called PstS-1 (identical to the well characterized 38 kDa antigen), PstS-2, and PstS-3.[33] The 38 kDa antigen PstS-1 was, alongside HSP65 and Ag85A, amongst the first antigens tested as DNA vaccine to protect mice against tuberculosis.[20] Results by Zhu and co-workers demonstrated strong cellular immune responses characterized by IFN-γ production and CD8-mediated cytotoxicity. Strikingly, the specificity of both CD4$^+$ and CD8$^+$ T cells was found to be different between *M. tuberculosis* infected and DNA vaccinated mice with respect to several epitopes. Vaccinated mice were significantly protected against *M. tuberculosis* challenge up to 12 weeks post-infection.[20] In contrast, although Tanghe et al, demonstrated high levels of Th1-type IL-2 and IFN-γ responses in mice vaccinated with DNA expressing either PstS-1, PstS-2 or PstS-3, no protection was observed in the PstS-1 vaccinated mice.[30] At present it is not possible to reconcile these conflicting results with PstS-1 as different mouse strains, different challenge routes, and different resting periods between vaccination and infection (2 vs. 10 weeks) were used. Vaccination with PstS-2 resulted in modest reduction in bacterial numbers in spleen whereas the PstS-3 vaccination caused significant and sustained protection both in lungs and spleens over a period of at least 12 weeks.[30] These results make PstS-3 a particular interesting vaccine antigen and it is hoped that it will be tested soon in other models.

ESAT-6

ESAT-6 is a low-molecular protein found in short-term culture filtrate of organisms of the *M. tuberculosis* complex (*M. tuberculosis, M. bovis, M. africanum*) and only a few other mycobacterial species (i.e., *M. kansasii* or *M. marinum*).[34,35] It is one of the most prominent targets of the cellular immune system in all hosts tested so far (mice, guinea pigs, humans, cattle).[36-39] Apart from its immunogenicity and relative specificity for tubercle bacilli, its main attraction is the presence of its gene in the RD1 region of the genome that is deleted in all strains of BCG. These attributes have made it an ideal candidate as a specific diagnostic reagent to distinguish between infected and BCG vaccinated individuals.[38] Nevertheless, due to its pronounced immunogenicity it has also been evaluated as a vaccine candidate. Besides being tested in classic protein-based vaccine approaches it has also been tested frequently as a DNA vaccine in mice. These diverse studies have resulted in widely variable efficacies. Whilst some studies demonstrated no, or only marginal protection, others demonstrated highly significant and strong protection.[22,23] Interestingly, the best protective effects were observed with DNA vaccine constructs where a tPA secretion signal peptide was included.[22-24,40] The ability of this antigen to protect other host species, like the guinea pig, against tuberculosis, is unknown at the time of the preparation of this manuscript.

The long list of antigens conferring some degree of protection against tuberculosis shows that many diverse proteins can protect against tuberculous infections. However, a number of antigens have also been described that did not confer any significant degree of protection (e.g., the 19 kDa protein Rv3763,[41,42] 16 kDa alpha-crystallin like chaperonin protein Rv2031c,[22] Ag85C Rv0129c,[29] PstS-2 Rv0932c,[30] Rv1796, Rv2428 or Rv2945c[42-44]).

DNA Vaccines as Post-Exposure Vaccines or Immunotherapeutic Agents

As we have pointed out earlier, there is also a need to develop novel tuberculosis vaccines that are effective in latently infected individuals (post-exposure vaccine), or are effective as immunotherapeutic reagents to assist conventional chemotherapy. Effective treatment with chemotherapy requires large doses of costly antibacterial drugs that have to be taken for at least 6 months, which is difficult to achieve in many developing countries. In addition, immuno-therapy could improve the treatment of multi-drug resistant tuberculosis. Addressing these objectives, Lowrie and colleagues demonstrated that DNA vaccines that were effective as pro-phylactic vaccines against tuberculosis, also had therapeutic effects against an established, chronic *M. tuberculosis* infection by reducing the bacterial burdens in both the lungs and spleens of treated mice.[45] They tested DNA vaccines expressing *M. leprae* HSP65, *M. tuberculosis* HSP70, ESAT-6, and another secreted antigen, MPT70 (see Table 1 for details of antigens) and could show strong therapeutic activity, mainly with HSP65 and MPT70. Interestingly, injection of DNA encoding IL-12 alone was as beneficial as injection of HSP65 or MPT70 although co-injection of IL-12 and HSP65 DNA resulted in reduced therapeutic efficacy.[45] The admin-istration of BCG had no effect in these experiments. The authors' experiments suggested that the therapeutic effect of HSP65 was due to a reprogramming of the immune responses charac-terized by a switch from IL-4 responses towards predominant IFN-γ responses.[45]

A frequent result of incomplete drug treatment of tuberculosis is the regrowth of bacteria that have persisted in a nonreplicating and physiologically drug-resistant form. Lowrie and colleagues used a mouse relapse model to show that HSP65 DNA vaccination given at the end of incomplete chemotherapy could prevent the corticosteroid-induced regrowth of *M. tubercu-losis* in the lungs of 8/8 mice and in the spleens of 6/8 mice.[45] Significantly, BCG was ineffec-tive in this experimental model of relapsing tuberculosis. These results raised the possibility that therapeutic DNA vaccination could be used as a post-exposure vaccine in man to elimi-nate latent or persistent infection prior to disease.

Turner et al[46] conducted similar immunotherapeutic vaccination experiments in mice with an Ag85A DNA vaccine. They were unable to demonstrate any effect of vaccination on the course of infection in the lung of aerosol-infected mice, but could demonstrate a reduction in bacterial loads in their spleens.

DNA Vaccines Protecting Against Mycobacterial Diseases Other than Tuberculosis

Mycobacteria are the causative agents of several other human diseases apart from tuberculo-sis. For example, *M. leprae* is the causative agent of leprosy in humans, *M. avium* is a common opportunistic pathogen of immuno-compromised individuals like AIDS patients, and buruli ulcer, an emerging health problem of developing countries is caused by *M. ulcerans*. DNA vaccines are being developed to protect against all three of these diseases. This is being facili-tated by the availability of suitable murine models to test vaccine efficacy. For example, DNA vaccination with *M. tuberculosis* Ag85B has been shown to impart protection against *M. leprae* growth on challenge of mouse footpads,[47] as did a DNA vaccine expressing the *M. leprae* 35 kDa antigen (ORF designation ML0841[47,48]). The degree of *M. leprae* growth reduction was similar to that achieved with BCG, which appears to confer protection against leprosy in man. DNA vaccination with the *M. avium* homologues of either the *M. leprae* 35 kDa antigen,[47,49] Ag85A,[50] or of HSP65[50] protected mice against *M. avium* infection to a degree similar to, but not exceeding, BCG vaccination. In addition, DNA vaccination with *M. tuberculosis* Ag85A protected mouse footpads against *M. ulcerans* infection almost as well as BCG vaccination.[51] Interestingly, "vaccination" of mice, either before or after *M. avium* infection using a construct expressing IL-18 reduced the bacterial loads and increased the survival period of treated mice.[52]

Strategies to Improve the Efficacy of Tuberculosis DNA Vaccines

As shown in Table 1 and emphasized throughout this chapter, protection levels with DNA vaccination against challenge with tuberculosis has been generally less effective than BCG vaccination alone. Our own DNA vaccination studies in cattle, which, unlike mice and guinea pigs, are natural hosts for *M. bovis*, have confirmed this observation. We have tested 5 different DNA vaccine constructs (*M. leprae* HSP65, *M. tuberculosis* HSP70, Apa, MPB83, MPB70), either individually or as cocktails, and, to date, we have not been able to protect cattle against *M. bovis* infection (Buddle, Vordermeier, Hewinson, unpublished data) despite the proven effectiveness of these vaccines in rodents.[32,52-54] Therefore, it is important to improve the efficacy of DNA vaccines against tuberculosis and several strategies have been applied to address this objective.

Use of Vaccine Cocktails

Several groups have applied DNA vaccines in pools of up to 4 different vaccines (e.g., ESAT-6, MPT43, Ag85B, or ESAT-6, MPT63, MPT64, KatG).[22,55] Some combinations gave improvements in protection compared to the individual vaccines. However, BCG was still the most protective vaccine in these studies (Table 2A). Interestingly, formulating Ag85B and MPT64 together with the cytokine GM-CSF did not further improve protection, despite increases in cellular immunity.[56]

Protein Modifications

DNA vaccines have been constructed to give either enhanced antigen secretion from mammalian cells (by the addition of the tPA secretion signal peptide[23,28]), or to give enhanced proteosome-dependent degradation of the endogenously produced proteins (by conjugation to ubiquitin).[57] Both strategies have resulted in improved protective efficacies (Table 2B), although the modified vaccines were still inferior to BCG.

Heterologous Prime-Boost Strategies

Heterologous prime-boost strategies have been adopted in the quest for improved TB vaccines, and sometimes resulted in protection superior to BCG vaccination. These protocols were based on the use of DNA vaccination to prime immune responses, followed by recombinant proteins, recombinant viruses, or BCG itself to boost the immune response. Examples of such experiments are summarized in Table 2C.

Boosting Ag85A DNA vaccinated mice with recombinant Ag85A resulted in improved protection compared to DNA or protein alone.[58] In these studies, two different adjuvants were used to deliver the protein boost: MPL and SBAAS2A. Boosting with the latter adjuvant resulted in the same level of protection as conferred by BCG vaccination, whilst the protection achieved by boosting with protein combined with MPL, although better than DNA alone, was not as good as BCG (Table 2C). This improved protective efficacy was characterized by increased IFN-γ and IL-2 responses.[58] Similarly, when we vaccinated cattle with HSP65 DNA and then boosted the responses with recombinant protein in incomplete Freund's adjuvant (IFA) there were increased and more homogenous IFN-γ and T cell proliferative responses and a more balanced IgG1/IgG2 ratio (Vordermeier and Hewinson, unpublished results). Challenge infections to assess the protective efficacy of this vaccine protocol in cattle are now underway (Table 2C).

McShane and co-workers[59] have used recombinant viruses to boost DNA primed responses. Mice that had been primed by 3 vaccinations with DNA expressing a fusion protein of ESAT-6 and another *M. tuberculosis* antigen, MPT63, were boosted with the same fusion protein expressed by recombinant modified vaccinia virus Ankara (MVA). This resulted in significantly

Table 2. Strategies to increase efficacies of tuberculosis DNA vaccines

A. Use of Vaccine Cocktails:

Antigen[a]	Model	Challenge[a]	Comments[b]	Ref.
ESAT-6, MPT64, Ag85B	Mouse	MT/air	Individual vaccines: 0.25-0.41; cocktail: 0.63	[40]
ESAT-6, MPT63, MPT64, KatG	Mouse	MT/air	Individual vaccines: 0.40-0.60; cocktail: 0.82	[22]
Ag85B, MPT64, GM-CSF	Mouse	MT/air	Increased in vitro proliferation and IFN-γ, but no improved protection	[56]

B. Modifications:

Antigen	Modifications	Model	Challenge	Comments	Ref.
ESAT-6, MPT64, KatG	tPa signal peptide	Mouse	MT/air	Better protection when signal peptide had been added e.g., tPa-ESAT-6: 0.72; ESAT-6: 0.41	[23]
Ag85B	tPa signal peptide	Mouse	MT/iv	Infection contained longer when signal peptide added	[28]
ESAT-6, MPT64	Ubiquitin fusions	Mouse	MT/air	Better protection depending on ubiquitin-construct used e.g., Ubiqitin-MPT64: 0.52; MPT64: 0.35	[57]

continued on next page

Table 2. Strategies to increase efficacies of tuberculosis DNA vaccines (Cont'd)

C. Prime-Boost Strategies:

Prime (DNA)	Boost	Model	Challenge	Comment	Ref.
Ag85A	Ag85A protein Adjuvant: MPL	Mouse	MT/iv	better than DNA or protein alone, not as good as BCG (DNA: 0.28; DNA/protein: 0.46)	[58]
Ag85A	Ag85A protein Adjuvant: SBAAS2A	Mouse	MT/iv	better than DNA or protein alone, as good as BCG (DNA: 0.53, DNA/protein: 1.03)	[58]
HSP65 (ML)	HSP65 (MT) protein Adjuvant: IFA	Cattle	NT	improved and more homogenous proliferative and IFN-γ responses[c]	
ESAT-6/MPT63 fusion	MVA-ESAT-6/MPT63	Mouse	MT/ip	Increase in IFN-γ responses, prime-boost as good as BCG DNA/MVA ca. 0.90 to 2.70	[59]
Ag85B	BCG	Mouse	MT/air	DNA/BCG: 1.67. Partially mediated by CD8+ T cells	[60]
HSP65 (ML), HSP70, Apa	BCG	Cattle	MB/it	Better protection than with BCG alone.(Mean pathology scores: Control or DNA: ca. 7; BCG: 3.2; DNA/BCG: 1.5)[d]	

[a]See Table 1 legend. [b] If not listed otherwise, results are given as relative lung protection value compared to BCG (see Table 1 legend). [c]Vordermeier and Hewinson, unpublished data. [d]Vordermeier, Hewinson and Buddle, unpublished data.

increased frequencies of CD4$^+$ IFN-γ secreting cells and protection equivalent to BCG vaccination (Table 2C).

Sequential immunization with an Ag85B expressing DNA vaccine followed by BCG boosting has been found to be more effective than BCG vaccination alone.[60] The protective effect of this heterologous prime boost protocol was superior to the effect seen in mice that were vaccinated twice with BCG ('homologous boost'). The improved protection was at least partially mediated by CD8$^+$ T cells (Table 1C). We have recently performed an experiment in which cattle were primed twice with a mixture of DNA plasmids expressing HSP65, HSP70 and Apa before BCG vaccination and found that the degree of pathology in these animals after *M. bovis* challenge was reduced compared to BCG vaccination alone. The DNA vaccination alone did not result in significant protection (Table 2C, Vordermeier, Hewinson and Buddle, unpublished results).

It is likely that advances will be made with additional strategies, such as optimizing CpG motifs,[61] identification of new antigens by sequence comparisons with mycobacterial genome sequences,[62,63] analysis of in vivo antigen expression and using epitope prediction algorithms.[64] Codon-optimized epitopes of novel antigens could be readily incorporated into existing DNA vaccines.

Characterization of Immune Responses after DNA Vaccination

It is not known with certainty what immune responses equate with protection in tuberculosis. As highlighted in Table 1, DNA vaccination generally induces strong antigen-specific T cell proliferative and type 1 cytokine responses and IFN-γ and class I restricted CD8$^+$ CTL responses tend to be prominent. The level of IFN-γ produced early after infection is certainly one parameter observed in protective immunity resulting from DNA vaccination of mice.[65,66] The breadth of the profile of epitopes recognized by both CD4$^+$ and CD8$^+$ T cells may be another. For example, CD4 T cells from PstS-1 DNA vaccinated BALB/c mice recognized epitopes on 7 different peptides compared to 5 recognized following *M. tuberculosis* infection;[20,67] CD4$^+$ T cells from *M. tuberculosis* infected mice recognized a single Ag85A derived epitope whereas 8 were recognized in DNA vaccinated mice.[68] Similarly, the CD8$^+$ T cell epitope repertoire of DNA vaccinated mice was also strikingly different and broader than that of *M. tuberculosis* infected mice (Ag85A: 3 *vs.* 0 epitopes, PstS-1: 3 *vs.* 1 epitope, in DNA vaccinated and *M. tuberculosis* infected mice, respectively).[20,67,68] The induction, through DNA vaccination, of T cells that are specific for epitopes that are sub-dominant or cryptic during natural infection, can have a significant impact on protective immunity. For example, it has been recently shown that effective protection could be obtained after vaccination with a peptide comprising a sub-dominant epitope of ESAT-6 that gave only negligible responses during *M. tuberculosis* infection, whereas vaccination with a peptide that was a dominant epitope of ESAT-6 during infection was not protective.[69]

Several studies have attempted to define the T cell subsets responsible for protecting mice after DNA vaccination. D'Souza and co-workers[70] vaccinated a range of CD4, β2 microglobulin and IFN-γ gene-disrupted mouse strains with an Ag85A DNA vaccine. They demonstrated that mice lacking CD4$^+$ T cells or unable to produce IFN-γ could not be protected against tuberculosis whereas β2-microglobulin-knockout mice effectively controlled the disease when vaccinated with the DNA vaccine. These experiments therefore confirmed the importance of IFN-γ produced by CD4$^+$ T cells in providing protective immunity against tuberculous infection in mice. The authors concluded that such cells mediate the protective effect of this vaccine independently of CD8$^+$ T cells. However, their experiments did not rule out a role for CD8$^+$ T cells in Ag85A mediated protection. Mice with an MHC H-2^b were used and no Ag85A-specific CD8$^+$ T cell responses were demonstrable, even in wild-type mice.[70] This experiment should therefore be repeated in mice of H-2^d background where strong CD8$^+$ T cell responses were reported after Ag85A vaccination.

The importance of CD8$^+$ T cells that both secrete IFN-γ and lyse *M. tuberculosis* infected cells in vitro, has been demonstrated in a series of experiments analyzing T cell clones prepared from *M. leprae* HSP65 DNA or J774-HSP65 vaccinated mice.[71-73] These CD8⁺CD44ʰⁱ activated/memory T cells were able to adoptively transfer protection into naïve recipients. IFN-γ producing, cytolytic CD4⁺CD44ʰⁱ clones were also able to transfer protection, although they were less potent than the CD8⁺CD44ʰⁱ cells. In contrast, CD8⁺CD44ʰⁱ T cells that produced IFN-γ but were not cytolytic were unable to transfer protection. Furthermore, adoptive transfer of protection correlated with the ability of the clones to use the cytotoxic granule mediated pathway to lyse infected targets; in other systems, T cells that lyse targets by apoptosis may also contribute to mycobacterial killing.[74]

A key question to be addressed is how long protective immunity will persist after DNA vaccination. The rest periods between vaccination and mycobacterial infection, in most cases, do not exceed 10 weeks, and are more frequently around 4 – 6 weeks. However, a study by Silva and colleagues indicated that mice were still protected after long periods post-DNA vaccination prior to challenge.[73] They vaccinated mice with a DNA vaccine expressing mycobacterial HSP65 and challenged them after rest periods of 1, 4, 8, and 15 months. Encouragingly, they demonstrated that mice were significantly protected even if they were challenged after a 15 months rest period. The same was observed for BCG vaccinated mice but the degree of protection after either DNA or BCG vaccination was lower than expected at all intervals. However, CD44ʰⁱ T cells that produced high levels of IFN-γ and were cytolytic were found in these mice throughout the 15 months rest period in the DNA vaccinated mice. It is encouraging to demonstrate such long-lived memory in rodent models, but it will obviously be important to determine how DNA-vaccine induced memory responses can be sustained in target species. We have addressed this recently in cattle by studying the kinetics of the immune response following immunization with DNA encoding the *M. leprae* HSP65. In these studies we determined T cell proliferation and IFN-γ production after vaccination and demonstrated potent proliferative T cell and IFN-γ responses in vitro 2 weeks after the final DNA vaccination (Fig. 2). However, after a 6 weeks rest period, we detected only low frequencies of IFN-γ producing cells and low levels of in vitro produced IFN-γ, whereas we were still able to demonstrate strong HSP65-specific T cell proliferation. This seemingly paradoxical finding is completely in line with current concepts of memory T cell development in mice and humans.[75-77] The frequency of IFN-γ producing *effector cells* declines with the decrease in antigen stimulus over time, to be replaced with low frequencies of *effector memory* cells producing cytokines like IFN-γ, and *central memory* cells. It is likely that the latter subset of T cells were those that proliferated but did not produce IFN-γ after the prolonged rest periods. Consistent with this hypothesis, IFN-γ producing effector cells were induced within 3 days of the subcutaneous injection of HSP65 protein (as judged by the amount of and frequency of IFN-γ cells found in the draining lymph nodes of these cows, Fig. 2). Therefore DNA vaccination against tuberculosis in rodent models as well as in a target species is able to induce long-lasting memory responses of the expected type. In conclusion, DNA vaccination against mycobacterial infections has been a potent tool to define protective antigens and to study aspects of protective immunity. It has now reached a stage of development where studies in target species are opportune. In particular we look forward to the exploitation in target host species of strategies like heterologous-prime boost protocols that have improved vaccine efficacies compared to BCG and to the immunotherapeutic applications of DNA vaccines in which role BCG vaccination is not effective.

DNA Vaccines Against Bacterial Diseases Other than Those Caused by Mycobacterial Species

Bacillus anthracis

Bacillus anthracis is the causative agent of anthrax, a potentially lethal gastrointestinal and respiratory disease of domesticated and wild animals, as well as humans. Virulent strains of *B. anthracis* are characterized by their expression of a polyglutamic acid capsule and the production of a protein toxin. The anthrax toxin has three components (protective antigen, lethal factor and

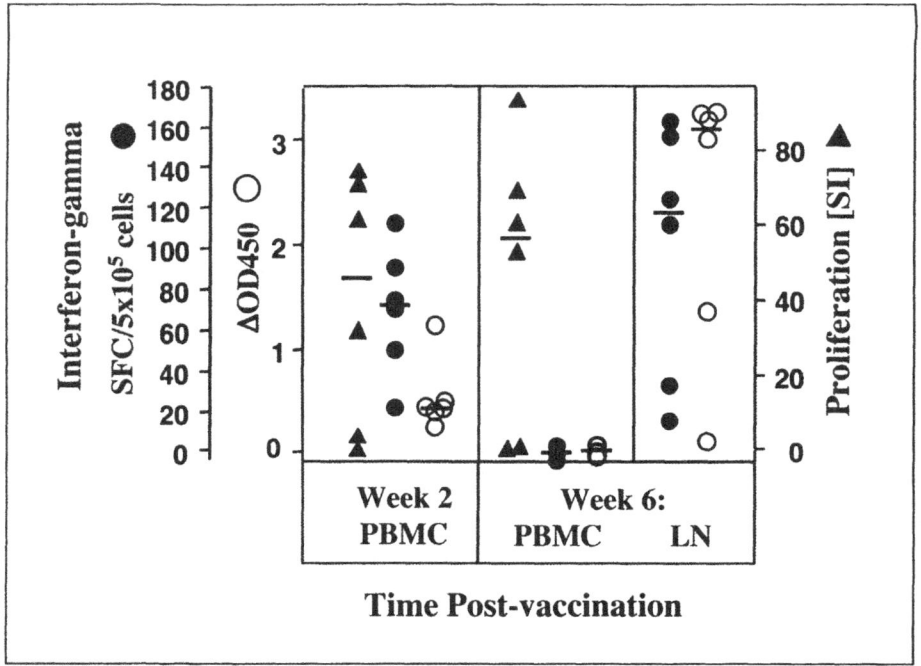

Figure 2. T cell memory responses in draining lymph nodes in *M. leprae* HSP65 DNA vaccinated cattle. Six cattle were vaccinated at weeks 0 and 3 with 1 mg of a *M. leprae* HSP65 expressing DNA vaccine.[17,19] Blood samples were taken 2 weeks after the second vaccination (week 5 of the experiment) and PBMC were prepared and tested for IFN-γ production and proliferative responses. Recombinant HSP65 protein was injected subcutaneously into the necks of these animals 4 days before further blood sampling, and post-mortem examination at week 9 to obtain the draining superficial lymph nodes. PBMC and lymph node cells were prepared and tested as above.

edema factor), although toxicity can be mediated experimentally using a combination of protective antigen (PA) and lethal factor (LF) alone. The PA of anthrax toxin binds to the target cell permitting access of the lethal and edema factors to the cytosol. The current licensed vaccine against anthrax contains PA adsorbed onto aluminum hydroxide. Antibodies to PA prevent binding to the target cell and confer protection from anthrax. To our knowledge there has only been one report of a DNA immunization approach to anthrax vaccination.[78] Mice immunized with a plasmid encoding the immunogenic portion of PA generated splenocytes that secreted IFN-γ and IL-4, consistent with the generation of a mixed Th1/Th2-like response. Immunized mice produced a serum IgG response to PA able to neutralize the biological activity of the protein in vitro, and were protected from lethal challenge with a combination of PA and LF. The level of protection varied from 75 – 100 %, depending on the dose of lethal factor administered. Complete protection to five lethal doses of toxin was obtained. It is hoped that a DNA vaccine against anthrax may be safer and more efficacious than the current licensed vaccine.

Borrelia burgdorferi

Borrelia burgdorferi is one of the members of the *Borrelia* genus of spirochetes responsible for Lyme disease; a chronic disease with arthritic and neurological sequellae if not diagnosed and treated adequately early in infection. The Outer Surface Protein A (OspA) of *B. burgdorferi* is the most frequent antigen component of vaccines under development for Lyme disease. Mice immunized with DNA expressing OspA generated a serum immunoglobulin response[79,80] that was dependent on both the dose of DNA administered and the promoter used to drive expression of OspA.[80] Interestingly, antibodies to OspA were generated following immunization with plasmid containing OspA in the reverse orientation,[80] suggesting that the upstream control region of OspA was active in eukaryotic cells; a fact confirmed in vitro.[80] All isotypes of murine IgG were generated to OspA DNA immunization, consistent with the proliferative response of spleen and lymph node T cells to OspA and the production of IFN-γ.[80] Vaccinated mice were protected from spirochete challenge.[79,80] Immune serum was demonstrated to inhibit growth of *B. burgdorferi* in culture,[79] as well as to confer protection to SCID mice against the onset of clinical arthritis following spirochete challenge.[80] Compared with OspA, DNA vaccination with OspB stimulated a lower titer of serum immunoglobulin and failed to protect mice from challenge, correlating with the inability of the immune serum to inhibit spirochete growth in culture.[79]

Brucella abortus

Brucella sp. are gram-negative facultative intracellular coccobacilli that infect both animals and humans. *Brucella* may enter the body via the skin, respiratory tract, or digestive tract. Once there, they can enter the blood and the lymphatics where they multiply inside phagocytes and eventually cause bacteraemia. Symptoms of Brucellosis vary from patient to patient but can include high fever, chills, and sweating. The four species of this genus that can infect humans are principally named after the animal in which they are most commonly found: *B. abortus* (cattle), *B. suis* (swine), *B. melitensis* (goats), *B. canis* (dogs). *B. abortus* causes spontaneous abortion and infertility in cattle and is the most common species associated with Brucellosis.

There is currently only one report of a DNA vaccination approach to Brucellosis.[81] In that study, mice were vaccinated once with DNA expressing the ribosomal protein L7/L12, the immunodominant protein of *B. abortus* identified in primed cattle.[82] Mice responded to vaccination by splenocyte proliferation and a serum IgG response specific for L7/L12. When challenged 28 days later, vaccinated mice had 0.47 to 1.26 log less bacteria in their spleens compared with controls, depending on the time post-infection. Thirty days post-infection, the protection induced by DNA vaccination was equivalent to that achieved with the live Brucella vaccine—*B. abortus* strain 19.

Chlamydia spp.

There are four species of Chlamydia recognized to-date: *Chlamydia pecorum, Chlamydia pneumoniae, Chlamydia psittaci,* and *Chlamydia trachomatis.* All are obligate intracellular pathogens, causing disease primarily through induction of the host immune response. This feature has hampered efforts to develop a protective vaccine against chlamydial infection. Humans are the predominant host for *C. pneumoniae* and *C. trachomatis,* whereas *C. pecorum* and *C. psittaci* can infect a variety of hosts, including humans. At least *C. trachomatis* and *C. psittaci* can be subdivided into biovariants, depending on both the disease they cause and the host they infect. The four biovariants of *C. trachomatis* cause trachoma or sexually transmitted infections in humans, mouse pneumonitis, or infections of swine, respectively. Similarly, four biovariants of *C. psittaci* have been defined, responsible for infections in guinea pigs, cats, ruminants, or birds, respectively. Most chlamydial species can infect the respiratory tract, although *C. pneumoniae* is the most common agent of chlamydial pneumonia, a potentially life-threatening condition with systemic sequellae. Infection of birds and farm mammals with Chlamydia can result in significant economic losses, as well as presenting an occupational zoonotic hazard. The prevalence of Chlamydia, together with the emergence of antibiotic-resistant isolates, provides the impetus to develop vaccines against chlamydial disease. Candidate antigens for incorporation into a vaccine, include the Major Outer Membrane Protein (MOMP), a cysteine rich outer membrane protein (Omp2), Heat Shock Protein 60 (HSP60), and ADP/ATP translocase (Npt1Cp). Determination of the genome sequences of *C. trachomatis* and *C. pneumoniae*[55,83] has resulted in opportunities for the empirical testing of open reading frames (ORFs) in the context of DNA immunization.[84]

C. pneumoniae

Intranasal vaccination of C57Bl/6 mice with a DNA plasmid encoding the HSP60 of *C. pneumoniae* did not generate an anti-HSP60 serum IgG response. However, seven days after intranasal infection with *C. pneumoniae*, the vaccinated mice had 5 – 20 fold less bacteria in their lungs as well as milder pneumonia, compared with controls.[85] In contrast, intradermal immunization with the same plasmid elicited a serum IgG response to HSP60 but failed to protect mice from intranasal challenge.[85] Mice immunized with the HSP60 plasmid and subsequently challenged, showed enhanced expression of IFN-γ in both the lung and from cultured splenocytes, compared with nonimmunized challenged controls. Protection was further enhanced by the co-administration of plasmid expressing IFN-γ or IL-12, but not GM-CSF, at the time of vaccination. A protective role for IFN-γ in this model was further demonstrated using IFN-γ receptor deficient mice. Protection was also shown to be dependent on CD4+ and CD8+ T cells. Induction of CD4+ cells by DNA immunization in the absence of CD8+ cells led to a worsening of the outcome following challenge.

A similar intranasal challenge model against the BALB/c background was used to screen eight ORFs of *C. pneumoniae* expressed individually as DNA vaccines.[84] Two constructs expressing the ORFs encoding MOMP and Npt1Cp gave protection, whereas those expressing ytfF, GltL, pml116, DnaK, ndk, and dagA did not. The utility of this approach was revealed in the confirmation of MOMP as a protective antigen in chlamydial infections,[86-91] as well as the identification of a novel target antigen, Npt1Cp.

A recent study of intramuscular DNA immunization with plasmids encoding HSP60, MOMP, or Omp2 confirmed the lack of correlation between a serum immunoglobulin response and protection against intranasal challenge with *C. pneumoniae*.[92] This study is noteworthy, in that it looked at the effect of immunization with a cocktail of all three plasmids. Although a proliferative response was seen to each individual antigen, no significant protection was obtained with the cocktail. In fact, significant protection was demonstrated only following immunization with DNA encoding MOMP and HSP60, and then in only one experiment.

C. psittaci

Two studies describe the efficacy of DNA immunization for *C. psittaci* infection of turkeys.[86,87] In both, animals were immunized with DNA encoding the MOMP of an avian *C. psittaci* strain (Serovar A). Two inoculation strategies were evaluated: combined parenteral and mucosal delivery, and gene-gun based epidermal delivery. Both modes of delivery elicited T and B cell responses, as evidenced by the production of anti-MOMP serum immunoglobulin and enhanced proliferation of peripheral blood lymphocytes. All animals that received the DNA vaccine were protected against a high dose aerosol challenge with *C. psittaci*. Protection was expressed in terms of a reduction or absence of signs of clinical disease, and of nasal shedding and tissue replication of organisms. No difference was observed in the relative protection afforded by the different routes of immunization. Antibody responses in vaccinated animals were weak and variable. Protection occurred in all vaccinated turkeys that had a serological response before challenge. The best protection was seen in turkeys that did not mount a secondary antibody response to challenge. Both modes of immunization generated antibodies of IgM, IgG and IgA isotype. Although IgG was the predominant isotype, levels of anti-MOMP IgG did not correlate with the level of protection, supporting observations made with *C. pneumoniae* in mice.[85,92] The immunogenicity and efficacy of intraepidermal vaccination was dependent on the size of the gold beads used for gene-gun delivery and the quantity of plasmid DNA delivered.

C. trachomatis

The majority of studies of DNA vaccination for *C. trachomatis* have used the MOMP gene product as the vaccine antigen and the biovariant of *C. trachomatis* that causes mouse pneumonitis (MoPn) as the challenge organism.[88-91] These studies reported variable levels of induction of an anti-MOMP serum immunoglobulin response. Where serum IgG responses were detected there was a bias to induction of the IgG2a isotype suggesting induction of a more Th1-like response. No serum IgA response was detected in the two studies where it was assayed.[88,90] The induction of cellular immune responses was similarly variable, although studies reported the induction of splenocyte proliferation,[90,91] IFN-γ production,[90] and cutaneous delayed-type hypersensitivity (DTH) reactions following injection of *C. trachomatis* elementary bodies, the infectious extracellular forms of the organism.[89,90] Induction of DTH was considered important since it has been reported to correlate with protection in the MoPn model.[93] Measurement of the relative levels of expression of IL-10 and IFN-γ further supported the notion that DNA immunization with MOMP favored a Th1-type response.[90] Following MOMP DNA vaccination, mice were most frequently challenged with MoPn intranasally. Generally, the levels of protection induced by MOMP DNA vaccination were encouraging; reaching as high as a 4 log reduction in lung bacterial burden[88] and significant protection from weight loss.[89,90] The level of protection was shown to depend on the route of immunization and the dose of DNA used.[88,90] Interestingly, protection in the lung was also obtained when an auxotroph of Salmonella was used to deliver the MOMP DNA plasmid orally.[88]

In one study, three different strains of vaccinated mice were challenged via the physiologically relevant genital route.[91] In this case DNA immunization induced weak cellular and humoral responses. No anti-MOMP vaginal antibodies were detected and vaccination had no effect on vaginal shedding of MoPn.[91] The failure of DNA vaccination in this case could be attributable to the route of challenge, dose of DNA, or even the expression plasmid used. Two other *C. trachomatis* antigens have been evaluated in the MoPn model: cytosine triphosphate (CTP) and serine-threonine kinase (STK). Protection was reported with the use of STK[94] but not CTP.[89]

Only one study has been published using a human strain of *C. trachomatis*: serovar L2, responsible for the invasive sexually transmitted disease, lymphogranuloma venereum. The MOMP antigen was used again but humoral responses were weak to absent, even when the DNA plasmid had been engineered for increased expression of MOMP protein in vitro. Mice vaccinated by either the intramuscular, intradermal or intranasal route were reported to resist intranasal challenge with *C. trachomatis* longer than controls, although the data were not analyzed statistically and all vaccinated mice eventually succumbed to infection.[95]

Clostridium tetani

Two studies report the evaluation of DNA immunization for protection against tetanus toxin; the potent neurotoxin synthesized by the anaerobic bacterium *Clostridium tetani*.[96,97] In both studies, BALB/c mice were immunized intramuscularly with plasmids encoding the nontoxic C-terminal domain of tetanus toxin (fragment C). Anti-fragment C serum immunoglobulin responses were obtained, as well as proliferative responses by cultured splenocytes. Evaluation of the IgG subclasses and cytokines produced by the splenocyte cultures suggested that DNA immunization induced a Th1-like response. In contrast, immunization with tetanus toxoid or a polypeptide of fragment C induced a Th2- like response. In most cases, the serum immunoglobulin response following DNA immunization was sufficient to protect 70 – 100 % of mice from lethal challenge with tetanus toxin. The level of protection was dependent on the amount of DNA used for immunization and the challenge dose. Whilst encouraging, the level of protection conferred by DNA immunization was inferior to that achieved with conventional toxoid or a polypeptide of fragment C.

Corynebacterium pseudotuberculosis

Corynebacterium pseudotuberculosis is the causative agent of caseous lymphadenitis (CLA); a chronic infectious disease of sheep and goats that results in abscesses of the lymph nodes, and occasionally, the internal organs. *C. pseudotuberculosis* may also cause ulcerative lymphangitis, a mildly contagious disease of horses characterized by inflammation of the lymphatic vessels of the lower limbs. CLA is highly contagious and causes significant economic loss to affected premises. Infection of humans with *C. pseudotuberculosis* is rare and is most often associated with skinning infected animals or the consumption of unpasteurized milk from such animals.

A DNA vaccination trial against CLA was recently conducted in sheep[98] and used the novel approach of directing the antigen to sites of immune induction by expressing it as a fusion protein with bovine CTLA-4. This approach had previously been shown to enhance both the speed and magnitude of the immune response to human immunoglobulin.[99] The phospholipase D gene of *C. pseudotuberculosis* was genetically detoxified (ΔPLD) and cloned into two DNA plasmids, designed (and confirmed in vitro) to secrete ΔPLD with or without targeting to CD80/86-positive antigen-presenting cells. A significant antibody response was generated only in those sheep vaccinated with DNA expressing the targeted CTLA-4-ΔPLD fusion. Animals were challenged above the coronet with 10^6 *C. pseudotuberculosis* six weeks after the primary immunization and killed six weeks later. The lymph nodes draining the site of challenge were inspected post mortem for the presence of abscesses. A proportion of animals that received either ΔPLD construct as DNA vaccine were completely protected from abscess formation: 45 % in the case of the nontargeted vaccine; 70 % with the targeted vaccine. This compared favorably to the level of protection (90 %) obtained with a commercial vaccine comprised of formalin-inactivated toxins and PLD.

Enterotoxigenic Escherichia coli

Enterotoxigenic *Escherichia coli* (ETEC) is one of the most common causes of acute diarrhea in children in developing countries and in travelers who visit those areas. ETEC is also one of the most important pathogens of pigs, causing diarrhea, edema disease or colisepticaemia. Essential virulence factors for the pathogenicity of ETEC include the enterotoxins and the Colonisation Factor Antigens (CFAs), which comprise fimbrial and nonfimbrial adhesins responsible for the adhesion of ETEC to enterocyte receptors. Studies of DNA vaccination for ETEC are restricted to the CFA/I fimbriae of human ETEC serogroups (encoded by *cfaB*) and FaeG, the major protein subunit of K88 fimbriae expressed by a subset of pig ETEC serogroups. Mice given a single inoculation of DNA expressing the CFA/I fimbrial adhesin mounted a serum IgG response, although the antibodies were unable to agglutinate bacterial cells or inhibit the haemagglutination (HA) promoted by CFA/I-bearing ETEC.[100] The serological response to CFA/I DNA vaccination was subsequently shown to be predominantly of the IgG2a isotype, compared with the predominantly IgG1 response induced by immunization with purified CFA/I protein subunits.[101] This group went on to show that the magnitude, isotype, and specificity of the serological response to CFA/I was dependent on the number of DNA inocula administered[102] as well as the plasmid used. Fusion of the *cfaB* gene to the tissue plasminogen activator protein (tPA) signal sequence resulted in the secretion of CFA/I protein from transfected cells[103] and vaccination with this construct induced serum IgG capable of inhibiting the HA of CFA/I-positive bacteria.[103,104] Cellular responses to CFA/I protein were demonstrable with all plasmids, but not IgA in gut mucosa or feces.[104] In order to generate mucosal IgA mice were primed with CFA/I plasmid DNA, followed by intragastric inoculation with recombinant Salmonella expressing CFA/I protein. The combined regime of DNA prime-Salmonella boost, generated serum IgG and fecal IgA responses greater than achieved with either DNA or recombinant Salmonella alone.[105]

In the pig ETEC study,[106] mice immunized with DNA encoding the *faeG* gene of ETEC K88ab produced a serum IgG2a response. Three pregnant sows were subsequently immunized with 200µg, 400µg or 1128µg total DNA. Neither of the sows receiving the lower doses of DNA seroconverted, although an antibody response was reported in the colostrum of the first sow. A serum IgG response was detected in the sow that received the highest dose of DNA, although it aborted 48 hours after the second DNA inoculation. Antibody titers in the surviving offspring were reportedly higher than those of their respective mothers, although these data were not presented in the paper.

Francisella tularensis

Francisella tularensis is the causative agent of tularaemia. There are five forms of clinical tularaemia. The most common (80 % of cases) is ulceroglandular tularaemia with symptoms of fever, headache, lymphadenopathy and characteristic ulceration of the site of inoculation. Tularaemia carries a mortality rate of 5 – 15 %, even higher with the typhoidal form, but antibiotics lower this rate to about 1 %. Many wild animals and some domestic animals can harbor *F. tularensis*, although the rabbit is the species most often involved in transmission. Infection can be acquired through numerous routes. The most common routes include: inoculation of the skin or mucous membranes with blood or tissue while handling, dressing or skinning infected animals, contact with fluids from infected flies or ticks, the bite of infected ticks or handling or eating insufficiently cooked contaminated meat.

An attenuated *F. tularensis* live vaccine strain (LVS) is available for those most at risk of infection, so there has been little effort to develop a DNA vaccine for tularaemia. However, lethal infection of mice with *F. tularensis* LVS has been used as a model with which to study the protective host response to noneukaryotic DNA.[107] BALB/c mice were injected with 0.01 – 20 μg chromosal DNA isolated from *F. tularensis* and then challenged with different doses of *F. tularensis* at various time points after immunization. Protection was dependent on the dose of DNA administered, the challenge dose and time to- challenge, but not the route of immunization. Protection was shown to be dependent on the unmethylated state of the bacterial DNA. Synthetic unmethylated CpG-rich oligonucleotides (CpG ODN) are common in prokaryotic genomes but rare in eukaryotes. Their ability to induce the innate immune response is well documented, and as such, CpG ODN have attracted much interest as mediators of vaccine enhancement[108] (see section on *Listeria monocytogenes*). The importance of unmethylated CpG in *F. tularensis* DNA for the induction of short-term protection was demonstrated using CpG-containing ODN. Immunization with unmethylated CpG-ODN, but not methylated, conferred 100 % protection against lethal challenge with either *F. tularensis* or *L. monocytogenes*. Using IFN-γ gene-knockout (GKO) mice, the protection obtained with either unmethylated *F. tularensis* DNA or CpG-ODN was shown to be dependent on IFN-γ, T cells and critically, B cells. Importantly, long-term pathogen-specific immunity was demonstrated in mice that had survived a previous lethal challenge through immunization with bacterial DNA.

Helicobacter pylori

Helicobacter pylori colonizes the gastric mucosa of more than half the world's population, causing asymptomatic gastritis in the majority. However, up to 10 % of infected individuals develop clinical disease, ranging from gastric and duodenal ulcers to adenocarcinoma of the stomach and lymphoma of mucosa-associated lymphoid tissue. The current standard treatment for *H. pylori* infection consists of antibiotics administered with an inhibitor of gastric acid secretion. The increasing resistance of *H. pylori* to antibiotics makes vaccination an attractive alternative, although complete elimination of the organism may be undesirable given the possible association of *H. pylori* colonization with reduced incidence of esophageal diseases. The latter observation has likely tempered enthusiasm for *H. pylori* vaccination and, in fact, only one study has been published on a DNA approach to vaccination.[109] Immunization of mice with plasmid encoding either the Heat Shock Protein (Hsp) A or B of *H. pylori* stimulated an antigen-specific serum IgG and IgA response. Differences in the relative levels of IgG to IgA and of IgG2a to IgG1 isotypes were observed between the response to Hsp A and Hsp B. DNA vaccination conferred protection against *H. pylori* replication in the stomachs of mice infected three months after vaccination and examined six months after challenge. Vaccinated mice also had less antral gastritis, as determined histologically. The protection afforded by Hsp B was greater than with Hsp A by both criteria.

Leptospira interrogans

Leptospirosis is the most widespread bacterial infection in the world, affecting both humans and animals. Zoonotic infection of humans with *Leptospira interrogans* responds well to treatment, but rarely infection may cause the acute life-threatening (Weil's) disease, resulting in jaundice and severe damage to the body organs. The protection conferred to guinea pigs by DNA vaccination with the seroreactive P68 antigen[110] is, to our knowledge, the only published report of a DNA vaccine approach to leptospirosis. In that study, 77 % of vaccinated guinea pigs were protected from death and 85 % from pulmonary diffuse hemorrhaging (PDH), following challenge with *L. interrogans*. In contrast, 75 % of guinea pigs vaccinated with vector plasmid alone died and nine out of ten had PDH post mortem.

Listeria monocytogenes

Listeria monocytogenes is the only one of seven species of Listeria that is pathogenic to humans. Diarrhea can follow ingestion of organisms via contaminated food, although infection is rare in normal populations. However, in individuals made susceptible by age, pregnancy, or immunosuppression, infection can lead to encephalitis, meningitis, stillbirth, and high mortality.

Infection of mice with *L. monocytogenes* is an established model for studying host antibacterial defense mechanisms.[111] The most extensively studied protective antigen in murine listeriosis is Listeriolysin O (LLO). Protection of BALB/c mice against challenge with *L. monocytogenes* can be mediated by CD8[+] T cytotoxic lymphocytes that recognize an H-2K^d restricted epitope from amino acids 91 – 99 of LLO.[112] It is not surprising, therefore, that the majority of studies on DNA vaccination against *L. monocytogenes* have included this antigen. However, with the possible exception of Mycobacteria, the strategies adopted for DNA vaccination against *L. monocytogenes* are amongst the most diverse and novel.

The first publication of a DNA vaccination approach to *L. monocytogenes* described the novel use of an attenuated strain of Salmonella to deliver DNA vaccine plasmids via the oral route.[113] In this way, plasmids encoding truncated *L. monocytogenes* genes *actA* and *hly* (encoding LLO) were most likely transferred from the Salmonella organisms to host antigen-presenting cells, resulting in CD8[+] CTL activity, CD4[+] T cell proliferation and the release of IFN-γ, as well as serum IgG and IgA responses.[113] Mice vaccinated with recombinant Salmonella containing the *hly* plasmid, but not the *actA* plasmid, were protected from a 10 x LD$_{50}$ challenge with *L. monocytogenes*.

Despite the protective nature of the 91 – 99 epitope of LLO, vaccination of BALB/c mice with DNA encoding the epitope failed to induce an immune response or confer protection against challenge.[114] However, when the plasmid was reconstructed so that the codon usage of the epitope was optimized for expression in mice, a CTL response was induced, mediated by IFN-γ-secreting CD8[+] T lymphocytes.[114] Furthermore, immunization with the optimized 91 – 99 epitope plasmid gave partial protection against sublethal challenge with *L. monocytogenes*. This study emphasized that the optimization of codon usage to improve the efficiency of antigen gene translation in the target species is an important consideration in constructing DNA vaccines against bacterial pathogens. Using DNA constructs containing other epitopes of LLO, these workers demonstrated further that the magnitude of protection was related to the level of CTL induction.[115] In contrast to the protective efficacy observed in the studies using Salmonella as a surrogate vector for DNA vaccines described above, a study using the whole *hly* gene for intramuscular DNA vaccination found the immune response to be poor and unable to confer significant protection against challenge.[116] Since LLO encoded by *hly* is a cytotoxin, these workers reasoned that expression of whole LLO by antigen-presenting cells might result in auto-toxicity and failure to induce protective immunity. A variety of plasmids were therefore constructed, expressing either mutant versions of LLO with reduced toxicity, or engineered to secrete the wild-type or mutant LLO (by virtue of fusion to the murine tPA signal sequence). In adoptive transfer protection experiments, the mutant form of LLO gave a ~ 0.6 log improvement in protection, the secreted wildtype form a ~ 0.3 log improvement, and the secreted mutant form a ~ 1.9 log improvement compared with the nonsecreted wild-type form of LLO.[116] However, in the vaccinated animal itself, only the secreted mutant form of LLO conferred protection to challenge.

Gene-gun immunization using LLO and p60 of *L. monocytogenes* has been used to examine the ability of CpG ODN to act as an adjunct to DNA immunization, in addition to boosting or co-administration of plasmid encoding GM-CSF.[117] An optimal vaccination schedule utilized LLO with or without p60 plus co-injection of CpG ODN. With this schedule over 3 log protection was obtained, representing sterile elimination of *L. monocytogenes* from the spleen of challenged mice[117] and demonstrating that CpG ODN can improve sub-optimal vaccination protocols. In another study, administration of CpG ODN alone to mice was sufficient to confer 100 % protection against lethal challenge with *L. monocytogenes* three days later.[107]

Mycoplasma pulmonis

Mycoplasma are minute pleomorphic gram-negative micro-organisms without cell walls that are intermediate in some respects between viruses and bacteria. *Mycoplasma pneumoniae* is an important human pathogen and mycoplasmoses among farm animals are responsible for considerable economic losses. Murine respiratory mycoplasmosis caused by *M. pulmonis* is one of the most important pathologies for laboratory rodents and is comparable to the pathological manifestations of *M. pneumoniae* in humans and of other animal respiratory mycoplasmoses. The experimental infection of mice with *M. pulmonis* has been used as a model for two novel approaches to DNA vaccination.[118,119] The first was termed Expression Library Immunization (ELI), in which an expression library of *M. pulmonis* DNA was used to vaccinate mice in order to identify protective antigens in an empirical fashion. The small genome size of *M. pulmonis* was ideal to prove the principle behind this approach. The second, and only other report of DNA vaccination against Mycoplasma, was to our knowledge the first to demonstrate the therapeutic potential of DNA immunization.

Using the ELI approach, the entire *M. pulmonis* genome was represented in each reading frame by nine libraries of about 3000 members. Injection of two of the nine libraries intradermally into groups of mice resulted in serum antibody and DTH responses to *M. pulmonis* antigens. Complete protection to *M. pulmonis* challenge was obtained with either library.[118] A *Listeria monocytogenes* ELI library was also constructed and used a negative control. Protection against *L. monocytogenes* challenge was not examined. The advantage of the ELI approach is that a DNA vaccine can be constructed without any prior knowledge of the pathogen's biology or of the suitable target genes to use for immunization. However, the approach may only be suitable for pathogens with genomes of manageable size and ultimately the genes responsible for protection must be dissected from the library and tested individually.

The therapeutic potential of DNA immunization in bacterial infections was reported twice, first for *M. pulmonis*[119] and subsequently, for *Mycobacterium tuberculosis*.[45,46] The latter papers are reviewed earlier in this chapter. In the *M. pulmonis* study, cellular and humoral immune responses were generated in mice immunized with DNA encoding *M. pulmonis* antigens A7-1 or A8-1, administered either singly or in combination. For the protection study, mice received a combination of A7-1 and A8-1 either one week prior to *M. pulmonis* challenge (vaccination) or one week after (therapy). Further immunizations were administered two more times at two-week intervals after the first immunization. Mice were killed monthly after infection and the extent of disease determined histopathologically and by enumerating bacteria in tracheo-lung lavage. Following either vaccination regime, the number of *M. pulmonis* recovered by lavage decreased monthly until no bacteria were cultured at four months. The continued reduction in colony forming units (CFU) was accompanied by reduction in the severity of lung inflammation.

Pseudomonas aeruginosa

Pseudomonas aeruginosa is an opportunistic pathogen, commonly associated with cystic fibrosis patients, immunocompromised individuals, including AIDS patients, and humans and animals that have a breach in their skin barrier caused by burn or wounding. Infection with *P. aeruginosa* is a significant cause of death in domestic animals. A number of virulence factors are secreted by Pseudomonas. The most toxic protein produced by Pseudomonas is Exotoxin A, which shares its mode of action (ADP-ribosylation) with diphtheria toxin. Neutralization of Exotoxin A toxicity by DNA immunization was the aim of two recent studies in which the coding sequence for nontoxic mutant forms of *P. aeruginosa* Exotoxin A were cloned into DNA expression vectors.[120,121] Both studies reported a serum immunoglobulin response to Exotoxin A following DNA immunization. Analysis of the IgG isotype and cytokine secretion by splenocytes suggested that vaccination favored a Th1-type response.[121] DNA vaccinated mice in both studies were completely protected from the lethal effect of intraperitoneal injection with wild-type Exotoxin A, consistent with the demonstration of Exotoxin A neutralization by immune sera in vitro.[121]

While neutralization of Exotoxin A activity following DNA immunization would not be expected to prevent infection with *P. aeruginosa*, it might prevent certain pathology associated with infection. The coding sequence for a nontoxic mutant form of Exotoxin A could, therefore, be an important constituent of a multicomponent *P. aeruginosa* DNA vaccine. An additional candidate for such a vaccine cocktail could be the *oprF* gene of *P. aeruginosa*, encoding Outer Membrane Protein F. Immunization with this protein has been shown to confer some protection in a variety of rodent models of *P. aeruginosa* infection.[122-124] Mice immunized with a DNA plasmid encoding *oprF* generated an IgG1 serum response to both OprF and OprH of the homologous strain of *P. aeruginosa*, and to heterologous immunotypes.[125] The immune serum also mediated opsonophagocytic uptake of *P. aeruginosa* by peripheral blood mononuclear cells in culture. In a chronic pulmonary infection model, DNA vaccinated mice showed a significant reduction in the number of severe lung lesions and the number of mice with lungs yielding > 5000 colony forming units (CFU) of *P. aeruginosa* compared with controls.[125]

Salmonella typhi

Salmonella typhi causes typhoid fever, a severe systemic disease characterized by fever and abdominal symptoms, including hemorrhage, which is the principle cause of death in such infections. If untreated, typhoid fever has a 10 – 20 % mortality rate. *S. typhi* infects only humans and chimpanzees, making study of the infection difficult. However, under certain experimental conditions it is possible to infect animals such as mice and mimic some aspects of the natural infection. The murine model has been used to evaluate a DNA immunization approach to prevent *S. typhi* infection.[126] In these studies, BALB/c mice injected with DNA expressing the Outer Membrane Protein C (OmpC) Porin of *S. typhi* produced a serum IgG response specific to the protein. Responses were greater to OmpC expressed with its leader sequence than without. Since vaccination of mice with recombinant OmpC induces protection from *S. typhi* in mice,[127] the preliminary DNA immunization results were taken as encouraging, although to our knowledge, no further studies have been reported.

Staphylococcus aureus

Contamination of food by toxins produced by *Staphylococcus aureus* is the major cause of food-borne intoxication. Symptoms are associated with acute gastroenteritis and mortality is less than 1 %. Since *S. aureus* is part of the normal respiratory and skin flora of approximately 30 – 40 % of people collectively, contamination of food with *S. aureus* most commonly originates from the food-handler. More significantly, such transmission may occur nosocomially from hospital personnel to susceptible patients by direct contact. This is of particular concern with regard to the transmission of methicillin-resistant *S. aureus* (MRSA) to immunocompromised individuals. More than 90 % of MRSA isolates produce the penicillin-binding protein PBP2', encoded by the *mecA* gene and carried on a foreign region of DNA integrated into the chromosome of MRSA. The source of *mecA* is not known but it may originate from other genera of bacteria or from within the Staphylococci.[128] Possession of *mecA* and expression of PBP2' distinguishes MRSA from methicillin-susceptible isolates.

A DNA vaccination approach to prevent MRSA infection was reported recently.[129] In this study, DNA encoding the *mecA* gene was used to immunize BALB/c mice three times at two-week intervals. Immunization resulted in a serum immunoglobulin response to PBP2', demonstrated by immunoblotting and ELISA. The immune serum enhanced the phagocytosis of PBP2'-positive MRSA by peritoneal macrophages. Fourteen days after the final immunization, the mice were challenged intravenously with 10^8 CFU MRSA 1191. Eight days after challenge, the kidneys were removed and the bacterial load determined. Vaccination with *mecA* plasmid gave a 0.4 log reduction in kidney CFU compared with controls, although this difference was statistically significant. This preliminary study suggests that PBP2' might be a target molecule for vaccination against MRSA.

Streptococcus pneumoniae

Despite the introduction of antibiotics, infection with *Streptococcus pneumoniae* still causes significant human mortality and morbidity through the induction of pneumococcal pneumonia, septicaemia, meningitis and otitis media. Most clinical isolates of *S. pneumoniae* express capsular polysaccharide that protects the organism from phagocytosis and represents the major virulence determinant of the organism. Ninety chemically and antigenically distinct capsular polysaccharides have been identified, forming the basis of the serotypic classification of *S. pneumoniae*. Recent efforts at vaccination have focused predominantly on the production of conjugate vaccines in which polysaccharide antigens are linked to a protein carrier in order to generate protective antibodies in recipients. DNA immunization against *S. pneumoniae* has not been rigorously pursued, although one study has reported the protection of mice from lethal pneumococcal septicaemia following immunization with a plasmid expressing pneumococcal surface protein A (PspA).[130] PspA is expressed on the surface of all serotypes of *S. pneumoniae* and is required for full virulence. Only 30 % of BALB/c mice immunized with the plasmid expressing PspA produced a serum immunoglobulin response to PspA protein although the isotype of immunoglobulin was not reported. Interestingly, the generation of this response did not correlate with protection from challenge with 20 lethal doses of virulent *S. pneumoniae*. None of the mice generated a mucosal response to PspA, as determined serologically using saliva and feces. Immunization with the plasmid protected 40 % of mice from death for up to 150 hours after challenge. All of the control mice were dead before 100 hours following challenge. Protection was also expressed in a significant reduction in the number of *S. pneumoniae* cultured from the blood. The efficacy of DNA immunization was less than that achieved with PspA protein, which had been shown to protect > 90 % of mice from challenge with more than one serotype of *S. pneumoniae*.[131,132]

Yersinia spp.

The genus Yersinia contains three pathogenic species: *Yersinia enterocolitica*, *Yersinia pseudotuberculosis* and *Yersinia pestis*. The first two species cause food-borne illness. *Y. enterocolitica* causes gastroenteritis characterized by fever and abdominal pain often accompanied by diarrhea and/or fever. *Y. pseudotuberculosis* typically causes an acute mesenteric lymphadenitis. *Y. pestis* is the causative agent of 'the plague', one of the most devastating infectious diseases known to man. Even though the three species use different routes to infect their host and cause diseases of different severity, they share a common tropism for the lymphoid tissue and are able to resist the primary immune defense mechanisms of the host. Only one immunogenicity study has been reported using DNA encoding the V antigen of *Y. pestis*.[133] Two challenge studies have been reported for *Y. enterocolitica*,[134,135] the latter study also investigated the outcome of challenge with *Y. pseudotuberculosis*.[135]

Y. enterocolitica and Y. pseudotuberculosis

The Heat-shock protein 60 (HSP60) of *Y. enterocolitica* is a protective antigen in mouse models of yersiniosis and invokes both the humoral and cellular arms of the immune response.[136] Vaccination of mice with a DNA plasmid encoding HSP60 induced a Th1-type response characterized by an IgG2a serum antibody response, antigen-specific splenocyte proliferation and the liberation of IFN-γ, but not IL-4 or IL-10.[134,135] Mice immunized with the HSP60 DNA vaccine were protected from challenge with a lethal dose of *Y. enterocolitica* administered either intravenously[134,135] or orally.[135] For those mice infected orally, protection was expressed at the level of a reduction in bacterial burden in the spleen but not locally at the site of mucosal entry. By using knockout mice, protection was found to be dependent on both CD4$^+$ and CD8$^+$ T cells.[135]

Table 3.

Organism	Antigen	Host	Immune Response	Challenge	Protection	Readout	Ref(s)	Notes
Bacillus anthracis	PA	Mouse	Serum IgG IFN-γ, IL-4	Anthrax toxin (2 component)	75 – 100 %	Death	[78]	
Borrelia burgdorfori	OspA	Mouse	Serum IgG Splenocyte proliferation IFN-γ, IL-4	*B. burgdorfori* ZS7	100%	Spirochete recultivation	[80]	i
	OspA OspB	Mouse	Serum Ig	*B. burgdorfori* Sh-2-82; N40	100% None	Spirochete recultivation	[79]	
Brucella abortus	L7/L12	Mouse	Serum IgG Splenocyte proliferation	*B. abortus* S2308	0.5 – 1.3 log	Spleen CFU	[81]	
Chlamydia pneumoniae	HSP60	Mouse	Serum IgG Splenocyte proliferation IFN-γ	*C. pneumoniae* Kajaani	0.7 – 1.3 log	Lung IFU	[85]	ii
	MOMP Npt1$_{Cp}$	Mouse	Not determined	*C. pneumoniae* AR39	~0.6 – 0.9 log ~0.6 log	Lung IFU	[84]	
	HSP60 MOMP Omp2 Cocktail	Mouse	Serum Ig Splenocyte proliferation	*C. pneumoniae* Kajaani	1.5 log 0.6 – 1.2 log 0.9 log 1.0 log	Lung IFU	[92]	
Chlamydia psittaci	MOMP	Turkey	Serum IgM, IgG, IgA PBL proliferation	*C. psittaci* 84/55	Up to 100 %	Bacterial shedding	[86]	
Chlamydia trachomatis	MOMP	Mouse	Western blot	*C. trachomatis* L2	Claimed	Death	[87]	iii
	MOMP	Mouse	Serum IgG DTH Splenocyte proliferation	*C. trachomatis* MoPn	Up to 4 log None	Lung IFU Vaginal IFU	[88*-90] [91]	*iv
	CTP		IFN-γ, IL-10 Serum IgG		None 2 – 3 log	Lung IFU Lung IFU	[89] [94]	
	STK							

continued on next page

Table 3. Cont'd.

Organism	Antigen	Host	Immune Response	Challenge	Protection	Readout	Ref(s)	Notes
Clostridium tetani	tetC	Mouse	Serum IgG Splenocyte proliferation IFNγ	Tetanus toxoid	70 - 100 %	Death	[96]	v
	tetC	Mouse	Serum IgG Splenocyte proliferation IFNγ CTL	Tetanus toxoid	75 - 100 %	Death	[97]	
Corynebacterium pseudotuberculosis	Genetic-ally-detox-ified PLD	Sheep	Serum Ig	C. pseudotuberculosis C231	45 - 70 %	Abscess formation	[98]	vi
Enterotoxigenic Escherichia coli	CFA/I	Mouse	Serum IgG IgA	None			[100-105*]	*vii
	faeG	Mouse, pig	Serum IgG	None			[106]	
Francisella tularensis	Chromo-somal DNA	Mouse	IFNγ, IL-6, IL-12	F. tularensis LVS	Up to 100%	Death	[107]	viii
Helicobacter pylori	Hsp A Hsp B	Mouse	Serum IgG, IgA	H. pylori SS1	~0.7 log ~1.3 log	Gastric CFU	[109]	
Leptospira interrogans	P68	Guinea pig	Not examined	L. interrogans serovar lai	77%	Death	[110]	ix
Listeria monocytogenes	LLO Act A	Mouse	Serum IgG, IgA Splenocyte proliferation IFNγ, IL-4 CTL	L. monocytogenes	60 – 100% None	Death	[113]	x
	LLO epitope	Mouse	IFNγ CTL	L. monocytogenes	1.7 – 2.7 log	Organ CFU	[114]	xi

continued on next page

Table 3. Cont'd.

Organism	Antigen	Host	Immune Response	Challenge	Protection	Readout	Ref(s)	Notes
	LLO	Mouse	CTL	L. monocytogenes	0.6 – 1.9 log	Spleen CFU	[116]	xii
	CpG ODN	Mouse		L. monocytogenes	100 %	Death	[107]	xiii
	LLO, p60	Mouse	Serum IgG Splenocyte proliferation IFNγ CTL	L. monocytogenes	0 – 3 log	Spleen CFU	[117]	xiv
Mycoplasma pulmonis	Genomic library	Mouse	Serum Ig DTH Macrophage migration inhibition	M. pulmonis	4.0 log - complete	Lung CFU	[118]	xv
	A7-1 A8-1	Mouse	Serum Ig, Wetern blot Splenocyte proliferation DTH Macrophage migration inhibition	M. pulmonis	Complete	Lavage CFU	[119]	xvi
Pseudomonas aeruginosa	Non-toxic form of ToxA	Mouse	Serum IgG	Wild-type ToxA	100%	Death	[120]	
	oprF		IFNγ Serum IgG	P. aeruginosa	~75 – 100 % Sig. reduction	Death Lung lesions Lung CFU	[121] [125]	
Salmonella typhi	OmpC	Mouse	Serum IgG, Western blot	None			[126]	
Methicillin-resistant Staphylococcus aureus	mecA	Mouse	Serum Ig, Western blot	MRSA 1191	0.4 log	Kidney CFU	[129]	
Streptococcus pneumoniae	pspA	Mouse	Serum Ig	S. pneumoniae A66	40 %	Death Blood CFU	[130]	xvii

continued on next page

Table 3. Cont'd.

Organism	Antigen	Host	Immune Response	Challenge	Protection	Readout	Ref(s)	Notes
Yersinia enterocolitica	HSP60	Mouse	Serum IgG, IgA	Y. enterocolitica strain WA-314	2.3 – 3.0 log	Spleen CFU	[135]	
			Splenocyte proliferation IFNγ, IL-4, IL-10	Y. pseudotuberculosis Y. enterocolitica strain WA-314	None 2.1 – 3.6 log	Spleen CFU	[134]	xviii
Yersinia pestis	V antigen	Mouse	Serum IgA	None			[133]	xix

i   The influence of promoter type on the magnitude of the immune response examined.
ii   Effect of the co-administration of plasmids expressing cytokines examined.
iii   The influence of gold bead size and DNA quantity on the efficacy of gene-gun delivery examined.
iv   DNA plasmid delivered by recombinant Salmonella orally.
v   First report of a DNA vaccine able to protect against a bacterial toxin.
vi   Antigen expressed as a fusion with CTLA-4, in order to direct it to antigen-presenting cells.
vii   Heterologous prime-boost strategy used: DNA (prime)/recombinant Salmonella (boost).
viii   Protection induced by administration of bacterial chromosomal DNA alone.
ix   In Chinese.
x   Attenuated Salmonella used to deliver DNA plasmids via the oral route.
xi   Protection only seen when codon usage optimized.
xii   Protection improved by secretion and/or mutation of the antigen.
xiii   Nonspecific protection induced by administration of CpG ODN alone.
xiv   Immune response and/or protection influenced by co-administration of CpG ODN or DNA expressing GM-CSF.
xv   First report of Expression Library Immunization (ELI).
xvi   First report of a therapeutic effect of DNA immunization on a bacterial infection.
xvii   Proposed as a model for determining the optimum expression of bacterial genes for DNA immunisation.
xviii   Cytokine genes cloned into and expressed from the plasmid encoding the antigen.
xix   Gene-gun mediated delivery found to be superior to manual delivery.

Protection of mice against yersiniosis requires a Th1 type response, characterized by the interplay of T cells and macrophages, and the production of IFN-γ. Therefore, in an effort to augment the response of mice to DNA vaccination, the HSP60 DNA plasmid was engineered to co-express the gene for IL-2, IL-4 or IFN-γ.[134] In each case, the cytokine gene was separated from *hsp60* by an internal ribosomal entry sequences (IRES) to facilitate equimolar expression of both the cytokine and HSP60. Co-expression of IFN-γ with HSP60 enhanced both splenocyte proliferation and the production of IgG. Coexpression of IFN-γ, but not IL-2 or IL-4, also enhanced protection against challenge by a further 1.58 log, compared with the HSP60 plasmid alone.

Since the HSP60 is highly conserved across bacterial genera, mice vaccinated with the HSP60 of *Y. enterocolitica* were also challenged with a variety of different bacteria, including *Y. pseudotuberculosis*, to determine whether any cross-protection had been generated following vaccination. No cross-protection was observed, even to *Y. pseudotuberculosis*.[135]

Y. pestis

The protective V antigen of *Y. pestis*[137] was used recently in a DNA immunogenicity study,[133] primarily as a model to examine the relative merits of gene-gun immunization over conventional needle inoculation for a DNA vaccine encoding a bacterial subunit protein. As such, immunization was not followed by challenge with *Y. pestis*. Immunization of mice by either method generated a serological response to the V antigen, although gene-gun mediated delivery was found to be more efficient.

General Conclusions

It may be considered remarkable that in almost every reported instance where DNA vaccination has been tested for protection against bacterial disease it has indeed given protection. That this is so, irrespective of the nature of the bacterium, its lifestyle and its mode of causing disease testifies to the power and potential of this approach. Obviously there may be bacterial targets against which the approach has been tested and so far failed. Negative results tend not to get published. Yet it would clearly be wrong to suppose that the technique will only be useful against intracellular pathogens because of an inherent bias towards cell-mediated immunity. Antibody responses to DNA vaccination have given good protection by opsonizing against *S. aureus* and by neutralizing anthrax, tetanus and pseudomonas toxins in mouse models. If and when the approach does fail one may anticipate that this will be due either to a lack of suitable antigens or to their inaccessibility in the target rather than to an inability of DNA vaccination to generate an appropriate response. In any case, techniques to selectively enhance responses to DNA vaccination in different directions are being developed, playing on the greatest strength of DNA vaccination, its versatility and the ease with which the DNA and the consequent responses can be modified.

As yet few DNA vaccines against bacteria have been evaluated in the final target host species, but encouraging results have been observed against psittacosis in turkeys and against tuberculosis in cattle. The big question, of course, is whether DNA vaccination will give responses that are strong enough or long-lasting enough to be practical in the real target species. Whether the prime-boost strategy will remain the best approach for strong, long-lasting responses, or if newer, more sophisticated forms of DNA vaccination alone will eventually suffice remains to be seen.

One final note is due on therapeutic vaccination. Vaccines are not usually therapeutic and the implications of the observations that DNA vaccines could cure mycoplasma and tuberculosis infections in mice have yet to be fully explored. We do not fully understand how DNA vaccines do this, but at the very least they serve as proof of principle; even the established inadequate or detrimental immune responses in heavily infected animals can be turned around. Thus there is some hope that in future immune therapy against bacterial disease, whether based on DNA vaccination or not, will complement and reduce the use of antibacterial drugs.

The multi-drug-resistant bacteria threat will always be with us and DNA immune therapy deserves further investigation.

References

1. WHO, Tuberculosis. Fact Sheet No 104 (http://www.who.org.) 2000.
2. Maher D, Raviglione MC. The global epidemic of tuberculosis: A World Health Organization perspective. In: D. Schlossberg, ed. Tuberculosis and nontuberculous mycobacterial infections. Philadelphia, London, Toronto, Montreal, Sydney, Tokyo: W.B. Saunders Company, 1999:104-113.
3. Colditz GA et al. Efficacy of BCG vaccine in the prevention of tuberculosis. Meta- analysis of the published literature. JAMA 1994; 271(9):698-702.
4. Brewer TF. Preventing tuberculosis with bacillus Calmette-Guerin vaccine: a meta- analysis of the literature. Clin Infect Dis 2000; 31Suppl 3:S64-7.
5. Fine PE. The BCG story: lessons from the past and implications for the future. Rev Infect Dis 1989; 11Suppl 2:S353-9.
6. Rodrigues LC, Diwan VK, Wheeler JG. Protective effect of BCG against tuberculous meningitis and miliary tuberculosis: a meta-analysis. Int J Epidemiol, 1993; 22(6):1154-8.
7. Cosivi O et al. Epidemiology of *Mycobacterium bovis* infection in animals and humans, with particular reference to Africa. Rev Sci Tech 1995; 14(3):733-746.
8. Cosivi O et al. Zoonotic tuberculosis due to *Mycobacterium bovis* in developing countries. Emerg Infect Dis 1998; 4(1):59-70.
9. Krebs JR. Bovine Tuberculosis in cattle and badgers. London: Ministry of Agriculture, Fisheries and Food Publications, 1997.
10. Andersen P. TB vaccines: progress and problems. Trends Immunol 2001; 22(3):160-168.
11. Flynn JL, Ernst JD. Immune responses in tuberculosis. Curr Opin Immunol 2000; 12(4):432-436.
12. Chan J, Kaufmann SHE. Immune mechanisms of protection, In: B.R. Bloom, ed. Tuberculosis: Pathogenesis, protection, disease. Washington, DC: ASM Press, 1994:389-415.
13. Orme IM, Cooper AM. Cytokine/chemokine cascades in immunity to tuberculosis. Immunol Today 1999; 20(7):307-312.
14. Dannenberg AM Jr. Pathogenesis of pulmonary *Mycobacterium bovis* infection: basic principles established by the rabbit model. Tuberculosis 2001; 81(1-2):87-96.
15. Vordermeier HM et al. Increase of tuberculous infection in the organs of B cell deficient mice. Clin Exp Immunol 1996; 106(2):312-316.
16. Silva CL, Lowrie DB. A single mycobacterial protein (hsp 65) expressed by a transgenic antigen-presenting cell vaccinates mice against tuberculosis. Immunology 1994; 82(2):244-248.
17. Lowrie DB et al. Towards a DNA vaccine against tuberculosis. Vaccine 1994; 12(16):1537-1540.
18. Huygen K. et al. Immunogenicity and protective efficacy of a tuberculosis DNA vaccine. Nat Med 1996; 2(8):893-898.
19. Tascon RE et al. Vaccination against tuberculosis by DNA injection. Nat Med 1996; 2(8):888-892.
20. Zhu X et al. Functions and specificity of T cells following nucleic acid vaccination of mice against *Mycobacterium tuberculosis* infection. J Immunol 1997; 158(12):5921-5926.
21. Orme IM, Andersen P, Boom WH. T cell response to *Mycobacterium tuberculosis*. J Infect Dis 1993; 167(6):1481-1497.
22. Morris S et al. The immunogenicity of single and combination DNA vaccines against tuberculosis. Vaccine 2000; 18(20):2155-2163.
23. Li Z et al. Immunogenicity of DNA vaccines expressing tuberculosis proteins fused to tissue plasminogen activator signal sequences. Infect Immun 1999; 67(9):4780-4786.
24. Lowrie DB et al. Protection against tuberculosis by a plasmid DNA vaccine. Vaccine 1997; 15(8):834-838.
25. Turner OC et al. Lack of protection in mice and necrotizing bronchointerstitial pneumonia with bronchiolitis in guinea pigs immunized with vaccines directed against the hsp60 molecule of *Mycobacterium tuberculosis*. Infect Immun 2000; 68(6):3674-3679.
26. Svanholm C et al. Enhancement of antibody responses by DNA immunization using expression vectors mediating efficient antigen secretion. J Immunol Methods 1999; 228(1-2):121-130.
27. Ragno S et al. Protection of rats from adjuvant arthritis by immunization with naked DNA encoding for mycobacterial heat shock protein 65. Arthritis Rheum 1997; 40(2):277-283.
28. Baldwin SL et al. Immunogenicity and protective efficacy of DNA vaccines encoding secreted and nonsecreted forms of *Mycobacterium tuberculosis* Ag85A. Tuber Lung Dis 1999; 79(4):251-259.
29. Lozes E et al. Immunogenicity and efficacy of a tuberculosis DNA vaccine encoding the components of the secreted antigen 85 complex. Vaccine 1997; 15(8):830-833.

30. Tanghe A et al. Immunogenicity and protective efficacy of tuberculosis DNA vaccines encoding putative phosphate transport receptors. J Immunol 1999; 162(2):1113-1119.
31. Baldwin SL et al. Evaluation of new vaccines in the mouse and guinea pig model of tuberculosis. Infect Immun 1998; 66(6):2951-2959.
32. Vordermeier HM et al. Effective DNA vaccination of cattle with the mycobacterial antigens MPB83 and MPB70 does not compromise the specificity of the comparative intradermal tuberculin skin test. Vaccine 2000; 19(9-10):1246-1255.
33. Lefevre P et al. Three different putative phosphate transport receptors are encoded by the *Mycobacterium tuberculosis* genome and are present at the surface of *Mycobacterium bovis* BCG. J Bacteriol 1997; 179(9):2900-2906.
34. Sorensen AL et al. Purification and characterization of a low-molecular-mass T-cell antigen secreted by *Mycobacterium tuberculosis*. Infect Immun 1995; 63(5):1710-1717.
35. Harboe M et al. Evidence for occurrence of the ESAT-6 protein in *Mycobacterium tuberculosis* and virulent *Mycobacterium bovis* and for its absence in *Mycobacterium bovis* BCG. Infect Immun 1996. 64(1):16-22.
36. Andersen P et al. Recall of long-lived immunity to *Mycobacterium tuberculosis* infection in mice. J Immunol 1995; 154(7):3359-3372.
37. Pollock JM, Andersen P. The potential of the ESAT-6 antigen secreted by virulent mycobacteria for specific diagnosis of tuberculosis. J Infect Dis 1997; 175(5):1251-1254.
38. Vordermeier HM et al. Development of diagnostic reagents to differentiate between *Mycobacterium bovis* BCG vaccination and *M. bovis* infection in cattle. Clin Diagn Lab Immunol 1999; 6(5):675-682.
39. Lalvani A et al. Enumeration of T cells specific for RD1-encoded antigens suggests a high prevalence of latent *Mycobacterium tuberculosis* infection in healthy urban Indians. J Infect Dis 2001; 183(3):469-477.
40. Kamath AT et al. Differential protective efficacy of DNA vaccines expressing secreted proteins of *Mycobacterium tuberculosis*. Infect Immun 1999; 67(4):1702-1707.
41. Yeremeev VV et al. The 19-kD antigen and protective immunity in a murine model of tuberculosis. Clin Exp Immunol 2000; 120(2):274-279.
42. Erb KJ et al. Identification of potential CD8+ T-cell epitopes of the 19 kDa and AhpC proteins from *Mycobacterium tuberculosis*. No evidence for CD8+ T-cell priming against the identified peptides after DNA-vaccination of mice. Vaccine 1998; 16(7):692-697.
43. Garapin A et al. Mixed immune response induced in rodents by two naked DNA genes coding for mycobacterial glycosylated proteins. Vaccine 2001; 19(20-22):2830-2841.
44. Lefevre P et al. Cloning of the gene encoding a 22-kilodalton cell surface antigen of *Mycobacterium bovis* BCG and analysis of its potential for DNA vaccination against tuberculosis. Infect Immun 2000; 68(3):1040-7.
45. Lowrie DB et al. Therapy of tuberculosis in mice by DNA vaccination. Nature 1999; 400(6741):269-271.
46. Turner J et al. Effective preexposure tuberculosis vaccines fail to protect when they are given in an immunotherapeutic mode. Infect Immun 2000; 68(3):1706-1709.
47. Britton WJ et al. Immunoprophylaxis against *Mycobacterium leprae* infection with subunit vaccines. Lepr Rev 2000; 71 Suppl:S176-81.
48. Martin E et al. DNA encoding a single mycobacterial antigen protects against leprosy infection. Vaccine 2001; 19(11-12):1391-1396.
49. Martin E et al. Protection against virulent *Mycobacterium avium* infection following DNA vaccination with the 35-kilodalton antigen is accompanied by induction of gamma interferon-secreting CD4(+) T cells. Infect Immun 2000; 68(6):3090-3096.
50. Velaz-Faircloth M et al. Protection against *Mycobacterium avium* by DNA vaccines expressing mycobacterial antigens as fusion proteins with green fluorescent protein. Infect Immun 1999; 67(8):4243-4250.
51. Tanghe A et al. Protective efficacy of a DNA vaccine encoding antigen 85A from *Mycobacterium bovis* BCG against Buruli ulcer. Infect Immun 2001; 69(9):5403-5411.
52. Kim SH, Cho D, Kim TS. Induction of in vivo resistance to *Mycobacterium avium* infection by intramuscular injection with DNA encoding interleukin-18. Immunology 2001; 102(2):234-241.
53. Chambers MA et al. Vaccination of mice and cattle with plasmid DNA encoding the *Mycobacterium bovis* antigen MPB83. Clin Infect Dis 2000; 30 Suppl 3:S283-287.
54. Chambers MA et al. Vaccination of guinea pigs with DNA encoding the mycobacterial antigen MPB83 influences pulmonary pathology but not hematogenous spread following aerogenic infection with *Mycobacterium bovis*. Infect Immun 2002; 70:2159-65.
55. Kalman S et al. Comparative genomes of *Chlamydia pneumoniae* and *C. trachomatis*. Nat Genet 1999; 21(4):385-389.

56. Kamath AT et al. Co-immunization with DNA vaccines expressing granulocyte-macrophage colony-stimulating factor and mycobacterial secreted proteins enhances T-cell immunity, but not protective efficacy against *Mycobacterium tuberculosis*. Immunology 1999; 96(4):511-516.

57. Delogu G et al. DNA vaccination against tuberculosis: expression of a ubiquitinconjugated tuberculosis protein enhances antimycobacterial immunity. Infect Immun 2000; 68(6):3097-3102.

58. Tanghe A et al. Improved immunogenicity and protective efficacy of a tuberculosis DNA vaccine encoding Ag85 by protein boosting. Infect Immun 2001; 69(5):3041-3047.

59. McShane H et al. Enhanced immunogenicity of CD4(+) t-cell responses and protective efficacy of a DNA-modified vaccinia virus Ankara prime-boost vaccination regimen for murine tuberculosis. Infect Immun 2001; 69(2):681-686.

60. Feng CG et al. Priming by DNA immunization augments protective efficacy of *Mycobacterium bovis* Bacille Calmette-Guerin against tuberculosis. Infect Immun, 2001. 69(6):4174-4176.

61. Verthelyi D et al. Human peripheral blood cells differentially recognize and respond to two distinct CPG motifs. J Immunol 2001; 166(4):2372-2377.

62. Cole ST et al. Deciphering the biology of *Mycobacterium tuberculosis* from the complete genome sequence. Nature 1998; 393(6685):537-544.

63. Cole ST et al. Massive gene decay in the leprosy bacillus. Nature 2001; 409(6823):1007-1011.

64. Meister GE et al. Two novel T cell epitope prediction algorithms based on MHC-binding motifs; comparison of predicted and published epitopes from *Mycobacterium tuberculosis* and HIV protein sequences. Vaccine 1995; 13(6):581-591.

65. Lyadova IV et al. Intranasal BCG vaccination protects BALB/c mice against virulent *Mycobacterium bovis* and accelerates production of IFN-γ in their lungs. Clin Exp Immunol 2001; 125:274-279.

66. Kamath AT et al. Protective effect of DNA immunization against mycobacterial infection is associated with the early emergence of interferon-gamma (IFNgamma)- secreting lymphocytes. Clin Exp Immunol 2000; 120(3):476-482.

67. Zhu X et al. Specificity of CD8$^+$ T cells from subunit-vaccinated and infected H-2^b mice recognizing the 38 kDa antigen of *Mycobacterium tuberculosis*. Int Immunol 1997; 9(11):1669-1676.

68. Denis O et al. Vaccination with plasmid DNA encoding mycobacterial antigen 85A stimulates a CD4$^+$ and CD8$^+$ T-cell epitopic repertoire broader than that stimulated by *Mycobacterium tuberculosis* H37Rv infection. Infect Immun 1998; 66(4):1527-1533.

69. Olsen AW et al. Efficient protection against *Mycobacterium tuberculosis* by vaccination with a single subdominant epitope from the ESAT-6 antigen. Eur J Immunol 2000; 30(6):1724-1732.

70. D'Souza S et al. CD4$^+$ T cells contain *Mycobacterium tuberculosis* infection in the absence of CD8$^+$ T cells in mice vaccinated with DNA encoding Ag85A. Eur J Immunol 2000; 30(9):2455-2459.

71. Silva CL et al. Protection against tuberculosis by passive transfer with T-cell clones recognizing mycobacterial heat-shock protein 65. Immunology 1994; 83(3):341-346.

72. Silva CL et al. Characterization of T cells that confer a high degree of protective immunity against tuberculosis in mice after vaccination with tumor cells expressing mycobacterial hsp65. Infect Immun 1996; 64(7):2400-2407.

73. Silva CL et al. Characterization of the memory/activated T cells that mediate the long- lived host response against tuberculosis after bacillus Calmette-Guerin or DNA vaccination. Immunology 1999; 97(4):573-581.

74. Silva CL Lowrie DB. Identification and characterization of murine cytotoxic T cells that kill *Mycobacterium tuberculosis*. Infect Immun 2000; 68(6):3269- 3274.

75. Kim CH, Campbell Dj, Butcher EC. Nonpolarized memory T cells. Trends Immunol 2001; 22(10):527-530.

76. Sallusto F et al. Two subsets of memory T lymphocytes with distinct homing potentials and effector functions. Nature 1999; 401(6754):708-712.

77. Wang X, Mosmann T. In vivo priming of CD4 T cells that produce interleukin (IL)-2 but not IL-4 or interferon (IFN)-gamma, and can subsequently differentiate into IL-4- or IFN-gamma-secreting cells. J Exp Med 2001; 194(8):1069-1080.

78. Gu ML, Leppla SH, Klinman DM. Protection against anthrax toxin by vaccination with a DNA plasmid encoding anthrax protective antigen. Vaccine 1999; 17(4):340-344.

79. Luke CJ et al. An OspA-based DNA vaccine protects mice against infection with *Borrelia burgdorferi*. J Infect Dis 1997; 175(1):91-97.

80. Simon MM et al. Protective immunization with plasmid DNA containing the outer surface lipoprotein A gene of *Borrelia burgdorferi* is independent of an eukaryotic promoter. Eur J Immunol 1996; 26(12):2831-2840.

81. Kurar E, Splitter GA. Nucleic acid vaccination of *Brucella abortus* ribosomal *L7/L12* gene elicits immune response. Vaccine 1997; 15(17-18):1851-1857.

82. Brooks-Worrell BM, Splitter GA. Antigens of *Brucella abortus* S19 immunodominant for bovine lymphocytes as identified by one- and two-dimensional cellular immunoblotting. Infect Immun 1992; 60(6):2459-2464.
83. Stephens RS et al. Genome sequence of an obligate intracellular pathogen of humans: *Chlamydia trachomatis*. Science 1998; 282(5389):754-759.
84. Murdin AD et al. Use of a mouse lung challenge model to identify antigens protective against *Chlamydia pneumoniae* lung infection. J Infect Dis 2000; 181 Suppl 3:S544-551.
85. Svanholm C et al. Protective DNA immunization against *Chlamydia pneumoniae*. Scand J Immunol 2000; 51(4):345-353.
86. Vanrompay D et al. Turkeys are protected from infection with *Chlamydia psittaci* by plasmid DNA vaccination against the major outer membrane protein. Clin Exp Immunol 1999; 118(1):49-55.
87. Vanrompay D et al. Protection of turkeys against *Chlamydia psittaci* challenge by gene gun-based DNA immunizations. Vaccine 1999; 17(20-21):2628-2635.
88. Brunham RC, Zhang D. Transgene as vaccine for chlamydia. Am Heart J 1999; 138(5 Pt 2):S519-522.
89. Zhang D et al. DNA vaccination with the major outer-membrane protein gene induces acquired immunity to *Chlamydia trachomatis* (mouse pneumonitis) infection. J Infect Dis 1997; 176(4):1035-1040.
90. Zhang DJ et al. Characterization of immune responses following intramuscular DNA immunization with the MOMP gene of *Chlamydia trachomatis* mouse pneumonitis strain. Immunology 1999. 96(2):314-321.
91. Pal S et al. Vaccination of mice with DNA plasmids coding for the *Chlamydia trachomatis* major outer membrane protein elicits an immune response but fails to protect against a genital challenge. Vaccine 1999; 17(5):459-465.
92. Penttila T et al. Immunity to *Chlamydia pneumoniae* induced by vaccination with DNA vectors expressing a cytoplasmic protein (Hsp60) or outer membrane proteins (MOMP and Omp2). Vaccine 2000; 19(9-10):1256-1265.
93. Yang X, HayGlass KT, Brunham RC. Genetically determined differences in IL-10 and IFN-gamma responses correlate with clearance of *Chlamydia trachomatis* mouse pneumonitis infection. J Immunol 1996; 156(11):4338-4344.
94. Brunham RC et al. The potential for vaccine development against chlamydial infection and disease. J Infect Dis 2000; 181 Suppl 3:S538-43.
95. Strugnell RA et al., DNA vaccines for bacterial infections. Immunol Cell Biol 1997; 75(4):364-9.
96. Anderson R et al. Immune response in mice following immunization with DNA encoding fragment C of tetanus toxin. Infect Immun 1996; 64(8):3168-73.
97. Saikh KU et al. Are DNA-based vaccines useful for protection against secreted bacterial toxins? Tetanus toxin test case. Vaccine 1998; 16(9-10):1029-1036.
98. Chaplin PJ et al. Targeting improves the efficacy of a DNA vaccine against *Corynebacterium pseudotuberculosis* in sheep. Infect Immun 1999; 67(12):6434-6438.
99. Boyle JS, Brady JL, Lew AM. Enhanced responses to a DNA vaccine encoding a fusion antigen that is directed to sites of immune induction. Nature 1998; 392(6674):408-411.
100. Alves AM et al. Epitope specificity of antibodies raised against enterotoxigenic *Escherichia coli* CFA/I fimbriae in mice immunized with naked DNA. Vaccine 1998. 16(1):9-15.
101. Alves AM et al. Immunoglobulin G subclass responses in mice immunized with plasmid DNA encoding the CFA/I fimbria of enterotoxigenic *Escherichia coli*. Immunol Lett 1998; 62(3):145-149.
102. Alves AM et al. Antibody response in mice immunized with a plasmid DNA encoding the colonization factor antigen I of enterotoxigenic *Escherichia coli*. FEMS Immunol Med Microbiol 1999; 23(4):321-330.
103. Alves AM et al. DNA immunisation against the CFA/I fimbriae of enterotoxigenic *Escherichia coli* (ETEC). Vaccine 2000; 19(7-8):788-795.
104. Alves AM et al. New vaccine strategies against enterotoxigenic *Escherichia coli*. I: DNA vaccines against the CFA/I fimbrial adhesin. Braz J Med Biol Res 1999; 32(2):223-229.
105. Lasaro MO et al. New vaccine strategies against enterotoxigenic *Escherichia coli*. II: Enhanced systemic and secreted antibody responses against the CFA/I fimbriae by priming with DNA and boosting with a live recombinant Salmonella vaccine. Braz J Med Biol Res 1999; 32(2):241-6.
106. Turnes CG et al. DNA inoculation with a plasmid vector carrying the *faeG* adhesin gene of Escherichia coli K88ab induced immune responses in mice and pigs. Vaccine 1999; 17(15-16):2089-95.
107. Elkins KL et al., Bacterial DNA containing CpG motifs stimulates lymphocyte dependent protection of mice against lethal infection with intracellular bacteria. J Immunol 1999; 162(4):2291-2298.
108. Krieg AM, Davis HL Enhancing vaccines with immune stimulatory CpG DNA. Curr Opin Mol Ther 2001; 3(1):15-24.

109. Todoroki I et al. Suppressive effects of DNA vaccines encoding heat shock protein on *Helicobacter pylori*-induced gastritis in mice. Biochem Biophys Res Commun 2000; 277(1):159-163.

110. Dai B et al. [Immunoprotection in guinea pigs using DNA recombinant plasmid rpDJt and expressed protein P68 in *L. interrogans* serovar lai]. Hua Xi Yi Ke Da Xue Xue Bao 1998; 29(3):248-251.

111. North RJ, Dunn PL, Conlan JW. Murine listeriosis as a model of antimicrobial defense. Immunol Rev 1997; 158:27-36.

112. Harty JT, MJ Bevan. CD8⁺ T cells specific for a single nonamer epitope of *Listeria monocytogenes* are protective in vivo. J Exp Med, 1992. 175(6):1531- 1538.

113. Darji A et al., Oral somatic transgene vaccination using attenuated *S. typhimurium*. Cell 1997; 91(6):765-775.

114. Uchijima M et al. Optimization of codon usage of plasmid DNA vaccine is required for the effective MHC class I-restricted T cell responses against an intracellular bacterium. J Immunol 1998; 161(10):5594-5599.

115. Koide Y et al. [DNA vaccines for infections with intracellular bacteria]. Nippon Saikingaku Zasshi 1999; 54(4):773-793.

116. Cornell KA et al. Genetic immunization of mice against Listeria monocytogenes using plasmid DNA encoding listeriolysin O. J Immunol 1999; 163(1):322-329.

117. Fensterle J et al. Effective DNA vaccination against listeriosis by prime/boost inoculation with the gene gun. J Immunol 1999; 163(8):4510-4518.

118. Barry MA, Lai WC, Johnston SA. Protection against mycoplasma infection using expression-library immunization. Nature 1995; 377(6550):632-635.

119. Lai WC et al. Therapeutic effect of DNA immunization of genetically susceptible mice infected with virulent *Mycoplasma pulmonis*. J Immunol, 1997. 158(6):2513-2516.

120. Denis-Mize KS et al. Analysis of immunization with DNA encoding *Pseudomonas aeruginosa* exotoxin A. FEMS Immunol Med Microbiol 2000; 27(2):147-154.

121. Shiau JW et al. Mice immunized with DNA encoding a modified *Pseudomonas aeruginosa* exotoxin A develop protective immunity against exotoxin intoxication. Vaccine 2000. 19(9-10):1106-1112.

122. Matthews-Greer JM, Gilleland HE Jr. Outer membrane protein F (porin) preparation of *Pseudomonas aeruginosa* as a protective vaccine against heterologous immunotype strains in a burned mouse model. J Infect Dis 1987; 155(6):1282-1291.

123. Gilleland HE Jr et al. Use of a purified outer membrane protein F (porin) preparation of *Pseudomonas aeruginosa* as a protective vaccine in mice. Infect Immun 1984; 44(1):49-54.

124. Gilleland HE Jr, Gilleland LB, Matthews-Greer JM. Outer membrane protein F preparation of *Pseudomonas aeruginosa* as a vaccine against chronic pulmonary infection with heterologous immunotype strains in a rat model. Infect Immun, 1988. 56(5):1017-1022.

125. Price BM et al. Protection against *Pseudomonas aeruginosa* chronic lung infection in mice by genetic immunization against outer membrane protein F (OprF) of *P. aeruginosa*. Infect Immun 2001; 69(5):3510-3515.

126. Lopez-Macias C et al. Induction of antibodies against *Salmonella typhi* OmpC porin by naked DNA immunization. Ann N Y Acad Sci, 1995. 772:285-8.

127. Isibasi A et al. Role of porins from *Salmonella typhi* in the induction of protective immunity. Ann N Y Acad Sci 1994; 730:350-352.

128. Archer GL, Niemeyer DM. Origin and evolution of DNA associated with resistance to methicillin in staphylococci. Trends Microbiol 1994; 2(10):343-347.

129. Ohwada A et al. DNA vaccination by *mecA* sequence evokes an antibacterial immune response against methicillin-resistant *Staphylococcus aureus*. J Antimicrob Chemother 1999; 44(6):767-774.

130. McDaniel LS et al. Immunization with a plasmid expressing pneumococcal surface protein A (PspA) can elicit protection against fatal infection with *Streptococcus pneumoniae*. Gene Ther 1997; 4(4):375-377.

131. McDaniel LS et al. PspA, a surface protein of *Streptococcus pneumoniae*, is capable of eliciting protection against pneumococci of more than one capsular type. Infect Immun 1991; 59(1):222-228.

132. Tart RC et al. Truncated *Streptococcus pneumoniae* PspA molecules elicit cross-protective immunity against pneumococcal challenge in mice. J Infect Dis 1996; 173(2):380-386.

133. Bennett AM et al. Gene gun mediated vaccination is superior to manual delivery for immunisation with DNA vaccines expressing protective antigens from *Yersinia pestis* or Venezuelan Equine Encephalitis virus. Vaccine 1999; 18(7-8):588-596.

134. Hornef MW et al. DNA vaccination using coexpression of cytokine genes with bacterial gene encoding a 60-kDa heat shock protein. Med Microbiol Immunol (Berl) 2000; 189(2):97-104.

135. Noll A et al. DNA immunization confers systemic, but not mucosal, protection against enteroinvasive bacteria. Eur J Immunol 1999; 29(3):986-996.

136. Noll A et al. Protective role for heat shock protein-reactive alpha beta T cells in murine yersiniosis. Infect Immun 1994; 62(7):2784-2791.
137. Williamson ED et al. A new improved sub-unit vaccine for plague: the basis of protection. FEMS Immunol Med Microbiol 1995; 12(3-4):223-230.
138. Flynn JL, ChanJ. Immunology of tuberculosis. Annu Rev Immunol 2001; 19:93 – 129.
139. Tanghe A et al. Tuberculosis DNA vaccine encoding Ag85A is immunogenic and protective when administered by intramuscular needle injection but not by epidermal gene gun bombardment. Infect Immun 2000; 68(7):3854-3860.
140. Mollenkopf HJ et al. Protective efficacy against tuberculosis of ESAT-6 secreted by a live *Salmonella typhimurium* vaccine carrier strain and expressed by naked DNA. Vaccine 2001; 19(28-29):4028-4035.
141. Coler RN et al. Vaccination with the T cell antigen Mtb 8.4 protects against challenge with *Mycobacterium tuberculosis.* J Immunol 2001; 166(10):6227- 6235.
142. Dillon DC et al. Molecular characterization and human T cell responses to a member of a novel *Mycobacterium tuberculosis* mtb39 gene family. Infect Immun 1999; 67(6):2941 -2950.
143. Skeiky YA et al. T cell expression cloning of a *Mycobacterium tuberculosis* gene encoding a protective antigen associated with the early control of infection. J Immunol 2000; 165(12):7140-7149.
144. Delogu G, Brennan MJ Comparative immune response to PE and PE_PGRS antigens of *Mycobacterium tuberculosis.* Infect Immun 2001; 69(9):5606-5611.

CHAPTER 11

DNA Vaccines as Cancer Treatment Modalities

Ronald C. Kennedy, Michael H. Shearer and Robert K. Bright

Abstract

Therapeutic vaccination has recently regained its optimism as a potential immunologic based modality for the treatment of cancer patients. Cancer vaccine based clinical trials are being initiated at a rapid pace, yet work still must be performed to determine the best formulation of the vaccine, the optimal route for immunization, the most effective timing and course of therapy, and the best and most accurate methods for evaluating clinical efficacy. Early studies involving tumor vaccination employed whole tumor cells, fragments of tumor cells or protein containing lysates from tumor cells as the strategy. Limited results with these approaches have led investigators to develop the next generation of cancer vaccines using defined tumor-associated antigens. With the discovery of new tumor antigens for common human malignancies, the formulation of the vaccine has been a focus. Most strategies attempt to invoke a predominant cell-mediated immune response and virtually ignore the humoral aspect of the immune response. Yet, antibodies have been shown to play a role in tumor immunity and the currently most effective immuno-therapeutic modalities use monoclonal antibodies to target the tumor antigen. One form of a universal vaccination strategy that can induce both components of the immune responses is represented by plasmid DNA encoding a foreign transgene product that represents the target antigen for the induction of a specific immune response. This vaccination strategy referred to as DNA or genetic immunization has been employed to induce a variety of immune responses against antigens from a variety of sources, including infectious agents and allergens. When the foreign transgene expresses a tumor antigen, the potential exists that the DNA immunization strategy will invoke an anti-tumor immune response. This report will focus on the activities of DNA immunization that induce immune responses against tumor antigens and will discuss particular aspects of what immunologic correlates have been associated with protection against tumor progression in vivo. A discussion of specific systems will highlight the possible immunologic correlates of anti-tumor immunity and will demonstrate that DNA-based strategies represent a viable alternative for the development and testing of cancer vaccination strategies that target a particular tumor antigen.

Introduction

Cancer is a term used to describe a group of related diseases whereby uncontrolled cell growth and differentiation leads to the spread of cells throughout the body. In developed countries, such as the United States, cancer is the second leading cause of death behind only cardiovascular disease. Approximately 1.2 million cases of cancer will be diagnosed in the United States this year. The estimated number of deaths in the United States as the result of cancer will exceed 500,000 a year. It has been estimated that in the late 1990s, total cancer-associated health care expenditures reached $110 billion annually worldwide. Specifically in 1999, cancer

DNA Vaccines, edited by Hildegund C. J. Ertl. ©2003 Eurekah.com and Kluwer Academic / Plenum Publishers.

diagnosis and cancer related therapies generated approximately $9.5 billion in revenues world-wide. As a consequence, cancer and associated therapies represents one of the largest untapped disease markets in the world.

The treatment of cancer and the modalities employed are continually evolving. The current major strategies for treating cancer include surgery, radiation, and chemotherapy or combinations thereof. These treatment modalities tend to be specific for the type of cancer and are often based on the individual patient. Treatment or therapy is intended to eliminate the tumor and prevent further metastasis. Alternatively, the treatment should also make the patient more resistant to subsequent tumor formation. Problems associated with these three treatment strategies include lack of specificity and toxicity along with associated morbidity and mortality. Thus, new strategies focus on improving the specificity of the treatment, along with reducing the toxicity. One of these strategies involves targeting the immune system to specifically recognize tumor cells. This strategy is referred to as immunotherapy and can either invoke or target antibody-mediated or cell-mediated immune responses, or both. When the immune component is activated outside of the host and the activated form is administered to the patient, this is referred to as passive immunotherapy. If the activation of the immune system involves manipulating the host directly by the administration of substances that induce a specific immune response, this is referred to as active immunotherapy. Both forms of immunotherapy are under intense investigation as a means of cancer treatment and new approaches to harness and specifically target the immune system are being developed and tested.[1]

The concept of immunotherapy to treat cancer is not new and was described in the 1800s when it was noticed by physicians that bacterial infections, such as with the etiologic agent of syphilis sometimes caused the regression of tumors. In the early 20th century, William B. Coley began to treat cancer patients with bacterial extracts, referred to as Coley's toxins.[2] Within the foundation of immunotherapy, exposure to an infectious agent or toxic product was shown to activate the systemic immune response, a portion of which might be directed nonspecifically against the tumor. The mechanism(s) of this anti-tumor activity were unknown at that time; however, in retrospect these bacterial extracts most likely induced inflammatory responses and the production of cytokines with tumoricidal activities.

Humoral immunity is mediated by B-lymphocytes and these cells represent the effector cell type involved in the production of antibodies. Recognition of an antigen and the subsequent activation and differentiation of B cells into antibody secreting plasma cells and longer lasting memory B cells involve a number of other cell types in the process. This includes antigen presenting cells (APCs), such as, cells of the macrophage/monocyte lineage and dendritic cells. B-lymphocytes themselves can also function as APCs. A subset of T lymphocytes expressing the CD4 molecule, designated CD4[+] T helper cells, also are involved in the activation of B cells. There exists several means by which antibodies that target tumor antigens can destroy tumors. Antibodies with specificity for a tumor antigen can recognize the tumor and activate other effector mechanisms of the immune system. Thus, an antibody bound to its tumor antigen can be involved via the Fc portion of the antibody in a general tumoricidal phenomena referred to as antibody-dependent cell-mediated cytotoxicity (ADCC). In this scenario, receptors for the Fc region (FcR) that are expressed on phagocytic cells, such as macophages and monocytes, or on neutrophils or natural killer (NK) cells recognize the antibody complexed to the tumor antigen, internalize and destroy the tumor cells. A second mechanism relies on the complement system to invoke a complement-mediated cytotoxicity (CDC) reaction. In this instance, complement can bind to the Fc portion of the antibody-tumor cell complex, become activated and cause tumor cell lysis. Antibodies have also been demonstrated to induce a programmed cell death, referred to as apoptosis that results in DNA fragmentation and cell death. Other mechanisms include the potential of antibodies to bind to the tumor antigen on its cell surface, induce signaling pathways that remove the tumor antigen from the cell surface and target the antigen to internal lysozymal pathways for degradation and removal from the cell surface. The specific mechanism of cell death in this latter scenario remains to be identified, but represents a complex set of interactions involving the tumor cell antigen that is being degraded and the alteration at the tumor cell surface as the result of the tumor cell antigen removal. Finally, one can also use the antibody to deliver toxic agents to the tumor cell that will allow for its destruction. The antibodies that are induced against the tumor cell antigen can either be administered passively or can result from exposure to a cancer vaccination strategy whereby the antibodies are actively induced.

Cell-mediated immune responses are predominantly associated with the activation of effector T cell populations. The prevailing belief regarding the immunologic rejection of cancer is that T lymphocyte-mediated immunity is essential for the destruction of most solid tumors. T cells are typically divided into two major groups, the CD4$^+$ T cell that provides help for the induction of both humoral and cell-mediated immune responses and the CD8$^+$ T cells that become the major effector cells in the form of cytotoxic T lymphocytes (CTLs). It is the CD8$^+$ CTL response that can directly mediated tumor cell destruction. CD4$^+$ T cells become activated following recognition of peptide antigens that are usually derived from exogenous proteins. The exogenous protein is processed into peptides and presented in the context of class II major histocompatibility complex (MHC) antigens by APCs (described above). CTLs become activated primarily by endogenously expressed proteins and their peptides are processed and presented in the context of MHC class I molecules that are expressed on the surface of APCs or target cells. MHC class I molecules are normally expressed on all somatic cells, including epithelial cells that give rise to carcinomas of the prostate, breast and lung.

Classifications of Tumor Antigens

In order to target tumors with immunotherapeutic strategies, one must identify the antigens present on the tumor. Several classifications of cancer antigens expressed on tumor cells have been described. Early studies classified tumor antigens according to the expression on tumor and normal noncancerous cells. Antigens that were expressed on both tumor and normal cell types were characterized as tumor-associated antigens. In a number of instances the tumor-associated antigens were over expressed on the tumor cell over-expressed either at much lower levels or selectively expressed on only certain cell types. A number of tumor-associated antigens are used to target the passive immunotherapy strategies described above. A second early classification of tumor antigens was referred to as tumor-specific antigens. Tumor-specific antigens were only expressed on the tumor cell and were not expressed on normal cell types. Examples of tumor-specific antigens are the oncogenic viral antigens that resulted from viral infection leading to malignant cell transformation. This led to expression of a tumor- specific viral encoded antigen on tumor cells and not normal cells. A more recent classification of tumor antigens has been defined that employs several factors including T cell responses that can be induced.[3] In this classification scheme, recognition by T cells, expression on certain cell types and the source of the antigen is the basis for the different tumor antigens. The first class of tumor antigens is the cancer testis antigens. These antigens were originally identified based on their expression on melanomas and several other tumors, but they are not expressed in normal tissues except for the testis. Expansion of this class of tumor antigens could also include cellular differentiation antigens that are expressed on normal cell types early in ontogeny and then lose expression during development and maturation. Included in this tumor antigen group are MAGE, BAGE, GAGE, NY-ESO-1, and selected oncofetal proteins. A second group of antigens were identified as self-antigens expressed during melanocyte differentiation and were shown to be targets for autologous CTL responses in melanomas. These are referred to as melanocyte differentiation antigens. Including in this group of tumor antigens are MART-1, tyrosinase and tyrosinase-related proteins, gp100, and gp75/TRP-1. The third group of cancer antigens is defined by point mutations of constitutive expressed self cellular proteins. These include p53, Ras, and a mutated form of cyclin dependent kinase 4. The fourth and fifth groups based on this classification are represented by over-expressed self antigens and viral encoded antigens, respectively. The over-expressed self antigens include HER2/neu and also

MART-1 which also represents a melanocyte differentiation antigen. Examples of viral encoded antigens include Epstein Barr Virus with Burkitt's lymphoma, hepatitis B and C viruses with hepatocellular carcinoma, human papilloma viruses (HPV) with anal and cervical carcinoma, human T lymphotropic virus type 1 with T cell leukemia and human herpes virus 8 in Kaposi's sarcoma. This latter classification employs $CD8^+$ CTL responses as a means to assist in the cancer antigen characterization scheme. For the purpose of this review, we will emphasize the tumor-associated versus tumor-specific nature of the antigen that is being used to target the immunotherapy. With regards to immunologic modalities being employed to treat existing cancers that target specific tumor antigens, to date only antibody based strategies have been approved and licensed for human use. Below we will briefly discuss some of these antibody based treatment modalities in the context of how the humoral immune response can play a role in tumor immunity.

Breast Cancer and Herceptin

Invasive breast cancer represents the second leading cause of cancer-related deaths among women in the United States. It is estimated that over 40,000 deaths will be attributed to invasive breast cancer this year and over 190,000 new cases will be diagnosed. The major risk factors for developing breast cancer are age and family history. The overall incidence of the disease increases with age and approximately 75% of diagnosed breast cancer occurs in women over 50 years of age. There are several forms of breast cancer that are clinically recognized. A benign breast tumor is simply an abnormal growth that rarely spreads outside the breast. This form often occurs and is not usually life-threatening. Four different forms of malignant breast cancer have been characterized. Ductal carcinoma in situ (DCIS) is the earliest form of breast cancer and is completely confined to the ducts of the breast. This form is easily detected by mammograms and DCIS has the highest cure rate approaching 100%. A more advanced form is invasive ductal carcinoma whereby the malignant growth begins in the ducts and then spreads into the fatty tissue of the breast. This is the most common form of breast cancer and accounts for approximately 80% of all breast cancer cases. A third form of cancer is lobular carcinoma in situ. Although this is not a true form of breast cancer, it results in an increase incidence in forming breast cancer with 30% of these women going on to develop breast cancer. The final form of breast cancer is invasive lobular carcinoma. This starts in the lobes prior to metastasizing to distal sites and represents between 10 to 15% of all breast cancers. These four forms of malignant breast cancer and their associated tumor types are considered early forms and have been designated stages I and II. Stage III breast cancer is a latter form of the disease and usually presents clinically with a primary tumor measuring greater than 5 cm that has spread to the skin, chest wall, or lymph nodes. Stage III accounts for approximately 10% of the breast cancer diagnoses. Stage IV is the most aggressive and severe form of breast cancer and is considered metastatic breast cancer. This form has a median survival time following diagnosis of 18-24 months and is rarely curable.

In early stage disease, most breast cancer patients undergo surgery to remove the tumor from the breast and in a number of instances the axillary lymph nodes are removed and analyzed for the presence of the tumor cells. The range of surgery includes a lumpectomy to remove the lump and some surrounding tissue to a radical mastectomy whereby the breast, lymph nodes and chest muscles are removed. Often time surgery is followed by radiation and/ or chemotherapy regimens to improve the overall survival rate. There are numerous options available for the treatment of early stage disease. In metastatic breast cancer combination chemotherapy is the standard, or in instances where the tumor expresses a hormone receptor, such as, the estrogen receptor (ER), hormone therapy is often times combined with chemotherapy. It appears that women with ER-positive tumors have a consistent survival advantage over ER-negative tumors after recurrence. For women with no prior history of hormone therapy, tamoxifen, a nonsteroidal anti-estrogen that binds to ER on tumors is the present standard of care.

A large amount of scientific evidence indicates a role for the ErbB family of epidermal growth factor receptor (EGFR) tyrosine kinases in the induction of breast cancer. The ErbB family of EGFRs interacts with epidermal growth factors that result in the initiation of a signaling cascade of intracellular reactions with the activation of tyrosine kinase as a key step in the process. There are presently four identified ErbB family members, designated ErbB1-ErbB4. ErbB1 is the receptor for epidermal growth factor (EGF) and related proteins, whereas, ErbB3 and ErbB4 bind to related ligands termed heregulins. It is also apparent that both sets of ligands can activate all four ErbB family members via formation of heterodimers. This ligand-induced heterodimer formation results in the activation of tyrosine kinase domain expressed on the cytoplasmic tail of the EGFR and the subsequent phosphorylation of multiple tyrosine residues within the cytoplasmic tail. The activation of the EGFR tyrosine kinase and the subsequent phosphopeptide motifs lead to the recruitment of several intracellular signaling proteins and the induction of signaling cascades. Since these cascades control gene transcription and cell cycle entry, they ultimately affect cell growth and can mediate transformation events via uncontrolled proliferation. EGFR expressed on tumor cells also appears to play an important role in apoptosis, angiogenesis, and in metastasis. The latter two are important steps in tumor growth and pathogenesis. ErbB2 appears to represent a superior receptor for inducing signaling pathways potentially because of its intracellular trafficking pattern that differs from the other EGFRs. It is also over-expressed in 20-35% of breast cancer and predicts a poor relapse-free survival following treatment.

Herceptin was the first humanized monoclonal antibody approved for the treatment of breast cancer. Herceptin or Trastuzumab monotherapy is indicated for the treatment of metastatic breast cancer whereby the tumor cells express the HER2/neu protein and have received one or more chemotherapy regimens. The HER2 received its designation based on human epidermal growth factor receptor 2 and is analogous to ErbB2. Herceptin was first approved in the United States in 1998. The side effects of this treatment regimen include hypersensitivity and pulmonary involvement that have been associated with anaphylactic reactions. The mechanism by which Herceptin destroys ErbB2 expressing breast cancer cells remains to be clarified and may include ADCC and apoptosis. Alternatively, the binding of Herceptin to the ErbB2 molecule may involve the activation of intracellular signaling pathways, the endocytosis of ErbB2 from the cell surface and the targeting of the intracellular form of the receptor to lysosomal degradation pathways. Thus, Herceptin may remove ErbB2 from the surface of the tumor cell and prevent its re-expression. Binding of the native ligand, EGF, to ErbB2 on the other hand actives intracellular signaling pathways and ErbB2 can be re-expressed on the tumor cell surface. This is a clear indication where a passive immunotherapy that uses a humanized monoclonal antibody can provide beneficial effects for treating human cancer.

Non-Hodgkin's Lymphoma and Rituxan

The term lymphoma is used to describe cancers that are derived from the lymphatic system. The lymphatic system consists of the lymph nodes, the spleen, and the vessels that link the lymph nodes, termed the lymphatics. The lymphatic system is an integral part of the immune system. Hodgkin's disease is one form of lymphoma with all other forms being grouped into non-Hodgkin's lymphoma (NHL). In the United States, lymphomas are responsible for approximately 5% of all cancers. This represents about 62,000 new cases of lymphoma per year. Of these, NHL accounts for 55,000 of these cases and estimates anticipate that 49% of the patients diagnosed will succumb within five years. Individuals with NHL are usually diagnosed at an advanced stage, stage III or IV. Patients with low-grade disease have lymphomas that grow slowly and cause lesser symptoms. The intermediate or high-grade lymphomas are more aggressive with regards to growth and spread quickly. This aggressive form tends to cause severe symptoms.

In most instances, localized low-grade disease (approximately 10% of patients) is treated with regional radiation therapy. This offers long-term control of the disease, with rates of survival ranging 75% in individuals younger than 60 years of age. Disease recurrence rates range from about 45% at ten years. Chemotherapeutic regimens and their efficacy as a first line therapy in low-grade lymphoma have not yet been determined. Chemotherapy is primarily utilized if a patient has failed radiation therapy or if there was a relapse of disease. As with other types of cancer, there are no set schedules or regimens for treating NHL with chemotherapy. It is highly subjective in nature and depends on the individual patient, disease state, type of lymphoma, and symptoms. In advanced stages of disease, combination radiation and chemotherapy treatments are employed. Individuals who exhibit a recurrence are usually not cured by conventional treatments, such as, radiation and chemotherapy. Stem-cell transplantation has resulted in remission from recurring disease. However, this type of therapy has a pronounced number of side effects, including high rates of mortality associated with the transplantation.

Lymphomas, by their nature, represent an ideal form of cancer to target with immunotherapeutic modalities. In particular, non-Hodgkin's B cell lymphomas express a B cell differentiation marker, designated CD20. CD20 is a 33-37 kDa phosphoprotein that crosses the cell membrane and is expressed on the cell surface of pre-B cells, resting and activated B cells, but not on plasma cells. The CD20 molecule is involved in the regulation of B cell activation. Rituxan or Rituximab is a chimeric human/mouse monoclonal antibody that is used for the treatment of CD20-positive B cell NHL. Rituxan is presently being used as a second-line treatment modality; however, the use of monoclonal antibodies as passive immunotherapies to treated NHL as front-line treatments are receiving a lot of attention. Presently, Rituxan, Bexxar, Zevalin all target the CD20 molecule. Each of these antibody-based agents have different levels of activity and toxicity. Zevalin is the mouse monoclonal anti-CD20 parent of Rituxan that has been conjugated with a chelating agent, yttrium 90, which is a β-emitting radioisotope. Generically, Zevalin is known as ibritumomab tituxetan and utilizing the cytotoxic activities of the chelating agent by directly targeting to CD20 expressing B cells is capable of killing tumor cells. Bexxar is also a monoclonal anti-CD20 that is conjugated to the radiolabel iodine. [131]Iodine, which is a γ-emitting radioisotope is also targeted to CD20 expressing B cells. Consequently, these targeted passive immunotherapies tend to be more toxic than Rituxan. The radioactivity associated with Bexxar and Zevalin can suppress the bone marrow and thyroid gland and enhance the normal toxicities involving monoclonal antibody treatment. Compared to Herceptin, the side effects associated with Rituxan are more severe forms of hypersensitivity. Side effects included mortality involving cytokine release syndrome. However, NHL is an incurable disease and few other alternatives exist. The mechanism of tumor immunity with regards to Rituxan appears to be FcR mediated and involves ADCC-associated killing of tumor cells expressing CD20. Studies have demonstrated that modifications of the FcR binding sites on the Fc region of the human portion of Rituxan had adverse effects and diminished in vitro killing of CD20-positive B cells derived from lymphomas. It is unclear whether the activities of Bexxar and Zevalin on CD20-positive tumor cells is both the result of ADCC and cytotoxic effects of the radioisotope upon the tumor cell targeted to the CD20 molecule. These data further support a role for antibodies in direct tumorcidal effects and therapeutic benefits for the treatment of NHL. Studies have also considered using monoclonal antibodies to CD19, CD22, CD38, and CD40 for treating B cell lymphoma since these cell surface antigens are capable of generating transmembrane signals and activating signal transduction pathways in normal and neoplastic B cells.

Additional Clinical Trials Utilizing Antibodies as Passive and Active Immunotherapeutics

A humanized monoclonal antibody specific for vascular endothelial growth factor (VEGF) is undergoing phase III clinical trial in patients with colorectal cancer and in a combination modality with chemotherapy in nonsmall cell metastatic lung cancer. This antibody ingibits angiogenesis in tumors. Safety concerns were reported in the colorectal cancer trial and deaths were associated with sudden bleeding. It remains to be determined whether this antibody will be approved as a passive immunotherapy treatment in humans. Passive administration of monoclonal antibodies to tumor-associated antigens expressed on gastrointestinal carcinoma cells have been reported to achieve a clinical benefit in patients with resected Dukes' C colorectal carcinoma.[4] Active immunotherapy utilizing a monoclonal anti-idiotype that mimics the high molecular weight melanoma antigen has also been reported to induce anti-tumor immunity in melanoma patients.[5] Together these studies demonstrate that antibodies can play a role in human tumor immunity to a variety of cancers. The observed immunity in these trials may only be partial and result in improved well-being or a prolonged survival rate. Thus it becomes difficult to definitely state an absolute role for antibodies in tumor immunity and to completely rule out other factors and their involvement. However, with the prevailing belief being that the immunologic rejection of cancer is a T cell-mediated phenomena, one can not rule out a role for antibodies in mediating tumor immunity with the evidence described above.

A number of cancer vaccine approaches are in or entering clinical trials. A large majority of these approaches target the cell-mediated immune responses and specifically attempt to induce CD8+ CTL responses. These include in vitro expansion and adoptive transfer of tumor-specific CTLs in the presence or absence of cytokines, peptide tumor antigen *ex-vivo* pulsed autologous dendritic cells and vaccination strategies that target the induction of CTL responses, among others. The induction of CD8+ CTL responses in cancer patients has been described; however, the clinical effectiveness of these responses remains under study. Thus, cancer vaccination strategies should focus on modalities that have the potential to induce both humoral and cell-mediated immune responses to specific tumor antigens.

DNA Vaccination

DNA immunization represents a relatively new approach to vaccine and immunotherapeutic development.[6] The injection of plasmid DNA encoding a foreign gene of interest can result in the subsequent expression of the foreign gene products and the induction of an immune response within a host. This is relevant for both prophylactic and therapeutic vaccination strategies in that exposure risk increases the incidence of certain cancers and prophylactic vaccination strategies for these types of cancers when exposure risk is unavoidable may be warranted. In addition, therapeutic vaccination can be used to treat existing cancers and potentially reduce the tumor burden as the result of targeting a specific immune response. DNA immunization can induce both humoral and cell-mediated immune responses and has been used as a modality to vaccinate against infectious diseases and cancer.[7,8] Studies described in this report will focus on the use of DNA immunization strategies as a means of invoking specific immune responses against tumor antigens associated with selected cancers where animal model systems exist to evaluate the induction of anti-tumor immunity. Much has been learned from murine models of various relevant human cancer scenarios where the tumor antigens have been identified. Our efforts have focused on a viral encoded tumor-specific antigen from an oncogenic DNA virus, namely simian virus 40 (SV40), and an encoded oncoprotein, designated as the large tumor antigen (Tag).

SV40 and Its Relevance to Human Infections and Cancer

SV40 is a polyomavirus of rhesus monkey (*Macaca mulatta*) origin. This DNA virus was initially discovered as a contaminant of poliovirus vaccines in 1960 that were used to immunize millions of people from 1955 to 1963.[9] SV40 is classified in the virus family Papoviridae within the genus *Polyomavirus*. This is based on the size and morphology of its icosahedral capsid and the size of its double-stranded DNA. SV40 is closely related to two other polyomaviruses that are of human origin and designed JC virus and BK virus. At a genomic level, these three viruses share approximately 70% homology. SV40 is an oncogenic virus that induces tumors in rodents and transforms a variety of cell types from a number of different species, including human, in vitro. SV40 is perhaps one of the best studied viruses that has been used in discerning the mechanisms of molecular processes in eukaryotic cells. It represents one of the premier molecular tools used to study cellular transformation events. The nonstructural Tag gene encoded by the SV40 genome is involved in viral replication and is expressed early in the DNA replication cycle. Large Tag also represents a viral oncoprotein that is responsible for cellular transformation and immortalization of the various cell types (see below). Three structural proteins are encoded for by the SV40 genome that make up the viral capsid proteins. These proteins, designated VP1, VP2, and VP3 are involved in capsid formation. VP1 is the capsid protein containing 362 amino acids that forms the pentameric capsomeres on the surface of the viral particle. Infection of a permissive host results in the production of infectious viral particles composed of these capsid proteins.[10]

SV40 is a natural infection of nonhuman primates, its permissive host. Most infections are benign and demonstrate no pathology. However, evidence is increasing that SV40 infection of immunocompromised nonhuman primates can be pathogenic and result in malignancies. SV40-induced progressive multifocal leukoencephalopathy, a demyelinating disease and astrocytomas have been reported in macaques that had been chemically immunocompromised or immunocompromised by a prior experimental infection with simian immunodeficiency virus.[11,12] SV40 can also infect a variety of human cell types and although humans were initially considered a semi-permissive host, neutralizing antibodies have been detect in both individuals exposed and not exposed to contaminated polio vaccines. SV40 DNA and the oncoprotein SV40 Tag have been detected in a number of human tumors, including mesotheliomas, osteosarcomas, and ependymomas. Although the evidence is increasing, the role of SV40 infection of humans in the development of cancer remains controversial.[13] It is noteworthy that a large number of patients with malignant pleural mesothelioma (MPM) appear to contain SV40 DNA sequences in their tumors including those encoding SV40 Tag which binds to the human tumor suppressor gene products, p53 and pRb. Overall, MPM is difficult to manage either surgically or with chemotherapy and little alternatives if any exist for treatment in the management of this disease. Clinical trials are being proposed to treat MPM patients with a cancer vaccination immunotherapy strategy that employs SV40 Tag.[14] Thus, the prevailing evidence could support a role of SV40 in human malignancies and the use of SV40 Tag as a viral-encoded tumor-specific antigen for the development of active immunotherapy to treat human cancers, such as MPM.

SV40 Tag and Its Role as a Target Antigen for Active Immunotherapy

Within its genome, SV40 encodes two early nonstructural proteins that share the 82 amino terminal amino acids. These are referred to as large and small T-antigen (Tag and tag, respectively). The production of these two nonstructural proteins utilizes a mechanism of alternative splicing of viral transcripts. The large SV40 Tag is composed of 708 amino acids and represents a protein with multiple functions (Fig. 1). The Tag undergoes a number of post-translation modifications that impart its multi-functionality. First, SV40 Tag is an essential protein for SV40 viral replication. It is involved in stimulating host cells to enter the S phase of the cell cycle and initiate DNA synthesis. The ability of SV40 Tag to alter the host cell cycle allows for its transforming capacity and oncogenic potential. Specific sites on Tag allow it to bind cellular tumor suppressor proteins, namely p53, pRb, p107, and p130/pRb2, which are involved in controlling the host cell cycle. Tag is also a DNA binding protein, contains nuclear localization and Zn finger and Leu Zipper domain motifs, and exhibits helicase activity. With regards to cancer and oncogenesis, the transformation properties of SV40 impart its oncovirus properties. Among the various SV40 homologs, the Tag is relatively conserved (70%) between the human (JC and BK) viruses and an isolate from baboons (SA12). Differences within the Tag involve short nucleotide sequence inserts or deletions and the amino terminus are highly conserved among the various Tags. The major differences are found within the variable carboxyl terminus domain. Although SV40 Tag is namely localized in the nucleus, a small amount of Tag is expressed on the cell surface of transformed cells. Thus, SV40 Tag can represent a target for anti-tumor immunity and function as a surface expressed viral-encoded tumor-specific

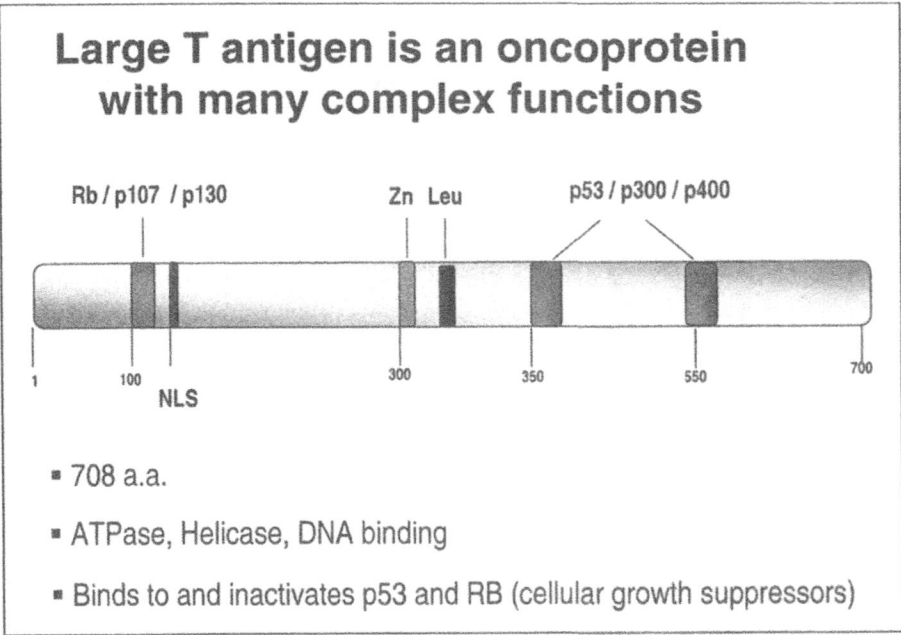

Figure 1. Schematic representation of the SV40 Tag. Sites of interaction of this multi-functional oncoprotein are depicted.

antigen. A number of studies support the notion that Tag is expressed on the surface of tumor cells and can function as an immunologic target. It was recently described that in humans with MPM, both antibodies to SV40 Tag epitopes and Tag specific CD8+ CTL's have been reported.[16] Our laboratory has been employing SV40 Tag as a model tumor antigen for evaluating immunologic strategies in developing cancer vaccines and as a means of dissecting the mechanism(s) of tumor, including DNA based vaccination strategies. The injection of SV40 transformed murine cells into immunocompetent mice induces the formation of tumors that express SV40 Tag on their cell surface[15] thus enabling us to study immune responses to Tag in well-defined rodent modes.

Animal Models for SV40 Tag Induced Tumor Immunity

One of the earliest animal models to show promise as a model for SV40 induced tumors were newborn hamsters. Depending on the site of inoculation with SV40, newborn hamsters were reported to develop fibrosarcomas, ependymomas, and mesotheliomas.[17,18,19,20] In addition, a number of these tumors that developed at the site of inoculation were capable of metastasizing to distal sites. Although these animal models showed extreme promise, their further development has been hindered by the lack and availability of hamster-specific immunologic reagents.

With regards to animal models, mice have served as the predominant animal model to evaluate both the in vivo effects of Tag expression and to evaluate SV40 Tag and its ability to induce tumor immunity. A number of transgenic strains of mice that utilized SV40 Tag expression under the control of a variety of promoters to demonstrate the oncogenic potential of Tag have been developed. In each instance, the transgenic mice showed extensive hyperplasia and tumor formation at the tissue and organ sites where SV40 Tag was targeted for expression. These transgenic mouse models have been used to dissect the functional domains of SV40 Tag and transformation events in vivo.[21-23] No information is presently available on the use of these models for studies involving the immunologic basis for tumor immunity or as tools for developing cancer vaccination modalities.

The best animal model system examined to explored SV40 Tag and in vivo tumor formation employs BALB/c mice inoculated with a tumorigenic syngeneic cell line. The cell line was derived from BALB/c mouse kidney fibroblasts that were transformed by SV40 in vitro.[24] This cell line is designated mKSA and intradermal or subcutaneous inoculation into BALB/c mice results in a visible solid tumor mass whose size can be measured as a means of quantitation. Eventually the size of the tumor mass results in a lethal tumor burden and one can also assess survival as a means of quantitation. Intraperitoneal inoculation of mKSA cells into BALB/c mice has also been used to evaluate in vivo tumor formation with the lethality of the tumor as a measuring endpoint. None of these routes of tumor inoculation appear to result in metastasis and these tumor models are examples of primary nonmetastatic solid tumors. Interestingly, inoculation of mice of non H-2^d haplotypes, the murine MHC of BALB/c mice with SV40-transformed syngeneic tumor cells (such as C57BL/6 mice injected with the SV40 transformed syngeneic kidney fibroblast BLK SV HD.2 cell line) fail to cause progressing tumors. Thus murine models employing mKSA as a tumor inoculum were limited to BALB/c mice and other H-2^d strains of mice. Inoculation of an F1 hybrid strain of mice (CB6/F1) that represented a cross between C57BL/6 mice and BALB/c mice were susceptible to tumor formation when inoculated only with mKSA and not the C57BL/6 transformed cell line. Other tumor model systems have employed nonSV40 transformed syngeneic tumor cell lines transfected with plasmids expressing SV40 Tag encoded genes. In these systems, SV40 Tag can be expressed on the surface of the tumor cell and employed as a target antigen for immunotherapy. One example is using an H-2^d mastocyoma cell line and transfecting it with SV40 Tag and isolating transfected cells that express SV40 Tag on their cell surface. Such transfected cell lines have been used to evaluate immunotherapy and mechanism(s) of tumor immunity employing SV40 Tag based strategies.[25]

More recently an experimental pulmonary metastasis murine model of SV40 tumors has been described that utilizes an intravenous injection of mKSA cells into BALB/c mice.[26] In this model, tumor cells can be recovered from the brain, spleen, lung, and kidneys of tumor bearing mice by primary organ cell culture. Immunofluorescence studies using antibodies to SV40 Tag confirmed that these isolated tumor cells expressed SV40 Tag on their cell surface and could represent a target for tumor vaccine based strategies. In order to quantitate the tumor burden in this model, the number of tumor lung foci were determined. This method either employed manual counting of the stained tumor cell foci from the lungs or used a computer-assisted method and an image analysis system.[27] In both quantitation systems, the lungs from inoculated mice were removed and stained with India Ink and Fekete's solution. The latter method was a major improvement for quantitation purposes in that both the number and size of the tumor lung foci in terms of pixel units could be determined. In addition, the subjective nature of manual counting was also removed when employing the image analysis and computer-assisted quantitation. Studies have suggested that the experimental pulmonary tumor model is a much more stringent model for evaluating immunotherapy when compared to the other nonmetastatic solid tumor models in BALB/c mice.[28] Based on the ability to accurately quantitate tumor burden, the metastatic parameters that can be evaluated and the more stringent nature of the system, the experimental pulmonary metastatis model has become the animal model of choice when comparing and evaluating SV40 Tag based immunotherapeutic modalities.

SV40 Tag Cancer Vaccination Studies and the Role of Antibodies in Tumor Immunity

Our efforts to develop and evaluate cancer vaccination strategies and determine the immunologic mechanism(s) of tumor immunity were initiated in the early 1980's. The system that was selected was SV40, since a number of our collaborators were examining the molecular mechanism(s) of cellular transformation and a number of reagents were available. The target tumor antigen that we selected was SV40 Tag and the systems to evaluate tumor immunity was BALB/c mice and mKSA cells. Our overriding hypothesis at that time was that antibodies would play a major role in tumor immunity and thus our cancer vaccination strategies initially relied on activation of CD4$^+$ T cells and the secretion of antibodies to SV40 Tag by activated B cells. In our thinking, it was these two cell types that were going to play a major role in SV40 tumor immunity and that vaccination strategies should target these two key players in the immune system. At this time and in our original hypothesis, CD8$^+$ T cells had little if any role in tumor immunity in this system other than to function in an ancillary role and perhaps mediate tumor immunity via nonantigen specific pathways. However, even during this stage of development and in this system in particular, the role of antibodies in SV40 Tag induced tumor immunity was controversial. Earlier studies had implicated CD8$^+$ CTLs as major players in SV40 Tag immunity and had characterized high and low CTL responding inbred stains of mice.[29] It was interesting to note that C57BL/6 mice were high CTL responses for SV40 Tag; whereas, BALB/c mice were low to nonresponding strains. These CTL responders were characterized in C57BL/6 mice and the epitopes and MHC restriction elements defined.[30] Investigators had determined that SV40 Tag extracts and irradiated tumor cells when immunized into BALB/c mice could prevent tumor formation, yet the mechanism of this tumor immunity was not determined and the role of antibodies was not examined.[31,32] From these studies, it was felt that CTL responses and cellular immunity played a major role in immunity and characterization of CTL responses in inbred strains of mice where no in vivo tumor challenge system was available continued. Although it was clear that CD8$^+$ CTL responses could be generated in some murine systems and that these effector cell types were extremely efficient at lysing SV40 Tag-expressing target cells in vitro, no direct evidence on their protective capacity in vivo was available. The current thinking at that time was that C57BL/6 mice generated CTL responses and no tumor could be induced by a challenge with a syngeneic SV40-transformed cell line therefore CTLs were responsible for tumor immunity. On the other hand, BALB/c mice generated no CTL responses and were susceptible to a challenge with a syngeneic SV40 transformed cell line again providing support the role for CTLs in SV40 Tag immunity. The role of antibodies was virtually ignored and our laboratory took a different tack at interpreting the above published results. It was our assessment that C57BL/6 mice failed to develop tumors following a challenge with syngeneic SV40-transformed cells because these cells were not tumorigenic in vivo. Alternatively, BALB/c mice developed tumors not because they failed to generate CTL responses, but rather they did not develop adequate antibody responses to SV40 Tag because the tumor cells were highly tumorigenic and weakly immunogenic. This gives some historical perspective with regards to the state of field at the time we initiated our cancer vaccination and mechanistic studies in the mid-1980's.[33] As new strategies for vaccination evolved, we incorporated these modalities into our systems to assess their ability to invoke anti-tumor immunity.

In the early 1990's, we started to examine additional strategies that might impart complete tumor immunity against SV40 Tag. Using a baculovirus expression system and the subunit concept approach, we evaluated recombinant SV40 Tag as a tumor vaccine approach in murine systems. We also predicted putative SV40 Tag B cells epitopes and employed a synthetic peptide conjugate vaccination strategy for SV40 Tag.[34] Our laboratory demonstrated that various SV40 Tag synthetic peptides when conjugated to carrier proteins could induce antibodies that recognized SV40 Tag. Additional synthetic peptides corresponding to the carboxyl terminus were capable of inducing partial to complete tumor immunity in BALB/c mice.[34] The protective SV40 Tag epitopes were mapped to the carboxyl terminus. Yet as with anti-idiotypic antibody vaccines, the tumor immunity induced by SV40 Tag synthetic peptides was only partial and not always complete.[33,35,36] This left us with the use of recombinant SV40 Tag as a means of inducing antibodies to SV40 Tag, CD4$^+$ T cell responses and complete tumor immunity. Indeed a number of comparative studies had demonstrated that recombinant SV40 Tag immunization resulted in complete tumor immunity in prophylactic murine solid tumor models.[36-38] Because the recombinant SV40 Tag immunization induced complete tumor immunity, we focused on this modality in an attempt to address mechanism(s) of tumor immunity within BALB/c mice. Clearly studies from our laboratory suggested that antibodies played a major role in this tumor immunity.[38] Based on in vitro studies, the putative mechanism of this antibody-based tumor immunity was determined to be ADCC. The ADCC-mediated immunity appeared to require CD32 (FcgR type II) effector cells and antibodies to SV40 Tag as this was the basis for the observed in vitro killing of SV40-expressing tumor cells. Complement and CDC did not appear to play a role in this in vitro tumor cell cytotoxicity, nor were NK cells involved in the direct killing of tumor cells. No CD8[+] CTL response was induced by the recombinant SV40 Tag immunization. Thus, we concluded that antibodies to SV40 Tag were involved in the observed tumor immunity.[38] In subsequent in vitro studies employing SV40 tumorigenic cells in culture, we ruled out that antibodies to SV40 Tag induced tumor cell apoptosis as a means of tumor cell death (unpublished observations). These studies supported the role of antibodies to SV40 Tag as a means of tumor immunity, yet any role for CD4[+] T cells induced by SV40 Tag immunization required additional clarification.

Tumor Immunity Mechanism Paradigm Switch Based on a DNA Vaccination Modality

Our laboratory had a bacterial plasmid that expressed SV40 Tag, designated pSV3neo which had be used by a variety of investigators for in vitro transfection and tumorigenesis studies (**Fig. 2**). As a lark, we purified DNA from pSV3neo, a commercially available plasmid expressing SV40 Tag and immunized BALB/c mice to determine whether we could detect an antibody response in their sera to SV40 Tag.[39] Mice were immunized a total of 4 times and sera was obtained following each immunization. Sera was screened by ELISA and no detectable antibodies to SV40 Tag was observed. Rather than terminate the experiment because of the negative antibody results, the pSV3neo immunized mice were challenged with a lethal dose of mKSA. To our surprise, all pSV3neo-vaccinated mice survived the lethal challenge while mice immunized with a control plasmid, pSV2neo (without the SV40 Tag insert) all succumbed to the tumor challenge. These experiments were repeated a number of times and in each instance pSV3neo immunization failed to induce significant antibodies to SV40 Tag, but generated tumor immunity to a solid tumor challenge.[40] Our laboratory characterized the immune response in the plasmid DNA-immunized mice and showed that CD8[+] MHC class I-restricted CTLs were induced and in the absence of antibodies to SV40 Tag were most likely responsible for the observed tumor immunity and the potential of a DNA vaccination approach to protect mice against SV40 induced tumors was described.[41] This was also relevant because DNA immunization could induce a CD8[+] CTL response in BALB/c mice, a previously considered nonresponding CTL strain for SV40 Tag. DNA vaccines to SV40 Tag have also been employed by other investigators who confirmed our results.[42] These investigations used a different inbred strain of mice as a challenge system and a different tumor cell challenge system to demonstrate in vivo anti-tumor immunity. Within our solid tumor model, the recombinant SV40 Tag and plasmid DNA approach gave us a unique system for evaluating a dichotomy of the immune response as a mechanism(s) for tumor immunity. Recombinant SV40 Tag induces antibodies to SV40 Tag and no CTL response, DNA immunization with pSV3neo generates little to no antibodies to SV40 Tag and a CD8[+] CTL response. Both immunization schemes generate CD4[+] T cell proliferative responses in vitro. Thus, this provides a unique opportunity to evaluate two different effector arms of the immune system and the mechanism(s) of tumor immunity to SV40 Tag immunization. We have subsequently generated a plasmid construct that employs the CMV promoter rather than the SV40 promoter with the SV40 Tag gene inserted. We have designated this plasmid pCMV Tag (see Fig. 2) and have shown that this plasmid can induce both antibodies and cell-mediated immune responses to SV40 Tag.

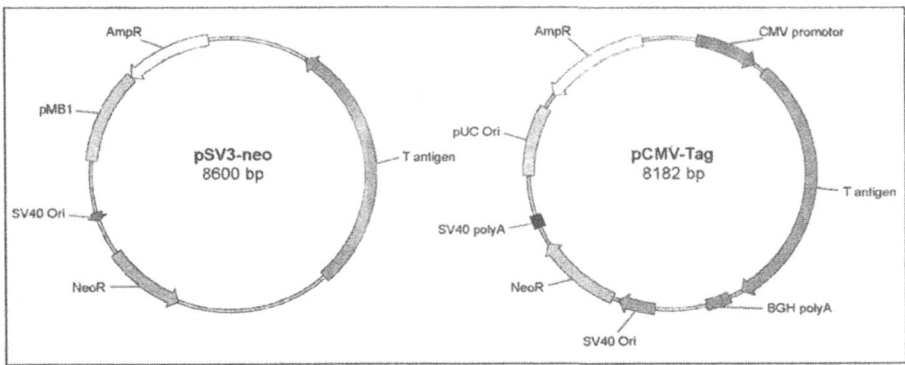

Figure 2. Plasmid DNA genetic maps of two different DNA vaccination strategies encoding SV40 Tag.

More Stringent Tumor Models Demonstrate That Antibodies Are More Effective in Providing Protective Immunity

As we discussed above, we have also developed a murine model for SV40 tumors that represents an experimental pulmonary metastasis scenario.[26] Rather than the solid nonmetastatic model that was utilized for our early studies, we have characterized the experimental pulmonary metastasis for its ability to evaluate a variety of cancer vaccination strategies. The advantages of the pulmonary metastasis model over the solid tumor challenge model are described in detail above. In one of our first studies, we compared the ability of recombinant SV40 Tag and pSV3neo DNA for their ability to prevent pulmonary metastasis following a challenge with SV40 tumorigenic cells. Recombinant SV40 Tag immunization afforded complete protection within the experimental pulmonary metastasis model; whereas, pSV3neo DNA injection resulted in only partial protection and in some instances, no protection when compared to control immunized mice.[28] This suggested that the experimental pulmonary metastasis model was more stringent with regards to evaluating cancer vaccination induced tumor immunity for SV40 Tag when compared to the nonmetastatic solid tumor model. Additionally, antibodies to SV40 Tag appear to be more effective than CTL responses in providing tumor immunity within this model. It remains to be determined what role CD4[+] T cell responses play in the tumor immunity induced within this animal model system. However, based on studies in the nonmetastatic solid tumor model, CD4[+] T cells should play a role both in the induction of antibodies to SV40 Tag, in promoting CD8[+] CTL responses and potentially as a direct effector cell whose role is critical in the killing of tumor cells in vivo.

Additional Studies on DNA Immunization to Target Tumor Antigen Specific Responses

As described above, B cell lymphomas express a number of cell surface markers that represent targets for immunotherapy. Of particular interest to DNA vaccination strategies are the idiotypic determinants associated with the surface immunoglobulin marker. DNA encoding the variable region sequences of an immunoglobulin molecule from the murine BCL1 lymphoma was used to generate anti-idiotype responses in approximately 50% of the mice. This anti-idiotype response recognized the BCL1 tumor cells.[44] Alternative constructs were generated that incorporated cytokine adjuvants, which when co-injected with the idiotype expressing plasmid DNA, increased the level of the anti-idiotype response. However, vaccinated mice did not survive challenge with the BCL1 tumor cells.[45] Other investigations demonstrated that a plasmid DNA expressing the immunoglobulin hypervariable region of the heavy chain could generate antibodies that would react with patients' tumor cells derived from

B-lymphoproliferative disorders.[46] Alternatively, DNA vaccines expressing a single chain immunoglobulin variable region fragment containing the idiotype when fused to the C fragment of tetanus toxoid induced enhanced anti-idiotype responses that recognized the lymphoma and provided protection in murine models of lymphoma and myeloma.[47,48]

A human carcinoembryonic antigen (CEA)-expressing syngeneic murine colon carcinoma cell line has been utilized to assess the ability of a DNA plasmid expressing human CEA to induce anti-tumor immunity. CEA is one of the most well characterized human tumor associated antigens and has been categorized as an oncofetal protein. CEA is a glycoprotein that is heterogeneous in nature and appears to be produced in high concentrations on certain human tumors, including the surface of adenocarcinoma. The plasmid DNA expressing human CEA induced both humoral and cell mediated immune responses in mice specific for CEA.[49] These studies did not include any tumor challenge studies. This group of investigators has also reported the induction of immune responses to CEA in nonhuman primates immunized by genetic vaccination.[50] These studies were performed in anticipation of a phase I human clinical trial employing DNA immunization in patients with colorectal carcinoma. A single DNA plasmid was constructed that expressed CEA and hepatitis B surface antigen (HBsAg) under the control of two separate CMV promoters. The HBsAg was included to provide an internal positive control for the efficacy of DNA immunization without the regard for breaking tolerance to the CEA self-antigen. Pig-tailed macaques were immunized by either intramuscular injection or particle bombardment of the skin. Both forms of immunization generated antibody and T cell proliferative responses to both HBsAg and CEA. One difference in immunization scheme was observed in the intramuscular injected group of animals where CEA specific IL-2 release and delayed type hypersensitivity was detected. No toxicity was observed as the result of the DNA immunization. In another study, protective immunity was observed in CEA transgenic mice that were immunized orally with plasmid DNA encoding CEA.[51] Tumor immunity was dependent on the activation of MHC class I restricted CD8[+] T cells. These activated T cells secreted INF-γ, IL-12 and GM-CSF. The transgenic mice were protected from a lethal challenge with murine MC38 colon carcinoma cells expressing CEA.

Based on these earlier studies, it became apparent that DNA immunization whereby the plasmid DNA expressed a tumor antigen could invoke both antibody and cell-mediated immune responses specific for the tumor antigen. More recently studies have described that utility of DNA vaccination can prevent in vivo tumor development and metastasis within animal models of human cancers. DNA vaccines expressing the rat proto neu tumor antigen prevented neu-expressing mammary tumor development in transgenic mice.[52] In this system, transgenic mice over-express the rat neu proto-oncogene which is under the control of the mouse mammary tumor virus promoter. Female transgenic mice spontaneously develop focal mammary tumors at 6 months of age. DNA immunized mice developed significantly fewer spontaneous tumors when compared to control immunized groups. Interestingly, protective tumor immunity could only be demonstrated in 3 month old mice, but not in 6 month old mice. These data suggested that the xenogeneic HER-2 expressed sequence can break immune tolerance to rat neu in these transgenic mice and induce protective immunity that impairs the neu oncogene driven progression of mammary carcinogenesis. DNA immunization with plasmids expressing p185 neu or Her-2/neu (ErbB-2) has also been reported to inhibited carcinogenesis in Her-2/neu transgenic mice or against challenge with Her-2/neu expressing murine tumors.[53-56] In these studies, DNA immunization caused a hemorrhagic necrosis of established cancer nests, in spite of the lack of a CTL response.[54] These investigators reported that antibodies to rat p185 neu were detected in DNA immunized animals. Other investigators reported that the protection induced by DNA immunization against Her-2/neu expressing murine tumors was dependent on CD4[+] T cells[56]

DNA vaccine strategies have also been employed as a means to treat melanomas. DNA expressing the Melanocyte Lineage-specific antigen, gp100, were reported to induce CTL and antibody responses to human gp100 and mediated protection against a lethal tumor challenge in mice.[57] The results demonstrated that DNA vaccinated mice were protected against a lethal challenge with the syngeneic B16 murine melanoma cell line expressing human gp100. This tumor immunity did not appear to result directly from the CTL response but rather was associated with the adoptive transfer of spleen-derived lymphocytes. Serum containing antibodies to gp100 appeared to play no role in anti-tumor immunity in this system. Partial protection and reduced melanoma formation was reported in mice immunized with DNA expressing gp100/pmel 17.[58] DNA vaccination of mice prior to B16 tumor cell challenge resulted in a 50% reduction of tumor size. When combined with DNA plasmids encoding GM-CSF, no increase in effectiveness was observed. However, control experiments did demonstrate that the plasmid encoding GM-CSF when administered alone led to a reduction in tumor size. Protective immunization against murine melanoma was also reported when the gp100 expressing DNA was incorporated into a liposome.[59] Intramuscular injection of the DNA and liposome generated both antibody to gp100 and CTL responses. Tumor immunity was again obtained in mice challenged with B16 melanoma cells. Combining this DNA vaccine with peptide vaccines specific for gp100 enhanced anti-tumor immunity as was demonstrated in a malignant melanoma mouse model employing C57BL/6 mice and the poorly immunogenic B16-FO murine melanoma cell line.[60] Depletion of T cell subsets indicated that the protective effects following DNA immunization was mediated by both CD4$^+$ and CD8$^+$ T cells. In the DNA and peptide combination regimen, tumor immunity appeared to be dependent on only the CD4$^+$ T cells. It was interesting to note that therapeutic effects were only observed with the combination regimen. The induction of immune responses in mice to both murine and human tyrosinase-related protein 2 (TRP-2) expressed in murine B16 and human melanomas has been reported as the result of plasmid DNA immunization.[61,62] Studies demonstrated that mice immunized with DNA encoding TRP-2 generated TRP-2 specific CTL responses and a delayed outgrowth of B16 melanoma. Employing cytokine co-delivery and biasing the immune responses towards Th1 by using IL-12 enhanced the anti-tumor efficacy of the DNA based melanoma vaccines.[61] Genetic immunization of mice with human TRP-2 resulted in a coat depigmentation that represents an autoimmune-mediated destruction of melanocytes and also provided significant protection against metastatic growth of the B16 melanoma.[62] Contrary to this study, anti-tumor immunity was observed as murine melanoma was eradicated without autoimmunity following DNA vaccination with TRP-2 expressing plasmids.[63] In this study, immunized mice rejected B16 tumors without the development of skin depigmentation or vitiligo, an autoimmune manifestation. Depletion experiments indicated that CD8$^+$ T cells and NK cells were critical for tumor immunity. Oral administration of a DNA vaccine encoding TRP-2 has also been demonstrated to protect from murine melanoma.[64] In this study, DNA immunization induced growth suppression and tumor rejections following a lethal challenge with a subclone of B16 melanoma cells. Tumor immunity was mediated by MHC class I restricted CD8$^+$ T cells that secreted INF-γ. These studies also demonstrated that peripheral T cell tolerance towards murine melanoma self-antigens can be broken by oral DNA immunization.

Similar to the situation with SV40 Tag and MPM, cervical carcinoma has been associated with infection by certain serotypes of HPV and HPV-transformed cells express a viral encoded oncoprotein designated E7. Thus, like SV40 Tag, HPV type 16 E7, represents a viral encoded tumor-specific antigen. Tumors expressing the HPV type 16 E7 oncoprotein and murine lung metastatic models have also been employed to demonstrate the ability of plasmid DNA construct expressing the E7 oncoprotein to induce tumor immunity.[65] In these studies, the investigators showed that DNA vaccination was capable of protecting mice against challenge with a more stringent subclone of tumor cells. The observed tumor immunity was dependent on E7-specific CD4$^+$ T cells secreting INF-γ. It was also apparent from these studies that IL-4 may play a detrimental role in the anti-tumor mediated effects by vaccination strategies other than DNA. These investigators concluded that DNA vaccines may provide superior tumor protection in this model system. Investigators have also described the induction of E7 specific CTL as the result of DNA vaccination with an HPV type 16 E7 expressing construct.[61,66] In one study, an E7 DNA vaccine was constructed whereby mutations in the two zinc-finger binding motifs resulted in the expression of a rapidly degraded E7 protein. When compared to the unmutated wild type E7 DNA vaccine that produced a more stable E7 protein, the mutant E7 DNA construct induced stronger E7 specific CTL response and better tumor protection presumably by more efficient antigen presentation.[66]

DNA vaccination whereby the HuD tumor antigen was expressed induced anti-tumor immunity in a murine model of small cell lung carcinoma.[67] There exists a clinically significant correlation between the presence of antibody against the paraneoplastic encephalomyelitis antigen, termed HuD and the limitation of tumor spread in patients with small cell lung carcinoma. In one study, a murine adenocarcinoma cell line stably transfected with HuD was used as a tumor challenge system to evaluate DNA immunization strategies that expressed HuD in mice. The observed anti-tumor immunity was associated with the induction of antibodies to HuD following DNA immunization. Prostate specific antigen (PSA) has also been utilized as a DNA based vaccination strategy.[68] Reports have described the induction of enhanced immune responses to DNA-PSA expressing vaccines when cytokine gene adjuvants were employed. These DNA based vaccines in conjunction with the cytokine gene adjuvants induced immune responses to PSA in nonhuman primate species.[69] Investigators demonstrated that both PSA humoral and cell-mediated immune responses were induced. Interestingly, in the nonhuman primate studies, the generation of a PSA specific immune response by DNA immunization was associated with breaking of tolerance to the PSA self-antigen.

Concluding Remarks

It is clear from the state of the art that DNA immunization represents a reasonable approach to develop targeted immunotherapy to treat cancer. This approach provides a number of advantages including ease of use, stability, and potential cost that allows one to rapidly develop and test a potential vaccination strategy once the gene of a tumor antigen has been cloned and identified. Yet problems are readily apparent and further studies are warranted. First when compared to other vaccination strategies, the DNA approach is relatively weak in its comparative ability to induce immune responses. A number of studies have shown that in comparison, a recombinant protein appears to be much more effective in the induction of tumor immunity in experimental metastatic animal model systems. Additionally, plasmid DNA immunization by itself fails to demonstrate complete protective immunity even in a prophylactic experimental challenge system. Enhancement of the immunogenicity of the plasmid DNA has utilized fusing additional genes to better target the DNA to APCs and/or improve in vivo stability. Alternatively, cytokines and cytokine gene adjuvants or other forms of adjuvants are necessary to enhance the immune response to the transgene tumor antigen product to observe protective tumor immunity. Finally the majority of the cancer models employed to evaluate protective tumor immunity use mice and mice appear to be highly responsive to DNA immunization when compared to larger species, such as, nonhuman primates. Several studies have utilized nonhuman primates to evaluate anti-tumor immune responses to plasmid DNA constructs expressing tumor antigens to provide preclinical information prior to initiating human clinical trials. As the field moves forward, the DNA immunization cancer field will tend to follow the lead and approaches of the DNA immunization strategies that target infectious agents as providing information for enhancing the immunogenicity and improving on anti-tumor immunity. The success of DNA cancer vaccination will depend on the cross fertilization and the improvement provided by investigations that target studies associated with infectious agents that cause human disease.

References

1. Dalgleish AG. Current problems in the development of specific immunotherapeutic approaches to cancer. J Clin Pathol 2001; 54:675-676.
2. Coley WB. The treatment of malignant tumors by repeated inoculations of erysipelas. With a report of ten original cases. 1893 (classical article) Clin Orthop 1991; 262:3-11.

3. Jager D, Jager E, Knuth A. Immune response to tumour antigens: implications for antigen specific immunotherapy of cancer. J Clin Pathol 2001; 54:669-674.
4. Riethmuller G, Schneider-Gadicke E, Schlimok G et al. Randomised trial of monoclonal antibody for adjuvant therapy of resected Dukes' C colorectal carcinoma. Lancet 1994; 343:1177-1183.
5. Mittelman A, Chen GZJ, Wong GY et al. Human high molecular weight melanoma associated antigen mimicry by mouse anti-idiotypic monoclonal antibody MK2-23: modulation of the immunogenicity in patients with malignant melanoma. Clin Cancer Res 1995; 1:705-713.
6. Weiner DB, Kennedy RC. DNA vaccines. Scientific American 1999; 281:50-57.
7. Benton PA, Kennedy RC. DNA vaccine strategies for the treatment of cancer. Curr Topics Microbiol Immunol 1998; 226:1-20.
8. Watts AM, Kennedy RC. DNA vaccination strategies against infectious diseases. Int J Parasitol 1999; 29:1149-1163.
9. Sweet BH, Hilleman MR. The vacuolating virus SV40. Proc Soc Exp Biol Med 1960; 105:420-427.
10. Butel JS. Simian virus 40. In: Webster RG, Granoff A, eds. Encyclopedia of virology. Vol 3. 1994:1322-1329.
11. Holmberg CA, Gribble DH, Takemoto KK et al. Isolation of simian virus 40 from rhesus monkeys (*Macaca mulatta*) with spontaneous progressive multifocal leukoencephalopathy. J Infect Dis 1977; 136:593-596.
12. Hurley JP, Ilyinskii PO, Horvath CJ et al. A malignant astrocytoma containing SV40 DNA in a macaque infected with simian immunodeficiency virus. J Med Primatol 1997; 26:172-180.
13. Butel JS, Lednicky JA. Cell and molecular biology of simian virus 40: Implications for human infections and disease. J Natl Cancer Inst 1999; 91:119-134.
14. Imperiale MJ, Pass HI, Sanda MG. Prospects for an SV40 vaccine. Semin. Cancer Biol 2001; 11:81-85.
15. Lewis AM, Cook J. A new role for DNA viruses early proteins in carcinogenesis. Science 1985; 227:15-20.
16. Bright RK, Kimchi ET, Shearer MH et al. SV40 Tag-specific cytotoxic T lymphocytes generated from the peripheral blood of cancer patients: evidence for Tag-specific immunity against malignant pleural mesothelioma in humans. Cancer Immunol Immunother 2002; 50:682-690.
17. Eddy BE, Borman GS, Berkeley WH et al. Tumors induced in hamsters by injection of rhesus monkey kidney extracts. Proc Soc Exp Biol Med 1961; 107:191-197.
18. Kirschstein RL, Gerber P. Ependymomas produced after intracerebral inoculation of SV40 into newborn hamsters. Nature 1962; 195:299-300.
19. Carbone M, Lewis AM, Matthews BJ et al. Characterization of hamster tumors induced by SV40 small tag deletion mutants as true histiocytic lymphomas. Cancer Res 1989; 49:1565-1571.
20. Cicala C, Pompetti F, Carbone M. SV40 induces mesotheliomas in hamsters. Am J Pathol 1993; 142:1524-1533.
21. Ewald D, Li M, Efrat S et al. Time sensitive reversal of hyperplasia in transgenic mice expressing SV40 T antigen. Science 1996; 273:1384-1386.
22. Sepulveda AR, Finegold MJ, Smith B et al. Development of a transgenic mouse system for analysis of stages in liver carcinogenesis using tissue specific expression of SV40 large tumor antigen control led by regulatory elements of the human alpha-1antitrypsin gene. Cancer Res 1989; 49:6108-6117.
23. Perez-Stable C, Altman NH, Mehta PP et al. Prostate cancer progression, metastasis, and gene expression in transgenic mice. Cancer Res 1997; 57:900-906.
24. Kit S, Kurimura T, Dubbs DR. Transplantable mouse tumor line induced by injection of SV40 transformed mouse kidney cells. Int J Cancer 1969; 4:384-392.
25. Watts AM, Bright RK, Kennedy RC. Antibody-Based Mechanisms of Tumor Immunity. In: Capra JD, Zannetti M, eds. The Antibodies. Vol 5. Harwood Academic Publishers, 1999:31-62.
26. Watts AM, Shearer MH, Pass HI et al. Development of an experimental murine pulmonary metastasis model incorporating a viral encoded tumor specific antigen. J Virol Meth 1997; 69:93-102.
27. Watts AM, Kennedy RC. Quantitation of tumor foci in an experimental murine tumor model using computer assisted video imaging. Anal Biochem 1998; 256:217-219.
28. Watts AM, Shearer MH, Pass HI et al.Comparison of SV40 large T-ag recombinant protein and DNA immunization in the induction of protective immunity from experimental pulmonary metastasis model. Cancer Immunol Immunother 1999; 47:343-351.
29. Knowles BB, Koncar M, Pfizenmaier K et al. Genetic control of the cytotoxic T cell response to SV40 tumor associated specific antigen. J Immunol 1979; 122:1798-1806.
30. Tevethia SS. Recognition of simian virus 40 T antigen by cytotoxic T lymphocytes. Mol Biol Med 1990; 7:83-96.
31. Anderson JL, Martin RG, Chang C et al. Nuclear preparations of SV40 transformed cells contain tumor specific transplantation antigen activity. Virology 1977; 76:420-426.

32. Law LW, Takemoto KK, Rogers MJ et al. Induction of simian virus 40 transplantation immunity in mice by SV40 transformed cells from various species. J Natl Cancer Inst 1997; 59:1523-1526.

33. Kennedy RC, Dreesman GR, Butel JS et al. Suppression of in vivo tumor formation induced by simian virus 40-transformed cells in mice receiving anti-idiotypic antibodies. J Exp Med 1985; 161:1432-1449.

34. Bright RK, Shearer MH, Kennedy RC. SV40 large tumor antigen associated synthetic peptides define native antigenic determinants and induce protective tumor immunity in mice. Mol Immunol 1994; 31:1077-1087.

35. Shearer MH, Lanford RE, Kennedy RC. Monoclonal anti-idiotypic antibodies induce antibody responses specific for simian virus 40 large tumor antigen. J Immunol 1990; 145:932-939.

36. Shearer MH, Bright RK, Kennedy RC. Comparison of humoral immune responses and tumor immunity in mice immunized with recombinant SV40 large tumor antigen and a monoclonal anti-idiotype. Cancer Res 1993; 53:5734-5739.

37. Shearer MH, Bright RK, Lanford RE et al. Immunization of mice with baculovirus derived recombinant SV40 large tumor antigen induces protective tumor immunity to a lethal challenge with SV40-transformed cells. Clin Exp Immunol 1993; 91:266-271.

38. Bright RK, Shearer MH, Kennedy RC. Immunization of BALB/c mice with recombinant SV40 large tumor antigen induces antibody-dependent cell-mediated cytotoxicity (ADCC) against SV40 transformed cells: An antibody based mechanism for tumor immunity. J Immunol 1994; 153:2064-2071.

39. Southern PJ, Berg P. Tranformation of mammalian cells to antibiotic resistance with a bacterial gene under control of the SV40 early region promoter. J Mol Appl Genet 1982; 1:327-341.

40. Bright RK, Shearer MH, Kennedy RC. Nucleic acid vaccination against viral induced tumors. Ann NY Acad Sci 1995; 772:241-251.

41. Bright RK, Beames B, Shearer MH et al. Protection against a lethal tumor challenge with simian virus 40 transformed cells by the direct injection of DNA encoding SV40 large tumor antigen. Cancer Res 1996; 56:1126-1130.

42. Schrimbeck R, Bohm W, Reimann J. DNA vaccination primes MHC class I restricted, simian virus 40 large tumor antigen specific CTL in H-2^d mice that reject syngeneic tumors. J Immunol 1996; 157:3550-3558.

43. Watts AM, Bright RK, Kennedy RC. DNA cancer vaccination strategies target SV40 large tumour antigen in a murine experimental metastasis model. Dev Biol Stand 2000; 104:143-147.

44. Hawkins RE, Winter G, Hamblin TJ et al. A genetic approach to idiotypic vaccination. J Immunother 1993; 14:273-278.

45. Stevenson FK, Zhu D, King CA et al. Idiotypic DNA vaccines against B-cell lymphoma. Immunol Rev 1995; 145:211-228.

46. Rinaldi M, Ria F, Parrella P et al. Antibodies elicited to naked DNA vaccination against complementary-determining region 3 hypervariable region of immunoglobulin heavy chain idiotypic determinants of B-lymphoproliferative disorders specifically react with patients' tumor cells. Cancer Res 2001; 61:1555-1562.

47. Spellerberg MB, Zhu D, Thompsett A et al. DNA vaccines against lymphoma: promotion of anti-idiotypic antibody responses induced by single chain FV genes by fusion to tetanus toxoid fragment C. J Immunol 1997; 159:1657-1652.

48. King CA, Spellerberg MB, Zhu D et al. DNA vaccines with single chain Fv fused to fragment C of tetanus toxin induce protective immunity against lymphoma and myeloma. Nat Med 1998; 4:1281-1286.

49. Conry RM, LoBuglio AF, Kantor J et al. Immune response to carcinoembryonic antigen polynucleotide vaccine. Cancer Res 1994; 54:1164-1168.

50. Conry RM, White SA, Fultz PN et al. Polynucleotide immunization of nonhuman primates against carcinoembryonic antigen. Clin Cancer Res 1998; 4:2903-2912.

51. Xiang R, Silletti S, Lode HN et al. Protective immunity against human carcinoembryonic antigen (CEA) induced by an oral DNA vaccine in CEA-transgenic mice. Clin Cancer Res 2001; 3S:856s-864s.

52. Pupa SM, Invernizzi AM, Forti S et al. Prevention of spontaneous neu-expressing mammary tumor development in mice transgenic for rat proto-neu by DNA vaccination. Gene Ther 2001; 8:75-79.

53. Amici A, Venanzi FM, Concetti A. Genetic immunization against neu/erbB2 transgenic breast cancer. Cancer Immunol Immunother 1998; 47:183-190.

54. Di Carlo E, Rovero S, Boggio K et al. Inhibition of mammary carcinogenesis by systemic interleukin 12 or p185neu DNA vaccination in Her-2/nue transgenic BALB/c mice. Clin Cancer Res 2001; 3S:830s-837s.

55. Wei WZ, Shi WP, Galy A et al. Protection against mammary tumor growth by vaccination with full-length modified human ErbB-2 DNA. Int J Cancer 1999; 81:748-754.
56. Foy TM, Bannink J, Sutherland RA et al. Vaccination with Her-2/neu DNA or protein subunits protects against growth of a Her-2/neu-expressing murine tumor. Vaccine 2001; 19:2598-2606.
57. Schreurs MWJ, de Boer AJ, Figdor CG et al. Genetic vaccination against the melanocyte lineage-specific antigen gp100 induces cytotoxic T lymphocyte mediated tumor protection. Cancer Res 1998; 58:2509-2514.
58. Nawrath M, Pavlovic J, Dummet R et al. Reduced melanoma tumor formation in mice immunized with DNA expressing the melanoma-specific antigen gp100/pmel17. Leukemia 1999; 13:S1:48-51.
59. Zhou WZ, Kaneda Y, Huang S et al. Protective immunization against melanoma by gp100 DNA-HVJ-liposome vaccine. Gene Ther 1999; 6:1768-1773.
60. Nawrath M, Pavlovic J, Moelling K. Synergistic effect of a combined DNA and peptide vaccine against gp100 in malignant melanoma mouse model. J Mol Med 2001; 79:133-142.
61. Tuting T, Gambotto A, DeLeo A et al. Induction of tumor antigen specific immunity using plasmid DNA immunization in mice. Cancer Gene Ther 1999; 73-80.
62. Steitz J, Bruck J, Steinbrink K et al. Genetic immunization of mice with human tyrosinase-related protein 2: implications for the immunotherapy of melanoma. Int J Cancer 2000; 86:89-94.
63. Bronte V, Apolloni E, Ronca R et al. Genetic vaccination with self tyrosinase-related protein 2 causes melanoma eradication but no vitiligo. Cancer Res 2000; 60:253-258.
64. Xiang R, Lode HN, Chao TH et al. An autologous oral DNA vaccine protects against murine melanoma. Proc Natl Acad Sci USA 2000; 97:5492-5497.
65. Chen CH, Wang TL, Ji H et al. Recombinant DNA vaccines protect against tumors that are resistant to recombinant vaccinia vaccine containing the same gene. Gene Ther 2001; 8:128-138.
66. Shi W, Bu P, Liu J et al. Human papillomavirus type 16 E7 DNA vaccine; mutation in the open reading frame of E7 enhances specific cytotoxic T lymphocyte induction and anti-tumor immunity. J Virol 1999; 73:7877-7881.
67. Ohwada A, Nagaoka I, Takahashi F et al. DNA vaccination against HuD antigen elicits antitumor activity in a small cell lung cancer murine model. Am J Respir Cell Mol Biol 1999; 21:37-43.
68. Kim JJ, Trivedi NN, Wilson DM et al. Molecular and immunological analysis of genetic prostate specific antigen (PSA) vaccine. Oncogene 1998; 17:3125-3135.
69. Kim JJ, Yang JS, Dang K et al. Engineering enhancement of immune responses to DNA based vaccine in a prostate cancer model in rhesus macaques through the use of cytokine gene adjuvants. Clin Cancer Res 2001; 3S:882s-889s.

CHAPTER 12

DNA Vaccines for Allergic Diseases

Kaw Yan Chua, Betina Wolfowicz and Patrick G. Holt

Abstract

DNA vaccines have prevented anaphylaxis in several models of murine allergy. The mechanisms of prevention are still unclear. In this chapter we describe different immunization protocols with a focus on those that may have induced immunoregulation. This analysis will unravel some of the elements that play key roles in the downregulation of atopic reactions.

Introduction

Allergy is characterized by skewed lymphocyte development. In atopic individuals, exposure to an otherwise innocuous environmental antigen drives the proliferation and differentiation of allergen specific CD4$^+$ and CD8$^+$ precursor T cells into interleukin (IL)-4, IL-5, and IL-13 secreting type 2 lymphocytes. It also represses their differentiation into type 1 lymphocytes. The lymphokines secreted by type 2 T helper (Th2) and type 2 cytolytic T (Tc2) cells up-regulate eosinophil, mast cell, basophil and IgE functions and are involved in the switching of B lymphocytes to IgE production.[1-3] When allergen binds IgE attached to the surface of receptor (R)$^+$ cells, it triggers the release of histamine and other mediators of pathogenesis. This is the common mechanism underlying asthma, rhinitis, hay fever and atopic dermatitis among other allergic diseases.[4] In nonallergic individuals, the responses to inhalant allergens are dominated by IL-12-driven differentiation of interferon (IFN)-γ-secreting Th1/Tc1 lymphocytes.[5] Th1/Tc1s are involved in phagocyte-mediated host defense, delayed type hypersensitivity, and in the induction of IgG2a secretion by B lymphocytes.[6]

Current immunotherapies for allergy aim at allergen desensitization. They consist of multiple applications of allergen extracts, at increasing dosage, over the course of a few years.

Allergen hyposensitization can be an effective strategy for some forms of atopy, as attested by the results of double-blind placebo-controlled trials. This therapy can result in a decrease in allergen skin test reactivity and a reduction of antigen specific lymphocyte proliferation.[7]

Recent data however suggest that once allergy is well established it may be difficult to eradicate.[8] Also, duration of efficacy correlates with duration of the treatment.[9] Moreover, the therapy is risky. Near-fatal reactions, although rare, have been described.[10] Indeed much work is focused now on the production of modified allergens that lose the ability to bind IgE, a major trigger in the allergic process.[11] Therefore a safe, long lasting, preventive vaccine may represent a better alternative than therapy.

The use of DNA vaccines for allergy prevention was conceived from the finding that naked plasmid DNA injections elicited a long-lasting antigen-specific response characterized in rodents by the production of IgG2a antibodies, CD8$^+$ and CD4$^+$ type 1 T cells.[12-16] It was postulated that DNA immunization might induce Th1 responses and/or concomitantly inhibit Th2 responses. Delivery of an IFN-γ gene directly into the pulmonary epithelium resulted in decreased pulmonary eosinophilia and airway hyperactivity.[17] Indeed, reduction of allergic responses by DNA vaccination has now been demonstrated in several animal models.

DNA Vaccines, edited by Hildegund C. J. Ertl. ©2003 Eurekah.com
and Kluwer Academic / Plenum Publishers.

The house dust mite allergen Der p 5 and β-galactosidase were the first to be described.[14-16] But DNA vaccines have also been used to decrease hypersensitivity to bee venom, ovalbumin (OVA), latex, and peanut in mice.[18-21]

DNA Vaccines for Allergy: An Overview

The vaccine produced by Hsu and coworkers consisted of a recombinant naked plasmid DNA carrying a house dust mite allergen.[15] These authors first established an allergy model in which intraperitoneal immunization of rats with this allergen, Der p 5, in alum, stimulated production of allergen-specific IgE. IgE production was accompanied by high levels of IL-4 and IL-5 secretion by splenic T cells, but not IFN-γ. In contrast, intramuscular injection of a full-length Der p 5 complementary DNA engineered into a eukaryotic expression vector, pCMVD, induced high levels of IgG2a and IgG1 anti-Der p 5 antibodies, but no IgE. Rats that had received pCMVD were then challenged by intraperitoneal injection of Der p 5 in alum. The level of specific IgE in pCMVD-pretreated rats was 90% lower than that of mock-treated animals. To determine if this preventive strategy could be used to modulate allergen-induced airway resistance, Der p 5 sensitized rats were challenged with aerosolized Der p 5 and pulmonary resistance was measured. In mock-immunized rats pulmonary resistance increased to 80% over baseline 20 min after challenge. In contrast, animals receiving pCMVD showed no significant changes over baseline.

Using naked plasmid DNA delivered intradermally, Raz and coworkers reported similar findings in a β-galactosidase model of allergy.[14] They demonstrated that mice immunized with a plasmid DNA encoding β-gal from *E. coli* (pCMV-LacZ) produced an IgG2a antibody response to β-gal, whereas protein in alum or in saline immunization induced IgG1 and IgE. Analysis of lymphokine secretion indicated that plasmid DNA induced IFN-γ but no IL-4 or IL-5, whereas β-galactosidase induced IL-4 and IL-5 but not IFN-γ. Plasmid-primed mice showed greatly reduced IgE antibody upon subsequent β-gal in alum intraperitoneal injection.

Oral DNA vaccination to prevent a common food allergy, hypersensitivity to peanuts, has been examined. The vaccine consisted of nanoparticles made by coacervation of plasmid encoding the peanut allergen Ara h2[21] and chitosan, a biocompatible vehicle. Chitosan is a nonimmunogenic, mucoadhesive polysaccharide derived from crustacean shells already in clinical use for controlled drug delivery. The chitosan protected the DNA from being rapidly degraded in the intestinal lumen, thus transfection of the stomach and intestinal epithelia was observed. Vaccinated mice produced Ara h2-specific secretory IgA and serum IgG2a. A reduction in allergen-specific IgE, allergen-induced anaphylaxis, decreased plasma histamine and vascular leakage were also observed.

One of the demands imposed on a DNA vaccine for allergy is that once an allergen specific Th1 scenario is set in motion, the effect is long lasting. Moreover, further allergen encounters must promote the Th1 pathway and inhibit Th2 cells. In this regard, Der p 5 allergen gene immunization induced long-lasting protection.[15] Rats that had received Der p 5 encoding plasmid showed reduction of antigen-induced airway hyperreactivity and histamine release into the broncheoalveolar lavage fluid 6 months after immunization. Recent results obtained by Jilek and coworkers[18] using the major antigen of honeybee venom, phospholipase A2 (PLA2), also suggest a persistent effect. Intradermal injection of naked plasmid DNA before challenge with PLA2 in alum, reduced PLA2 specific-IgE and IgG1 titers 7 months after sensitization, with concomitant rise in IgG2a and IgG3. Splenocytes recovered 5-6 months after the last DNA immunization exhibited a sustained IFN-γ and IL-10 secretion and reduced IL-4 production. PLA2 protein challenge of DNA vaccinated mice boosted IFN-γ and IL-10 secretion, suggesting the reactivation of quiescent memory lymphocytes.

It has been shown in the mouse model for atopic asthma that treatment with allergen genes after sensitization can down-regulate ongoing IgE responses in vivo. Balb/c mice were first sensitized with the mite allergen Der p 5 in alum, and they were subsequently injected with the Der p 5 encoding DNA pCMVD. It was observed that the level of specific IgE in treated mice was significantly reduced when compared to mock-treated animals.[16] Other allergen encoding plasmids are also being explored for therapeutic use. In the β-gal system, pCMV-LacZ treatment after priming with β-gal protein in alum resulted in a 75% decrease in the IgE antibody titer within 6 weeks.[22] In the PLA2 model, treatment of mice with plasmid DNA after PLA2 sensitization resulted in a high percentage (65%) of recovery from subsequent anaphylactic challenge.[18]

Established Th2 responses to other allergens may be more difficult to down-modulate, or may even be exacerbated by allergen DNA injections.[8,23] One option to enhance the efficacy of DNA vaccines is to exploit the adjuvanticity of Th1 skewing cytokines, such as IL-12, IL-18 or IFN-γ. It has been demonstrated that expression of IFN-γ in the lungs at the time of pulmonary challenge can decrease the allergic attack.[17] Mice were presensitized to the antigen conalbumin. The animals then aspirated a mixture of IFN-γ DNA/lipofectamine, that transfected the IFN-γ gene directly into the pulmonary mucosa. IFN-γ production, presumably mostly from the transfected gene, was detected in the broncho-alveolar lavage (BAL) fluid as early as 48h after treatment and peaked on day 3. If mice were challenged at this time, intratracheally, with conalbumin, eosinophil counts in the BAL fluid, airway pressure, and IL-4 and IL-5 were lower than in the sham-vaccinated group.[17] This work demonstrates that robust, local, and timely expression of a Th1 lymphokine may counteract a Th2 trigger. Similar results have been obtained by intranasal delivery, before challenge, of IL-12 encoding DNA, either naked or inside a fowl poxvirus vector.[24]

Systemic induction of IFN-γ may be sufficient to prevent local reactivity. To induce an IFN-γ-rich environment, mice have been immunized with the protein antigen OVA and free recombinant (r)IL-12.[25] This protocol enhanced T cell production of IFN-γ, but the IFN-γ production was not specific for OVA. Therefore the mice were treated with an OVA-IL-12 fusion protein. The mice were first primed with OVA in alum to induce a Th2-dominated response and then received the fusion protein. It was observed that OVA-IL-12 induced more significant increases in IFN-γ and serum IgG2a and decreases in IL-4 and serum IgE than the mixture of OVA and free IL-12.[25] To further the strategy, Maecker and collaborators switched to the use of plasmid instead of protein immunization and studied the effects of another powerful Th1 cytokine, IL-18.[19] Mice were injected in the muscle with vectors engineered with OVA cDNA alone, murine IL-18 alone, the two plasmids co-injected or a fusion construct encoding OVA and mouse IL-18. Either construct reduced the subsequent methacholine triggered-airway hyperreactivity (AHR), increased IFN-γ production and reduced OVA-specific IgE production. Similar to the results obtained for OVA-IL-12 protein vaccination, the OVA-IL-18 fusion DNA was the most effective under all those criteria. Moreover, only the OVA-IL-18, but not the OVA cDNA plasmid alone had therapeutic properties: it reversed established AHR in mice with the preexisting condition, reduced the number of eosinophils in the BAL fluid, increased allergen-specific IFN-γ and decreased antigen-specific IL-4 production. The authors suggest that co-expression of antigen and cytokine may focus the activity of IL-18 or IL-12 on OVA-specific B and T cells, thus also minimizing any bystander effects of free inflammatory cytokine.

It has recently been reported that the expression vector pSec TagA without inserted allergen sequences could inhibit Th2 responses to PLA 2 and OVA challenge.[18] Opposite results were obtained in the OVA model using a different vector[19] as well as in the Der p 5 model.[15] The protection conferred by pCMVD immunization was highly antigen-specific. Furthermore, animals immunized with pCMV devoid of Der p 5 encoding DNA were not protected against allergen sensitization and subsequent induction of airway reactivity. One possible explanation is that although the vector used in Jilek's work did not carry sequences for the endogenous production of antigen, antigen was being supplied from different sources: in the case of OVA, possibly as an ingredient of the animal feed. In the case of PLA2, many inflammatory stimuli induce the production of this endogenous phospholipase through the nuclear transcription factor kappa B (NF-κB) pathway.[26,27] Similar to inflammatory mediators, the plasmid vectors used for DNA immunization also induce this pathway (see relevant chapters in this book). Because of the highly conserved regions among the PLA2 enzymes of different species,[28] endogenous PLA2 may supply enough cross-reactive epitopes to prime specific T cells in a strong Th1 milieu. Alternatively, the vector may create a strong Th1 environment with truly antigen-independent effects.[18]

Safety First

The possibility of adverse effects has to be considered at the design stage, but no simple predictive parameter has yet been identified. In a worst case scenario a DNA vaccine may actually prime a Th2 response predisposing the host to anaphylaxis. That was the case for a model of peanut allergy using the major peanut allergen Ara h2.[23] Single or multiple injections of C3H/HeSn mice with naked pAra h2 DNA sensitized the animals against this allergen in a dose-dependent manner. The severity of the reactions ranged from scratching and rubbing around the nose and head, to decreased activity, increased respiratory rate, and ultimately convulsions and death in some individuals. The authors indicated that hypersensitivity induction was not unique to the peanut allergen gene since they could also induce anaphylaxis in this mouse strain to the ovomucoid antigen using the same methodology. When the vaccination and challenge protocol were tried in three different strains, C3H/HeSn, AKR and Balb/c mice, only C3H/HeSn animals developed anaphylaxis. Interestingly, the morbidity did not correlate with IL-4 or IL-5 increases unique to this strain, and it occurred despite slight enhancement of IFN-γ secretion and lack of IgE induction. The only parameter that was exclusive to the C3H mice was the high induction of IgG1 anti-Ara h2 antibodies. When C3H antisera containing these antibodies were transferred to naïve mice, they could mediate vasodilation upon antigen challenge of the host, in a typical positive passive cutaneous anaphylaxis (PCA) reaction.

As described in the previous section, when the Ara h2 peanut allergen was administered orally to AKR mice within chitosan nanoparticles, it induced protective immunity.[21] Whether this form of administration would make the DNA vaccine safe for C3H mice is unclear.

While PCA⁺ IgG1 antibodies were observed in food allergen models, they were not detected in inhalant allergen models. In the Der p 5 system for example, pCMVD protected mice also produced large amounts of anti-Der p 5 IgG1.[15] Also, it has been demonstrated that some parasite-derived antigens induce IgG1 antibodies incapable of mediating PCA.[29] The presence of these PCA⁻ IgG1 antibodies in IL-4-deficient mice indicates that they are induced in an IL-4 independent pathway. Reciprocally, in IL-12-deficient mice, or in IFN-γ-depleted mice, the same antigens induce higher levels of PCA⁺ IgG1 antibodies as well as lower levels of nonanaphylactic IgG1. These results indicate that IL-12 has a negative effect on the secretion of anaphylactic antibodies and a positive role on PCA⁻ IgG1 synthesis. The results also suggest that one subset of IgG1 antibodies is IL-4-dependent and can mediate allergic processes and another subset is promoted by IL-12-dependent pathways and lacks anaphylactic activity. According to the authors, the two types of IgG1 are derived from the same gene (no IgG1 is present in IgG1⁻/⁻ mice), and preliminary RT-PCR experiments have failed to detect alternative splicing. They suggest that perhaps the differences are posttranslational at the glycosylation level. It is not known whether both molecules fix complement; this information could be relevant since some of the proteins of the complement pathway have anaphylactic and chemotactic properties.

Taken together, reduced IgE alone is not predictive of protection, but increases in IgG1 are not predictive of allergy either. In fact some IgG1 secretion may result from increased IL-12 and IFN-γ production. Apparently not all the IgG1 can mediate anaphylaxis and the induction of PCA⁺ IgG1 seems to be antigen dependent.[29] However, major histocompatibility complex (MHC) alleles do not seem to determine susceptibility to allergy, at least not in the case of peanut allergy, since AKR and C3H mice share the same haplotype and yet AKR mice do not develop anaphylactic reactions.[23]

It is noteworthy that the opposite pattern emerges from the strain-dependent murine development of asthma: peanut allergy-resistant AKR mice are susceptible to asthma whereas peanut allergy-prone C3H mice are resistant to asthma.[30] Although the C3H mice used for the food allergy model (C3H/HeSn) and those used for the asthma model (C3H/HeJ) belong to different substrains, the effector mechanisms relevant to diverse atopic manifestations most likely differ. The genetic factors involved in the susceptibility to asthma among inbred mouse strains were studied by comparative microarray technology. Transcriptome analysis combined with single nucleotide polymorphism-based genotyping showed that asthma-susceptible AKR and A/J mice had a deficiency in a component of the complement pathway, C5a, whereas asthma-resistant C3H/HeJ and Balb/c animals had normal C5a expression.[30] Although C3H/HeJ mice are LPS-resistant, this trait was unrelated to their resistance to asthma. One of the protective mechanisms at work may be the induction of IL-12 secretion by the binding of C5a to its receptor on macrophages.

The route of vaccination may also tilt the balance from protective to harmful. It is likely that the degree of DNA sequestration varies with the delivery route, but protracted expression of even small amounts of allergenic protein may be undesirable.[31] Slater et al for example are developing a DNA vaccine to prevent latex allergy.[20] They have produced a construct that carries the sequence for Hev b 5 latex allergen in the p394 expression vector, a vector that uses the immediate early CMV promotor. The plasmid was injected subcutaneously at the base of the tail and its pattern of expression was analyzed. Hev b 5 RNA was observed at the injection site and in the lymph nodes, spleen, and lungs within 1 day after injection and persisted for at least 14 days. On day 14 it was also detected in the blood and tongue.[32] No Hev b 5 protein was detected. It is unclear whether these data reflect widespread tissue transfection, or, more likely, transcript expression by migratory surveillance cells, such as myeloid dendritic cells, monocytes, etc.

Many allergens such as honeybee venom PLA 2 or Der p 1, one of the major allergens in house dust mites, are enzymes. Endogenous production of an active enzyme after DNA vaccination may add unnecessary risks. For example, the sequence of Der p 1 is homologous to that of other cysteine proteases such as papain, actinidin, cathepsin B and cathepsin H.[33,34] The residues making up the active site are highly conserved. The enzymatic activity of Der p 1 has been confirmed by the demonstration that Der p 1 proteolytically cleaves CD25 and also the low-affinity receptor for human IgE, CD23.[35] It has been postulated that the enzymatic activity of Der p 1 has direct bearing on its allergenicity.[36] Native Der p 1 is initially synthesized in pre-/proenzyme form. During expression of native Der p 1 the pre- and prodomains are excised, and the mature form is secreted. The pre- and proenzyme sequences are required for proper folding and enzymatic bioactivity.[37,38] Therefore a Der p 1 vaccine lacking the pre/proenzyme regions can be safe. It can also be effective because these precursor regions are not part of the allergen, and native conformation is irrelevant to T cell priming.[39]

Allergen DNA could be architected to encode physiologically inactive immunogenic fragments. Our laboratory is using this strategy to design a DNA vaccine against Der p 1.[40] Three DNA constructs have been produced in which the pre- and proenzyme DNA sequences are deleted. The Der p 1 leader peptide sequence was replaced by the Der p 5 leader, a signal peptide that facilitates protein export to the endoplasmic reticulum. Construct Der p 5/1-116 contains the sequence corresponding to the N-terminal alpha-helical domain of Der p 1; Der p 5/114-222 corresponds to the β-sheet-rich, C-terminal domain of Der p 1, and Der p 5/1-222 corresponds to secreted mature Der p 1. The pEGFP-N3 reporter vector encoding green fluorescence protein was used to monitor protein expression in vivo by confocal microscopy. There was good correlation between the frequency and levels of DNA expression and antibody production. Fluorescence signal could be detected after vaccination with either Der p 5/1-116 or Der p 5/1-222, but only rarely with Der p 5/114-222. Expression of Der p 5/1-116 in muscle sections was the most frequently detected, as early as 6 h after injection, whereas Der p 5/1-222 protein was detected only after 2 weeks of antigen accumulation. The Der p 5/1-116 construct induced the highest anti-Der p 1 antibody response. The anti-Der p 1 response after Der p 5/1-222 injection was lower but highly significant, whereas no anti-Der p 1 antibodies were detected after Der p 5/114-222 immunization. Two injections of either Der p 5/1-116 or Der p 5/1-222 naked plasmid, three weeks apart, generated long lasting immunity (at least 5 months). As expected, all the anti-Der p 1 antisera induced by Der p 5/1-116 reacted with epitopes located in the N-terminal domain, but 80% of the anti-Der p 1 antisera induced by Der p 5/1-222 also reacted with the N-terminal domain. Interestingly, Der p 5/1-116 induced a mixture of IgG2a and IgG1 Der p 1-specific antibodies, but the anti- Der p 1 antibodies induced by Der p 5/1-222 were overwhelmingly IgG2a . Whether the IgG1 produced is IL-4- or IL-12-dependent has not yet been determined. These data seem to indicate a correlation between magnitude of expression and IgG subclass induction. IgG subclass differences may also reflect the secretory, cytoplasmic or membrane bound character of the antigen. It has been shown that plasmids expressing the membrane-bound, native form of influenza hemagglutinin after intramuscular injection preferentially induce IgG2a, whereas DNA immunization with a recombinant secreted form of the antigen predominantly induces IgG1.[41] It is possible that our observations correspond to the same phenomenon. In the hemagglutinin study, IgG1 secretion was IL-4-dependent but not DNA-dose dependent.

The delivery method also has to be carefully evaluated. DNA immunization by gene gun is inadequate for allergy DNA vaccines because it favors Th2 immunity.[42] Despite the use of the native membrane-bound hemagglutinin plasmid, gene gun vaccination elicited predominantly IL-4-dependent IgG1.[41]

Immunogenicity: Antigen Expression and Presentation

In this section we will discuss new information that has fueled the debate on whether there is a coupling of antigen expression, antigen presentation and immunostimulatory functions after DNA vaccination. This information is of particular relevance to the design of a vaccine with immunomodulatory functions.

The site of injection determines the major cellular target for antigen expression. This cell acts as antigen factory and temporary reservoir. After intramuscular immunization with dust mite allergen gene constructs accompanied by electroporation, there were increased protein levels in the muscle bundles up to 10-14 days, possibly due to antigen accumulation. Protein signal started to decrease after 2 weeks, the time when antibody response first became detectable, and no protein could be shown 3 to 4 weeks after injection.[40] The fate of plasmid after intramuscular injection has been analyzed by imaging of fluorescent tagged DNA.[43] It was shown that immediately after injection, the DNA localized mostly in the myotendinous junctions of the muscle, mostly in the extracellular space between the muscle cells and some inside the myocytes. One day later, most of the DNA between muscle cells was inside mononuclear cells (MNC), and also in the draining lymph nodes. The MNC held the DNA within cytoplasmic vesicles. The MNC were positive for the CD11b myeloid marker, and the costimulatory molecules CD80 and CD86. Antigen transcripts however (either β-gal or HIV gag) could only be detected in muscle cells by RT-PCR. The MNC were still transcript negative 2 days and 7 days after injection. Electroporation greatly enhanced the uptake of the DNA within the nuclei of muscle cells, but still the authors could not detect transcript expression within nonmuscle cells.[43] The lack of detection could be due to lack of sensitivity and the migratory nature of antigen presenting cells (APCs), but the possibility of tissue-specific antigen expression cannot be excluded.

Since professional APC's are sparse in the muscle, and myocytes lack proper costimulatory molecules necessary for immunogenicity, it could be argued that allergen proteins presented in the context of muscle cells induce tolerance.[44] However, the available evidence indicates that antigen is not presented by somatic tissues but by professional APCs of bone marrow origin.[45] Moreover, all the models that induced allergy protection by plasmid immunization have shown allergen-specific Th1 immunity.[14-22] These observations indicate that prophylaxis is not an anergy phenomenon but one of immune deviation.

After skin vaccination, most of the transgene expression occurs in keratinocytes,[46] but there is also abundant evidence showing dendritic cells (DCs) and macrophages containing both plasmid DNA and the transgene product.[47,48] Gene gun immunization resulted in direct transfection of DCs, and the antigen also localized in these cells 24 h after DNA vaccination.[48] The straightforward interpretation of these data is that transfection of DNA into the DCs resulted in antigen synthesis in these cells. However, since RT-PCR was not performed in some of these studies, the possibility that DCs also acquired antigen by cross priming cannot be ruled out. Transcript expression was shown in DCs that had migrated from explant cultures of skin after intradermal injection of a pCMV-HEL (hen egg lysozyme) model vaccine.[47]

DCs play an essential role in the immunogenicity of DNA vaccines. These cells emerge as the most potent inducers of effector T cell function. Perhaps DCs have a dual function in immunomodulation models, as APCs and as antigen manufacturers following DNA vaccination. It is likely that even if only a minor proportion of antigen synthesis is derived from DC factories, this could be an important contribution to immunoregulation.

A possible strategy to enhance the efficacy of DNA vaccines is the inclusion, in the plasmid, of organelle-targeting sequences that promote association of processed antigens to peptide presentation molecules, such as MHC class II,[49] MHC class I,[50] or CD1d.[51] It is likely that the success of this strategy depends on the direct transfection of professional APCs, and co-translation of the antigen and the fused targeting protein.[50] It is still unknown if this approach can bias immunity towards Th1 induction. The available data suggest that MHC class II or class I targeting does not correlate with Th1 cytokine production: when a DNA construct was engineered for efficient loading onto MHC class II molecules, intramuscular delivery of the plasmid induced protection against tumor challenge and CD4$^+$ T cells secreted IFN-γ. If the same plasmid was delivered intraperitoneally in a vaccinia vehicle though, the CD4$^+$ T cells secreted both IFN-γ and IL-4, and the vaccine was not protective. In this case, independent of the construct, the delivery method determined both the pattern of cytokine induction and the efficacy of the vaccine.[49] Targeting the MHC class I favors priming of CD8$^+$ T cells,[50] but these cells are also a potential source of IL-4 and IL-5.[3]

Alternatively, DNA sequences encoding Th1 cytokines may increase the success of an allergy vaccine. Maecker and coworkers showed how inoculation of a DNA fusion construct encoding OVA and IL-18 was more effective in preventing AHR than the mixture of OVA and IL-18 encoding DNA.[19] The cell responsible for AHR suppression was a CD8$^+$, IFN-γ secreting T cell. These results suggest a mechanism where a DC synthesized OVA and IL-18 and interacted with the T cells both via T cell receptor (TCR) and IL-18 receptors. However the results do not rule out the access of OVA-IL-18 protein into the APCs via cross priming.

These reports indicate that engineering additional modulatory sequences in the vector DNA may enhance the efficacy of DNA vaccines for allergy. The question of whether this strategy requires antigen synthesis (myocytes, epithelial or endothelial cells or professional APCs) and antigen presentation co-localization to the same cell is unresolved. The persistence of immunity, in particular long-lasting antibody responses, despite the temporary expression of antigen at the site of injection suggests that other cells, such as follicular DCs, may act as antigen reservoir.[51] The way antigen gets access to APCs may be of great relevance to antigen-specific, immunomodulation protocols. It is unknown whether addition of targeting, immunomodulatory, or signaling sequences will make the difference between efficacy or lack thereof.

Antigen Dose and Polarization of T Cells

The degree of polarization of T cells into Th1 and Th2 phenotypes reflects the nature of the antigenic and environmental stimuli to which the cells have been exposed. A factor frequently invoked in the Th1 versus Th2 development pathway is the antigen dose.[31,52] Low antigen doses injected intraperitoneally in alum favor Th2 responses.[31] On the other hand, low antigen doses administered orally induce lower IL-4 and higher IFN-γ secretion than high oral antigen doses.[52,53]

For DNA vaccination systems, the effect of the dose of expressed antigen in T cell polarization has not been fully investigated. Some evidence suggests that the quantity of expressed antigen plays a minor role in T cell polarization. For example, a vector with several CpG motifs had more Th1 adjuvanticity than a vector that expressed the DNA insert better, but had fewer CpG motifs. Expression, however, was determined in vitro.[54]

The data from Torres et al suggest that secretion of the antigen is more relevant to Th2 polarization than the quantity of expression.[41] Plasmids expressing the membrane-bound, native form of influenza hemagglutinin after intramuscular injection preferentially induced IgG2a. The secreted form of hemagglutinin induced IL-4 dependent IgG1. Similar amounts of gene transcripts were detected after cell transfection in vitro, for both antigen forms. Increase in the DNA dose did not change the response elicited by the membrane form in vivo. These results are similar to those obtained by Boyle et al, comparing the responses to DNA encoding secreted, cytoplasmic, and membrane bound OVA.[55] RT-PCR analysis showed similar levels of transcription for all immunogens. However, while cytoplasmic and membrane forms induced mostly IgG2a, secreted forms generated 10- to 100-fold higher IgG and 50-100-fold higher IgG1. The authors attribute the increase in humoral response to the higher availability of secreted antigen for B cell priming.

Additional results do not allow ruling out a role for the dose of antigen. When the level of hemagglutinin expressed was increased by gene gun DNA delivery, membrane-bound antigen elicited Th2 responses.[56] Data obtained for two Der p 1 allergen gene constructs also showed that the construct that resulted in higher antigen expression induced elevated IgG1 responses. It has yet to be determined if this IgG1 is IL-4 dependent.[40]

The influence of the route of immunization on the quality of the response is also antigen-dependent. Intradermal injection of plasmid DNA was superior to intramuscular immunization for the induction of IL-4 and IgG1 antibodies to OVA or HEL, but not for human immunoglobin.[57] A β-gal construct injected intradermally on the other hand induced potent Th1 responses.[14] It has been shown that in vivo different tissues may transcribe DNA with different efficiency, as in the case of myocytes and myeloid cells described above.[43] Also, whereas mononuclear cells in the muscle are scarce, Langerhans cells and other professional APC are enriched in the dermis. Intradermal injections may result in direct APCs transfection more frequently. Endogenous antigen synthesis by APCs versus soluble antigen uptake may induce different APC maturation pathways.

Effector Mechanisms of Allergy Protection

Both CD4+ and CD8+ T cells have been shown to respond to DNA vaccination.[14,15,19] In the Der p 5 model in which the IgE response to Der p 5 in alum was inhibited by previous intramuscular vaccination with a pCMVD construct carrying the full length Der p 5 cDNA, this effect could be adoptively transferred with CD8+ T cells from treated mice. Similarly, the cells responsible for protection in the OVA-IL-18 DNA vaccination model were also reported to be CD8+ T cells that produced IFN-γ: treatment of mice with neutralizing anti-IFN-γ monoclonal antibodies (mAb) or with depleting anti-CD8 mAb abolished the protective effect.[19] In the β-gal model on the other hand, inhibition of IgE production could be transferred with both CD4+ and CD8+ cells.[14]

The therapeutic effects of plasmid DNA vaccination imply a shift from Th2 to Th1 immunity. The immunostimulatory characteristics of plasmid DNA have led to the hypothesis that DNA vaccines protect against allergy by priming a Th1 or a Tc1 allergen specific responses.[14-16] Antigen is processed for presentation to CD8+ and CD4+ T cells. Antigen uptake and processing stimulate IL-12 secretion. IL-12 in turn induces IFN-γ secretion, which promotes differentiation of Th1 or Tc1 cells. These cells may in turn exert a positive feedback on antigen primed B lymphocytes to reinforce the polarization of the response. The molecules responsible for switching on a set of cytokines are starting to be unraveled. It has been postulated that (IL-12 R-β2) chain-mediated stimulation was sufficient to induce Th1 cell differentiation.[58] Several recent findings call this into question. Among others, IL-12 deficient mice, that fail to produce IFN-γ in response to mycobacterial infection, can develop Th1 responses to adenoviral infection even in the absence of IL-18.[59] Also, ectopic expression of IL-12R-β 2 in committed Th2 cells does not affect the production of IL-4.[60]

Recent research has shown that a T-box signaling protein expressed in T lymphocytes (T-bet) commits T cells to the Th1 lineage.[61] T-bet expression correlated with IFN-γ expression in Th1, natural killer (NK) and activated B cells. More striking was the finding that transduction of this gene into already polarized effector Th2 cells redirected their commitment towards a Th1 secretory pathway, inducing the synthesis of IFN-γ and repressing that of IL-4 and IL-5. It is also noteworthy that a corresponding series of regulatory proteins with transcription factor activity, including in particular GATA-3 and cMaf, have been implicated in Th2 commitment.[62]

A recent report suggests that IL-10 secreted by regulatory T cells may be responsible for down-regulating both Th1 and Th2 responses.[63] These regulatory T cells are neither Th1 nor Th2, and they inhibit IgE secretion independently of their ability to produce IFN-γ. Myeloid DCs that cannot produce IL-12 can be stimulated to secrete IL-10. The relative importance of IL-10, IL-12, and IFN-γ to allergy protection has not been fully elucidated yet.

Future Directions

It is clear that more mechanistic studies will be necessary for the rational design of allergy vaccines. Priming DC polarization may hold the key to long lasting allergy protection. As new DC markers are identified, the heterogeneity of DCs becomes evident. DC subsets can also be separated by their proliferative and developmental response to lymphokines. Many issues remain to be addressed: the potential of different populations of DCs to secrete IL-12 or if they are terminally differentiated, the role of antigen or DNA in this differentiation, the capacity for DNA / antigen internalization by DC subsets,[64] and whether the ability to induce IL-12 depends on this event. Most data support indiscriminate binding of DNA to the cell surface.[65] In most cases no sequence-specific receptor has been identified. However, oligodeoxynucleotides consisting of different immunostimulatory motifs elicit different cytokine secretion patterns from peripheral blood mononuclear cells (PBMC).[66] The mechanisms that drive functional distinctions in vivo are still poorly understood.

The most effective age for immunization will also have to be determined. Interestingly, the DC1 phenotype could not be generated from newborn precursor (p) DCs. Neonatal pDCs from cord blood were cultured with IL-4 and GM-CSF and then stimulated by lipopolysaccharide (LPS), CD40 ligation, or poly(I:C). None of these protocols induced IL-12 secretion, due to impaired transcription of IL-12(p35) mRNA. IL-12(p40) gene expression was not altered. The addition of rIFN-γ to LPS-stimulated newborn DCs restored their expression of IL-12(p35) and their synthesis of IL-12 (p70). Neonatal DCs were less efficient than adult DCs in the induction of IFN-γ production by adult CD4⁺ T cells. This defect was corrected by the addition of rIL-12.[67]

When dealing with inhalant allergens, it will be important to characterize the subpopulations of DCs in the human respiratory tract. It has recently been shown that DCs from the respiratory tract selectively prime Th2 responses. However, if they are precultured with GM-CSF and TNF-α or CD40L they can prime Th1 responses.[68] On the other hand, protection from skin or food allergies may require skin or oral delivery of genes. It is also of interest to note that the nasal mucosa contains a large population of cells with both DC1 and DC2 phenotypes[69] which respond vigorously to allergenic challenge; nasal antigen delivery is known to favor selective downregulation of IgE responses,[70] and hence targeted gene delivery at this site may have unique effects.

B lymphocytes also contribute to their microenvironment. A DNA vaccine that induces antibodies to secreted antigens may result in complement activation. Constituents of the complement pathway have both protective and susceptibility roles in the asthmatic process, which may not be confined to the effector phase of asthma.[30]

It is now becoming clear that B cells not only specialize in antibody production and efficient antigen presentation, but also condition their environment through the array of secreted cytokines. B cells can induce Th2 responses.[71] Naïve antigen-specific B cells stimulate CD4$^+$ T cells, naïve or primed, to secrete high levels of IL-4 only in the presence of the cognate antigen. But as reported more recently, B lymphocytes can also differentiate into, IFN-γ producing cells if their Th1 counterparts educate them.[72] Generation of these effector B cells, Be1 and Be2, was done in vitro, by co-culture of transgenic naïve B cells with transgenic effector Th1 or Th2. Effector B cells could also be isolated from mice infected with two types of microorganisms. Be1 were recovered from *Toxoplasma gondii* infected mice, a parasite that induces a predominant Th1 response. *Heligmosomoides polygyrus* infection, which induces a predominant Th2 response, resulted in Be2 recovery from infected mice. The role of antigen on Be1 or Be2 differentiation is unclear. This moves IFN-γ and IL-4 secretion out of the exclusive realm of T cells and brings another player into consideration.

Different strategies are currently being studied for raising the immunogenicity of DNA vaccines. One of them is the prime-boosting approach. Ramsay and coworkers used consecutive immunization with naked DNA followed by a fowl poxvirus recombinant, both encoding hemagglutinin (HA), to raise the amount of systemic anti-HA antibodies.[24] Similarly, a DNA vaccine that primes a low-level but persistent Th1 immunity, may be greatly enhanced by boosting with a different vehicle.

A rapidly evolving area of research is rational oligodeoxynucleotide (ODN) design to be used as an adjuvant for protein vaccines. Different motifs stimulate mouse or human cells, different cell types, and different cytokine secretion. There is so far little available data comparing the efficacy of ODN as adjuvant to that of plasmid vaccines for the prevention of allergy. Also, the safety and feasibility of an effective ODN formulation needs to be evaluated.

Given the relevance of IL-10 as an immunomodulatory cytokine, it will be important to determine how to stimulate the IL-10 pathway without concomitant Th2 cytokine production. Although IL-10 has an inhibitory role in IL-4 production, it affects neither IL-5 nor IL-13 activities. Both IL-10 and IFN-γ may be required for immunotherapy of established disease. Another important cytokine in the allergy prevention field is GM-CSF. Since "first line of response" pDC1 are enriched in GM-CSF receptors, strategies that incorporate this cytokine either directly, as plasmid DNA, or in fusion constructs may also increase allergy DNA vaccine efficacy. Both GM-CSF and IL-10 are T cell products, and their secretion is usually accompanied by other lymphokines. Therefore a passive approach may be better suited for the purpose than the natural induction of these cytokines. Other cytokines that either target APCs or T cells may also be effective.

Insights will be gained from an understanding of the molecular differences between Th1 and Th2. Creative manipulation of signaling molecules may in the future be used to control T cell differentiation.[61] A DNA vaccine encoding Th1 differentiation molecules may hold promise, if targeting T cells in this way does not trigger immune deficiency. Current immunotherapeutic strategies aim at this indirectly, by the choice of antigen, vehicle, dose and route of administration. All of these factors are of relevance in determining the type of response. It is of interest, for example, that CD8$^+$ γδ T cells present in the digestive tract seem refractory to Th2 differentiation. For these cells, the default pathway is the production of IFN-γ, even in the presence of IL-4.[73] This is of particular clinical interest since oral immunization is a practical approach to human vaccination.

The safety of engineered DNA and delivery vehicles must be determined. Manipulation of DNA constructs may maximize protective phenotypes while minimizing side effects.

There is evidence that at birth, virtually all children exhibit low-level Th2 polarized responses to environmental antigens present in the maternal environment. Immune deviation is therefore an important mechanism capable of redirecting these responses towards the Th1 cytokine phenotype in protection against allergy. A DNA vaccine coding for a major allergen able to promote an allergen specific Th1 response and downregulate the Th2 response upon subsequent antigen exposure may need to be administered early in life in order to achieve high efficacy for prevention of allergy.[74]

References

1. Romagnani S. Regulation of the development of type 2 T-helper cells in allergy. Curr Opin Immunol 1994; L6:838-846.
2. Abbas AK, Murphy KM, Sher A. Functional diversity of helper T lymphocytes. Nature 1996; 383:787-793.
3. Kemeny, D. CD8⁺ T cells in atopic disease. Curr Op Immunol 1998; 10:628.
4. Chang TW. The pharmacological basis of anti-IgE therapy. Nature Biotechnology 2000; 18:157-162.
5. Wierenga E, Snoek M, de Groot C et al. Evidence for compartmentalization of functional subsets of CD4+ T lymphocytes in atopic patients. J Immunol 1990; 144:4651.
6. Sher A, Coffman RL. Regulation of immunity to parasites by T cells and T cell derived cytokines. Ann Rev Immunol 1992; 10:385-409.
7. Ohashi Y, Nakai Y, Tanaka A et al. Ten-year follow-up study of allergen-specific immunoglobulin E and immunoglobulin G4, soluble interlieukin-2 receptor, interleukin-4, soluble intercellular adhesion molecule-1 in serum of patients on immunotherapy for perennial allergic rhinitis. Scand J Immunol 1998; 47:167.
8. Hansen G, Berry G, DeKruyff RH et al. Allergen-specific Th1 cells fail to counterbalance Th2 cell-induced airway hyper reactivity but cause severe airway inflammation. J Clin Invest 1999; 103:175-183.
9. Des Roches A, Paradis L, Knami J et al. Immunotherapy with a standardized Dermatophagoides pteronyssimus extract. V. Duration of the efficacy of immunotherapy after its cessation. Allergy 1996; 51:430.
10. Wolf BL, Hamilton RG. Near-fatal anaphylaxis after Hymenoptera venom immunotherapy. J Allergy Clin Immunol 1998; 102:527-8.
11. Schramm G, Kahlert H, Suck R et al. "Allergen engineering": variants of the timothy grass pollen allergen Phl p 5b with reduced IgE-binding capacity but conserved T cell reactivity. J Immunol 1999; 162:2406-2414.
12. Wolff JA, Malone RW, Williams P et al. Direct gene transfer into mouse muscle in vivo. Science 1990; 247:1465-1468.
13. Raz E, Carson DA, Parker SE et al. Intradermal gene immunization: the possible role of DNA uptake in the induction of cellular immunity to viruses. Proc Natl Acad Sci USA 1994; 91:9519-9523.
14. Raz E, Tighe, H, Sato, Y et al. Preferential induction of a Th1 immune response and inhibition of specific IgE antibody formation by plasmid DNA immunization. Proc Natl Acad Sci USA. 1996; 93:5141-5145.
15. Hsu CH, Chua KY, Tao MH et al. Immunoprophylaxis of allergen-induced immunoglobulin E synthesis and airway hyper responsiveness in vivo by genetic immunization. Nat Med 1996; 2:540.
16. Hsu CH, Chua KY, Tao MH et al. Inhibition of an in vivo allergen-specific IgE response in mice by direct gene transfer. Int Immunol 1996; 8:1405-1411.
17. Li XM, Chopra RK, Chou TY et al. Mucosal IFN-gamma gene transfer inhibits pulmonary allergic responses in mice. J Immunol 1996; 157:3216-3219.
18. Jilek S, Barbey C, Spertini F et al. Antigen-Independent Suppression of the Allergic Immune Response to Bee Venom Phospholipase A(2) by DNA Vaccination in CBA/J Mice. J Immunol 2001; 166:3612-3621.
19. Maecker HT, Hansen G, Walter DM et al. Vaccination with allergen-IL-18 fusion DNA protects against, and reverses established, airway hyperreactivity in a murine asthma model. J Immunol 2001; 166:959-65.
20. Slater JE, Zhang YJ, Arthur-Smith A et al. A DNA vaccine inhibits IgE responses to the latex allergen Hev b 5 in mice. J Allergy Clin Immunol 1997; 99:S504.
21. Roy K, Mao HQ, Huang SK et al. Oral gene delivery with chitosan—DNA nanoparticles generates immunologic protection in a murine model of peanut allergy. Nat Med 1999; 5:387.
22. Raz E, Spiegelberg HL. Deviation of the allergic IgE to an IgG response by gene immunotherapy. Int Rev Immunol 1999; 18:271-89.
23. Li X-m, Huang C-K, Schofield BH et al. Strain-dependent induction of allergic sensitization caused by peanut allergen DNA immunization in mice. J Immunol 1999; 162:3045-3052.
24. Ramsay, A, Kent SJ, Strugnell, RA et al. Genetic vaccination strategies for enhanced cellular, humoral and mucosal immunity. Imm Reviews 1999; 171:27-44.
25. TS Kim, RH DeKruyff, R Rupper et al. An ovalbumin-IL-12 fusion protein is more effective than ovalbumin plus free recombinant IL-12 in inducing a T helper cell type 1-dominated immune response and inhibiting antigen-specific IgE production. J Immunol 1997; 158:4137.
26. Ribardo DA, Crowe SE, Kuhl KR et al. Prostaglandin levels in stimulated macrophages are controlled by phospholipase A2-activating protein and by activation of phospholipase C and D. J Biol Chem 2001; 276:5467-75.

27. Walker G, Kunz D, Pignat W et al. Suppression by cyclosporin A of interleukin 1 beta-induced expression of group II phospholipase A2 in rat renal mesangial cells. Br J Pharmacol 1997; 121:787-793.
28. Rogers MV, Henkle KJ, Herrmann V et al. Evidence that a 16-kilodalton integral membrane protein antigen from Schistosoma japonicum adult worms is a type A2 phospholipase. Infect Immun 1991; 59:1442-7.
29. Faquim-Mauro EL, Coffman RL, Abrahamsohn IA et al. Cutting Edge: Mouse IgG1 antibodies comprise two functionally distinct types that are differentially regulated by IL-4 and IL-12. J Immunol 1999; 163:3572-3576.
30. Karp CL, Grupe A, Schadt E et al. Identification of complement factor 5 as a susceptibility locus for experimental allergic asthma. Nature Immunol 2000; 1:221-226.
31. Kolbe L, Heusser C, Kölsch E. Antigen dose-dependent regulation of Bε-memory cell expression. Int Arch Allergy Appl Immunol 1991; 95:202-206.
32. Slater JE, Paupore E, Zhang YT et al. The latex allergen Hev b 5 transcript is widely distributed after subcutaneous injection in BALB/c mice of its DNA vaccine. J Allergy Clin Immunol. 1998; 102:469-75.
33. Chua K, Stewart G, Thomas W et al. Sequence analysis of cDNA coding for a major house dust mite allergen, Der p 1. Homology with cysteine proteases. J Exp Med 1988; 167:175-182.
34. Chua K-Y, Kehal PK, Thomas, WR et al. High-frequency binding of IgE to the Der p allergen expressed in yeast. J Allergy Clin Immunol. 1992; 89:95-102.
35. Hewitt CR, Brown AP, Hart, BJ et al. A major house dust mite allergen disrupts the immunoglobulin E network by selectively cleaving CD23: innate protection by antiproteases. J Exp Med. 1995; 182:1537-1544.
36. Wan H, Winton HL, Soeller C et al. Der p 1 facilitates transepithelial allergen delivery by disruption of tight junctions. J Clin Invest 1999; 104:123.
37. Jacquet A, Haumont M, Massaer M et al. Biochemical and immunological characterization of a recombinant precursor form of the house dust mite allergen Der p 1 produced by Drosophila cells. Clin Exp Allergy 2000; 30:677-684.
38. Takahashi K, Takai T, Yasuhara T et al. Production of enzymatically and immunologically active Der f 1 in Escherichia coli. Int Arch Allergy Immunol. 2000; 122:108-114.
39. Korematsu S, Tanaka Y, Hosoi S et al. C8/119S mutation of major mite allergen Derf-2 leads to degenerate secondary structure and molecular polymerization and induces potent and exclusive Th1 cell differentiation. J Immuinol 2000; 165:2895- 2902.
40. Liew, LN, Wolfowicz, BS, Chua KY. Differential immunogenicity of DNA carrying Der p 1 gene fragments (manuscript in preparation).
41. Torres CAT, Yang K, Mustafa F et al. DNA immunization: effect of secretion of DNA-expresssed hemagglutinins on antibody responses. Vaccine 2000; 18:805-814.
42. Feltquate DM, Heaney S, Webster RG et al. Different T-helper cell types and antibody isotypes generated by saline and gene gun DNA immunization. J Immunol 1997; 158:2278-2284.
43. Dupuis M, Denis-Mize K, Woo C et al. Distribution of DNA vaccines determines their immunogenicity after intramuscular injection in mice. J Immunol 2000; 165:2850-2858.
44. Rolland J, O'Hehir R. Immunotherapy of allergy: anergy, deletion and immune deviation. Curr Opin Immunol 1998; 10:640-645.
45. Corr MA, von Damm A, Lee DJ et al. In vivo priming by DNA injection occurs predominantly by antigen transfer. J Immunol 1999; 163:4721-4727.
46. Porgador A, Irvine KR, Iwasaki A et al. Predominant role of directly transfected dendritic cells in antigen presentation to CD8+ T cells after gene gun immunization. J Exp Med 1998; 188:1075-1083.
47. Bouloc A, Walker P, Grivel JC et al. Immunization through dermal delivery of protein-encoding DNA: a role for migratory dendritic cells. Eur J Immunol 1999; 29:446.
48. Condon C, Watkins SC, Celluzzi CM et al. DNA-based immunization by in vivo transfection of dendritic cells. Nat Med 1996; 2:1122.
49. Chen CH, Wang TL, Ji H et al. Recombinant DNA vaccines protect against tumors that are resistant to recombinant vaccinia vaccines containing the same gene. Gene Ther 2001; 8:128-138.
50. Rodriguez F, Shang J, Whitton JL. DNA immunization: ubiquitination of a viral protein enhances cytotoxic T-lymphocyte induction and antiviral protection but abrogates antibody induction. J Virol 1997; 71:8497-8503.
51. MacLennan ICM. Germinal centers. Annu Rev Immunol 1994; 12:117-139.
52. Sato MN, Carvalho AF, Silva AO et al. Low dose of orally administered antigen down-regulates the helper type 2-response in a murine model of dust mite hypersensitivity. Immunol 1999; 98:338-344.
53. Alpan O, Rudomen G, Matzinger P. The role of dendritic cells, B cells, and M cells in gut-oriented immune responses. J Immunol 2001; 166:4843-4852.

54. Tighe H, Corr M, Roman M et al. Gene vaccination: plasmid DNA is more than just a blueprint. Imm. Today 1998; 19:89-97.
55. Boyle JS, Koniaras C, Lew AM. Influence of cellular location of expressed antigen on the efficacy of DNA vaccination: cytotoxic T lymphocyte and antibody responses are suboptimal when antigen is cytoplasmic after intramuscular DNA immunization. Int Immunol 1997; 9:1897-1906.
56. Robinson H, Pertmer TM. DNA vaccines for viral infections: basic studies and applications. Adv Virus Res 2000; 55:1-74.
57. Boyle JS, Silva A, Brady et al. DNA immunization: induction of higher avidity antibody and effect of route on T cell cytotoxicity. Proc Natl Acad Sci USA 1997; 94:14626-14631.
58. Rogge, L, Barberis-Maino L, Biffi M et al. Selective expression of an interleukin-12 receptor component by human T helper cells. J Exp Med 1997; 185:825-831.
59. Xing Z, Zganiacz A, Wang J et al. IL-12-independent Th1-type immune responses to respiratory viral infection: requirement of IL-18 for IFN-gamma release in the lung but not for the differentiation of viral-reactive Th1-type lymphocytes. J Immunol 2000; 164:2575-2584.
60. Heath VL, Showe L, Crain C et al. Cutting edge: Ectopic expression of the IL-12 receptor-β2 in developing and committed Th2 cells does not affect the production of IL-4 or induce the production of IFN-γ. J Immunol 2000; 164:2861-2865.
61. Szabo S, Kim ST, Costa G et al. A novel transcription factor, T-bet, directs Th1 lineage commitment. Cell 2000; 100:655-669.
62. O'Garra A, Arai N. The molecular basis of T helper 1 and T helper 2 cell differentiation. Trends Cell Biol 2000; 10:542-550.
63. Cottrez F, Hurst SD, Coffman RL et al. T regulatory cells 1 inhibit a Th2-specific response in vivo. J Immunol 2000; 165:4848-4853.
64. Randolph GJ, Inaba K, Robbiani DF et al. Differentiation of phagocytic monocytes into lymph node dendritic cells in vivo. Immunity 1999; 11:753-61.
65. Krieg AM, Hartmann G, Yi A-K. Mechanism of action of CpG DNA. Curr Top Microbiol Immunol 2000; 247:1-21.
66. Verthelyi D, Ishii KJ, Gusel M et al. Human Peripheral blood cells differentially recognize and respond to two distinct CpG motifs. J Immunol 2001; 166:2372-2377.
67. Goriely S, Vincart B, Stordeur P et al. Deficient IL-12(p35) gene expression by dendritic cells derived from neonatal monocytes. J Immunol 2001; 166:2141-6.
68. Stumbles PA, Thomas JA, Pimm CL et al. Resting respiratory tract Dendritic cells preferentially stimulate Th2 responses and require obligatory cytokine signals for induction of Th1 immunity. J ExpMed 1998; 188:2019-2031.
69. Jahnsen F, Moloney ED, Hogan T et al. Rapid dendritic cell recruitment to the bronchial mucosa of atopic asthmatics in response to local allergen challenge. Thorax (submitted).
70. Van Halteren AG, vander Cammen MJ, Cooper D et al. Regulation of antigen-specific IgE, IgG1 and mast cell responses to ingested allergen by mucosal tolerance induction. J Immunol 1997; 159:3009-15.
71. Macaulay AE, DeKruyff RH, Goodnow CC et al. Antigen-specific B cells preferentially induce CD4$^+$ T cells to produce IL-4. J Immunol 1997; 158:4171-4179.
72. Harris DP, Haynes L, Sayles PC et al. Reciprocal regulation of polarized cytokine production by effector B and T cells. Nature Immunol 2000; 1:475-482.
73. Yin Z, Zhang D-H, Welte T et al. Dominance of IL-12 over IL-4 in $\gamma\delta$ T cell differentiation leads to default production of IFN-γ: failure to down-regulate IL-12 receptor β_2-chain expression. J Immunol 2000; 164:3056-3064.
74. Holt PG, Macaubas C, Stumbles PA et al. The role of allergy in the development of asthma. Nature 1999; 402:B12-B17.

CHAPTER 13

Immune Responses in Gene Transfer for Genetic Disorders

Denise E. Sabatino and Katherine A. High

Introduction

Gene transfer is a novel area of therapeutics in which a nucleic acid sequence is the active agent. Transferred in via a gene delivery vehicle, or vector, the donated gene sequence, referred to as a transgene, is usually introduced into a single target tissue in the host, where it directs synthesis of a missing or therapeutic protein. Gene transfer strategies are being evaluated in clinical trials for a variety of genetic and acquired disorders; in the setting of genetic disorders, the transgene generally encodes a protein that is missing or defective in the host and to which the recipient may therefore not be fully tolerant. As work in the field of gene transfer has progressed, it has become clear that avoiding or suppressing an unwanted immune response to the transgene product will be critical to successful gene therapy for genetic disease. In this sense, then, the focus of this chapter differs from that of others in this volume; instead of stimulating an adaptive immune response, we wish to avoid it altogether.

Analysis of immune responses to therapeutic proteins used to replace missing proteins in the setting of genetic disorders provides a framework for investigation of responses in the setting of gene transfer, but the problems presented by gene transfer are considerably more complex than those presented by simple protein replacement (Table I). For example, the immune response to the vector itself can influence the response to the transgene product; thus, adenoviral and plasmid vectors can trigger innate and adaptive immune responses that alter the local cytokine milieu in which transgene expression occurs. Second generation vectors are fully deleted for viral coding sequences, such as adeno-associated virus (AAV) and helper-dependent adenoviral (Ad) vectors, have been developed in an attempt to eliminate this factor. Other factors such as the identity of the target tissue and the host's degree of tolerance to the product of the donated gene also play a role. The interplay of these factors is not yet well understood, but there are a number of studies in the literature that at least bear on the question. These will be reviewed and areas requiring further study will be highlighted. We will make frequent use of hemophilia as a model for studying immune response to the product of a donated gene; the disease has been treated using both protein and gene transfer approaches, and it is thus possible to compare responses in these settings.

Tolerance in the Setting of Genetic Deficiency States

During fetal development and throughout postnatal life, self-reactive T cell clones are deleted or anergized. In the setting of genetic disease, an individual may not express epitopes recognized by T cells specific to the wild-type protein. These T cells mature, and on encounter with the antigen in an inflammatory setting differentiate into effector cells. The likelihood that this will occur in gene therapy depends to some extent on the underlying mutation; some types of mutations (e.g., some missense mutations) will result in full tolerance to the wild-type

DNA Vaccines, edited by Hildegund C. J. Ertl. ©2003 Eurekah.com and Kluwer Academic / Plenum Publishers.

Table 1. Factors that influence immune response to a protein in protein replacement therapy and in therapy by gene transfer

Protein Replacement Therapy
    Underlying mutation
    Genetically determined characteristics of the immune system
       (e.g., TAP, proteosomes, MHC class I and class II molecules)
    Infection, inflammation or massive tissue injury at the time of initial antigen exposure

Gene Transfer
    All of the above
    Type of vector
    Dose of vector
    Route of administration of vector
    Tissue specificity of the promoter in the vector cassette
    Tolerance established prior to vector administration

protein, whereas others (e.g., complete gene deletion) will be associated with an absence of any tolerance to the encoded protein. Hemophilia B, a bleeding disorder caused by an absence of functional blood coagulation Factor IX, illustrates this point well. The disease is currently treated by infusion of protein concentrates of Factor IX. The overall incidence of formation of inhibitory antibodies in response to infused protein is ~3%, but rises to ~20% for those with gene deletions, and falls to nearly zero for individuals with missense mutations.[1,2]

Antigen Processing and Presentation in Gene Therapy vs. Protein Replacement Therapy

In both gene transfer and protein replacement approaches to treatment of genetic deficiency states, the immune system is exposed to a protein that may be perceived as foreign. If the protein circulates, it will be taken up by antigen presenting cells (APCs) and presented preferentially through major histocompatibility complex (MHC) class II molecules (Fig. 1). Through cross-presentation the antigen may also have access to MHC class I pathways. Two important differences between protein replacement therapy and gene transfer approaches may influence the immune response to the therapeutic protein. First, in gene transfer the therapeutic gene product is synthesized endogenously; there may thus be more extensive presentation through class I pathways than would be the case for an infused protein. Second, the immune system is poised to respond to invading microbial insults, so that, depending on the nature of the viral or nonviral vector, cytokines may be released that influence the quality of the T cell response to the transgene product. These differences should be borne in mind when analyzing immune response to the protein (or transgene product) in the setting of genetic deficiency states.

Factors that Influence Immune Response to Transgene Product in Gene Transfer for Genetic Disease

Role of Vector in Determining Immune Response to Transgene Product

A complexity of gene transfer as a therapeutic modality is its multicomponent nature. Every gene transfer strategy is characterized by a specific combination of vector, transgene, and target tissue, and each of these specific combinations may trigger a unique response to the transgene product on the part of the immune system. The analysis of these responses is still at an early stage, but some basic findings have emerged from studies carried out in the past few years. Clearly, the vector itself, and the nature of the immune response to it, can have a profound

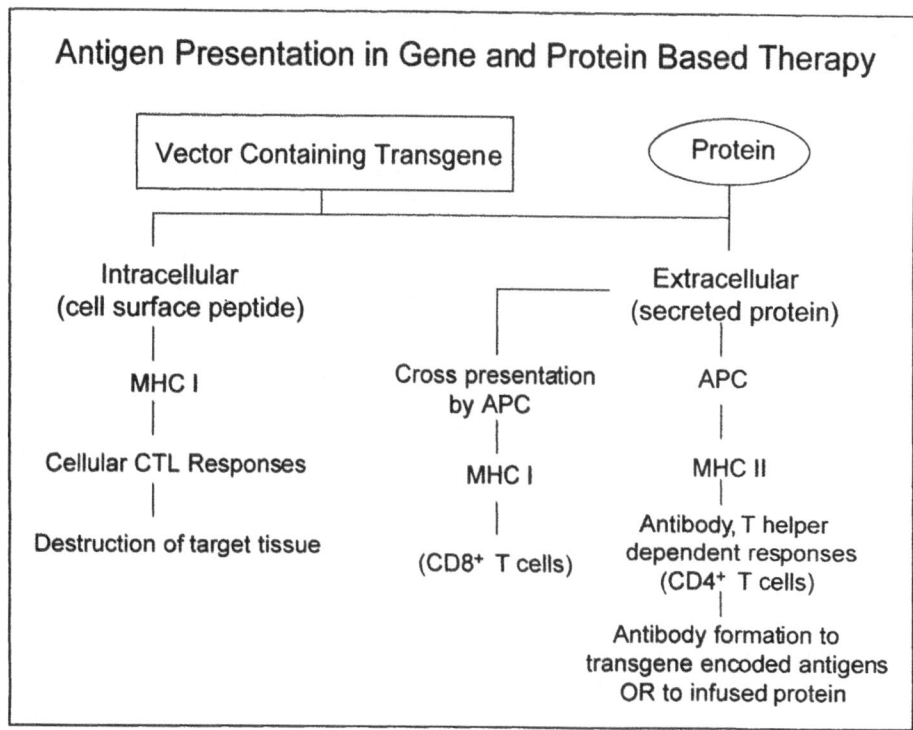

Figure 1. Antigen presentation in gene and protein based therapy. Antigen processing and presentation differ based on endogenously synthesized protein (gene based therapy) or protein that is produced exogenously and then infused (protein replacement therapy). CTL responses may occur in response to protein produced intracellularly, although this is not always the case. The factors listed in Table 1 influence whether the transgene encoded protein elicits a CTL response.

influence on the immune response to the transgene product. This has now been illustrated for a number of transgenes. Thus, a large body of work demonstrates that injection of adenoviral vectors into lung, liver, muscle, and joints leads to an inflammatory infiltrate, destruction of the target cells, and loss of transgene expression at the site of injection. These events are mediated by antigen-specific MHC class I restricted cytotoxic T lymphocyte (CTL) responses to the transgene product and to expressed viral proteins. Activation of CD4+ T cells to both viral protein and transgene product antigens also occurs, facilitating a fully competent CTL response.[3-8] Injection of AAV vectors encoding identical transgenes, however, shows stable transgene expression and an absence of CTL (or CD4+ T helper cell) response to the transgene product.[9,10] Jooss et al, in an attempt to elucidate the mechanisms underlying the vector-dependent differences in the immune response to the transgene product, carried out a series of experiments in which mice were injected with either adenoviral or AAV vector encoding the bacterial protein β-galactosidase. Intramuscular injection of AAV-lacZ into one hindlimb, followed later by Ad-lacZ injection in the other, resulted in CD4+ and CD8+ infiltration into the AAV-injected limb, accompanied by loss of transgene expression, suggesting that AAV-transduced skeletal muscle can be a suitable target for antigen-specific CTL once activated. Adoptive transfer experiments in which either dendritic cells, macrophages, or B cells from an Ad-lacZ injected mouse were infused into an AAV-lacZ injected animal, demonstrated that only dendritic cells were capable of eliciting a strong T cell response to AAV-lacZ transduced muscle fibers. Adoptive transfer of APCs from an animal transduced with an empty adenoviral vector did not cause any diminution of lacZ expression, suggesting that presentation of endogenously produced antigen was critical for the immune response. In in vitro transduction experiments, the authors demonstrated that APCs are readily transduced by adenoviral vectors but that they are relatively resistant to transduction by AAV vectors. They concluded that this difference in APC transduction was an important factor in the difference in immune response to the transgene product when these two different vectors are introduced into skeletal muscle. In these experiments, the transgene was under the control of a cytomegalovirus promoter; an interesting test of their hypothesis would involve transduction with vectors that direct tissue-specific expression (vide infra), since this would eliminate the role of expression in the APC. It is likely that cross presentation, resulting in access to class I pathways in APCs even without direct antigen synthesis by these cells, may result in a similar outcome despite the absence of expression of the donated gene in APCs.[11]

Fields et al[10] carried out a detailed analysis of immune response to the human F.IX (hF.IX) protein in C57Bl/6 normal and hemophilic mice injected at intramuscular sites with either an AAV, an adenoviral, or a plasmid vector expressing hF.IX under the control of a CMV promoter. Transduction with the AAV vector showed a modest Th2 response to the transgene product, with low levels of IL-10 secretion by lymphocytes harvested from injected mice and stimulated with hF.IX, and a hF.IX-antibody profile consistent with a Th2 driven response, i.e., a predominantly IgG1 response. Stimulation with hF.IX of lymphocytes from Ad-F.IX injected mice showed cytokine profiles consistent with activation of both Th1 and Th2 subsets, and the hF.IX-specific antibody profiles reflected this, with similar levels of IgG1, IgG2b, and IgG2c. Consistent with other studies, a strong F.IX-specific CTL response was seen in the Ad-hF.IX injected mice, while no CTL response to hF.IX was seen with lymphocytes harvested from the AAV-hF.IX-injected mice. Studies with plasmid injected mice, although less extensive, showed a strong anti-hF.IX antibody response and inflammatory infiltrates similar to those observed following adenoviral vector injection (Fig. 2). In follow-up studies, Fields et al[12] also characterized the immune response to murine F.IX following both protein infusion or gene transfer in hemophilia B knockout mice. These animals have an extensive deletion of the murine F.IX gene and no detectable F.IX transcripts.[13] Intravenous infusion of recombinant murine F.IX protein resulted in the formation of inhibitory antibodies, predominantly of the IgG1 subclass, consistent with a Th2-driven humoral immune response; intramuscular injection of an AAV vector encoding murine F.IX under the control of the CMV promoter resulted in an essentially identical profile. (Of note, hemostatically normal mice, presumably tolerant to murine F.IX, did not develop inhibitory antibodies to the protein following intramuscular injection of AAV-mF.IX, underscoring the role of tolerance to the transgene product, vide infra). Thus, the response to F.IX antigens following AAV-mediated gene transfer is comparable to that seen with protein infusion of the antigen. This is at first glance puzzling since F.IX is a neoantigen synthesized in the setting of what is essentially a viral infection. It may be that the delay in transgene expression that is characteristic of AAV-mediated gene transfer[14] results in a temporal separation of response to the vector and response to the transgene product, so that the immune response to F.IX is not influenced by the antecedent response to the viral vector. This hypothesis can be tested using a genetic "switch", in which transgene expression requires a small molecule partner that is administered orally.[15,16] In such a system, the immune response to the vector could run its course before the transgene product was expressed. Such a temporal separation of responses to the vector and the transgene product may be a key to abrogating unwanted immune responses to the latter.

Although the foregoing discussion has focused on adaptive immune responses to the transgene product, it is likely that innate immune responses triggered by the viral vector itself may influence the response to the transgene product. Delineation of these innate immune responses is still at an early stage, and specific effects on immune response to the transgene product have not been defined. The most comprehensive studies carried out thus far have examined innate immune responses to adenoviral vectors; these have been most extensively studied because they

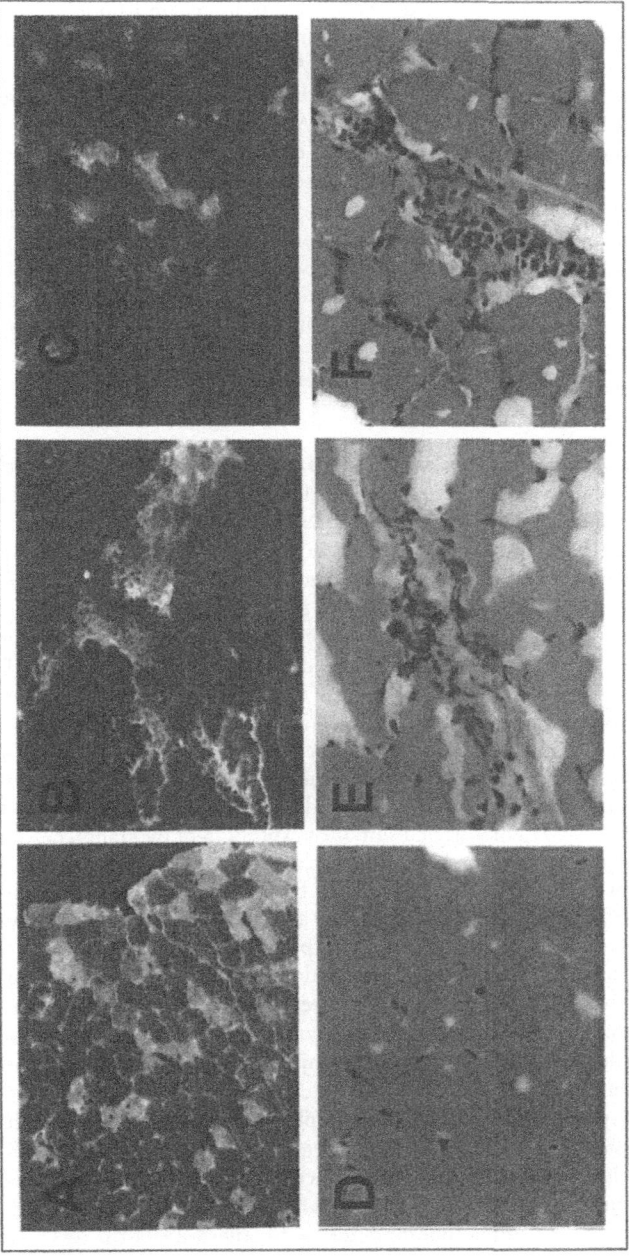

Figure 2. Histology of muscle tissue after hFIX vector administration. Cross-sections of AAV (A, D), adenovirus (B, E) and plasmid injected muscle (C, F) two weeks after vector administration into C57BL/6 mice are shown. Each vector contains a hFIX transgene. Panels A-C are immunofluorescent staining of hFIX (original manification 200x). Panels D-F are H & E staining that demonstrate the presence of inflammatory infiltrates (original magnification: 400x).

trigger a marked response associated with clinical symptoms. In studies in which nonhuman primates were injected via the portal vein with a dose (7.5 x 10^{12} vector genome [vg]/kg) of first generation adenoviral vector just below that which causes severe morbidity in these animals, early (2 hr after vector infusion) systemic release of high levels of IL-6 was documented.[17] Findings were similar in animals injected with UV-irradiated vector, demonstrating that viral gene expression was not required to trigger IL-6 release. Biodistribution studies with fluorescently labeled vector at the same early time points showed vector uptake in Kupffer cells of the liver and in macrophages and dendritic cells in the spleen, leading the authors to hypothesize that IL-6 release occurs as a consequence of activation of APCs coincident with vector uptake. The effect(s) of APC activation and cytokine release on the ensuing acquired immune response to the transgene product merits further investigation. An important question that is not yet resolved is whether fully deleted adenoviral vectors (so-called gutted adenoviral vectors) will trigger a similar innate immune response. The experiments cited above, using UV-irradiated vector, would suggest that the capsid proteins alone are adequate to trigger the innate immune response, a conclusion supported by other studies in the literature.[18] However, one cannot assume that irradiated vector is completely devoid of viral protein expression. Similarly, the fact that fever, flu-like symptoms, and thrombocytopenia developed in a human subject treated with a fully deleted adenoviral vector[19] would support the notion that viral gene expression is not required to elicit these responses, but again, such a conclusion would require careful analysis of the vector for contamination with helper adenovirus, so that this issue must be regarded as not yet settled.

Influence of Vector Dose on Immune Response to the Transgene Product

A goal of pre-clinical investigation in gene transfer is to define parameters of dosing and route of administration that confer efficacy while insuring safety. Typically, studies in animals are carried out at a range of doses, and these studies are used to define a safe starting dose in human studies. In the setting of gene transfer for genetic disease, this type of investigation has led to the finding, for some combinations of vector, transgene, and target tissue, that the likelihood of harmful immune responses to the transgene product increases with increasing dose of vector. In the case of AAV-mediated, muscle-directed gene transfer for hemophilia B, the first indication of this dose dependence was in a study of hemophilic dogs injected with a range of doses of an AAV vector expressing canine F.IX.[20] All five dogs in the study showed correction of clotting assays, but the animal treated at the highest dose (8.5×10^{12} vg/kg, 2×10^{12} vg/site) also developed an inhibitory antibody to canine Factor IX, which first appeared two weeks after injection, peaked 5-6 weeks after injection, and disappeared altogether by week 10 (Fig. 3). The animal subsequently demonstrated sustained expression of the transgene, demonstrating that the transient antibody response had not been accompanied by loss of transgene expression or substantial loss of transgene-expressing cells. Besides receiving the highest dose of vector, this animal also suffered from a skin infection at the time of injection; thus it was unclear whether the dose of vector or some other factor predisposed to antibody formation. In follow-up studies, hemophilic dogs with the same mutation underwent intramuscular injection of similar doses of vector. A hemophilic dog injected with vector using the same parameters outlined above developed a transient noninhibitory antibody response and subsequently showed sustained expression of the transgene. Another animal injected with a slightly higher dose in vg/kg (1.1×10^{13} vg/kg) but a six-fold higher dose/site (1.2×10^{13} vg/site) developed a high-titer, long-lasting inhibitory antibody response.[21] The inhibitory antibodies eventually disappeared but Factor IX expression was never documented, suggesting that the transduced cells had been destroyed. This conclusion was supported by evidence of a strong T cell proliferative response to canine Factor IX in peripheral blood mononuclear cells (PBMCs) harvested from this animal. (Additional studies showed that noninhibitory antibodies were of the IgG2 subclass but that animals with inhibitory antibodies showed both IgG1 and IgG2 subclasses [note that these do not correspond to antibody subclasses in humans or mice]). Similarly, in experiments where dogs with a different underlying mutation (a nonsense mutation in the 5' end of the canine F.IX gene[22]) were treated via mesenteric vein infusion with an AAV vector expressing canine F.IX under the control of a liver-specific promoter, there was a dose-dependence to the likelihood of inhibitor formation, with animals treated with 1×10^{12} vg/kg expressing high levels of canine F.IX and showing no evidence of antibody formation (Fig. 3), while a dog

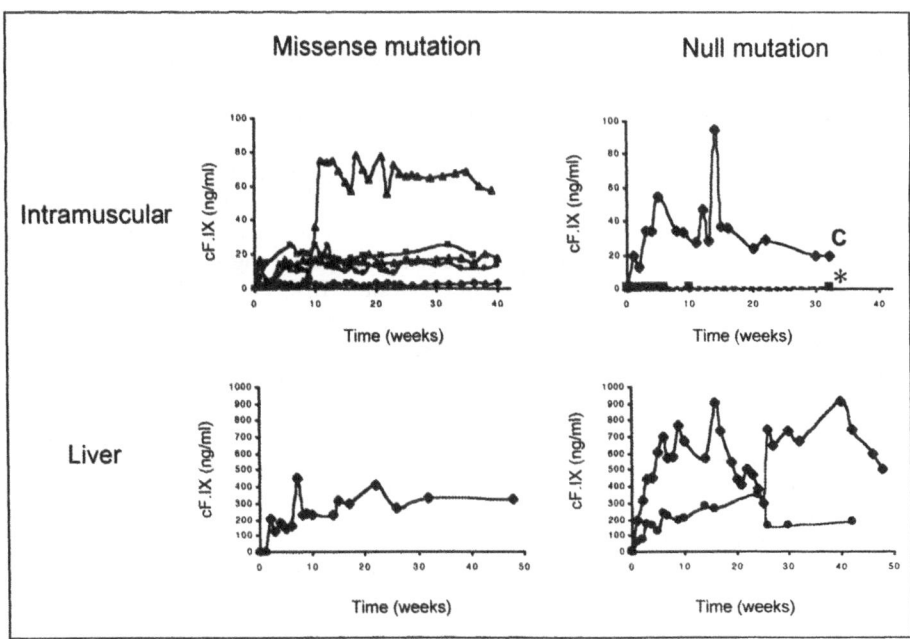

Figure 3. Levels cF.IX in hemophilia B dogs after administration of AAV-cF.IX. The levels of cF.IX after gene transfer are shown in hemophilia B dogs with a missense mutation of a null mutation. Dogs with each mutation were treated with intramuscular or liver directed administration of AAV-cF.IX. Each graph shows the levels or cF.IX (ng/ml) over time (weeks). The solid lines indicated that no inhibitory antibodies were observed. The broken lines identified by * indicate that inhibitory antibodies were observed (n=2). The C indicated that the animal was treated with immunosuppressive drug cyclophosphamide biweekly beginning on the day of vector administration through week 6.

treated with 3×10^{12} vg/kg developed a high titer inhibitor to F.IX.[23] This result must be interpreted cautiously, since the dog treated at the highest dose also had liver disease secondary to iron overload associated with (an unrelated) hereditary hemolytic anemia, but nonetheless underscores the importance of caution during dose escalation in human trials of gene transfer for genetic disease.

Experience with a variety of transgenes and vectors increasingly supports the notion of a dose-dependence to the likelihood of immune response to the transgene product, but has not yet yielded substantial insights into the potential mechanisms for these findings. Hypotheses that need to be explored include:

1. whether risk of forming inhibitory antibodies is related to total number of viral particles injected. This may relate to issues of innate immunity to viral particles (vide supra). The finding that alternate serotypes of AAV can result in much higher levels of transgene expression per injected particle dose should facilitate investigation of this possibility.

2. whether some contaminant in the viral prep acts as an adjuvant.

3. whether more efficient transduction of APCs occurs at higher doses, although this would not explain results seen with a tissue-specific promoter.

4. whether the only parameter of importance is local levels of transgene (antigen) production. The mechanism in this case may involve better access to class I presentation pathways in APCs (cross presentation) at higher antigen levels.

Role of Underlying Mutation

Of interest, it is clear in the case of gene transfer in genetic disease that the underlying mutation, and the resulting degree of tolerance to the transgene product, plays some role in defining doses of vector that can be administered without triggering antibody formation. Thus normal mice can be injected intramuscularly with doses of 4 x 10^{12} vg/kg of AAV-murine F.IX without developing antibodies, whereas hemophilia B mice, with a large deletion of the F.IX gene, readily develop inhibitory antibodies at the same dose.[24] Similarly, the experiments described in the foregoing paragraph, carried out in dogs with a missense mutation in the canine F.IX gene, defined parameters of dose/kg and dose/site that did not trigger inhibitory antibody formation following intramuscular injection of AAV-canine F.IX (Fig. 3). However, when the same parameters were used to perform intramuscular injections of dogs with a nonsense mutation at the 5' end of the canine F.IX gene,[22] the dogs developed high-titer, long-lasting inhibitory antibodies, which occurred even when the dose/site and the dose/kg were lowered to one-third the doses used in the missense mutation dogs (1.2 x 10^{12} vg/kg and 6.3 x 10^{11} vg/site) (Fig. 3).[25] Clearly, the availability of a wider range of animal models of human disease, that more comprehensively reflect the types of mutations that can be encountered in human disease, would facilitate evaluation of the role of this parameter.

Route of Administration of Vector

As gene transfer strategies are developed, it is becoming clear that the route of administration can dramatically affect the immune response. Several studies have directly compared different routes of administration using the same vector. A single dose of AAV-Ovalbumin was administered into normal C57BL/6 mice using four different routes: intraperitoneal, intravenous, subcutaneous, and intramuscular. An anti-ovalbumin and anti-AAV antibody response was observed with all modes of delivery while a strong cytotoxic T cell response specific to ovalbumin was observed only with intraperitoneal, subcutaneous and intravenous routes. This suggests that some routes of delivery may facilitate a cellular immune response more efficiently than others while a humoral response may be less dependent on the route of administration.[26]

Since Factor IX can be expressed from several tissues (i.e., liver or muscle), comparisons of different delivery routes have been made with F.IX in animal models. AAV-hF.IX using either a chimeric CMV/Moloney murine leukemia virus promoter/enhancer or CMV enhancer/β-actin promoter delivered intravenously or via the portal vein into liver in immune competent mice express high levels of hF.IX without developing anti-hF.IX antibodies while intramuscular administration reliably elicits an anti-hF.IX antibody response resulting in no detectable hF.IX.[27,28] These patterns of immune response are observed at low doses of AAV-hF.IX.[27] AAV-hF.IX with a ubiquitously expressed eukaryotic promoter (EF1α) delivered by splenic capsule injection into several strains of normal mice (CD1, C3H, C57Bl/6, BALB/c), demonstrated that only one strain developed antibodies to the hF.IX (CD1) given through this route while all strains develop anti-hF.IX antibodies to intramuscularly injected AAV-hF.IX.[14,29] This again shows differences in immune response based on route of administration but also emphasizes that there can be differences in immune response among different strains of mice. Those mice that did not develop antibodies to the liver delivered vector remain tolerant or only develop low titer antibodies when challenged with hF.IX in complete Freund's adjuvant.[29] Hemophilia B dogs that have a null mutation[22] and develop antibodies to intramuscularly administered AAV-CMV-cF.IX do not develop anti-cF.IX antibodies to AAV-ApoE/hAATenh-cFIX (a liver-specific promoter) administered via the portal or mesenteric vein (Fig. 3).[23]

While it is unclear why there are different immune responses to intramuscular and hepatic gene transfer, it may be related to the cellular components of each tissue. There is evidence that the liver has cells that may facilitate the induction of tolerance. For example, liver sinusoidal endothelial cells have been found to induce antigen specific CD8+ T cell tolerance.[30]

Tissue Specificity of the Promoter in the Vector Cassette

Another factor that has been demonstrated to affect the immune response to the transgene product is the tissue specificity of the promoter that is used to direct expression of the transgene. Many gene transfer studies utilize ubiquitous viral (e.g., cytomegalovirus [CMV]) or eukaryotic (e.g., phosphoglycerate kinase 1 [PGK]) promoters, which are expressed at high levels in many tissues. While in many cases the vector is used to target a particular tissue, the vector is potentially able to transduce multiple cell types including dendritic cells and macrophages that then express the transgene and present the antigen to the immune system. While APCs can also take up antigen and present it to the immune system, transduction and expression of a transgene in APCs efficiently enhances the activation of the immune response. This results in an immune response to the transgene that eliminates the effectiveness of the transgene product. There is evidence that adenoviruses, plasmid vectors and retroviruses are able to transduce immature dendritic cells by binding to Toll-like receptors. This drives maturation of dendritic cells that express the antigen and activate T and B cells. AAV vectors fail to efficiently transduce dendritic cells.[9,31] Jooss et al presented evidence that AAV enters the dendritic cell but does not result in expression of the transgene. Another hypothesis may be that AAV fails to induce maturation of the dendritic cell. This is an area that requires further investigation.

Work in mouse models of muscular dystrophy underscores the importance of the tissue specificity of the promoter. In these mouse models the number of antigen presenting cells in dystrophic muscle is significantly higher than in healthy muscle, making tissue specific, i.e., muscle specific expression, critical to minimizing the immune response to the transgene. In a mouse model of limb-girdle muscular dystrophy, in which γ-sarcoglycan (γSG) is deleted, it was observed that AAV-CMV-γSG had low levels of expression and induced a humoral immune response to the γSG but there was no evidence of a CTL response.[32] The same construct with the muscle specific creatine kinase promoter resulted in high levels of stable expression and no evidence for a humoral response.[32] Whether the same findings would be seen in healthy muscle tissue is not yet clear.

In the Duchenne muscular dystrophy mouse model (mdx), an adenovirus expressing β-galactosidase either under the control of the CMV promoter or the muscle specific promoter CK6 (a version of the mouse muscle creatine kinase promoter) demonstrated similar results.[11] A CTL response to β-galactosidase was detected with the CMV vector but not the CK6 vector; however, a CTL response to the adenovirus vector caused a loss of transgene expression over time.[11]

Models of gene transfer into liver have also demonstrated the importance of tissue specific expression despite suggestions that the liver itself may be more prone to tolerance induction. Studies using an adenoviral vector demonstrated that liver specific promoters such as the hAAT or the mouse albumin promoter did not generate an immune response to an α1 antitrypsin transgene product while a vector with the ubiquitously expressed phosphoglycerate kinase 1 (PGK) promoter resulted in formation of antibodies to the hAAT protein.[33] Whether these findings can be generalized to other vectors is unclear; the powerful innate immune response to the adenoviral vector (vide supra) on the one hand, and the altered immune cell population in dystrophic muscle, on the other hand, may result in circumstances where the tissue-specificity of the promoter in the vector assumes special importance. Further studies will be required to clarify this point.

Tolerance Regimens Prior to Vector Administration

As an alternative approach to preventing an immune response to the transgene product, it may be possible to tolerize an individual to the transgene protein prior to administering a gene therapy. There has been significant interest in inducing immune tolerance for autoimmune diseases and a growing interest in addressing tolerance to transgene products as effective gene transfer becomes feasible. The concept of tolerance is to induce hyporesponsiveness to the antigen by suppressing T cell mediated or antibody mediated responses to a specific antigen.

Oral tolerance has been extensively studied for use in autoimmune diseases including multiple sclerosis, experimental allergic encephalomyelitis, and rheumatoid arthritis. Several proposed mechanisms of oral tolerance depend on the dose of antigen. Upon oral administration of antigen, the antigen interacts with gut-associated lymphoid tissue that contains Peyer's patches consisting of M cells, macrophages, dendritic cells, and B and T lymphocytes. At low doses of antigen, the Th2 cells and Th3 cells are induced to secrete IL-4, IL-10 and TGF-β that are cytokines that are associated with immune suppressive responses. At high doses of antigen, some antigen may enter the circulation and result in clonal deletion or anergy of Th1 and Th2 cells.[34]

There has been success in using oral tolerance in animal models of autoimmune disease including experimental allergic encephalomyelitis and collagen-induced arthritis;[35] however, it has been difficult to induce oral tolerance for autoimmune disease in humans. The effectiveness of oral tolerance may depend on the disease, dose of antigen and the type of protein that is administered. Failure to induce tolerance in human autoimmune disorders may relate to the difficulty in overcoming or tolerizing the immune response to an antigen when it has already developed. Genetic diseases such as hemophilia that may be treated with a protein therapy or gene therapy may be less difficult as models for inducing oral tolerance, since the individual could be tolerized to the protein prior to administration of protein or gene therapy. Studies in oral administration of human Factor IX in neonatal hemophilia B and wild type mice have demonstrated that it may be possible to induce tolerance to human F.IX prior to administration of gene therapy.[36] Studies of oral administration of human F.IX in adult wild type mice or hemophilia B mice show that the immune response to human F.IX administered in complete Freund's adjuvant is significantly decreased.[37]

Another approach to induce tolerance to a specific antigen has been to use a nasal administration route. Nasal administration of antigen utilizes administration of short peptides with T cell epitopes, preferably immunodominant epitopes, which are able to move across the epithelium more easily than full-length proteins. This has been demonstrated in animal models for autoimmune diseases associated with CD4+ T cell responses such as collagen induced arthritis[38] as well as diseases characterized by antibody mediated responses such as experimental myasthenia gravis..[39]

To induce specific T cell tolerance, systemic administration of anti-CD4 monoclonal antibody and a chimeric immunoglobulin containing the immunodominant epitope of a transgene peptide has been studied in a mouse model. BALB/c mice that received anti-CD4 and an immunoglobulin with influenza virus hemaglutatinin treatment and were then given a gene therapy using intramuscular administered AAV-hemagluatinin did not develop an immune response to the antigen. T cells isolated from lymph nodes did not respond to hemagluatinin in in vitro stimulation assays; these results support T cell tolerance induction. However, it is not clear whether this tolerance is long term in mice.[40]

Role of Immunosuppression at the Time of Vector Administration

Another strategy that has been used to avoid an immune response following gene transfer is administration of immune suppressive therapy at the time of vector administration. Ideally, an immune suppressive approach would be specific for the transgene antigen, however, many of these agents act nonspecifically to impede the immune response.

Various transient immunosuppressive regimens were compared in a model system for gene therapy for hemophilia B. Some of these agents, used extensively in organ transplantation, act by inhibiting T or B cell proliferation; others block differentiation of T or B cells. Cyclosporin A, FK506 and cyclophosphamide were each administered separately at the time an AAV-hF.IX vector was injected at four intramuscular sites in Hemophilia B mice with a large mF.IX deletion. Previous studies with these mice demonstrated that within 2 to 3 weeks after vector administration the mice develop anti-hF.IX neutralizing antibodies. Cyclosporin A, which is a general immune suppressive drug, had no effect on eliminating anti-hF.IX antibodies and caused toxicity that was sometimes lethal. FK506 (Tacrolimus), which acts to inhibit T cell activation, prevented antibody formation to the transgene product during administration of the drug. However, once the drug was discontinued at six weeks post vector injection, neutralizing hF.IX antibodies developed and the levels of hF.IX in the circulation decreased. Cyclophosphamide (Cytoxan), a DNA alkylating agent that is cytotoxic to dividing cells, was also studied in this model. In clinical practice this drug is used extensively for myeloblation prior to bone marrow transplantation and for immune suppression during organ transplanation or treatment of autoimmune diseases. Older literature had documented that transient immunosuppression at the time of neoantigen exposure in mice could result in tolerance. The mechanism of tolerance induction by cyclophosphamide in this setting is unclear, since evidence exists supporting both clonal deletion of antigen specific lymphocytes and induction of T cell anergy.[41] By administering cyclophosphamide every two weeks beginning on the day of vector injection and continuing up to six weeks post vector injection, most of the mice expressed the transgene and showed long term a correction of clotting function.[24] Notably, antibodies to AAV were observed after vector administration in these mice so that cyclophosphamide treatment did not prevent an immune response to the vector capsid. Cyclophosphamide treatment at the time of intramuscular administration of AAV vector was also studied in a hemophilia B dog model that has a null mutation. Without immune suppression treatment these dogs develop neutralizing antibodies to the canine F.IX transgene. When cyclophosphamide was administered at the time of AAV-cF.IX injection, cF.IX antibodies did not develop and long term cF.IX expression was ensued (Fig. 3).[25]

Strategies using agents that block the costimulatory signals that initiate the immune response were also analyzed in the setting of gene transfer in the hemophilia B mouse model. This approach interferes with the CD40-CD40Ligand (CD40L) costimulatory pathway, which is important for the antigen specific activation of T cells by antigen presenting cells, by using anti-CD40L antibody to block the pathway and, therefore, T cell activation. These signals are also required for CD4$^+$ T helper dependent B cell activation (Fig. 4). The treatment was administered beginning three days before intramuscular AAV vector administration extending through day 9 after AAV injection. Following AAV-mediated transfer in the hemophilia B mouse model system, anti-CD40L treatment resulted in immune responses that were delayed or weakened compared to untreated animals, but the treatment did not result in stable gene expression.[24] Similar results were obtained in studies using AAV-hemaglutinin (HA) or Ad-HA administration in skeletal muscle in BALB/c mice with anti-CD40L treatment. T cells from T cell receptor (TCR)-HA mice that express α/β TCR specific for an influenza virus HA peptide were adoptively transferred into the BALB/c mice prior to administration of virus vector and anti-CD40L treatment. While anti-CD40L treated mice did not have any transgene expression from the Ad-HA vector, AAV-HA injected mice showed an improvement in duration of expression although this was eventually lost. This may have been due to the inability to silence new antigen specific T cells.[42]

Interference with the B7-CD28 costimulatory pathway may be another approach that can disrupt the activation of T cells (Fig. 4). The effect of the CTLA-4-Ig soluble fusion protein, which interferes with the B7-CD28 pathway, was tested in the hemophilia B gene transfer model by administering antibody beginning three days before intramuscular injection of AAV-hF.IX and continuing through day 9 after AAV injection. This treatment resulted in transgene expression and functional correction of the clotting activity. However, the immune suppression was transient since the transgene expression diminished and hF.IX antibodies appeared over time.[24] The CTLA-4-Ig has also been used to determine if the immune response to adenovirus vectors can be blocked.[43] These vectors generate CTL immune responses to the vector and cells containing the vector that result in destruction of the transduced cells as well as activation of B cells that results in development of neutralizing antibodies. An E1-deleted adenovirus vector containing a CMV promoter-lacZ gene was administered to lung or to liver while CTLA-4-Ig was administered on days 0 to 4. In the lung transgene expression was observed through day 14 after vector administration and then transgene expression was almost

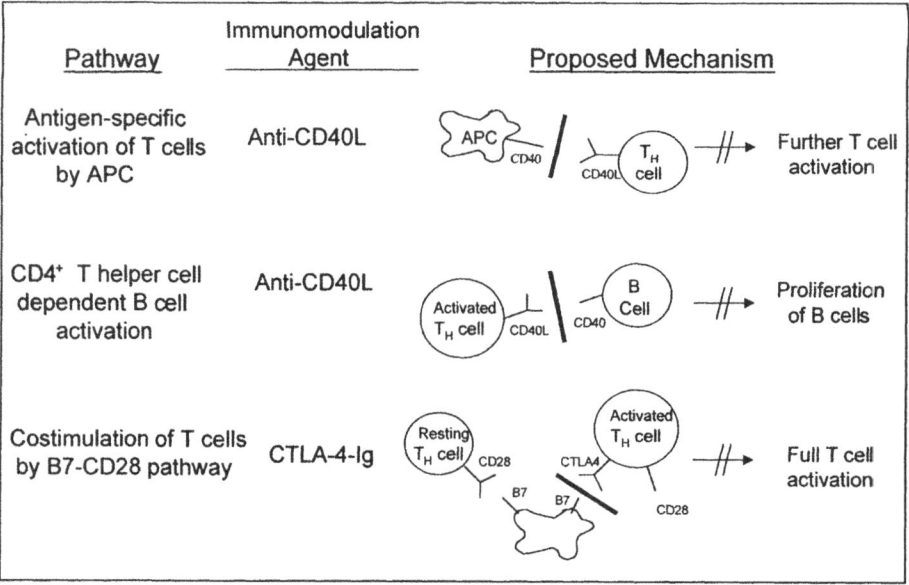

Figure 4. Proposed mechanisms of imunomodulation strategies.

undetectable by day 28. Since low levels of neutralizing antibody formation were observed after adenovirus administration in the lung, readministration of vector using a Ad-CMV/ β-actin-alkaline phosphatase was possible 4 to 6 weeks after the initial vector administration. The liver targeted vector developed neutralizing Ad-antibodies but transgene expression was detectable through 28 days after vector administration, which was longer than transgene expression observed in the lung. These studies suggest that CTLA-4-Ig reduced T cell activation.[43] While suppressing the immune system at the time of vector administration may be sufficient to obtain long term transgene expression, further studies will be required to determine the treatment schedule that will lead to success using these types of approaches.

References

1. Giannelli F, Green PM, Sommer SS et al. Haemophilia B: database of point mutations and short additions and deletions, 7th edition. Nucl Acids Res 1997; 25:133-135.
2. High KA. Factor IX: Molecular structure, epitopes, and mutations associated with inhibitor formation. Aledort LM, Hoyer LW, Lusher JM et al, eds. Vol. 386. New York: Plenum Press, 1995:79-86.
3. Tripathy SK, Black HB, Goldwasser E et al. Immune responses to transgene-encoded proteins limit the stability of gene expression after injection of replication-defective adenovirus vectors. Nature Med 1996; 2:545-550.
4. Yang Y, Ertl HCJ, Wilson JM. MHC class I-restricted cytotoxic T lymphotyces to viral antigens destroy hepatocytes in mice infected with E1-deleted recombinant adenoviruses. Immunity 1994; 1:433-442.
5. Yang Y, Xiang Z, Ertl HCJ et al. Cellular and humoral immune responses to viral antigens create barriers to lung-directed gene therapy with recombinant adenoviruses. J Virol 1995; 69:2004-2015.
6. Yang Y, Su Q, Wilson JM. Role of viral antigens in destructive cellular immune responses to adenovirus vector-transduced cells in mouse lungs. J Virol 1996; 70:7209-7212.
7. Yang Y, Haecker SE, Su Q et al. Immunology of gene therapy with adenoviral vectors in mouse skeletal muscle. Human Mol Genet 1996; 5:1703-1712.
8. Zsengeller ZK, Wert SE, Hull WM et al. Persistence of replication-deficient adenovirus-mediated gene transfer in lungs of immune-deficient (nu/nu) mice. Human Gene Ther 1995; 6:457-467.
9. Jooss K, Yang Y, K.J. F et al. Transduction of dendritic cells by DNA viral vectors directs the immune response to transgene products in muscle fibers. J Virol 1998; 72:4212-4223.

10. Fields PA, Kowalczyk DW, Arruda VR et al. Role of vector in activation of T cell subsets in immune responses against the secreted transgene product factor IX. Mol Ther 2000; 1:225-235.
11. Hartigan-O'Connor D, Kirk CJ, Crawford R et al. Immune evasion by muscle-specific gene expressino in dystrophic muscle. Mol Ther 2001; 4:525-533.
12. Fields P, Armstrong EA, J.N. H et al. Intravenous administration of E1, E3-deleted adenoviral vector induces tolerance to a Factor IX transgene in C57Bl/6 mice. Gene Ther 2001; 8:354-361.
13. Lin HF, Maeda N, Smithies O et al. A coagulation factor IX-deficient mouse model for human hemophilia B. Blood 1997; 90:3962-3966.
14. Herzog R, Hagstrom N, Kung S et al. Stable gene transfer and expression of human FIX following intramuscular injection of recombinant AAV. Proc Natl Acad Sci USA 1997; 94:5804-5809.
15. Furth PA, St. Onge L, Boger H et al. Temporal control of gene expression in transgenic mice by a tetracycline-responsive promoter. Proc Natl Acad Sci USA 1994; 91:9302-9306.
16. Kistner A, Gossen M, Zimmerman F et al. Doxycycline-mediated quantitative and tissue-specific control of gene expression in transgenic mice. Proc Natl Acad Sci USA 1996; 93:10933-10938.
17. Schnell MA, Zhang Y, Tazelaar J et al. Activation of innate immunity of nonhuman primates following intraportal administration of adenoviral vectors. Mol Ther 2001; 3:708-722.
18. Kafri T, Morgan D, Krahl T et al. Cellular immune response to adenovrial vector infected cells does not require de novo viral gene expression: implications for gene therapy. Proc Natl Acad Sci USA 1998; 95:11377-11387.
19. Fang X, Andreason G, Hariharan M et al., Pre-clinical efficacy and safety studies of a gutless adenovirus vector (MaxADFVIII) for treatment of hemophilia A., XVIII ISTH Meeting, Paris, France, 2001, p. Abstract #OC2490.
20. Herzog RW, Yang E, Couto L et al. Long-term correction of canine hemophilia B by AAV-mediated gene transfer of blood coagulation factor IX. Nature Med 1999; 5:56-63.
21. Herzog RW, Fields PA, Arruda VR et al. Characterization of B and T cell responses against factor IX in AAV vector-based gene therapy for canine hemophilia B. Mole Ther 2000; 1:S27.
22. Mauser AE, Whitney KM, Lothrop CD Jr. A deletion mutation causes hemophilia B in Lhasa Apso dogs. Blood 1996; 88:3451-3455.
23. Mount JD, Herzog RW, Tilson DM et al. Sustained phenotypic correction of hemophilia B dogs with a factor IX null mutation by liver-directed gene therapy. Blood 2002; in press.
24. Fields PA, Arruda VR, Armstrong E et al. Risk and prevention of anti-factor IX formation in AAV-mediated gene transfer in context of a large factor IX gene deletion. Mol Ther 2001; 4:201-210.
25. Herzog RW, Mount JD, Arruda VR et al. Muscle-directed gene transfer and transient immune suppression result in sustained partial correction of canine hemophilia B caused by a null mutation. Mol Ther 2001; 4:192-199.
26. Brockstedt DG, Podsakoff GM, Fong L et al. Induction of immunity to antigens expressed by recombinant adeno-associated virus depends on the route of administration. Clin Immunol 1999; 92:67-75.
27. Ge Y, Powell S, Van Roey M et al. Factors influencing the development of an anti-factor IX (FIX) immune response following administration of adeno-associated virus-FIX. Blood 2001; 97:3733-3737.
28. Nathwani AC, Davidoff A, Hanawa H et al. Factors influencing in vivo transduction by recombinant adeno-associated viral vectors expressing the human IX cDNA. Blood 2001; 97:1258-1265.
29. Mingozzi F, Arruda VR, Liu Y-L et al. Induction of immunological tolerance to a coagulation factor antigen by hepatic gene transfer. Blood 2001; 98(Suppl):694a.
30. Limmer A, Ohl J, Kurts C et al. Efficient presentation of exogenous antigen by liver endothelial cells to CD8+ T cells results in antigen-specific T-cell tolerance. Nat Med 2000; 6:1348-1354.
31. Chang LJ, He J. Retroviral vectors for gene therapy of AIDS and cancer. Curr Opin Mol Ther 2001; 3:468-475.
32. Cordier L, Gao G-P, Hack AA et al. Muscle-specific promoters may be necessary for adeno-associated virus-mediated gene transfer in the treatment of muscular dystrophies. Hum Gene Ther 2001; 12:205-215.
33. Pastore L, Morral N, Zhou H et al. Use of a liver-specific promoter reduces immune response to the transgene in adenoviral vectors. Hum Gene Ther 1999; 10:1773-1781.
34. Weiner HL. Oral tolerance: immune mechanisms and treatment of autoimmune diseases. Immunol Today 1997; 18:335-343.
35. Khare SD, Krco CJ, Griffiths MM et al. Oral administration of an immunodominant human collagen peptide modulates collagen-induced arthritis. J Immunol 1995; 155:3653-3659.
36. Sabatino DE, Arruda VR, Liu Y-l et al. Generation of whey acidic protein promoter-human factor IX transgenic mice for studies of oral tolerance to human factor IX in neonatal hemophilia B mice. Blood 2001; 98(Suppl):825a.

37. Alpan O, Kamala T, Velander W et al. Milky Way: An inexpensive and easy method of inducing tolerance to human F.IX. Blood 2001; 98 (Suppl):825a.
38. Staines NA, Harper N, Ward FJ et al. Mucosal tolerance and suppression of collagen-induced arthritis (CIA) induced by nasal inhalation of synthetic peptide 184-198 of bovine type II collagen (CII) expressing a dominant T cell epitope. Clin Exp Immunol 1996; 103:368-375.
39. Karachunski PI, Ostlie NS, Okita DK et al. Prevention of experimental myasthenia gravis by nasal administration of synthetic acetylcholine receptor T epitope sequences. J Clin Invest 1997; 100:3027-3035.
40. Lanoue A, Bona C, von Boehmer H et al. Conditions that induce tolerance in mature CD4+ T cells. J Exp Med 1997; 185:405-414.
41. Mayumi H, Umesue M, Nomoto K. Cyclophosphide -induced immological tolerance; an overview. Immunobiology 1996; 195:129-139.
42. Sarukhan A, Camugli S, Gjata B et al. Successful interference with cellular immune responses to immunogenic proteins encoded by recombinant viral vectors. J Virol 2001; 75:269-277.
43. Jooss K, Turka L, Wilson JM. Blunting of immune responses to adenoviral vectors in mouse liver and lung with CTLa4Ig. Gene Ther 1998; 5:309-319.

CHAPTER 14

The Use of DNA Vaccines for Neonatal/Early Life Childhood Immunization

Jiri Kovarik, Xavier Martinez and Claire-Anne Siegrist

Abstract

DNA vaccines could represent a major advancement in the development of novel antigen-delivery systems to be used in early life, although the demonstration of their efficacy and safety in human adults must be awaited before any prediction on the future of such vaccines in early life can be made. Their capacity to efficiently induce adult-like neonatal T helper (Th)1-type and cytolytic T lymphocyte (CTL) responses to encoded antigens in several animal species could be particularly beneficial in view of the preferential Th2-polarization and weak CTL activity frequently observed at this young age. In contrast, their frequent failure to induce early life vaccine antibody responses above those elicited by live attenuated/adjuvanted protein vaccines, remains a significant limitation. This feature calls for much more complex vaccine strategies (i.e., prime-boost approaches), that may not prove feasible in countries that most need novel DNA vaccines. Similarly, the expectation that the in vivo antigen production would allow DNA vaccines to readily escape from the inhibitory influence of maternal antibodies has not been confirmed, despite the reported success in a few experimental settings. Notably, considerable heterogeneity in the quantity and quality of early life T and B cell responses to DNA immunization has been observed. These are reviewed here, underlining that a better understanding of the respective roles of the routes and methods of administration, the doses and types of antigen(s) and constructs, the co-administration of antigens or immunomodulators, the ages of recipients and the titers of pre-existing maternal antibodies is clearly needed. DNA immunization models in various animal species should thus be used to progressively identify the essential parameters and characteristics of the novel DNA vaccine candidates which may be considered for use in early life.

The Need for Novel Antigen-Delivery Systems for Early Life Immunization

The Susceptibility of Neonates to Infections: What Is the Vaccine Need?

The prevention of severe infections by routine vaccination represents worldwide the most promising approach to decreasing the very high mortality rate in newborns and young infants. Particularly during the first 12 months of life, infants are likely to develop severe or persistent infectious diseases. Thus the major goal for novel early life vaccines, in addition to optimal safety, is the rapid induction of protective and long lasting immunity against the target pathogen(s). There is a rather large variety of pathogens that can cause severe infections in early life, including bacterial, viral, parasitic or fungal organisms or their products.[1,2] Nevertheless, a rather limited number of viral and bacterial agents, whose spectrum varies depending upon the level of industrialization, cause the majority of early deaths and are thus primary targets for

DNA Vaccines, edited by Hildegund C. J. Ertl. ©2003 Eurekah.com
and Kluwer Academic / Plenum Publishers.

vaccine development. Routine immunization against some common infant pathogens, such as pertussis, haemophilus influenzae type B (Hib), poliomyelitis, hepatitis B, tuberculosis and measles, is already applied in many countries and represents an effective method of prevention. However, it frequently requires either repeat vaccine administration or delayed immunization, as with live-attenuated measles vaccines that remain poorly immunogenic before the age of 9 months, in addition to being most susceptible to inhibition by antibodies of maternal origin.[3] The immune defense against other early infant pathogens, such as Respiratory Syncytial Virus (RSV), would clearly benefit from vaccines capable of rapidly inducing strong and sustained neutralizing neonatal antibody responses. But clearly the main expectation of the application of DNA vaccines in early life would be to generate rapid protection against important pathogens, such as malaria, tuberculosis and HIV, which require the induction of strong cellular immune responses.

What Are the Immunological Challenges to be Met by DNA Vaccines in Early Life?

A) Limitations of Th1 Responses in Early Life

The increased susceptibility of neonates and young infants to infectious disease is a direct consequence of the immaturity of their immune system. The effectiveness of major components of innate immunity, such as neutrophils, the complement system, natural killer (NK) cells, monocytes/macrophage and dendritic cells (DCs), appears reduced in newborns and infants.[4] Of particular importance, a limited production of interleukin (IL)-12 by early life antigen presenting cells (APCs) was identified in mice[5-7] and confirmed in human infants,[8,9] which could have a main impact on the development of subsequent adaptive responses. As a probable consequence of differences in early life APC activation properties, neonatal and infant T cells seem to follow a different pathway of differentiation, to that of adult T cells. In infants, cord blood derived T cells are characterized by spontaneous in vitro production of IL-4 when compared to their adult counterparts[10] and rather low levels of interferon (IFN)-γ are produced by neonatal T cells.[11] This preferential differentiation into Th2-type populations of early life T cells is similarly observed in mice : T cells derived from neonatal mice produce less IFN-γ and more IL-4/IL-5 than adult T cells.[12]

As could be expected, limitations of early life T- cell responses to pathogens also apply to vaccine responses. When restimulated with measles antigen, measles vaccine-specific T cells obtained from human infants were shown to secrete less IFN-γ than T cells harvested from adults.[13] Similarly, an initial preferential Th2 polarization of human infant responses to acellular pertussis vaccine was demonstrated by Rowe et al.[14] In contrast, studies demonstrated the capacity of infant T cells to raise both Th1-like (IFN-γ and IL-2) and Th2-like (IL-5) responses after 3 doses of pertussis vaccine.[15,16] It was thus concluded that T cell responses to acellular pertussis vaccine are characterized by mixed Th0 or Th1/Th2 responses in infants, whereas a preferential Th2 cytokine pattern is observed in mice.[17] Indeed, the preferential Th2-type cytokine production of vaccine-specific T cells is often more marked in mice, which secrete very high levels of IL-4/IL-5 and little or no IFN-γ in response to live viral, peptide or protein vaccines,[7,18] than in human infants. This early life Th2 polarization of CD4 T cells goes along with an impaired induction of CTL responses to most, but not all vaccines.[7,18,19] Certain vaccines such as BCG, however, elicit preferential IFN-γ production both in human neonates[20] and in mice.[18] Thus, in humans as in mice, certain vaccines may elicit early life Th1 responses. However, none of the infant studies yet address the important question of whether the Th1/Th2 polarization of T cell responses elicited in infants is similar—or not—to the one induced in immunologically mature adults. This question requires comparison of T cell responses to a given vaccine after neonatal/adult immunization, and such studies have only been initiated recently and are still ongoing. To add a level of complexity to the issue, observations made initially in mice[12,21,22] and recently confirmed in infants[14,23] suggest, that a secondary loss of

Th1 immunity and the preferential induction of Th2 memory cells may occur after early life vaccination. Thus, the first major challenge for novel vaccines to be used in early life is the rapid induction of strong, adult-like Th1 effector and memory responses.

B) Limitations of CTL Responses in Early Life

In parallel to the limited Th1 responses, the efficiency in the generation of CTL responses after human immunodeficiency virus (HIV) and RSV infections appears to be age dependent, although some CTL activity may be detected within the first weeks of life following infections in human infants.[24,25] Furthermore, measles-specific CTL have been observed following immunization of 9-month-old infants.[26] However, comprehensive age-dependent studies of neonatal or infant CTL responses to vaccines (which may be quite different from those elicited by chronic infection) are not yet available. The limited generation of Th1-type responses, in association with weak CTL activity, is likely to contribute to the increased early life susceptibility to intracellular pathogens, and the early induction of CTL responses represents a second major challenge for novel early life vaccines.

C) Limitations of Antibody Responses in Early Life

In addition to the limitations of T cell responses, infant B cell responses to infections show significant age dependent limitations, such as reduced antibody responses and limited antibody repertoire diversity, possibly associated with lower antibody avidity.[27,28] In comparison to adults, human infant antibody responses to infections are characterized by a strong predominance of IgG1 and IgG3 subclasses, whereas the generation of IgG2 antibodies remains weak during the first 18 months of life.[29] This lack of IgG2 production could reflect the preferential induction of Th2 versus Th1 responses,[30] although correlation between antibody isotypes and cytokine production has not yet been established as solidly as in mice. These limitations, together with the inability in young infants to respond to most T-independent antigens (i.e., bacterial cell wall polysaccharides), suggest the existence of a significant delay in the maturation of the B cell compartment.

As expected, the magnitude of vaccine-specific antibody responses in early life is largely age dependent.[31] Similarly, an age dependent increase of antibody responses was observed after the immunization of 1-, 2-, 3-week-old or adult mice with various vaccine antigens.[18] The antibody responses of young mice are not only reduced in their final titers, but a delay of approximately 2 to 3 weeks can also be observed after the immunization of 1-week-old mice. The delayed immune responses observed in 1-week old mice could thus, based on all available parameters, correspond to the immune maturation of human newborns.[31] Factors responsible for this delay have not yet been characterized in detail, although recent observations in infant mice demonstrate delays in both the induction of the early peak of antibody secreting plasma cells and in the establishment of the bone marrow pool of plasmocytes.[32] The preferential secretion of IgG1 and the lack or decrease of IgG2a subclass responses to various vaccine antigens observed in young mice, are likely reflecting the Th2-type polarization of T cells.[18] Whether this absence of IgG2a responses in young mice correlates with the reduced IgG2 levels observed in human neonates can only be implied. Lastly, antibody responses to most polysaccharides vaccines are limited in infant mice as they are in human infants and toddlers.[33]

D) Inhibition of Antibody Responses by Maternal Antibodies

In addition to T- and B cell immaturity that infant vaccines have to overcome, antigen-specific maternal antibodies (MatAb) can interfere with vaccine induced immune responses. Although MatAb can protect neonates and infants from early infection, they progressively wane below protective thresholds, leaving the infant susceptible to infections—usually before vaccine induced antibody responses have been been raised to protective levels. Inhibitory effects on antibody responses by MatAb were observed long ago with live as well as with nonlive vaccines,[31] but the precise mechanisms responsible for this inhibition have only recently been investigated in more detail. The importance of several factors, such as the titer of

MatAb present at the time of immunization, the epitope specificity of the inhibition of B cell responses, and the lack of inhibition of T cell responses by MatAb has been demonstrated in a variety of early life murine vaccination models and recently confirmed in human infants.[3,13,34-36] Therefore, the main challenge for novel vaccines is to be able to overcome inhibition of vaccine antibody responses by pre-existing MatAb, regardless of the magnitude of such titers.

Immune Responses after Neonatal DNA Vaccination

Neonatal Tolerance or Immunity?

The issue of neonatal tolerance induction by plasmid DNA vaccination was raised by the observation, that the immunization of 1 to 5 day old BALB/c mice with a plasmid encoding the circumsporozoite protein (CSP) protein of *plasmodium* (p.) yoelii induced CSP-epitope specific B and T cell tolerance to subsequent pCSP challenge.[37] These results were in apparent contrast to many other published reports where neonatal and young mice had been successfully immunized by plasmid DNA encoding various antigens. Briefly, CTL responses and protection to subsequent pathogen challenge induced by DNA vaccination of newborn mice were reported in the Cas-Br-M murine leukemia virus model[38] and by vaccination with plasmids encoding nucleoprotein (NP) and hemagglutinin (HA) in the influenza A virus model.[39,40] The immunization of neonatal mice with plasmids encoding NP of lymphocytic choriomeningitis virus (LCMV),[41] rabies virus glycoprotein D and measles virus HA[42] induced T cell responses[43] and antibody responses were detected after influenza virus HA,[40] rabies virus glycoprotein D[43] and influenza A virus WSN HA plasmid DNA immunization of newborn mice.[44] Furthermore, antibodies against influenza virus HA obtained after intramuscular (i.m.) or gene gun (g.g.) DNA vaccination of one-day-old mice, were shown to be protective against subsequent challenge.[45]

Based on Burnet's clonal deletion theory,[46] neonatal tolerance, as first described after the in utero or neonatal exposure of mice to allogenic spleen cells,[47] develops as a consequence of the elimination of antigen-specific lymphocytes that are considered as self-specific. Neonatal tolerance has long been made responsible for the obvious difficulties in mounting effective and protective immune responses to various antigens in newborn mice. However, numerous reports have since shown that when tolerizing experimental conditions were slightly modified, newborn mice could mount significant T cell responses. This included providing adult professional APCs and increased costimulation,[48] reducing the inoculation dose of a live replicative leukemia virus,[49] co-administration of incomplete Freunds' adjuvant (IFA)[22,50] or immunization at 7 days of age with a panel of peptide, protein or even attenuated live vaccines.[18] Therefore, the induction of either antigen-specific immunity or tolerance in neonatal mice seems to depend on the type and amount of antigen and adjuvant used for immunization and on the age of the recipient. Precisely the same rules appear to apply for DNA vaccines. In the *p. yoelii* pCSP model, 2 day-old mice were reported particularly susceptible to tolerance induction, whereas no tolerance was observed in mice older than 7 days.[37] Furthermore, tolerance induction in 1 and 2 day-old mice was dependent on the dose of pCSP. High doses (10 µg and 100 µg) favored tolerance induction, whereas a low dose (1 µg) of plasmid clearly reduced susceptibility to tolerance.[51] Also, no tolerance was observed even in neonatal mice when other malaria antigens were delivered by the same plasmid vector.[51] Perhaps the fact that CSP, as soluble protein, is weakly or not immunogenic in neonatal mice, could explain at least in some part the reason for tolerance induction.[51] Interestingly, tolerance induction observed after immunization with the pCSP was overcome by the co-administration of a plasmid encoding granulocyte-macrophage colony stimulating factor (GM-CSF).[52]

Is tolerance induction an issue for neonatal human immunization with DNA vaccines? Human neonates are much more immune-competent than newborn mice. This is suggested by the apparent lack of immune consequences after neonatal thymectomy and by the lack of reports on "true" immune tolerance in human neonates.[31] Nevertheless, a detailed correlation between the developmental stages of both immune systems is difficult to firmly establish. The responsiveness of other animal species to neonatal DNA immunization, particularly nonhuman primates, could thus give important indications as to the potential risk of human neonates to tolerance induction by plasmid immunization. Only a few reports on immune responses to genetic immunization are available in species other than mice. Nevertheless, antibody responses to DNA vaccination with the pre-S/S large envelope protein L of duck hepatitis B virus were induced in 3-day-old ducklings,[53] while one-day-old piglets raised antibody responses to pseudorabies virus glycoprotein D delivered as DNA vaccine.[54] Furthermore, 3-day-old lambs were able to mount antigen specific T- as well as antibody responses to herpesvirus-1 glycoprotein D plasmid immunization.[55] In primates, T and B cell responses were detectable after immunization of newborn baboons with influenza HA and NP plasmids[56,57] and antigen-specific immunity to genetic immunization with HIV and Hepatitis B virus surface antigens was demonstrated in chimpanzees vaccinated as neonates.[58,59] Thus, although only few neonatal immunization studies have yet been performed in species other than mice, the induction of neonatal tolerance reported after DNA immunization seems to be an exception rather than the rule, as suggested recently.[60]

Factors Influencing the Type of Immune Response after Neonatal DNA Administration

The Route and Method of DNA Administration

A DNA vaccine can be administered by various routes and to many different sites. The most frequently used routes to deliver DNA vaccines include the i.m., intradermal (i.d.), subcutaneuos (s.c.), and g.g., but intravenous, intraperitoneal and various mucosal routes are also under investigation. In mice, and also in rhesus monkeys, quantitative as well as qualitative differences of the immune response to hepatitis B surface antigen (HBsAg) antigen were detected, depending on the route of immunization used.[61] Most comparative studies were conducted with plasmid DNA administered by either the i.d., i.m. or by the g.g route in mice. Whereas the efficacy to generate CTL responses seems comparable by all three methods, antibody responses differed significantly. The i.d. delivery of plasmid DNA encoding ovalbumin (OVA), human Ig, influenza virus HA or HBsAg resulted in higher IgG1/IgG2a ratios than i.m. delivery.[61-63] Similar to the i.d. route, the administration of plasmid DNA encoding various vaccine antigens by g.g. lead to higher IgG1/IgG2a ratios than i.m. immunization,[61,64,65] and this difference was apparently not due to the reduced dose of DNA administered by g.g.[64] All these observations point to fundamental differences in the immune responses generated as a consequence of the delivery of plasmid DNA either into the skin or into the muscle. In fact, the excision of target skin shortly (10min) after immunization with g.g. highly reduced or even abrogated antibody and CTL responses to various plasmid DNA constructs, whereas the rapid excision of the target tissue in the muscle did not affect immune responsiveness.[66]

In early life immunization models, the importance of the route of administration might be even more complex, and further studies are needed to address this issue. When comparing the IgG isotype distribution after g.g. or i.m. influenza virus HA plasmid DNA administration in either adult or 1-day-old mice, the increased IgG1/IgG2a ratios observed after g.g. administration in adult mice were not observed in 1-day-old mice, where either exclusive IgG1, exclusive IgG2a or both IgG1 and IgG2a response were detected.[65] In our experiments, the i.m. immunization of 1 week-old BALB/c mice with 3 different DNA vaccines led to a similar IgG subclass distribution and cytokine secretion profiles as in adults:[42] immunization with Sendai virus pNP and measles virus pHA led to higher IgG2a than IgG1 levels and preferential IFN-γ secretion by antigen-specific T cells, whereas immunization with a plasmid encoding fragment C of tetanus toxoid resulted in equal IgG2a and IgG1 titers and similar IFN-γ and IL-5 responses in both age groups.[42] To exactly what extent the method of delivery of DNA vaccines may affect early life responses could be worth assessing in nonhuman primates, as this could be an important issue in the development of DNA vaccines for use in early life.

Induction of Early Life Th1/CTL Responses: Influence of Immunostimulatory CpG-Motifs in the Plasmid Backbone

Following immunization with plasmid DNA, adult animals generate strong Th1 and CTL responses to the encoded antigen in most experimental models used.[67] The induction of adult-like Th1 and CTL responses by DNA vaccines has also been reported in early life, as demonstrated after DNA vaccination of newborn mice in the Cas-Br-M murine leukemia virus model,[38] with plasmids encoding influenza NP and HA,[39,40,68] LCMV NP,[41] rabies virus glycoprotein,[43] measles virus HA or Sendai virus NP.[42]

There are several characteristics of the DNA vaccines that may account for this effect. Apart from the prolonged in vivo antigen synthesis and presentation of antigen, the bacterial plasmid backbone itself contains immunostimulatory motifs that can provide significant adjuvant activity for the encoded antigen.[69,70] Nonmethylated CpG-motifs, that are found in higher frequency in bacterial DNA as compared to vertebrate DNA, can activate APC probably by signaling through the Toll like receptor 9.[71] As a result, a large variety of cytokines, costimulatory and MHC molecules, as well as anti-apoptotic and survival factors, are produced that lead to a general activation of the immune system.[72]

Synthetic bacterial oligodeoxynucleotides containing one or several CpG-motifs are extremely potent adjuvants that can induce Th1 and CTL responses when co-administered with peptide, protein and even live viral vaccines in newborn and young mice.[7,73] It is thus likely, that at least part of the capacity of DNA vaccine to induce Th1 and CTL activity in early life is related to the presence of these immunostimulatory CpG-motifs on the plasmid backbone. This hypothesis is supported by our observation that the simple addition of CpG-rich oligodeoxynucleotides (ODN) to attenuated measles virus vaccine can induce CTL activity in mice immunized i.m. at one week of age (Fig. 1). A similar CTL activity is detectable after immunization with the DNA plasmid encoding measles HA, whereas no CTL lysis is observed after vaccination with measles virus vaccine alone (Fig. 1). Thus, immunostimulatory CpG-motifs present in the DNA plasmid backbone seem to be, at least to a certain extent, directly responsible for the Th1- and CTL driving capacity of DNA vaccines. However, presence of CpG-ODN in a plasmid backbone is not sufficient to guarantee the induction of adult-like Th1/Th2 neonatal responses[19] and comparisons on the influence of the number/quality of immunostimulatory motifs on neonatal responses have not yet been reported. This would be worth assessing under the few experimental conditions where neonatal T cell responses to DNA vaccines were found to differ from those elicited in adults.

The Form and Type of the Encoded Antigen: A Distinct Influence in Early Life?

DNA vaccines encode for antigens (mostly proteins or parts thereof) that contain one or several well-characterized epitopes, that mediate protective immunity against the infectious agent. Because DNA vaccines provide a relatively easy system to select and modify the cDNA encoding for the antigen, a rather large variety of DNA plasmid constructs, encoding for the same antigen, are now used for vaccination in preclinical models. Vaccine inserts can be designed to encode for a protein that is produced by the target cells in either a membrane bound (m-) form (if containing an anchor sequence), a secreted (s-) form or a cytoplasmic (c-) form (lacking a signal sequence). However, some important differences in the immune responses generated by these constructs are beginning to surface, that seem to be a direct consequence of this cellular localization of the antigen produced. Irrespective of the route and method used to deliver the DNA vaccine, s-antigens seem to generate higher antibody responses compared to m-antigens.[74-76]

Qualitative differences in B cell responses can be detected even more frequently than quantitative changes. In general, s-antigens induced higher IgG1 and lower IgG2a titers when directly compared with the same antigens encoded in their m-form.[75-81] These differences were observed in adult mice of different genetic backgrounds, in a variety of immunization models

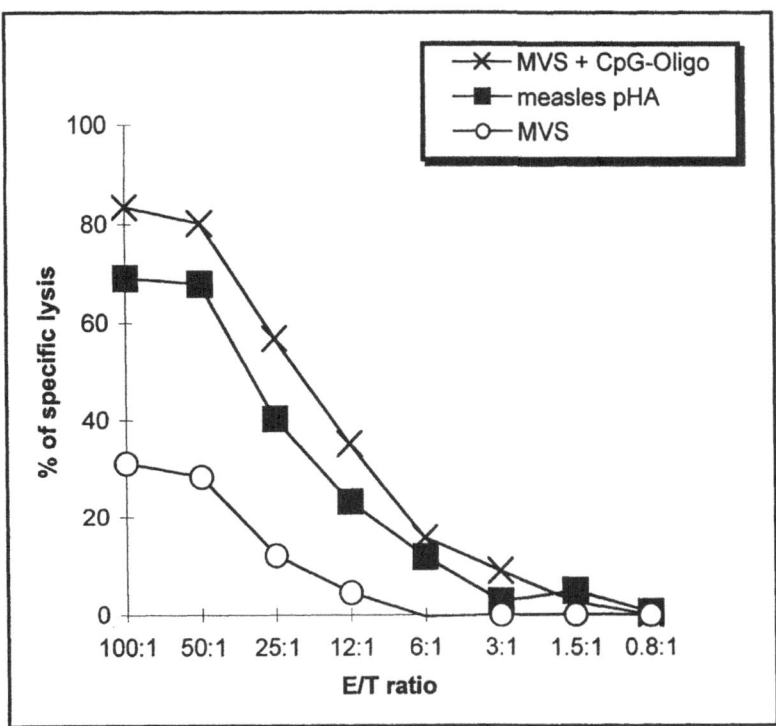

Figure 1. Lytic activity of restimulated CTL harvested from immunized mice at different effector/target ratios, measured by a standard chromium-release assay on HA-transfected P815 target cells. One week old BALB/c (6 to 8 mice per group) mice were immunized with measles virus vaccine (MVS: $5x10^5CCID_{50}$),[18] measles virus HA DNA vaccine (measles pHA: 100μg)[42] or measles virus vaccine in co-administration with a CpG-oligodeoxynucleotide (MVS + CpG-Oligo: 100μg).[7] Three weeks later, spleen cells were harvested and restimulated in vitro with measles virus HA CTL peptide 544-552 and the CTL assay was performed after 7 days of culture.

and irrespective of the administration route used. As with the m-antigens, c-antigens induce predominantly IgG2a responses[75,82] or a balanced IgG1/IgG2a isotype distribution.[78] Whether these differences could be considered as a consequence of priming different Th cell subsets, or whether altering the localization of the antigen affected the ability to prime B cells, remains an open issue.[75]

In models where CTL responses were measured after administration of s-, m- and c-forms of DNA vaccines, all three types of constructs were able to generate similarly strong CTL activity.[75,79,82] Nevertheless, in the tick-borne encephalitis virus envelope protein model,[77] bovine herpesvirus-1 gD model[80] and in the Herpes simplex virus gD model,[83] s-products tended to generate Th2-type cytokine responses when compared to the constructs encoding for the m-form. A direct role of IL-4 in the generation of IgG1/IgG2a predominance after s-form DNA vaccination was suggested by experiments using IL-4 knockout mice. In the absence of IL-4, the elevated IgG1 levels induced by the s-forms of influenza and measles virus pHA were dramatically reduced, whereas a significant increase of the IgG2a levels was noted.[81]

There are only a few reports that investigate the influence of the cellular localization of the antigen after DNA vaccination of neonatal or young mice. Yet the characteristic of the s-form antigens to increase the IgG1/IgG2a ratio and to promote Th2-type priming, could represent a potential disadvantage in the construction of plasmids to be used in neonates.[18] In order to investigate this in detail, we compared immune responses in 1 week-old BALB/c mice after i.m. immunization with measles virus HA, encoded as either m- or s-form. In the latter, about 10% of the protein produced is secreted into the medium after in vitro transfection of target cells.[79] In contrast to the m-form that induced exclusively Th1-type responses (i.e., IFN-γ production after in vitro restimulation of T cell and high IgG2a/IgG1 titers in the serum), immunization with the s-form of pHA generated antigen-specific IL-5 responses, in addition to IFN-γ secretion,[19] as summarized in Table 1. Interestingly, IL-5 was only detectable after priming of 1 week-old mice and was not detectable after immunization of adult mice with the pHA-construct. Furthermore, measles s-pHA immunization resulted in an IgG1 predominance in young mice, but more balanced IgG1 and IgG2a titers in adult BALB/c mice (Table 1). Thus, whereas immune responses in adult mice were not (cytokine profile), or only a little (IgG isotype distribution) changed, the modification of m-HA into a s-HA resulted in a clear shift towards Th2-type responses in young mice. Such an age-dependent priming of different Th-subsets by DNA vaccines does not only occur in response to measles HA, but apparently also in response to influenza A virus pHA immunization.[40] Whereas T cells from BALB/c mice primed in adulthood produced exclusively IFN-γ, T cells from mice vaccinated as neonates not only secreted less IFN-γ but also generated IL-4 responses.[40] The exact reasons for these age-dependent variations of T cell responses to m- or s-antigens remain unknown, but they are most likely related to differences at the level of APC to T cell interactions triggered by immunization. Regardless of the mechanism, these observations imply that the induction of adult-like Th1/Th2 responses by early life DNA immunization cannot be anticipated from studies in adults. This calls for detailed preclinical studies of candidate DNA vaccines in early life immunization models.

Induction of Adult-Like Antibody Responses by DNA Vaccines: Is This an Improvement over Conventional Vaccines?

After DNA vaccination of adult animals, considerable antibody responses (which however, frequently remain inferior to those elicited by conventional vaccines) have usually been detected in various immunization models. In order to investigate the capacity of DNA vaccines to raise antibody responses in young individuals, we assessed antigen-specific total IgG responses induced by a plasmids encoding measles HA, tetanus toxin fragment C and Sendai virus NP in 1 or 2 week old BALB/c mice. With these DNA vaccines, equivalent antibody responses with similar endpoint titers were reached in young as well as adult mice.[42] Therefore, at least with the constructs investigated, DNA vaccines seemed capable of inducing adult-like antibody responses in young mice. This could represent a potential advantage to other subunit vaccines, where antibody responses in young mice remain inferior to those observed in adults.[18,84] However, antibody responses elicited by these DNA vaccines in 1-2 week old mice were not higher than those induced by conventional vaccines in control infant mice, and remained at least 10 fold lower than those induced in adults by the corresponding conventional vaccines.[18,84,85] This is illustrated for measles vaccines in Figure 2. Although DNA-HA immunization elicited similar IgG anti-HA titers in 1 week old and adult mice, antibodies remained lower, or equal, to those elicited by conventional vaccines such as live attenuated measles virus vaccine MVS or recombinant canarypox virus vaccine ALVAC-HA. Thus, DNA immunization failed to enhance early life antibody responses over those elicited by conventional vaccines. In some cases, DNA vaccinated neonatal mice even raised clearly inferior antigen-specific IgG titers compared to mice immunized with inactivated virus, such as following immunization with inactivated herpes simplex virus[86] or attenuated rabies virus.[87] Similarly, we observed much weaker antibody levels after delivery of a RSV-F protein DNA vaccine to 2 week old mice in comparison to the response to alum-adjuvanted RSV-F protein.[88] In contrast, a DNA vaccine encoding HBsAg was recently shown to generate similar anti-HBs titers as alum-adjuvanted recombinant HBsAg in 1 week old BALB/c mice[89] and influenza virus pHA immunization induced somewhat higher anti-HA antibody responses than UV-inactivated influenza virus in newborn mice.[44] These contrasting results could indicate that certain

Table 1. *Influence of the cellular localization of measles virus HA produced by DNA vaccines on antigen-specific immune responses after i.m. immunization of BALB/c mice*

Antigen Form	IFN-γ	IL-5	IgG Profile
1-week-old mice			
membrane bound	++	0	IgG1 < IgG2a
secreted	++	++	IgG1 > IgG2a
Adult mice			
membrane bound	++	0	IgG1 < IgG2a
secreted	++	0	IgG1 = IgG2a

antigens, possibly the less immunogenic ones in early life, may benefit from the fact that the prolonged in vivo production of antigen by DNA vaccines allows ongoing maturation of the immune system to occur and eventually enhance antibody responses. Thus, although DNA vaccines seem to possess a rather weak capacity to raise antibody responses in adult individuals, they could prove potentially interesting for early life immunization to certain antigens, providing optimal antigenic constructs allowing prolonged expression are used.

Could Early Life DNA Vaccination Elicit Life-Long Responses?

To be useful for early life vaccination, the immunological requirements for a novel antigen delivery system should be to induce adult-like (protective) sustained antibody responses, as well as Th1-type and memory CTL responses. There are many reports indicating that DNA vaccines (encoding for various antigens) are very good inducers of memory CTL responses in adult mice.[90] Furthermore, CTL responses could also be measured after HIV-antigen DNA immunization of adult chimpanzees[91] and human adults.[92,93] CTL were also detectable in healthy adult volunteers after injection of several doses of *p. falciparum* CSP protein DNA vaccine.[94] Not only in adult but also in newborn mice, NP-specific memory CTL were detectable even one year post LCMV-DNA immunization.[95] In this model, antigen-specific antibody responses were still detectable 6 months after vaccination, suggesting a long-lived humoral response to NP.[95] These observations, that DNA vaccination has the potential to induce persistent B and memory T cell responses for up to 18 months, was recently confirmed by Bot et al after immunizing newborn baboons with plasmids encoding influenza HA and NP.[57] Therefore, DNA vaccines represent an antigen-delivery system that appears to meet the requirements for induction of sustained vaccine responses after early life immunization. This in itself, may represent a significant advantage over conventional vaccines requiring repeated booster administration.

DNA Vaccines and Inhibition of Responses by Maternal Antibodies

Transport of pathogen- or vaccine-specific IgG from mother to offspring is important for passive protection against a number of early life pathogens. Such antibodies of maternal origin readily bind to invading bacteria or viruses and facilitate their elimination via neutralization, opsonization and/or complement activation. However, MatAb can just as well bind to vaccine antigens and limit infant vaccine responses. This has been observed with most if not all types of currently used vaccines, although some vaccines or delivery systems might be less susceptible to MatAb inhibition than others.[31] Recent studies, in both experimental models and clinical settings, have progressively identified the main determinants of MatAb mediated inhibition of infant responses to conventional vaccines and shed some light on what appeared initially to be

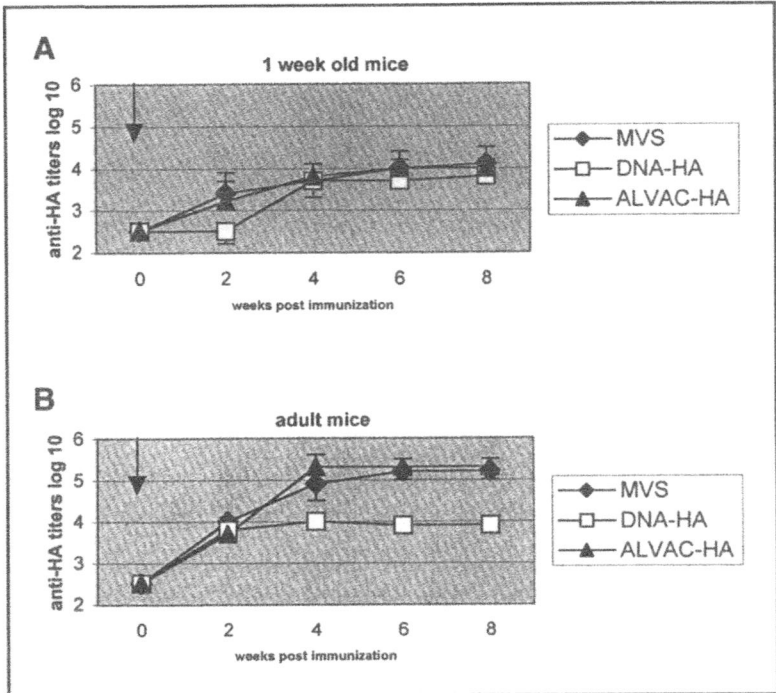

Figure 2. Measles-HA specific total IgG titers (log_{10}) measured by ELISA after the immunization of either 1 week old (A) or adult (B) BALB/c mice with measles virus vaccine (MVS: $5x10^5CCID_{50}$),[18] measles virus HA DNA vaccine (measles pHA: 100μg)[42] or a recombinant canarypox viral vector expressing measles-HA.[18] Indicated at each time point are mean antibody titers of each group containing 5 to 8 mice.

a large series of frequently contradictory results. Factors identified as essential for the outcome of immunization with conventional vaccines in the presence of maternal antibodies include A) the relative ratio of MatAb to vaccine antigen, B) the B cell epitope specificity of antigens used in mothers/infants, and C) the influence of MatAb on T cell responses.

A) The Influence of the Relative Ratio of MatAb to Vaccine Antigen

The influence of the titer of MatAb on the outcome of infant vaccination has long been recognized and was the basis for postponing measles vaccination until the end of the first year of life. It was, however, recently observed that either reducing MatAb titers or increasing the vaccine antigen dose was often all that was required to circumvent the inhibitory influence of MatAb.[34,36,96,97] This led to the conclusion that the main determinant of MatAb mediated inhibition was the relative ratio of MatAb to vaccine antigen. Due to the administration of similar doses of vaccine antigen to all infants, it is the titer of MatAb present at time of immunization that essentially determines the degree of MatAb-mediated inhibition of vaccine-induced antibody responses in early life. This is best explained by the formation of antigen-antibody complexes, whose immunological outcome depends on whether antigens or antibodies are in excess following immunization in the presence of pre-existing maternal antibodies. Hence, all factors that have an effect on the magnitude of titers of MatAb at the site of vaccine administration also affect the immunization outcome. These factors include: a) the type of antigens used to induce MatAb in mothers, which may be more or less immunogenic, b) the age at first infant immunization, taking into account that acquisition of MatAb in rodents requires suckling such that antibody titers only raise progressively during the first days of life and that MatAb decline thereafter with a half-life of approximately 4 weeks, c) the infant immunization regimen, including number of doses and intervals between doses, and d) the use of the immunization route, as mucosal surfaces are largely shielded from MatAb IgG.

B) The B Cell Epitope Specificity of Antigens Used in Mothers/Infants

Many reports suggest that MatAb-mediated inhibition of B cell responses occurs through the masking of immunodominant epitopes resulting from the formation of antigen-antibody complexes between vaccine antigens and MatAb, which prevents the access of infant B cells to epitopes on the vaccine antigen. Thus, at a given MatAb titer, escape from MatAb inhibition is more likely to occur if antigen delivery systems with distinct immunodominant B cell epitopes are used in mothers and infants rather than following use of the same delivery system.[36]

C) The Influence of MatAb on T Cell Responses

We recently demonstrated in murine models that pre-existing MatAb essentially affect B cell responses and that successful priming of CD4[+] and CD8[+] T cells does occur even in the presence of high titers of MatAb. Accordingly, measles-specific MatAb did not inhibit infant CD4 T cell responses to measles vaccine[13] and successful priming to hepatitis A was observed despite the presence of MatAb.[35] This factor plays an essential role in the outcome of multiple dose immunization regimen.

Why could DNA vaccines be less susceptible to the inhibitory effects of MatAb than live viral, bacterial or peptide or protein vaccines? It is known that the transfection of target cells by DNA vaccines can lead to the in vivo production of vaccine antigen over a period of several weeks.[98] It was therefore hoped that in vivo antigen production would allow B cell epitopes to escape from binding by MatAb, and that prolonged antigen delivery would allow the time period at which MatAb persist at inhibitory levels to be exceeded. These potential advantages of DNA immunization thus raised substantial hopes that the inhibitory effects of MatAb would be more easily evaded with this delivery system than with any other subunit vaccine.

Contradictory Results in Various Murine Models

There have been contradictory results published from various experimental models on the influence of MatAb on antibody responses of newborn or young animals to DNA vaccines (summarized in Table 2). We suggest these contradictory results essentially reflect the use of different models and experimental conditions and postulate here, that the variation of MatAb titers at time/site of antigen delivery, the use of similar/distinct antigen delivery systems in mothers and pups and the noninhibition of T cell responses despite MatAb by DNA vaccines, could emerge as the most important determinants for the effect of MatAb on immune responses to DNA vaccines, as previously demonstrated for conventional vaccines.

Table 2 readily illustrates that whenever T cell responses to neonatal DNA immunization were assessed, no inhibition was reported, even under experimental conditions where antibody responses were completely inhibited by pre-existing MatAb.[34,68,89,99] This strongly suggests that although vaccine antigens produced following DNA immunization may remain inaccessible to infant B cells, they are readily taken up (presumably partly in the form of MatAb-vaccine antigen complexes) and presented by neonatal APCs. This process is thus similar to that observed following administration of conventional vaccines.

Our own reports[34] demonstrated that although antibody responses induced by measles virus HA DNA immunization were not inhibited by intermediate titers of MatAb, raised either by DNA or live attenuated measles virus immunization, higher MatAb titers completely inhibited antibody responses to DNA-HA. This led to the conclusion that the ratio of MatAb to vaccine antigen remains an essential determinant of the impact of MatAb on infant responses to DNA vaccines, resulting in either detectable or absent antibody responses. The few reports showing a lack of inhibition of MatAb on antibody responses to DNA immunization

Table 2. *Comparison of the influence of maternal antibodies on immune responses induced by DNA vaccines in a variety of experimental models using neonatal and/or young (< 2 weeks old) animals*

Reference	Species	Age at Immun.	MatAb Against	MatAb Titers	Encoded Antigen	Immun. Route	Ab Responses	T Cell Responses
(68)	BALB/c mice	< 24h old	influenza virus	pups: >100 µg/ml	HA, membrane	i.m. and g.g.	inhibited	not inhibited
	BALB/c mice	< 24h old	influenza virus		NP, internal	i.m. and g.g.	not inhibited	not inhibited
(86)	BALB/c mice	< 24h old	HSV-1	mothers: 27 µg/ml	gB, membrane	s.c.	not inhibited	not inhibited
(87)	C3H/He mice	< 48h old	rabies virus	n.d.	glycoprotein, membrane	s.c.	not inhibited	n.d.
(89)	BALB/c mice	1,3 and 7 days old	HBsAg	pups: 4.0 $\log_{10}$	sAg, membrane	i.m.	inhibited	not inhibited
(41)	BALB/c mice	< 2 weeks old	LCMV	n.d.	NP, internal	i.m.	not inhibited	not inhibited
(99)	BALB/c mice	2 weeks old	influenza virus	pups: 40 µg/ml	HA, membrane	i.m.	inhibited	not inhibited
(34)	BALB/c mice	2 weeks old	pHA	pups: 4.0 $\log_{10}$	HA, membrane	i.m.	not inhibited	not inhibited
	BALB/c mice	2 weeks old	MVS	pups: 5.0 $\log_{10}$	HA, membrane	i.m.	inhibited	
	BALB/c mice	2 weeks old	MVS	pups: 3.5 $\log_{10}$	HA, membrane	i.m.	not inhibited	not inhibited
(54, 100)	piglets	1 day old	pseudorabies virus	n.d.	gD, membrane	i.m.	inhibited	n.d.
(55)	lambs	3 days old	bovine herpesvirus-1	mothers: "high"	gD, membrane	i.d.	not inhibited	n.d.

HSV-1: herpes simplex virus-1
MVS: attenuated measles virus Schwarz-strain
n.d.: not determined

do not contradict this conclusion. In two of these reports, DNA immunization with plasmids encoding for internal viral antigens was used. It is expected that internal antigens such as the influenza[68] or LCMV[41] NPs are less readily accessible to maternal B cells than epitopes at the viral surface. This likely results in the induction of relatively lower titers of NP-specific antibodies subsequently transferred, than for surface viral antigens. This hypothesis could be confirmed by determination of NP-specific antibody titers present in pups of influenza or LCMV infected mothers at time of immunization, which was unfortunately not assessed in either report.

There are 2 other reports of successful induction of antibody responses following immunization with DNA vaccines encoding surface viral antigens—one in pups of herpes simplex virus (HSV)-1 infected mothers and the other in pups injected with hyperimmune serum raised by rabies virus infection.[86,87] As MatAb titers were not assessed at the time of immunization in these pups, it cannot be excluded that this successful induction of antibody responses could be ascribed to the presence of relatively low MatAb titers in these 24-48 h old pups. To define whether DNA immunization represents a true advantage over immunization with conventional vaccines for the escape from MatAb inhibition of antibody responses, it should be assessed whether the experimental conditions used do result in the inhibition of antibody responses to conventional vaccines (such as adjuvanted inactivated virus). This was performed in the HSV-1 immunization model, where B cell responses to UV-inactivated HSV virus, but not to the plasmid immunization, were completely inhibited by MatAb.[86] This suggests that, under certain experimental conditions, DNA immunization could better overcome the MatAb-mediated inhibition of antibody responses than conventional vaccines. It would be of significant interest to define whether this is due to a particularly prolonged in vivo antigen production or to other yet unknown properties.

In many other murine studies (Table 2), early life B cell responses to DNA vaccines were indeed clearly inhibited by MatAb,[89] even when the antibodies were induced by viral infection.[34,68,99] Likewise, in one day-old piglets from sows vaccinated against pseudorabies (PRV), antibody responses to PVR virus gD DNA vaccine were inhibited.[54,100] Although the interpretation of these various results is rather difficult, we would hypothesize that, as with all other conventional vaccines, the ratio of antigen-specific MatAb to vaccine antigen and possibly the persistence of vaccine antigen production following DNA immunization, seem to determine whether a DNA vaccine administered to newborns/young individuals would be able to bypass the inhibitory effects of MatAb or not.

Apart from the routine strategy of delivering several booster immunizations to overcome the initial B cell inhibition mediated by MatAb, the delivery of antigens by the mucosal route is currently being investigated. This delivery method has the important advantage in that the mucosal tissue is less accessible to MatAb. As it is possible to deliver not only certain conventional vaccines but also DNA vaccines by mucosal administration,[101] this method could represent a way of bypassing the MatAb-mediated inhibitory effect on vaccine-specific antibody responses. However, weak antigen-specific immune responses are generally observed after mucosal delivery in comparison to i.m. and i.d. delivery of DNA vaccines. Therefore, apart from developing adjuvants and live delivery systems to optimize the mucosal delivery of DNA vaccines, this strategy may not prove successful.

In conclusion, the capacity of DNA immunization to induce significant antibody responses, despite the presence of high titers of MatAb, currently requires the use of an immunization regimen including the administration of repeat vaccine doses, as most conventional infant vaccines.

Optimizing DNA Vaccination Strategies for Neonates

The susceptibility to early life infectious diseases can be reduced in many cases by the induction of sufficiently high titers of neutralizing antibodies. In view of the limited B cell responses in infants, vaccines or delivery systems that induce stronger antibody responses than the currently used "conventional" vaccines would be needed. Hence, various strategies to increase antibody responses to DNA vaccines are under investigation in early life experimental models.

DNA Prime/Protein Boost Strategies

To use DNA vaccines with their strong Th1 and CTL-driving capacity to prime early life vaccine responses, followed by other delivery systems such as protein or viral vaccines that are able to induce stronger antibody responses seems an attractive strategy. The importance of the initial use of formulations capable of eliciting neonatal Th1- and CTL responses should be underlined, because even strong Th1-driving formulations such as CpG-ODN or DNA vaccines failed to redirect Th2-type vaccine responses previously established in young mice.[7,42] In accordance to many observations reported in adult animals, booster immunization with a conventional vaccine (i.e., attenuated measles virus vaccine) 3 weeks after DNA-HA priming of 1 week old BALB/c mice resulted in anti-HA total IgG final titers that were more than 10 fold higher than titers induced by prime/boost immunization with DNA-HA alone (Fig. 3A). The high IgG2a/IgG1 subclass distribution observed after the measles virus vaccine boost indicated that DNA-HA primed B cells were efficiently reactivated by the recall immunization (Fig. 3B). These recent observations confirm previous results from a RSV early life immunization model.[88] Furthermore, CTL-peptide specific precursor frequencies (1/7399 vs 1/5157 of spleen cells) were similarly high in mice primed at 1 week of age with DNA-HA and boosted with either DNA-HA or live attenuated measles virus vaccine. This is also in agreement with the expansion of CTL responses reported in various adult immunization models by heterologous prime-boost strategies.[102] Thus, by boosting plasmid DNA primed individuals with a heterologous rather than a homologous formulation, a certain synergy might be achieved to broaden the T cell and B cell vaccine responses in early life.

Combined Vaccination with DNA and Other Delivery Systems

Potential limitations to heterologous prime-boost strategies in early life do, however, exist. In addition to logistic issues, which may preclude their use in the field, prime-boost strategies would require several months for completion of a multiple dose regimen, which may not be compatible with the need for rapid protection. One potentially interesting approach could thus be to simultaneously administer two different vaccines or delivery systems coding for the same antigen. When a DNA plasmid encoding measles HA was administered to 1 week old BALB/c mice with a single dose of attenuated measles virus vaccine, HA-specific IgG responses were earlier and higher than following DNA-HA immunization alone (Fig. 4A). In contrast to an almost exclusive IgG1 response to the attenuated measles virus vaccine and an almost exclusive IgG2a response to DNA-HA, the simultaneous administration of the two vaccines induced an equivalent IgG1 and IgG2a subclass distribution as measured 5 weeks post-immunization (Fig. 4B). The secretion of both IFN-γ and IL-5, after in vitro restimulation of measles specific splenocytes from mice immunized with the vaccine mixture, suggested that Th1-type priming had been induced by the DNA vaccine, and also that Th2-type priming had been triggered by the measles virus component of the preparation (Fig. 4C). However, despite this induction of Th2-type cytokines, the co-delivery of attenuated measles virus vaccine with pHA did not abolish or decrease HA-specific CTL responses (Fig. 4D) induced by the DNA component of the vaccine. Similar observations were made with the co-delivery of a plasmid DNA encoding for RSV-F protein and purified, aluminum hydroxide adsorbed RSV-F protein.[88] Therefore, combining DNA vaccines with other delivery systems that induce stronger antibody responses could represent a strategy to enhance and accelerate early life B cell responses, while conserving the advantages of the DNA vaccine to efficiently prime CTL responses.

Other Approaches

Aluminum salts, when co-administered with DNA vaccines, were shown to increase antigen-specific antibody responses dramatically in mice and considerably, but less efficiently, in nonhuman primates.[103] Interestingly, the co-administration of DNA constructs with these

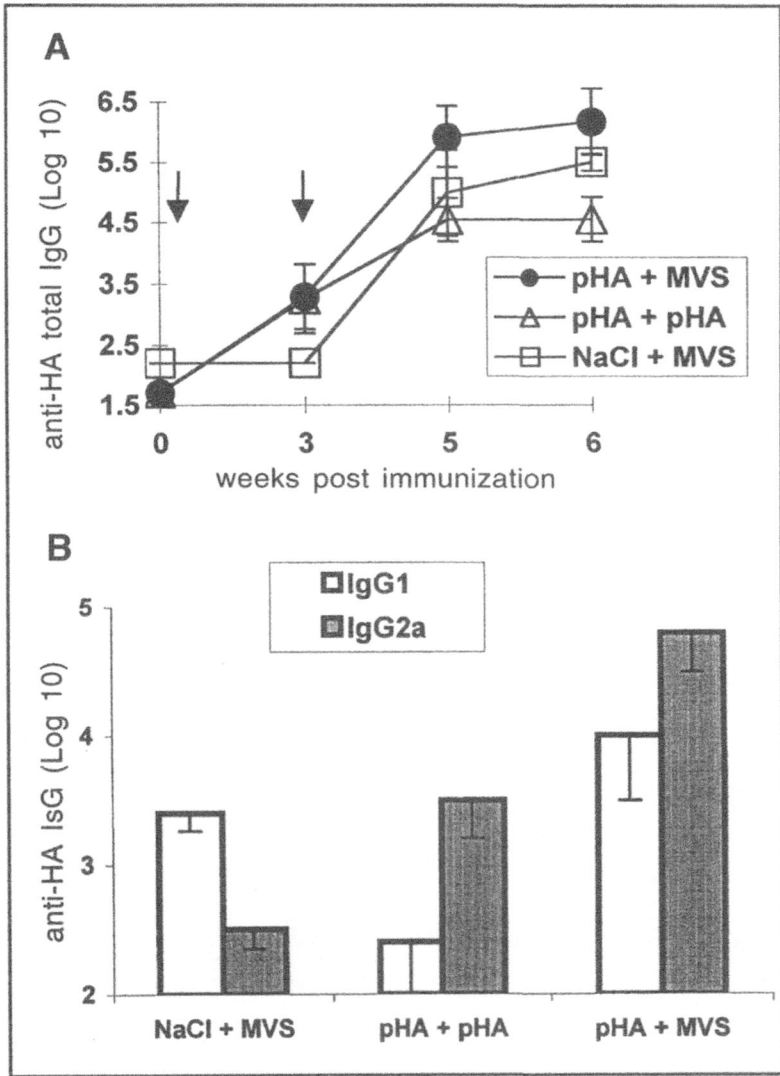

Figure 3. (A) Antibody responses after booster immunization with measles virus vaccine (MVS) or measles HA DNA vaccine (pHA) in BALB/c mice primed at one week of age with measles HA DNA vaccine (or saline for controls). Measles-HA specific total IgG titers ($\log_{10}$) of each group containing 5 to 8 mice are indicated for each time point. Immunizations are indicated as arrows. (B) Antibody subclass distribution in these groups of mice 3 weeks post booster immunization.

salts, that are know to polarize immune responses towards the Th2 phenotype, apparently did not affect the isotype distribution and did not abrogate cell-mediated immunity induced by the DNA vaccine.[103] Similarly, an increase in antibody responses (but not T cell responses) was obtained by a preparation of DNA vaccines with cationic lipid formulations, such as Vaxfectin.[104] In our experiments, the addition of Al(OH)₃ did not significantly affect antibody and T cell responses after immunization of 1 week old BALB/c mice with measles virus pHA.

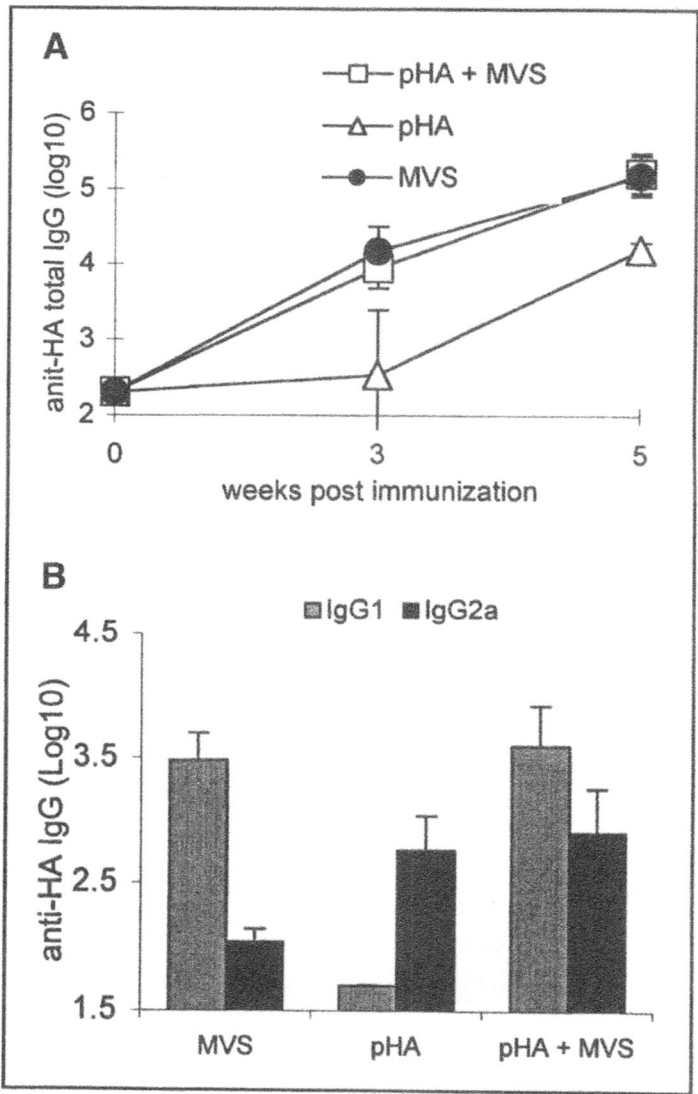

Figure 4. Influence of the co-administration of measles HA DNA vaccine (pHA) together with measles virus vaccine (MVS) in 1 week old BALB/c mice. (A) group means of measles-HA specific total IgG titers ($\log_{10}$) after immunization. (B) anti-HA IgG subclass distribution measured 5 weeks post immunization.

continued on next page

   The co-delivery of antigen, together with "endogenous" adjuvants such as cytokines, chemokines or other molecules encoded in the same or distinct plasmids, could be an attractive approach to increase the immunogenicity and efficacy of DNA vaccines. Very few data exist on the success of the co-delivery of immunostimulatory molecules in early life models. The co-delivery of the IL-12 gene[19,65] and the IFN-γ gene,[65] together with either measles or influenza HA plasmids, resulted in increased Th1-type immune responses in neonatal and one-week-old

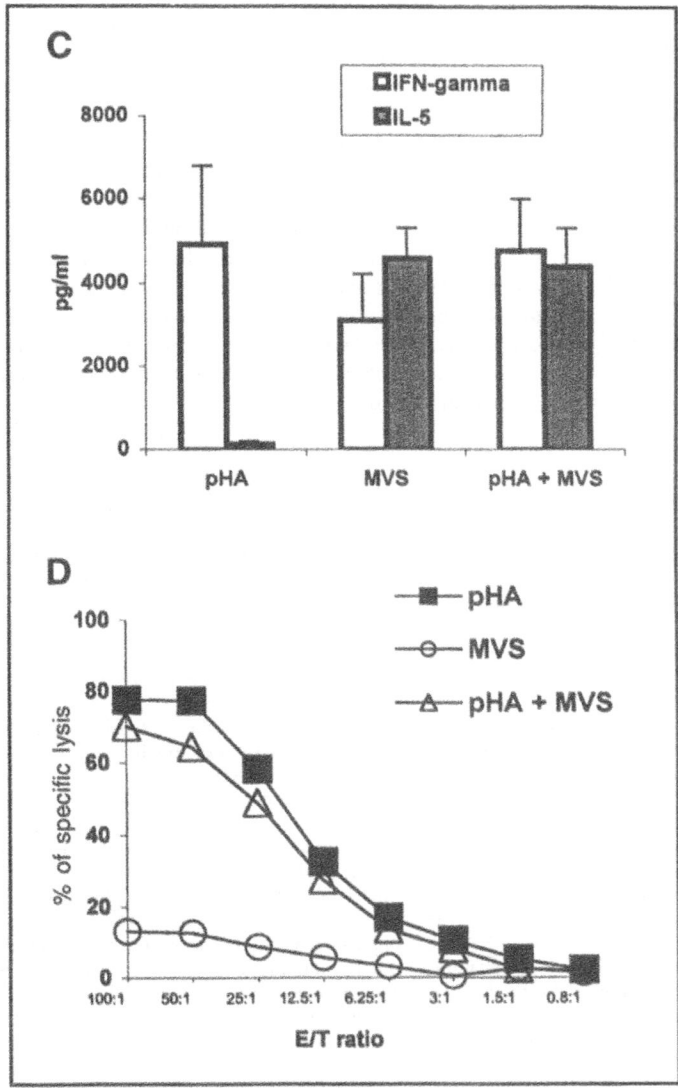

Figure 4 (con'd). (C) cytokine (IFN-γ and IL-5) content of supernatants from spleen cells harvested 5 weeks post immunization and restimulated with antigen (measles virus) for 48 h in vitro. Indicated are mean titers of 5 to 8 mice per group, measured by sandwich ELISA.[18] (D) lytic activity of in vitro restimulated CTL obtained from mice 5 weeks post immunization, measured by a standard chromium-release assay on HA-transfected P815 target cells.

mice. In adult mice, the administration of plasmid DNA encoding for a fusion product of the antigen (human Ig) with L-selectin or CTLA4, dramatically increased antigen-specific antibody responses and proliferative T cell responses.[105] Whether such targeting of the vaccine antigen to a site where immune responses are induced, might be of benefit in early life immunizations, remains to be investigated.

Conclusions and Perspectives

DNA vaccines, in contrast to most subunit vaccines, have the capacity to drive adult-like Th1 and CTL responses to most (although not all) encoded antigens in neonatal and young mice. However, DNA vaccines do not generate stronger neonatal antibody responses than conventional vaccines and they remain susceptible to the inhibitory influence of high titers of MatAb. This feature calls for more complex vaccine strategies that may not prove feasible in countries that most need such DNA vaccines. As considerable heterogeneity in the quantity and quality of early life T and B cell responses to DNA immunization has been observed, a better understanding of the respective roles of the routes and methods of administration, the doses and types of antigen(s) and constructs, the co-administration of antigens or immunomodulators, the ages of recipients and the titers of pre-existingmaternal antibodies are clearly needed. Thus, after an initial period of merely descriptive studies whose main results are reviewed here, immunization models in various animal species should now be used to progressively identify the essential parameters and characteristics of the novel DNA vaccine candidates which are to be considered for use in early life.

References

1. Wright PF. Infectious diseases in early life in industrialized countries. Vaccine 1998; 16:1355-1359.
2. Mulholland K. Serious infections in young infants in developing countries. Vaccine 1998; 16:1360-1362.
3. Gans HA et al. Deficiency of the humoral immune response to measles vaccine in infants immunized at age 6 months. JAMA 1998; 280:527-532.
4. Kovarik J, Siegrist CA. Immunity in early life. Immunol Today 1998; 19:150-152.
5. Donckier V et al. IL-12 prevents neonatal induction of transplantation tolerance in mice. Eur J Immunol 1998; 28:1426-1430.
6. Arulanandam BP, Van Cleave VH, Metzger DW. IL-12 is a potent neonatal vaccine adjuvant. Eur J Immunol 1999; 29:256-264.
7. Kovarik J et al. CpG oligodeoxynucleotides can circumvent the Th2 polarization of neonatal responses to vaccines but may fail to fully redirect Th2 responses established by neonatal priming. J Immunol 1999; 162:1611-1617.
8. Goriely S et al. Deficient IL-12 (p35) gene expression by dendritic cells derived from neonatal monocytes. J Immunol 2001; 166:2141-2146.
9. Chougnet C et al. 2000; Influence of human immunodeficiency virus-infected maternal environment on development of infant interleukin-12 production. J Infect Dis 2001; 181:1590-1597.
10. Delespesse G et al. Maturation of human neonatal CD4+ and CD8+ T lymphocytes into Th1/Th2 effectors. Vaccine 1998; 16:1415-1419.
11. Hodge S, Hodge G, Flower R et al. Cord blood leucocyte expression of functionally significant molecules involved in the regulation of cellular immunity. Scand J Immunol 2001; 53:72-78.
12. Adkins B, Hamilton K. Freshly isolated, murine neonatal T cells produce IL-4 in response to anti-CD3 stimulation. J Immunol 1992; 149:3448-3455.
13. Gans HA et al. IL-12, IFN-gamma, and T cell proliferation to measles in immunized infants. J Immunol 1999; 162:5569-5575.
14. Rowe J et al. Antigen-specific responses to diphtheria-tetanus-acellular pertussis vaccine in human infants are initially Th2 polarized. Infect Immun 2000; 68:3873-3877.
15. Ausiello CM et al. Vaccine- and antigen-dependent type 1 and type 2 cytokine induction after primary vaccination of infants with whole-cell or acellular pertussis vaccines. Infect Immun 1997; 65:2168-2174.
16. Zepp F et al. Pertussis-specific cell-mediated immunity in infants after vaccination with a tricomponent acellular pertussis vaccine. Infect Immun 1996; 64:4078-4084.
17. Ryan M, Gothefors L, Storsaeter J et al. Bordetella pertussis-specific Th1/Th2 cells generated following respiratory infection or immunization with an acellular vaccine: comparison of the T cell cytokine profiles in infants and mice. Dev Biol Stand 1997; 89:297-305.
18. Barrios C et al. Neonatal and early life immune responses to various forms of vaccine antigens qualitatively differ from adult responses: predominance of a Th2-biased pattern which persists after adult boosting. Eur J Immunol 1996; 26:1489-1496.
19. Kovarik J et al. Limitations of in vivo IL-12 supplementation strategies to induce Th1 early life responses to model viral and bacterial vaccine antigens. Virology 2000; 268:122-131.

20. Marchant A et al. Newborns develop a Th1-type immune response to Mycobacterium bovis bacillus Calmette-Guerin vaccination. J Immunol 1999; 163:2249-2255.
21. Chen N, Field EH. Enhanced type 2 and diminished type 1 cytokines in neonatal tolerance. Transplantation 1995; 59:933-941.
22. Singh RR, Hahn BH, Sercarz EE. Neonatal peptide exposure can prime T cells and, upon subsequent immunization, induce their immune deviation: implications for antibody vs. T cell-mediated autoimmunity. J Exp Med 1996; 183:1613-1621.
23. Rowe J et al. Heterogeneity in diphtheria-tetanus-acellular pertussis vaccine-specific cellular immunity during infancy: relationship to variations in the kinetics of postnatal maturation of systemic th1 function. J Infect Dis 2001; 184:80-88.
24. Wasik TJ et al. Diminished HIV-specific CTL activity is associated with lower type 1 and enhanced type 2 responses to HIV-specific peptides during perinatal HIV infection. J Immunol 1997; 158:6029-6036.
25. Chiba Y et al. Development of cell-mediated cytotoxic immunity to respiratory syncytial virus in human infants following naturally acquired infection. J Med Virol 1989; 28:133-139.
26. Jaye A et al. Ex vivo analysis of cytotoxic T lymphocytes to measles antigens during infection and after vaccination in Gambian children. J Clin Invest 1998; 102:1969-1977.
27. Schroeder HW Jr. et al. Developmental regulation of the human antibody repertoire. Ann N Y Acad Sci 1995; 764:242-260.
28. Lucas AH, Azmi FH, Mink CM et al. Age-dependent V region expression in the human antibody response to the Haemophilus influenzae type b polysaccharide. J Immunol 1993; 150:2056-2061.
29. Plebani A et al. Serum IgG subclass concentrations in healthy subjects at different age: age normal percentile charts. Eur J Pediatr 1989; 149:164-167.
30. Jelonek MT et al. Comparison of naturally acquired and vaccine-induced antibodies to Haemophilus influenzae type b capsular polysaccharide. Infect Immun 1993; 61:5345-5350.
31. Siegrist C. Neonatal and early life vaccinology. Vaccine 2001; 19:3331-3346.
32. Pihlgren M et al. Delayed and deficient establishment of the long-term bone marrow plasma cell pool during early life. Eur J Immunol 2001; 31:939-946.
33. Kovarik J et al. Adjuvant effects of CpG oligodeoxynucleotides on responses against T-independent type 2 antigens. Immunology 2001; 102:67-76.
34. Siegrist CA et al. Influence of maternal antibodies on vaccine responses: inhibition of antibody but not T cell responses allows successful early prime-boost strategies in mice. Eur J Immunol 1998; 28:4138-4148.
35. Dagan R et al. Immunization against hepatitis A in the first year of life: priming despite the presence of maternal antibody. Pediatr Infect Dis J 2000; 19:1045-1052.
36. Siegrist CA et al. Determinants of infant responses to vaccines in presence of maternal antibodies. Vaccine 1998; 16:1409-1414.
37. Mor G et al. Induction of neonatal tolerance by plasmid DNA vaccination of mice. J Clin Invest 1996; 98:2700-2705.
38. Sarzotti M et al. Induction of cytotoxic T cell responses in newborn mice by DNA immunization. Vaccine 1997; 15:795-797.
39. Bot A, Bot S, Garcia-Sastre A et al. DNA immunization of newborn mice with a plasmid-expressing nucleoprotein of influenza virus. Viral Immunol 1996; 9:207-210.
40. Bot A et al. Induction of humoral and cellular immunity against influenza virus by immunization of newborn mice with a plasmid bearing a hemagglutinin gene. Int Immunol 1997; 9:1641-1650.
41. Hassett DE, Zhang J, Whitton JL. Neonatal DNA immunization with a plasmid encoding an internal viral protein is effective in the presence of maternal antibodies and protects against subsequent viral challenge. J Virol 1997; 71:7881-7888.
42. Martinez X et al. DNA immunization circumvents deficient induction of T helper type 1 and cytotoxic T lymphocyte responses in neonates and during early life. Proc Natl Acad Sci USA 1997; 94:8726-8731.
43. Wang Y, Xiang Z, Pasquini S et al. Immune response to neonatal genetic immunization. Virology 1997; 228:278-284.
44. Antohi S, Bot A, Manfield L et al. The reactivity pattern of hemagglutinin-specific clonotypes from mice immunized as neonates or adults with naked DNA. Int Immunol 1998; 10:663-668.
45. Pertmer TM, Robinson HL. Studies on antibody responses following neonatal immunization with influenza hemagglutinin DNA or protein. Virology 1999; 257:406-414.
46. Burnet FM. The clonal seletion theory of aquired immunity. Cambridge University Press. 1953.
47. Billingham RE, Brent L, Medawar PB. Actively acquired tolerance of foreign cells. Nature 1953; 172:603-606.

48. Ridge JP, Fuchs EJ, Matzinger P. Neonatal tolerance revisited: turning on newborn T cells with dendritic cells. Science 1996; 271:1723-1726.
49. Sarzotti M, Robbins DS, Hoffman PM. Induction of protective CTL responses in newborn mice by a murine retrovirus. Science 1996; 271:1726-1728.
50. Forsthuber T, Yip HC, Lehmann PV. Induction of TH1 and TH2 immunity in neonatal mice. Science 1996; 271:1728-1730.
51. Ichino M et al. Factors associated with the development of neonatal tolerance after the administration of a plasmid DNA vaccine. J Immunol 1999; 162:3814-3818.
52. Ishii KJ, Weiss WR, Klinman DM. Prevention of neonatal tolerance by a plasmid encoding granulocyte-macrophage colony stimulating factor. Vaccine 1999; 18:703-710.
53. Rollier C et al. Early life humoral response of ducks to DNA immunization against hepadnavirus large envelope protein. Vaccine 2000; 18:3091-3096.
54. Monteil M et al. Genetic immunization of seronegative one-day-old piglets against pseudorabies induces neutralizing antibodies but not protection and is ineffective in piglets from immune dams. Vet Res 1996; 27:443-452.
55. Van Drunen Littel-van den Hurk S. et al. Immunization of neonates with DNA encoding a bovine herpesvirus glycoprotein is effective in the presence of maternal antibodies. Viral Immunol 1999; 12:67-77.
56. Bot A et al. Induction of antibody response by DNA immunization of newborn baboons against influenza virus. Viral Immunol 1999; 12:91-96.
57. Bot A et al. Induction of immunological memory in baboons primed with DNA vaccine as neonates. Vaccine 2001; 19:1960-1967.
58. Bagarazzi ML et al. Safety and immunogenicity of intramuscular and intravaginal delivery of HIV-1 DNA constructs to infant chimpanzees. J Med Primatol 1997; 26:27-33.
59. Prince AM, Whalen R, Brotman B. Successful nucleic acid based immunization of newborn chimpanzees against hepatitis B virus. Vaccine 1997; 15:916-919.
60. Bot A. DNA vaccination and the immune responsiveness of neonates. Int Rev Immunol 2000; 19:221-245.
61. McCluskie MJ et al. Route and method of delivery of DNA vaccine influence immune responses in mice and nonhuman primates. Mol Med 1999; 5:287-300.
62. Boyle JS, Silva A, Brady JL et al. DNA immunization: induction of higher avidity antibody and effect of route on T cell cytotoxicity. Proc Natl Acad Sci USA 1997; 94:14626-14631.
63. Haensler J et al. Intradermal DNA immunization by using jet-injectors in mice and monkeys. Vaccine 1999; 17:628-638.
64. Feltquate DM, Heaney S, Webster RG et al. Different T helper cell types and antibody isotypes generated by saline and gene gun DNA immunization. J Immunol 1997; 158:2278-2284.
65. Pertmer TM, Oran AE, Madorin CA et al. Th1 genetic adjuvants modulate immune responses in neonates. Vaccine 2001; 19:1764-1771.
66. Torres CA, Iwasaki A, Barber BH et al. Differential dependence on target site tissue for gene gun and intramuscular DNA immunizations. J Immunol 1997; 158:4529-4532.
67. Ulmer JB, Sadoff JC, Liu MA. DNA vaccines. Curr Opin Immunol 1997; 8:531-536.
68. Pertmer TM et al. DNA vaccines for influenza virus: differential effects of maternal antibody on immune responses to hemagglutinin and nucleoprotein. J Virol 2000; 74:7787-7793.
69. Krieg AM et al. CpG motifs in bacterial DNA trigger direct B-cell activation. Nature 1995; 374:546-549.
70. Sato Y et al. Immunostimulatory DNA sequences necessary for effective intradermal gene immunization. Science 1996; 273:352-354.
71. Hemmi H et al. A Toll-like receptor recognizes bacterial DNA. Nature 2000; 408: 740-745.
72. Wagner H. Bacterial CpG DNA activates immune cells to signal infectious danger. Adv Immunol 1999; 73:329-368.
73. Brazolot Millan CL et al. CpG DNA can induce strong Th1 humoral and cell-mediated immune responses against hepatitis B surface antigen in young mice. Proc Natl Acad Sci USA 1998; 95:15553-15558.
74. Inchauspe G et al. Plasmid DNA expressing a secreted or a nonsecreted form of hepatitis C virus nucleocapsid: comparative studies of antibody and T-helper responses following genetic immunization. DNA Cell Biol 1997; 16:185-195.
75. Boyle JS, Koniaras C, Lew AM. Influence of cellular location of expressed antigen on the efficacy of DNA vaccination: cytotoxic T lymphocyte and antibody responses are suboptimal when antigen is cytoplasmic after intramuscular DNA immunization. Int Immunol 1997; 9:1897-1906.

76. Scheiblhofer S et al. Removal of the circumsporozoite protein (CSP) glycosylphosphatidylinositol signal sequence from a CSP DNA vaccine enhances induction of CSP-specific Th2 type immune responses and improvesprotection against malaria infection. Eur J Immunol 2001; 31:692-698.
77. Aberle JH et al. A DNA immunization model study with constructs expressing the tick-borne encephalitis virus envelope protein E in different physical forms. J Immunol 1999; 163:6756-6761.
78. Haddad D et al. Differential induction of immunoglobulin G subclasses by immunization with DNA vectors containing or lacking a signal sequence. Immunol Lett 1998; 61:201-204.
79. Cardoso AI et al. Immunization with plasmid DNA encoding for the measles virus hemagglutinin and nucleoprotein leads to humoral and cell-mediated immunity. Virology 1996; 225:293-299.
80. Lewis PJ, van Drunen Littel-van den H, Babiuk LA. Altering the cellular location of an antigen expressed by a DNA-based vaccine modulates the immune response. J Virol 1999; 73:10214-10223.
81. Torres CA, Yang K, Mustafa F et al. DNA immunization: effect of secretion of DNA-expressed hemagglutinins on antibody responses. Vaccine 1999; 18:805-814.
82. Cardoso AI et al. Measles virus DNA vaccination: antibody isotype is determined by the method of immunization and by the nature of both the antigen and the coimmunized antigen. J Virol 1998; 72:2516-2518.
83. Higgins TJ et al. Plasmid DNA-expressed secreted and nonsecreted forms of herpes simplex virus glycoprotein D2 induce different types of immune responses. J Infect Dis 2000; 182:1311-1320.
84. Barrios C et al. Partial correction of the TH2/TH1 imbalance in neonatal murine responses to vaccine antigens through selective adjuvant effects. Eur J Immunol 1996; 26:2666-2670.
85. Siegrist CA et al. Induction of neonatal TH1 and CTL responses by live viral vaccines: a role for replication patterns within antigen presenting cells? Vaccine 1998; 16:1473-1478.
86. Manickan E, Yu Z, Rouse BT. DNA immunization of neonates induces immunity despite the presence ofmaternal antibody. J Clin Invest 1997; 100:2371-2375.
87. Wang Y, Xiang Z, Pasquini S et al. Effect of passive immunization or maternally transferred immunity on the antibody response to a genetic vaccine to rabies virus. J Virol 1998; 72:1790-1796.
88. Martinez X et al. Combining DNA and protein vaccines for early life immunization against respiratory syncytial virus in mice. Eur J Immunol 1999; 29:3390-3400.
89. Weeratna RD et al. Priming of immune responses to hepatitis B surface antigen in young mice immunized in the presence of maternally derived antibodies. FEMS Immunol Med Microbiol 2001; 30:241-247.
90. Tighe H, Corr M, Roman M et al. Gene vaccination: plasmid DNA is more than just a blueprint. Immunol Today 1998; 19:89-97.
91. Boyer JD et al. Protection of chimpanzees from high-dose heterologous HIV-1 challenge by DNA vaccination. Nat Med 1997; 3:526-532.
92. Calarota S et al. Cellular cytotoxic response induced by DNA vaccination in HIV-1-infected patients. Lancet 1998; 351:1320-1325.
93. Calarota SA et al. Immune responses in asymptomatic HIV-1-infected patients after HIV-DNA immunization followed by highly active antiretroviral treatment. J Immunol 1999; 163:2330-2338.
94. Wang R et al. Induction of antigen-specific cytotoxic T lymphocytes in humans by a malaria DNA vaccine. Science 1998; 282:476-480.
95. Hassett DE, Zhang J, Slifka M et al. Immune responses following neonatal DNA vaccination are long-lived, abundant, and qualitatively similar to those induced by conventional immunization. J Virol 2000; 74:2620-2627.
96. Cutts FT et al. Immunogenicity of high-titre AIK-C or Edmonston-Zagreb vaccines in 3.5-month-old infants, and of medium- or high-titre Edmonston-Zagreb vaccine in 6-month-old infants, in Kinshasa, Zaire. Vaccine 1994; 12:1311-1316.
97. Dagan R et al. High-dose inactivated Hepatitis A vaccine (HD-HAV-VAC) in infants: Comparison of response in the presence versus absence of maternally-derived antibodies (Mat-Ab). Proceedings of the 38th Annual ICAAC. 1998.
98. Boyle CM, Robinson HL. Basic mechanisms of DNA-raised antibody responses to intramuscular and gene gun immunizations. DNA Cell Biol 2000; 19:157-165.
99. Radu DL et al. Effect ofmaternal antibodies on influenza virus-specific immune response elicited by inactivated virus and naked DNA. Scand J Immunol 2001; 53:475-482.
100. Le Potier MF, Monteil M, Houdayer C et al. Study of the delivery of the gD gene of pseudorabies virus to one-day-old piglets by adenovirus or plasmid DNA as ways to by-pass the inhibition of immune response by colostral antibodies. Vet Microbiol 1997; 55:75-80.
101. McCluskie MJ, Davis HL. Novel strategies using DNA for the induction of mucosal immunity. Crit Rev Immunol 1999; 19:303-329.
102. Schneider J et al. Induction of CD8+ T cells using heterologous prime-boost immunisation strategies. Immunol Rev 1999; 170:29-38.

103. Ulmer JB et al. Enhancement of DNA vaccine potency using conventional aluminum adjuvants. Vaccine 1999; 18:18-28.
104. Reyes L et al. Vaxfectin enhances antigen specific antibody titers and maintains Th1 type immune responses to plasmid DNA immunization. Vaccine 2001; 19:3778-3786.
105. Boyle JS, Brady JL, Lew AM. Enhanced responses to a DNA vaccine encoding a fusion antigen that is directed to sites of immune induction. Nature 1998; 392:408-411.

DNA Delivery with Attenuated Intracellular Bacteria

Joachim Fensterle and Stefan H. E. Kaufmann

Intracellular Bacteria: From the Pathogen to the Vaccine—An Overview

For centuries, intracellular bacteria have been a major cause of death globally. Notably, *Mycobacterium tuberculosis* killed more than 2 million individuals in 1999 worldwide, a number which was only exceeded by HIV.[1]

Intracellular bacteria are defined by their capability to survive and live inside eukaryotic host cells. Therefore, they comprise different unrelated species of bacteria that have developed diverse strategies to survive within this compartment. This includes the genus *Mycobacterium*, gram-positive genera like *Listeria*, and gram-negative genera like *Salmonella* and *Shigella*.

Obviously, efficacious vaccines against intracellular pathogens would be highly desirable. How can vaccines be developed for this threatening group of bacterial pathogens and how can one even consider using these bacteria as vaccine carriers? Due to the intracellular life style of these bacteria, T cells are critical mediators of the acquired immune response. Killed bacteria as vaccines are in most cases insufficient stimulators of T cells, although the first ever vaccine against intracellular bacteria, the typhoid vaccine developed 1897 by A. Wright and D. Sample, was based on this principle.[2]

Therefore, a frequently used first step in vaccine development against intracellular bacteria is to isolate attenuated strains of the bacteria that fail to cause the disease and assess their capacity to protect against infection with the virulent strain. Without doubt, *M. tuberculosis* is the most threatening pathogen of this group and consequently, an attenuated vaccine strain against *M. tuberculosis* was the first live vaccine available against intracellular bacteria. This vaccine strain is derived from the bovine pathogen *Mycobacterium bovis*, a close relative of *M. tuberculosis*. In 1909, A. Calmette and C. Guérin began to attenuate virulent *M. bovis* isolates by more than 200 in vitro passages.[3] This paragon live vaccine strain called *M. bovis* Bacillus Calmette-Guérin (BCG) is avirulent for humans and can protect against infection with *M. tuberculosis* though to a varying degree. Despite its poor efficacy, *M. bovis* BCG is still the only vaccine available against tuberculosis.

Attenuation by in vitro passage of virulent strains is time consuming, especially for the slow-growing mycobacteria. Consequently, the next generation of attenuated bacteria was generated by chemical or transposon mutagenesis. Using the latter technique, Hoiseth and Stocker developed an attenuated strain of *Salmonella typhimurium* with a defect in the synthesis of aromatic amino acids.[4]

Although mutagenesis accelerated the generation of attenuated bacterial vaccine strains, this technique still relies on the fact that genes involved in pathogenicity or auxotrophy are affected arbitrarily and extensive testing of a large number of different clones is required in order to finally identify suitable mutants. Furthermore, attenuated strains generated by

DNA Vaccines, edited by Hildegund C. J. Ertl. ©2003 Eurekah.com and Kluwer Academic / Plenum Publishers.

nondirected mutagenesis still bear the risk that mutations can revert to the virulent wild type strain. With advances in microbial genetics and genetic engineering, directed deletion of genes essential for pathogenicity or auxotrophy has become feasible. An example is the generation of an attenuated strain of *Listeria monocytogenes*, a food-born pathogen. One factor responsible for pathogenicity is the protein ActA, which promotes bacterial spreading from cell to cell.[5] The targeted disruption of the virulence gene encoding for ActA resulted in a highly attenuated *L. monocytogenes* strain constituting a potential vaccine candidate.[6]

Why should these attenuated intracellular bacteria be useful as heterologous vaccines when they cannot protect against the corresponding disease in some instances? The answer to this question lies in the unique microbiological and immunological features of this group of bacteria. An essential requisite for an effective vaccine is the induction of a potent immune response by antigen presenting cells (APCs).[7] These cells constitute the major targets for intracellular bacteria. Moreover, infection of APCs with most intracellular bacteria results in the activation of these cells and subsequent induction of a robust immune response.[8] Consequently, in the case of many intracellular bacteria, infection induces an immune response, without the need for additional adjuvants. However, the simple induction of an immune response is not the end of the game; the quality of the immune response is equally decisive. To fight against infections with intracellular bacteria, protozoa and viruses or against cancer, the vaccine must induce a T-helper 1 (Th1) immune response.[8] The Th1 response is characterized by IFN-γ secreting $CD4^+$ T-cells and the generation of cytotoxic T-lymphocytes (CTL).[9] Both, IFN-γ and CTL play a pivotal role in fighting infections with intracellular pathogens and cancer. IFN-γ activates professional phagocytes like macrophages to kill the engulfed pathogens, and CTL kill cells that fail to control their predators.

Indeed, the induction of a sustained Th1 response is the hallmark of infection with intracellular bacteria.[10] Although unique microbial factors are involved in the generation of the primary cytokine milieu and APC activation after infection (see below), pattern recognition receptors (PRR) on APC and other cells of the innate immune system turn out to be of central importance.[11] Recent focus increasingly turned towards a subfamily of PRR, the Toll-like receptors (TLR), a receptor family related to the Toll receptor of Drosophila.[12] As summarized in Figure 1, members of this family recognize different bacterial agents, including lipopolysaccharides (LPS) abundant in the cell wall of Gram-negative bacteria,[13,14] cell wall components of gram-positive bacteria,[15-18] hypomethylated CpG motifs on bacterial DNA,[19] and bacterial flagellae.[20] Recognition of bacterial agents by TLR results in signaling via the IL-1 receptor pathway and finally the activation of the cell.[21]

In summary, the targeting of intracellular bacteria to APCs, the possibility to apply them orally and their capacity to induce a Th1 response are intriguing features of this group of bacteria that qualify them as attractive carriers for heterologous antigens. Combined with the advantages of DNA immunization, intracellular bacteria may offer promising novel avenues towards the design of future vaccines against still untreatable or unpreventable diseases notably tuberculosis, AIDS and cancer.

DNA Delivery versus Naked DNA Vaccination—The Pros and Cons

The discoveries that naked DNA injected into the mouse muscle is expressed by host cells,[22] and that naked DNA injection can elicit an immune response[23] were major breakthroughs in modern vaccine research. Soon after the initial experiments, several groups demonstrated in animal models that naked DNA vaccination can protect against a variety of pathogens and cancer.[24] However, confirmation in clinical trials of the promising results in small animal models turned out to be difficult.[25] Major drawbacks include the high amount of DNA needed for injection and the low seroconversion rate of vaccinees. In contrast, conventional vaccination with heterologous live bacterial carriers allows easy application via the oral route and some carrier strains have already been approved for clinical use.[26] However, bacterially expressed

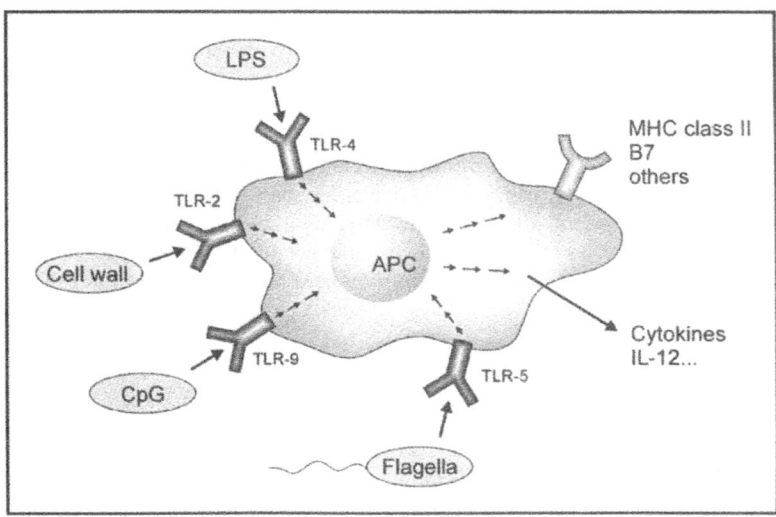

Figure 1. Activation of phagocytes and cells of the innate immune system by bacterial products via Toll-like receptors (TLR). Bacterial products are sensed via TLR and subsequently, signaling through the IL-1 signaling pathway mediates cytokine secretion and upregulation of activation markers. LPS: lipopolysaccharides, CpG motifs: immunostimulatory, hypomethylated CpG sequences on bacterial DNA. See text for further details.

antigens can have different structural features compared to proteins expressed by eukaryotic cells and, depending on the carrier strain, induction of CTL is often relatively poor.

Combination of naked DNA vaccines with intracellular bacteria as carriers can open a new area of vaccine development, combining the advantages of both techniques, namely: (I) the protein is still produced by the host cell, (II) CTL, as well as Th1 CD4$^+$ T-cell responses, are efficiently induced, (III) the vaccine can be administered orally, (IV) the vaccine is directly targeted to the APCs and (V) the vaccine is easily produced. The problem, however, remains, that the bacterial vaccine carrier will induce an immune response directed against itself. Although this reaction is a prerequisite for the intrinsic adjuvanticity of bacterial vaccines, it poses several problems. Repeated vaccinations or previous contact with the homologous or a related pathogen could focus the immune response against the carrier and impair the response against the antigen encoded by the DNA vaccine.

Studies considering the problem of preexisting immunity to the carrier produced somehow conflicting results.[27-35] Taken together, the animal studies suggest that antibody responses are diminished in animals immunized with pathogens to which they have preexisting immunity, whereas immunization can still be successful if the animals are challenged with the pathogen corresponding to the heterologous antigen. Nevertheless, these studies suggest that the problem can be solved by the appropriate choice of a carrier strain expressing minimal cross-reactivity to human pathogens, and with optimal expression of the heterologous antigen, which is easily achieved by DNA delivery. However, only clinical trials with different bacterial DNA carriers can ultimately provide the correct answer.

Probably, the optimal schedule will not consist of a single vaccination with naked DNA, DNA delivered by bacteria or heterologous live vaccines. Recent work suggests that the combination of different vaccine carriers in a heterologous prime/boost schedule yields superior results[36] and therefore it appears necessary to optimize each technique and to develop the best-suited prime/boost protocol combining the advantages of different types of vaccines.

Intracellular Bacteria: How Do They Survive?

Intracellular bacteria have the capacity to reside and survive within eukaryotic host cells. In order to reach their preferred target cells and to survive in macrophages, which are specialized to kill engulfed pathogens, they have developed distinct strategies. To understand how DNA delivery by these bacteria functions, some basic knowledge about the life cycle of intracellular bacteria is required. Figure 2 summarizes major techniques different bacteria employ to infect phagocytic cells and how they manage to survive inside these cells.

In the following, the microbiology and infection biology of three representative members of the group of intracellular bacteria will be discussed in greater detail. These species, namely *L. monocytogenes*, *Shigella flexneri* and *S. typhimurium* have already been used as DNA carriers. The main emphasis will be placed on *L. monocytogenes*, as it is the best-studied microorganism of this group.

Listeria monocytogenes

The food-born pathogen *L. monocytogenes*, a nonsporulating, gram-positive bacterium is the causative agent of listeriosis that causes serious infections in elderly and immunosuppressed individuals but poses only a minor risk for immunocompetent individuals. Yet, *L. monocytogenes* infection is a problem for pregnant women, and in about 50% of affected females, infection of the fetus leads to abortion and death of the fetus.[37] Overall, some 100 individuals in the US succumb to listeriosis annually. In addition, the recall of contaminated food products can be extremely costly and easily reach the level of hundreds of millions US Dollars.

Although the overall clinical significance of *Listeria* is low, *L. monocytogenes* had a major impact on basic research. Use of *L. monocytogenes* as a model organism dates back to 1962, when G.B. Mackaness introduced the mouse model of listeriosis.[38] Since then, numerous studies have taken advantage of this model system, and today *L. monocytogenes* is probably the best studied intracellular pathogen from the aspect of immunology and infection biology.

Which features render *Listeria* such an attractive model organism? First of all, *L. monocytogenes* is easy to grow and the risk of laboratory infection is small. Listeriosis in mice (in contrast to human infections) progresses rapidly and results in acute disease, allowing fast evaluation of the immune response and the efficacy of vaccination schedules. Finally, *L. monocytogenes* spreads from cell to cell without leaving the cell and hence is restricted to the intracellular compartment.[39]

Molecular Biology of Listerial Infection

The principal manner of listerial uptake is via the oral route. Ingested *Listeria* reach the blood circulation via the intestine. Only a few hours after the initial systemic phase, the bacteria are filtered out in their target organs, spleen and liver, and completely cleared from the blood stream. *L. monocytogenes* can then further disseminate to the brain stem and the placenta.[40] *Listeria* can infect, propagate in and disseminate to many cell types, especially APCs, which renders them attractive vaccine carriers. The molecular mechanisms underlying uptake and dissemination between cells will be described in the following paragraph.

As a first step of bacterial invasion of nonphagocytic cells the cytoskeleton is manipulated by the bacteria to promote their entry. Two bacterial proteins have been identified so far, internalin (or InlA) and InlB, which are involved in this process.[41] The entry of *Listeria* into human epithelial cell lines is promoted by the interaction of internalin with E-cadherin, which is expressed on epithelial cells at adherence junctions.[42,43] However, internalin does not interact with murine E-cadherin. Recently, using transgenic mice expressing human E-cadherin, the in vivo role of internalin for epithelial cell entry in the intestine was revealed.[44] Two cellular counterparts of InlB have been identified, namely a poorly characterized surface molecule, gC1qR[45] and, the Met receptor tyrosine kinase.[46] Binding of InlA and InlB to mammalian cells leads to signaling via the Met receptor and possibly other receptors, and finally to rearrangement of the actin cytoskeleton.[47] *Listeria* mutations causing defects in internalin or InlB block the uptake of *Listeria* in many cell types, but not the uptake into dendritic cells (DCs).[48]

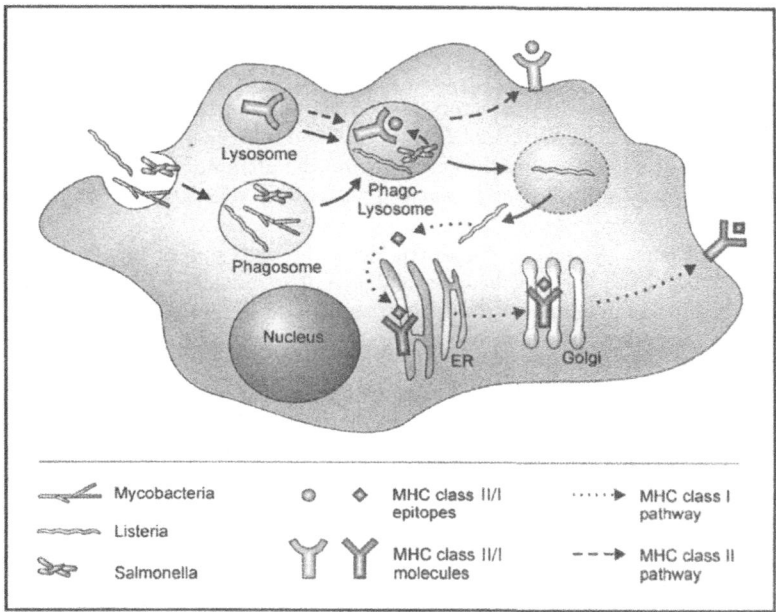

Figure 2. Uptake and localization of intracellular bacteria in professional antigen presenting cells (APCs) and access of bacterial products to the MHC class I/II presentation pathway. *Mycobacteria, Salmonella* or *Listeria* are phagocytized by APCs. In the phagolysosome, bacterial products have access to the MHC class II presentation pathway and can induce CD4+ T-cells. To allow bacterial survival, *Mycobacteria* block phagosome—lysosome fusion, *Salmonella* modify the antibacterial phagolysosomal properties and *Listeria* escape into the cytosol by lysing the membrane. In the cytosol, listerial products have access to the MHC class I presentation pathway. *Shigella* also have the capability to escape out of the phagolysosome and thus have similar access to the MHC class I presentation pathway as *Listeria*.

Within the cell, *Listeria* egresses from the phagosome into the cytosol with the help of two virulence factors, listeriolysin (LLO) and the broad host range phospholipase, phosphatidyl-inositol-specific phospholipase C (PI-PLC), encoded by the plcA gene.[49] LLO, a hemolysin with an acidic pH optimum of pH 5.5, forms pores in the membrane of *Listeria*-containing vesicles,[50] which, together with the action of PI-PLC allows bacterial escape into the cytoplasm. LLO is also responsible for the expression of several surface molecules and the induction of chemokines via NF-κB.[51] This protein therefore could be essential not only for virulence, but also for induction of an efficient immune response. LLO is harmful in the cytoplasm due to its pore forming activity. As recently shown, the amino acid sequence of LLO comprises a so-called PEST sequence, which is responsible for rapid degradation of the protein by host proteases.[52] Once this sequence is removed, LLO remains active and infected cells die quickly from listerial infection.

Once in the cytoplasm, *L. monocytogenes* propagates and moves around using a sophisticated strategy: the listerial protein ActA, which is expressed in the cytoplasm causes the polymerisation of actin to the tail of the bacteria which then pushes the pathogen through the cell.[5] Using this actin motor, *Listeria* are pushed through the cellular membrane and thus reach adjacent cells.[41] This process allows *Listeria* to infect neighboring cells without the need to leave the cellular compartment. On their way to the next cell, *Listeria* are entrapped in vesicles with double membranes with properties different from phagosomes. Via the action of the second phospholipase C, phosphatidylcholine-specific phospholipase C (PC-PLC) encoded by the *plcB* gene, the bacteria escape from this compartment into the cytoplasm.[53] The role of

LLO in cell-to-cell spread has been controversial, due to the acidic pH optimum of LLO and the requirement of this protein for primary infection. Using nonhemolytic *Listeria* covered with LLO, Gedde et al convincingly demonstrated that LLO is required for escape from the double-membrane vacuole.[54]

Several well-described virulence factors (and certainly also additional unknown factors) are required for the intracellular lifestyle of *Listeria*. InlA and InlB are indispensable for infection, while LLO and PLBC are required to escape from the phagosome and the double membrane vacuole. Finally, spreading of *L. monocytogenes* within cells is mainly promoted by the ActA protein.

As described later, knowledge of essential factors for each step of cellular infection and intracellular dissemination allows the rational design of attenuated strains that are optimized for DNA delivery and induction of the appropriate immune response.

Immunology of Listerial Infection

The most important part of the host response for vaccination is the acquired immune response. Early studies basically concentrated on the strength of the cellular and humoral response induced by a given vaccine candidate. However, it has become increasingly clear that the type of immune response is equally important for vaccination.[8] Therefore, it is important to consider the different phases of the immune response, including the early, nonspecific response, for design of the next-generation vaccines.

As already mentioned, one of the most important findings for vaccine development in the last years was the discovery of two distinct types of CD4$^+$ T-helper cells, namely Th1 and Th2 cells.[9] These cells can be distinguished by the cytokines they produce. Th1 cells secrete predominantly IL-2 and IFN-γ, whereas Th2 cells secrete mainly IL-4, IL-5 and IL-10. For protection against most diseases against which vaccines are not available so far, induction of a Th1 response is pivotal.[8] Accordingly, this chapter will focus on the factors promoting the Th1 type immune response induced by *Listeria*.

During natural infection with *L. monocytogenes*, the first line of defense comprises neutrophils.[55] These cells are attracted by infected hepatocytes undergoing apoptosis during infection and destroy the vast majority of bacteria.[56] Depletion of neutrophils severely impairs resistance to *L. monocytogenes* due to the uncontrolled growth of bacteria at the initial site of infection.[57] However, the role of neutrophils in the induction of the T cell response appears negligible. Rather the interplay between macrophages and natural killer (NK) cells appears crucial in the early phase of infection and the subsequent activation of T cells.[58] Upon listerial infection, macrophages produce TNF-α and IL-12, both of which stimulate NK cells to secrete IFN-γ.[59] Experimentally infected animals die rapidly if any of these cytokines is absent.[60-65] Subsequently, IFN-γ produced by NK cells activates macrophages, allowing them to kill *L. monocytogenes* organisms, which still reside in the phagosome. MHC class II molecules are also upregulated. Besides the already mentioned cytokines, IL-1 and IL-6 participate in the early cytokine response and act mainly by recruiting neutrophils.[66-68] A recent study in knockout animals revealed an essential role of an unusual cytokine, Eta-1.[69] It was found that animals deficient in Eta-1 have a severely impaired capacity to mount Th1 responses, and are therefore extremely sensitive to infection with *L. monocytogenes*.[69] It is interesting to note that prior to this study, the role of this cytokine, which was already described in 1989,[70,71] was incompletely understood,[72] and only knockout animals revealed its important function in antibacterial immunity. It is therefore likely that the interplay of cytokines in listeriosis still bears some surprises.

Activation of professional APCs at the early stage of infection leads to the presentation of epitopes derived from listerial proteins by MHC class I and MHC class II molecules.[73-75] This links innate immunity to the adaptive T cell response against *Listeria*. Primed CD4$^+$ and CD8$^+$ T-cells induce rapid clearance of *L. monocytogenes* from the infected host.[76]

The critical T cells for clearing listerial infection are CD8$^+$ T-cells.[77] Early research suggested a role for the classical effector molecules of CD8$^+$ T cells, namely perforin and granzyme,[78] as well as IFN-γ.[64] More recently, Harty and colleagues demonstrated that TNF-α produced by CD8[+] T cells also plays an important role in protection.[79,80] Another hallmark of immunity against *L. monocytogenes* was elucidated by Pamer and collegues using the MHC class 1 tetramer technique, which allows the quantitative analysis of the CD8[+] T cell response.[81,82] This group demonstrated the induction of high numbers of memory CD8[+] T cells, which form the secondary CD8[+] T cell response and cause rapid clearance of *L. monocytogenes* challenge infection. For further details, the interested reader is referred to an excellent review by Busch and Pamer.[76]

Beside CD8[+] T cells, *L. monocytogenes* also induces CD4[+] T cells and, to a minor extent, specific antibodies. Recent research suggests that CD4[+] T cells[83,84] and specific antibodies can protect against primary infection.[85]

The major task of a next generation vaccine is the efficient establishment of a memory pool of T cells and the induction of a Th1 type response; intracellular bacteria fulfill these demands effectively. The first step is the early generation of a cytokine milieu promoting the Th1 immune response. As described previously, *L. monocytogenes* infection causes early secretion of IFN-γ, IL-12 and Eta-1. IFN-γ counteracts the production of IL-4, while IL-12 and Eta-1 promote the activation of CD4[+] T cells of the Th1 phenotype. In addition, several bacterial factors (LPS for gram-negative bacteria, lipoteichoic acids for gram-positive bacteria and hypomethylated CpG motifs from DNA) fully activate professional APCs (see Figure 1) through binding to TLR[86] and upregulating costimulatory molecules like B7-1, B7-2, VLA-4, ICAM-1, CD40 and also MHC class I and MHC class II molecules.[87]

Rational Design of *L. Monocytogenes* Vaccine Strains

The first step in the development of a live vaccine is the generation of an attenuated strain, which can be applied safely but has preserved the capacity to induce a protective immune response efficiently. In the early days of vaccinology, attenuated strains were used as vaccines without further fine characterization of the attenuation. These vaccines bore the risk of reverting to virulence. A modern vaccine must possess a stable genotype incapable of reverting to virulence.

Most attenuations of intracellular bacteria to date affect essential metabolic pathways, rendering the bacteria dependent on particular nutrients (auxotrophy). For example, Alexander and coworkers obtained a highly attenuated mutant of *L. monocytogenes* using transposon mutagenesis disrupting phenylalanine synthesis,[88] and Rouquette et al described an attenuated strain with a mutation in iron uptake.[89] These strains grow well in culture medium containing aromatic acids or iron, whereas in vivo these nutrients are limiting.[88-90] Targeted attenuation has been used in a number of bacterial strains, including vaccines accepted for clinical use. However, most strains were created by transposon mutagenesis or carry point mutations thus bearing the risk of reversion. New listerial vaccine strains based on mutations in metabolic pathways need to carry gene deletions to minimize the risk of reversion.

The general growth inhibition in vivo of auxotrophic bacteria can be disadvantageous with regard to vaccine efficacy. These bacteria normally persist for a short time in the host. Therefore, only low concentrations of antigen are available for the immune system. With the increasing knowledge of infection biology, rational design of vaccine strains has become possible. To date, several groups have developed attenuated strains through deletion of key virulence factors essential for pathogenicity, but not influencing immunogenicity. For example, an attenuated *Listeria* strain should not be mutated in the listeriolysin gene, as this would lead to reduced antigen presentation to CD8[+] T-cells.[91] One of the most straightforward approaches to listerial attenuation is the deletion of the *actA* gene.[92] These mutants still possess the ability to invade APCs, replicate in these cells and efficiently induce both CD4[+] and CD8[+] T cells. However, as they lost their capability to spread to adjacent cells they are highly attenuated. Several groups have used attenuated *L. monocytogenes* strains based on this principle and demonstrated induction of an effective immune response against heterologous antigens expressed by the listerial carriers themselves[93,94] or encoded by naked DNA constructs delivered by the listerial carriers.[6] Another virulence factor suitable for attenuation is the phospholipase PC-PLC involved in the escape from the double-membrane vesicles. Indeed, *L. monocytogenes* strains carrying a deletion of *plcB* and *actA* were found to be highly attenuated but still potent vaccine strains for DNA delivery.[6]

In summary, various targets for attenuation of *L. monocytogenes* exist. For use in humans, a vaccine strain must encompass at least two independent deletions of virulence factors and/or metabolic enzymes to ensure a minimal risk of reversion.

Shigella flexneri

Shigella are gram-negative, rod-like bacteria which are responsible for several intestinal diseases ranging from mild diarrhea to acute inflammatory bowel disease, bacillary dysentery. The disease mainly affects impoverished people and especially young children of low economic status.[95] A vaccine against *Shigella* has a high priority, as antibiotic treatments are not affordable in the developing world.[96]

Although *Shigella* belongs to a different genus than *Listeria*, both have developed comparable survival strategies: phagosomal escape and actin driven cell-to-cell spread. Yet *Shigella* only transiently inhabits its host cells and shigellosis rarely becomes systemic; infection is normally restricted to the intestine.[97]

Molecular Biology of Shigella Infection

Shigella enter the host via the oral route. As a first step, epithelial cells of the colon are invaded. The reason for the specificity of *Shigella* for colonic epithelial cells is unknown, but it is very likely that colon-specific adhesion factors are involved.

Which factors are responsible for the uptake of *Shigella* into epithelial cells and how do the bacteria escape the phagolysosome?

Most virulence factors of *Shigella* are found on a 220 kb virulence plasmid with a complex pattern of regulation also encompassing chromosomal genes.[98] The central component for virulence encoded on this plasmid is a type III secretion system, which also exists in pathogenic *E. coli* species and in *Salmonella enterica*.[99] The basic flagella-like-structure is composed of at least five proteins encoded by the mxi operon localized on the pathogenicity island (PAI) of the plasmid. The secreton encompasses a needle-like structure composed of a channel of 2-3 nm, through which secreted proteins can pass.[100] Contact with epithelial cells induces expression of the type III secretion system. The first proteins essential for infection of epithelial cells are IpaB, IpaC and IpaD. The current model holds that IpaB and IpaD control the flux through the type III secretion system by forming a complex on the top of the needle-like structure. Upon contact with their target cell, IpaB and IpaC are inserted into the eukaryotic cell membrane and inject bacterial proteins into the cytoplasm.[101,102] Furthermore, the interaction of IpaC with the eukaryotic cell initiates remodeling of the actin skeleton to allow bacterial uptake into the cell.[103] IpaC induces signaling via Cdc42 and Rac, resulting in the local assembly of actin. Yet, the exact mechanisms of actin polymerization induced by IpaC for bacterial entry are more sophisticated and have been reviewed extensively elsewhere.[104] IpaA, which is injected through the type III secreton into the cytoplasm activates the cellular protein vinculin and thus induces the bundling and partial depolymerization of actin.[105,106] Ultimately these interactions promote the uptake of *S. flexneri* into epithelial cells via membrane ruffles.

Once within the cellular vacuole, *Shigella* lyses the surrounding membrane via the action of IpaB and IpaC and escapes into the cytosol.[107,108] Like *Listeria*, *Shigella* has the capacity to move within the cytosol and to infect neighboring cells via the polar assembly of actin. Actin assembly is mediated by a single protein anchored in the outer membrane of the bacterium, IcsA, also called VirG.[109] Again, host cell proteins like N-WASP and Arp2/3 are recruited via IcsA, which ultimately cause actin assembly.[110-112] Several other virulence factors are also involved in virulence of *Shigella*, either directly or by regulating expression of virulence genes.

Immunology of Shigella Infection

The major hallmark of *S. flexneri* infection is the massive inflammation of the infected areas accompanied by partial tissue destruction by the exaggerated host response. Does the bacterium have any benefit from inducing such a massive response ultimately causing its eradication; a process Sansonetti has termed "a case of fatal attraction?"[104]

Similar to other enteroinvasive pathogens, the initial events of infection are focused on M cells, which overlay lymphoid structures in the intestinal mucosa. The follicle-associated epithelium (FAE), where M cells are located, lacks mucus, glycocalyx and brush borders.[113] Hence, it provides an accessible way into the intestinal tissue. However, using this port of entry, the pathogen is immediately exposed to numerous host cells of the immune system, notably neutrophils and macrophages. Therefore *Shigella* requires mechanisms that allow escape from antibacterial activities of these cells. Even though *Shigella* can egress from the phagolysosome into the cytoplasm, the central strategy of *Shigella* is induction of apoptosis of macrophages and escape into subepithelial tissues.[114]

IpaB is responsible for macrophage apoptosis.[115] Through binding to caspase 1, IpaB initiates the caspase-cascade ultimately causing apoptosis.[116] At the same time, caspase 1 activation causes release of two pro-inflammatory cytokines, IL-1β and IL-18. The release of these cytokines seems to be not accidental, as experimental infusion of IL-1 receptor antagonist not only reduces inflammation, but also inhibits further invasion of the bacteria.[117] More recently, the roles of IL-1 and IL-18 in shigellosis were elucidated in greater detail.[118] Indeed, IL-1 seems to be involved in tissue destruction allowing bacterial dissemination, whereas IL-18 is responsible for IFN-γ induction, which finally promotes control of *Shigella* infection.[119] Central to host defense are IFN-γ and immune activation via bacterial LPS. Indeed, blocking of CD14, a surface receptor delivering LPS to TLR4,[21] severely impairs host resistance against shigellosis.[119-121]

In addition, epithelial cells secrete large amounts of IL-8, a potent proinflammatory cytokine and chemoattractant for neutrophils.[122,123] Invasion of large numbers of neutrophils further contributes to tissue destruction. At the same time, neutrophils engulf and kill *Shigella*. Even though the bacteria do not survive within these cells, blocking of neutrophil immigration decreases tissue destruction and bacterial invasion.[124]

In summary, *Shigella* infection is primarily characterized by massive inflammation caused by IL-1 and IL-8, whereas control of infection is mediated primarily via IL-18, IFN-γ and the activation of phagocytes by LPS.

Rational Design of a *Shigella* Vaccine

Shigella appears to be a more problematic vaccine carrier candidate than *Listeria*. This is mainly due to the massive inflammation and tissue destruction induced at the very beginning of infection. Therefore, metabolic attenuation is certainly insufficient and virulence attenuation which ameliorates infection is critically required.

A number of attenuated *Shigella* strains exist, some of which have already been tested in human trials for safety and efficacy. Most strains carry more then one mutation with the majority encompassing mutations within the aromatic amino acid or nucleotide synthesis pathways.[125-134] Mutation of the gene responsible for cell-to-cell spreading, *icsA*, efficiently attenuates the bacteria and has been frequently combined with auxotrophic mutations[127,128,135-138] However, these strains still cannot be applied orally because high doses result in significant inflammation, fever and diarrhea. Therefore, further attenuation affecting factors responsible for inflammation are needed with additional mutations in the endotoxin synthesis further lowering the inflammatory potential.[139]

Taken together, the generation of attenuated *Shigella* strains is possible, but more complicated as compared to *Listeria*, because of its profound inflammatory potential.

Salmonella enteritidis

Salmonella enteritidis is a gram-negative rod comprising a multitude of serovars with different host specificity.[140] The most important human serovar is *S. enteritidis* serovar Typhi (*S. typhi*), the causative agent of typhoid fever. Several other serovars of *S. enteritidis* cause gastrointestinal infections in humans. The best-studied serovar is *S. enteritidis* serovar Typhimurium (*S. typhimurium*), which causes gastroenteritis in man and typhoid in mice. Hence it provides a useful model for human typhoid. *Salmonella* infections are highly prevalent globally and cause infections in children and elderly people. Accordingly, control of *S. enteritidis* infection in humans and farm animals is of major importance for global health.[141,142] *Salmonella* infections can be treated by chemotherapy. However, emergence of drug resistance increasingly demands the development of vaccination strategies.[143]

Salmonella has developed a strategy different from that employed by *Listeria* and *Shigella* to escape antibacterial activities of phagocytes: it modifies the phagolysosome to allow its survival and its propagation within this compartment.[144]

Molecular Biology of *S. typhimurium* Infection

S. typhimurium is an enteropathogen which normally enters the host by the oral route. After passage through the acidic stomach, which is facilitated by the products of several acid tolerance (*atr*) genes,[145,146] *S. typhimurium* reaches the intestine where it colonizes the mucosal tissue and traverses the intestinal epithelium.[147] Most genes responsible for virulence of *S. typhimurium* are localized in two pathogenicity islands called *Salmonella* pathogenicity island (SPI) 1 and SPI2.[148] Central to pathogenicity are two inducible type III secretion systems resembling the type III system of *Shigella*.[99,149] Upon colonization of the intestinal mucosa, which is mainly mediated by fimbrial genes,[150-153] the SPI1 encoded type III secretion apparatus is expressed.[154] Several factors are important for the structure and functioning of this system including: InvA, InvG, PrgH and PrgK, which are structural components of the needle complex,[155] InvJ, SpaO and SipD, which are necessary for maintaining and regulating the secretion process[156,157] and SipB, SipC and SipD, which are required for the translocation of secreted proteins into the host cell cytosol.[158,159]

One of the first proteins secreted into the host cell belongs to a family originally designated "*Salmonella* Outer Proteins", SopE.[160] By activating Cdc42, SopE promotes reorganization of the actin cytoskeleton via Rac[161] and translocation of NF-κB and activation of p38 and JNK via MEK kinases.[162] However, deletion of the *sopE* gene does not cause marked inactivation. A recently identified protein, SopE2, which is homologous to SopE, possibly has a similar function and *sopE2* negative mutants are highly attenuated.[163] The resulting membrane ruffles induced in host cells at the contact site are essential for uptake of *S. typhimurium*. Depending on the host and serovar, *Salmonella* either exclusively invades via M-cells or it infects enterocytes overlying domed villi; this results in differential severity of disease.[147] However, bacterial invasion is not essential for the severity of intestinal disease and only the minority of bacteria invade intestinal cells.[147]

Other translocated proteins, like SopA, C, D and in particular SopB, are responsible for the extensive fluid release into the intestinal lumen. SopB (also called SigD), an inositol phosphate phosphatase,[164,165] hydrolyses inositol phosphate, which can block closure of chloride channels through several signaling events.[166,167]

Certain *Salmonella* infections are not restricted to the intestine and then cause typhoid-like disease. *Salmonella* attracts phagocytes, and a small proportion uses them as a vehicle for systemic dissemination. Even though *Salmonella* can induce apoptosis in macrophages via SipB,[168] this process remains limited so that *Salmonella* can misuse these cells for systemic dissemination. As a consequence, *Salmonella* must survive within professional phagocytes. The virulence factors central for survival within phagocytes are located in SPI2. This chromosomal region encompasses another type III secretion system that is pivotal for survival within the phagosome.[169] Several virulence factors, which are secreted by this secreton modulate the properties of the phagosomal compartment within the APCs.[170] These include SpiC, which modifies phagosome-lysosome maturation,[171] PmrA- regulated pags and SpvA-D, which allow survival at low pH, promote resistance against attack with cationic peptides and reactive oxygen intermediates and promote growth under oxygen and nutrient-limiting conditions.[172-176]

Much effort has been made to understand the complex regulation of virulence factors and several key regulator genes have been identified.[177] Numerous attenuated strains of *S. enteritidis* have become available encompassing mutations in virulence gene regulators with major impact on attenuation and immunogenicity of which we will only discuss the two most important ones. Certainly one of the most critical regulators of expression and repression of virulence, including SPI1 and SPI2 encoded factors, is PhoP-PhoQ. This two-component system senses extracellular cation levels and activates the expression of PhoP activated genes (pags).[178] Another central regulator of SPI1 and SPI2 expression in *S. typhimurium* is EnvZ/OmpR.[179] This two-component system seems pivotal for expression of the SPI1 encoded type III secretion system. Numerous other factors are involved in the regulation of virulence genes and this complex network ultimately regulates expression of virulence genes by sensing environmental conditions. For *Salmonella* organisms, which spend part of their lives within the host and part in the sewage, this sophisticated regulation of virulence gene expression allows efficient use of the latter during infection and avoids a prompt immune response against antigens, which are not needed at early stages of infection.

Immunology of *Salmonella* Infection

Both *Salmonella* and *Shigella* misuse the immune response for disrupting the intestinal mucosa to promote widespread dissemination throughout this tissue. However, the inflammatory response induced by *Salmonella* is less severe and obviously regulated differently.

The first contact between *Salmonella* and the immune system occurs after passage of the intestinal epithelium, which causes attraction of phagocytes.[180] Some bacteria misuse macrophages and neutrophils as transporters to distant tissues.[147] Even though *Salmonella* is principally able to survive within macrophages, intracellular survival and killing remains in a delicate balance and strongly depends on the stage of macrophage activation. IFN-γ and TNF-α activate macrophages and induce a whole array of antibacterial capacities.[181,182] Another important factor for defense against *Salmonella* infection is the expression of a functional Nramp1 molecule, which functions as a cation channel.[183-186] Depending on the activation stage of the macrophage, the presence of a functional Nramp1 molecule and the way of bacterial uptake, *Salmonella* either survives in phagocytes or is killed.

Central to the induction of the massive host response early in infection is LPS.[187,188] As depicted in Figure 1, LPS is sensed via specific PRR, notably CD14 and TLR–4.[189] As a result, a variety of proinflammatory cytokines (including TNF-α, IL-1, IL-6, IL-12 and IL18) and chemokines are produced, which cause massive infiltration. This cytokine milieu, especially IL-12 and IL-18, promotes Th1 polarization of the adaptive immune response[190,191] and prompts production of IFN-γ by NK cells, macrophages and other cells.[190,192-196]

It is not surprising that *S. typhimurium* infection induces a profound Th1 response, including high titers of *Salmonella* specific IgG2a and IgG3 antibodies with opsonizing activity. *Salmonella* usually resides in the phagosome and does not escape into the cytosol, hence the observation that *S. typhimurium* infection induces CD8$^+$ T cells was quite surprising[197] and the mechanism of CD8$^+$ T cell-priming remained obscure until recently, when it was shown that *S. typhimurium* induces apoptosis in infected macrophages. Resulting apoptotic bodies are engulfed by adjacent DCs, which subsequently present these antigens in the context of MHC class I molecules.[198,199]

The roles of the different T cell subsets and antibodies in clearing primary and secondary *S. typhimurium* infection have been reviewed elsewhere.[200] In short, mainly CD4$^+$ T cells are responsible for clearing Salmonella infection whereas the relative contribution of antibodies and CD8$^+$ T cells remains controversial. Yet, the capacity of *S. typhimurium* to potently stimulate CD4$^+$ T cells, CD8$^+$ T cells and opsonizing antibodies renders them promising vaccine carriers. Moreover, most attenuated strains of *S. typhimurium* persist for several weeks in mice. For a vaccine strain, this long lasting persistence is attractive, as it promotes induction of memory T cells.

Rational Design of *Salmonella* Vaccines

Although a variety of attenuated *Salmonella* strains have already been generated.[201] many highly attenuated strains were found to exhibit only poor immunogenicity. For example, the disruption of the PhoP/Q regulator system not only reduces virulence, but also immunogenicity.[202,203] Similarly and even more obviously, disruption of the key regulator of SPI1 genes, HilA, almost completely abrogates immunogenicity.[179,203]

The best established attenuation of *Salmonella* is the block in the synthesis of aromatic amino acids.[204] The *aroA* mutation in *S. typhimurium* produces a highly attenuated strain, that is still immunogenic. Another candidate gene for attenuation of *Salmonella* is *sopB*. This mutation results in reduced severity of diarrhea and PMN infiltration and should only slightly influence antigenicity.[164] In contrast, mutations in the type III secretion system markedly affect immunogenicity. Hence mutations in these virulence genes should be avoided whereas nutrient attenuations appear more promising.

The existing attenuated strains of *S. typhimurium* and *S. typhi* do not yet represent optimal vaccine strains. However, the growing knowledge about *Salmonella* makes it very probable that potent and safe vaccine strains will become available in the near future.

DNA Delivery with Intracellular Bacteria

For many years, intracellular bacteria have been candidate carriers for heterologous antigens in animal models and in human trials.[201,205-207] The paper of Sizemore et al in 1995, demonstrating DNA delivery with live bacteria, unfolded this fertile field.[208] The next sections will summarize the work on DNA delivery with attenuated *Shigella*, *Listeria* and *Salmonella* strains.

The Starting Point: DNA Delivery with S. Flexneri

The first DNA delivery experiments with viable bacteria were performed with auxotrophic *S. flexneri* containing a deletion in the *asd* gene.[208] In vivo, this strain fails to synthesize the cell wall component diaminopimelinic acid (DAP) and thus is lysed upon arrival in the host cell cytoplasm.[209] To demonstrate successful DNA delivery and expression in vitro and in vivo, the model antigen β-galactosidase, encoded by a eukaryotic expression plasmid was used.[208] Furthermore, intranasal immunization of mice with the auxotrophic strain induced both, humoral and cellular immunity, which however was incompletely characterized.[210] Later, successful induction of a Th1 response was demonstrated, encompassing specific CD8+ T-cells after immunization of mice with a *S. flexneri* strain carrying genes encoding for antigens of the measles virus.[211] The strength of the T cell response could be boosted by additional vaccinations with the same strain. In this particular situation at least, the immune response to the carrier did not dominate over that to the encoded antigen. Even if this observation cannot be extrapolated to other bacteria and other antigens, it suggests that in principle consecutive immunizations with the same strain are possible.

In May 1995, Powell and coworkers filed a comprehensive patent (claiming the method of DNA delivery for *Shigella* and a variety of other bacteria), using *S. flexneri* strains carrying *aroA* and *virG* mutations as well as strains with *asd* mutations for successful DNA delivery of a β-galactosidase construct in vitro.[212,213] In the same year, another group published that invasive *S. flexneri* and *E. coli* strains could be used to transfect mammalian cell lines in vitro.[214] Using appropriate plasmids, stable integration of the foreign gene into the mammalian chromosome was achieved (see below).

More recently, Anderson et al compared the efficacy of heterologous antigen expression or DNA delivery by an attenuated *S. flexneri* strain with that of naked DNA immunization.[215] These authors used an engineered *S. flexneri* 2a strain with deletions in the *guaBA* gene (auxotrophy for guanin biosynthesis) and the *virG* gene[128] and the carboxy-terminal portion of tetanus toxoid (TT) as antigen. They compared the TT specific antibody response after two consecutive intranasal immunizations of guinea pigs with the *S. flexneri* constructs or two i.m. immunizations with naked DNA. Overall, both *S. flexneri* strains induced a robust antibody response with 2 log higher titers compared to naked DNA immunization. In all cases, the response was Th1 biased as the major antibody isotype was IgG2a.[215] However, only vaccination with the *S. flexneri* construct expressing the TT polypeptide induced neutralizing antibodies, with titers more then 4 fold lower compared to the commercially available adsorbed TT vaccine.[215] On the first glance this study suggests a superior efficacy of protein expressing vaccine strains over DNA delivery or naked DNA vaccination. However, induction of a humoral immune response is not the major scope of DNA vaccination and T cell responses were not examined in this study. Therefore, additional experiments are needed to compare different modes of vaccination with emphasis on the Th1 response.

Getting Gram-Positive: DNA Delivery with L. Monocytogenes

In the first study using *L. monocytogenes* as DNA carrier, Dietrich et al used a sophisticated strategy for optimal DNA delivery.[6] The vaccine carrier contained engineered deletions of the *mpl*, *actA* and *plcB* genes, disabling both its intra- and intercellular movement. In addition, the plasmid containing the *gfp* reporter gene encoded a phage lysine under the control of the *actA* promotor. Therefore, this construct is lysed upon arrival in the cytosol and the plasmid is released.[6] In vitro, this attenuated suicide *L. monocytogenes* strain efficiently delivered the reporter gene into a murine macrophage cell line. Interestingly, lysis of the carrier strain via the phage lysine resulted in higher transfection efficiency than lysis of intracellular bacteria by treatment with antibiotics.[6] Furthermore, this *L. monocytogenes* carrier strain could deliver the reporter gene to primary bone marrow derived macrophages[216] and human DCs in vitro.[217,218] After intraperitoneal infection of mice, successful transfer was observed in peritoneal macrophages, demonstrating the feasibility of this system in vivo.[216]

Despite extensive in vitro and in vivo demonstrations of successful plasmid transfer, the use of this carrier strain for efficient induction of protective immune responses has not been described, yet. However, in vitro delivery into a murine macrophage cell line with this strain carrying a plasmid encoding for ovalbumin resulted in efficient presentation of ovalbumin derived peptides in the context of MHC class I molecules.[6]

DNA Delivery with S. Typhimurium

Use of intracellular bacteria that escape into the cytosol for DNA delivery, is straightforward. In contrast, the possibility to deliver DNA with bacteria like *S. typhimurium* appears surprising at first sight because these bacteria reside in the phagosome and do not escape into the cytosol. The DNA released into the phagosome has to be transported into the cytosol and subsequently into the nucleus. However, experiments with attenuated *S. typhimurium* as a carrier resulted in successful DNA delivery.[219] Originally, the commercially available eukaryotic expression plasmid pCMV-beta was used, in which the β-gal gene was substituted by genes encoding for LLO or ActA. LLO not only constitutes an important virulence factor for *L. monocytogenes*, but also an immunodominant antigen.[73,220] In contrast, experimental infection of mice does not induce ActA specific CD8+ T cells although a response can be induced by immunizing mice with antigen pulsed DCs.[221] Mice were immunized orally with the attenuated strain *S. typhimurium aroA*- carrying the ActA or LLO encoding DNA vectors.[219] After immunization, high levels of antibodies, as well as CD4+ and CD8+ T cells were detectable with a Th1 bias. In addition, vaccination with *S. typhimurium* constructs carrying the plasmid encoding LLO protected mice from lethal challenge with virulent *L. monocytogenes*.[219] Yet, the plasmid used in this study was suboptimal for DNA delivery, because plasmids with pUC origin are rapidly lost in *S. typhimurium*, even under selective conditions.[222]

Using a similar system with the nonmodified pCMV-beta plasmid, Paglia et al induced a Th1 response after prime/boost vaccination comprising high antibody titers, as well as specific CD4+ and CD8+ T cells.[223] Furthermore, after 3 consecutive vaccinations with a 15-day interval, the *S. typhimurium* construct carrying the pCMV-beta plasmid protected mice against challenge with β-galactosidase expressing tumor cells.[223]

In two other studies with different experimental settings, DNA delivery with *S. typhimurium* was compared with naked DNA immunization.[224,225] In both systems, DNA delivery by *S. typhimurium* seemed to be superior over naked DNA vaccination. Using TT as antigen, DNA delivery and heterologous antigen expression by *S. typhimurium* as well as naked DNA vaccination induced a comparable CD4+ T cell response, whereas antibody titers were markedly higher using DNA delivery or heterologous expression by *S. typhimurium* as compared to naked DNA vaccination.[224] Furthermore, nasal and oral application of the attenuated *S. typhimurium* strain with a eukaryotic expression plasmid encoding for the nucleoprotein of measles virus caused superior mucosal immune responses in the lung than intramuscular (i.m.) injection of naked DNA.[225] Another study comparing CD8+ T cell responses after naked DNA immunization and DNA delivery by *S. typhimurium* using plasmids encoding for the human immunodeficiency virus gp140 protein[226] revealed similar frequencies of CD8+ T cells. Taken together, DNA delivery with *S. typhimurium* seems to be at least equally effective as naked DNA vaccination in inducing CD4+ and CD8+ T cells, whereas DNA delivery by *S. typhimurium* induces higher titers of systemic and mucosal antibodies.

These studies have used *S. typhimurium* as a carrier strain, which does not cause systemic infection in humans (see above). Fennelly et al employed the *S. typhi* Ty21a strain approved as a live oral vaccine in humans, as a carrier for a DNA vaccine encoding for the measles virus nucleoprotein.[211] After intraperetoneal application, the *S. typhi* construct induced CD8+ T cells as efficiently as intranasally applied *S. flexneri* constructs in mice. Although *S. typhi* is unable to cause systemic infection after oral application in mice and hence fails to induce a satisfactory immune response by the oral route, this experiment suggests that attenuated *S. typhi* strains can be used for DNA delivery in humans as well.

Remaining Problems and Strategies to Enhance Efficacy

Despite the encouraging results in small animal models, DNA delivery by bacterial carriers still faces problems. One of the major concerns, affecting both DNA delivery with intracellular bacteria and naked DNA vaccination, is the risk of chromosomal integration of the foreign DNA. Integration is problematic for two major reasons: 1.) Stable integration of plasmids encoding for intact tumor antigens could transform the target cell into a tumor cell and 2.) integration of the strong constitutive promoters (e.g., the cytomegalovirus (CMV) promotor) used for antigen expression next to a proto-oncogene could lead to the expression of the latter.

There are several strategies to minimize these problems: oncogenes can be modified to become nononcogenic antigens with minimal risk of reversion to the native antigen. Furthermore, antigen modification minimizes the risk of homologous recombination, if homologous sequences are present in the eukaryotic chromosome. Use of tissue specific promoters, e.g., promotors only active in APCs, may circumvent the second problem. However, these promotors possess reduced transcriptional activity compared to strong viral promotors, like the CMV promotor.

Is DNA integration a real problem in practice? For naked DNA vaccination, several studies have addressed this question and mostly failed to detect integration events in vivo.[227-231] However, rare nonhomologous recombination events resulting in the loss of antigen expression are difficult to detect and may underestimate the integration rate. Using enteroinvasive *E. coli* or *L. monocytogenes* as DNA delivery vehicles, two independent studies demonstrated stable chromosomal integration in vitro.[6,214] Another study used direct intrasplenic injection of plasmid constructs encoding rearranged immunoglobulins and detected stable integration of the vector into the host chromosome.[232] Although the plasmid vector employed in this study encompassed a large gene with a high degree of homology to chromosomal sequences thus enhancing the risk of integration, this work shows that principally stable integration is possible.

Consequently, integration of DNA into the host chromosome can occur, even though the integration rate seems to be extremely low and only detectable in special experimental settings. The risk of integration can be lowered by reducing the amount of DNA per injection. Therefore, vaccination with DNA delivery seems favorable over i.m. injection of naked DNA because lower amounts of DNA can be used and specific cells are targeted. In any case, integration of foreign DNA remains a major concern for DNA vaccines and must be carefully investigated before initiating broad clinical trials.

From a theoretical standpoint, cytoplasmic release of DNA from attenuated suicide strains of *S. flexneri* or *L. monocytogenes* should yield superior results as compared to DNA delivery to the phagosome by *Salmonella* constructs. However, the feasibility of DNA delivery by *S. typhimurium* has been proven, even though this strain does not reach the cytoplasm. The possibility needs to be evaluated whether apoptotic blebs induced in *Salmonella* infected macrophages can deliver DNA constructs to bystander APC.[198] Potential mechanisms for the poorly understood plasmid transport from the phagosome to the cytoplasm have been discussed.[233] In vitro experiments suggest improved MHC class I presentation by *S. typhimurium* constructs, which escape the phagosome.[234] *S. typhimurium* can be endowed with phagosomal escape mechanisms by secretion of listerial LLO via the *E. coli* hemolysin secretion apparatus.[235] This strain induces breaks in the phagosomal membrane, resulting in a partial cytoplasmic release of secreted proteins with enhanced CD8$^+$ T cell induction after oral vaccination of mice.[236] It is likely that the enhanced capacity of this strain to deliver DNA in vitro can result in higher levels of CD8$^+$ T cells in vivo.[234]

A frequent problem of bacterial vectors for DNA delivery is plasmid stability. Indeed, recombinant bacteria can rapidly lose plasmids that are not optimized even under selective conditions, or the transgene can be lost due to recombination.[237,238] The latter event should be of minor concern for DNA delivery, because the antigen is not expressed in the bacterial carrier. Provided the number of repetitive elements remains small, selective deletion of the antigen encoding part should not be advantageous for the bacterial carrier. In order to enhance plasmid stability, several strategies are currently pursued: either the plasmid carries a gene, which prevents bacterial disintegration, or the plasmid contains an essential gene for bacterial growth, which is deleted in the bacterial chromosome.[239]

Finally, directed bacterial disintegration upon arrival in the cytoplasm might be favorable, as demonstrated for *L. monocytogenes* in vitro.[6] This approach is currently being adapted for *S. typhimurium* using inducible expression of phage lysins resulting in efficient release of native plasmids in vitro.[240] However, these in vitro data do not necessarily reflect an improved immune response in vivo. For example, the induction of a potent and long lasting memory response probably requires persistence and availability of the antigen for a protracted period of time.[241-243] Bacterial disintegration could reduce the duration of antigen availability for the immune system. Hence, the impact of bacterial disintegration on the strength of the primary immune response, as well as long-term immunologic memory, must be analyzed carefully.

Outlook

DNA delivery represents an attractive combination of DNA vaccination with live vaccine carriers. Five years after the first description of successful vaccination with a *Shigella* based DNA delivery construct, the feasibility and power of this new technique has been well documented for *S. flexneri*, *L. monocytogenes* and *S. typhimurium* in small animal models. Provided these results can be verified in clinical trials, DNA delivery by live intracellular bacteria can become an important milestone towards the development of efficacious novel vaccines.

Still, a considerable amount of basic work is needed: the efficacy of DNA delivery needs to be improved, carriers with optimal modifications for attenuation and immunogenicity must be developed and the efficiency of different carrier system must be compared.

The future of vaccinology will not be based on an isolated technique, but rather require a combination of different strategies using a heterologous prime/boost protocol. Therefore, it is imperative to optimize each single technology, including DNA delivery with attenuated intracellular bacteria.

References

1. The world health report 1999. Making a difference. World Health Organization, Genf, 1999.
2. Wright AE, Sample D. Remarks on vaccination against typhoid fever. Brit Med J 1897; 1:256-259.
3. Calmette A, Guérin C. Sur quelques propriétés du bacilles tuberculeux d'origine bovine cultivé sur la bile de boef glycérine. Compt R Acad Sci 1909; 149:716.
4. Hoiseth SK, Stocker BA. Aromatic-dependent *Salmonella typhimurium* are nonvirulent and effective as live vaccines. Nature 1981; 291:238-239.
5. Kocks C, Gouin E, Tabouret M et al. *Listeria monocytogenes* induced actin assembly requires the actA gene product, a surface protein. Cell 1992; 68:521-531.
6. Dietrich G, Bubert A, Gentschev I et al. Delivery of antigen-encoding plasmid DNA into the cytosol of macrophages by attenuated suicide *Listeria monocytogenes*. Nat Biotechnol 1998; 16:181-185.
7. Germain RN. The biochemistry and cell biology of antigen presentation by MHC class I and class II molecules. Implications for development of combination vaccines. Ann N Y Acad Sci 1995; 754:114-125.
8. Kaufmann SHE, Fensterle J, Hess J. The need for a novel generation of vaccines. Immunobiology 1999; 201:272-282.
9. Abbas AK, Murphy KM, Sher A. Functional diversity of helper T lymphocytes. Nature 1996; 383:787-793.
10. Kaufmann SHE. Immunity to Intracellular bacteria. In: Paul WE, ed. Fundamental Immunology. Philadelphia: Lipincott-Raven Publishers, 1999:1335-1371.
11. Medzhitov R, Janeway C Jr. Innate immune recognition: mechanisms and pathways. Immunol Rev 2000; 173:89-97.
12. Imler JL, Hoffmann JA. Toll and Toll-like proteins: an ancient family of receptors signaling infection. Rev Immunogenet 2000; 2:294-304.
13. Qureshi ST, Lariviere L, Leveque G et al. Endotoxin-tolerant mice have mutations in Toll-like receptor 4 (Tlr4). J Exp Med 1999; 189:615-625.
14. Poltorak A, He X, Smirnova I et al. Defective LPS signaling in C3H/HeJ and C57BL/10ScCr mice: mutations in Tlr4 gene. Science 1998; 282:2085-2088.
15. Ozinsky A, Underhill DM, Fontenot JD et al. The repertoire for pattern recognition of pathogens by the innate immune system is defined by cooperation between toll-like receptors. Proc Natl Acad Sci USA 2000; 97:13766-13771.
16. Flo TH, Halaas O, Lien E et al. Human toll-like receptor 2 mediates monocyte activation by *Listeria monocytogenes*, but not by group B streptococci or lipopolysaccharide. J Immunol 2000; 164:2064-2069.
17. Underhill DM, Ozinsky A, Smith KD et al. Toll-like receptor-2 mediates mycobacteria-induced proinflammatory signaling in macrophages. Proc Natl Acad Sci USA 1999; 96:14459-14463.
18. Takeuchi O, Hoshino K, Kawai T et al. Differential roles of TLR2 and TLR4 in recognition of gram-negative and gram-positive bacterial cell wall components. Immunity 1999; 11:443-451.
19. Hemmi H, Takeuchi O, Kawai T et al. A Toll-like receptor recognizes bacterial DNA. Nature 2000; 408:740-745.
20. Hayashi F, Smith KD, Ozinsky A et al. The innate immune response to bacterial flagellin is mediated by Toll-like receptor 5. Nature 2001; 410:1099-1103.
21. Muzio M, Mantovani A. Toll-like receptors. Microbes Infect 2000; 2:251-255.
22. Wolff JA, Malone RW, Williams P et al. Direct gene transfer into mouse muscle in vivo. Science 1990; 247:1465-1468.
23. Ulmer JB, Donnelly JJ, Parker SE et al. Heterologous protection against influenza by injection of DNA encoding a viral protein. Science 1993; 259:1745-1749.
24. Gurunathan S, Klinman DM, Seder RA. DNA vaccines: immunology, application, and optimization. Annu Rev Immunol 2000; 18:927-974.
25. Le TP, Coonan KM, Hedstrom RC et al. Safety, tolerability and humoral immune responses after intramuscular administration of a malaria DNA vaccine to healthy adult volunteers. Vaccine 2000; 18:1893-1901.
26. Hess J, Kaufmann SHE. Live antigen carriers as tools for improved anti-tuberculosis vaccines. FEMS Immunol Med Microbiol 1999; 23:165-173.
27. Ugozzoli M, Santos G, Donnelly J et al. Potency of a genetically detoxified mucosal adjuvant derived from the heat-labile enterotoxin of *Escherichia coli* (LTK63) is not adversely affected by the presence of preexisting immunity to the adjuvant. J Infect Dis 2001; 183:351-354.
28. Kohler JJ, Pathangey LB, Gillespie SR et al. Effect of preexisting immunity to *Salmonella* on the immune response to recombinant *Salmonella* enterica serovar *typhimurium* expressing a Porphyromonas gingivalis hemagglutinin. Infect Immun 2000; 68:3116-3120.

29. Mandl S, Hix L, Andino R. Preexisting immunity to poliovirus does not impair the efficacy of recombinant poliovirus vaccine vectors. J Virol 2001; 75:622-627.
30. Roberts M, Bacon A, Li J et al. Prior immunity to homologous and heterologous *Salmonella* serotypes suppresses local and systemic anti-fragment C antibody responses and protection from tetanus toxin in mice immunized with *Salmonella* strains expressing fragment C. Infect Immun 1999; 67:3810-3815.
31. Belyakov IM, Moss B, Strober W et al. Mucosal vaccination overcomes the barrier to recombinant vaccinia immunization caused by preexisting poxvirus immunity. Proc Natl Acad Sci USA 1999; 96:4512-4517.
32. McClain DJ, Pittman PR, Ramsburg HH et al. Immunologic interference from sequential administration of live attenuated alphavirus vaccines. J Infect Dis 1998; 177:634-641.
33. Bergquist C, Lagergard T, Holmgren J. Anticarrier immunity suppresses the antibody response to polysaccharide antigens after intranasal immunization with the polysaccharide-protein conjugate. Infect Immun 1997; 65:1579-1583.
34. Barington T, Skettrup M, Juul L et al. Nonepitope-specific suppression of the antibody response to Haemophilus influenzae type b conjugate vaccines by preimmunization with vaccine components. Infect Immun 1993; 61:432-438.
35. Barington T, Kristensen K, Henrichsen J et al. Influence of prevaccination immunity on the human B-lymphocyte response to a Haemophilus influenzae type b conjugate vaccine. Infect Immun 1991; 59:1057-1064.
36. Ramshaw IA, Ramsay AJ. The prime-boost strategy: exciting prospects for improved vaccination. Immunol Today 2000; 21:163-165.
37. Kaufmann SHE. Listeriosis: new findings—current concern. Microbial Pathogen 1988; 5:225-231.
38. Mackaness GB. Cellular resistance to infection. J Exp Med 1962; 116:381-389.
39. Cossart P, Mengaud J. *Listeria monocytogenes*. A model system for the molecular study of intracellular parasitism. Mol Biol Med 1989; 6:463-474.
40. Cossart P, Lecuit M. Interactions of *Listeria monocytogenes* with mammalian cells during entry and actin-based movement: bacterial factors, cellular ligands and signaling. Embo J 1998; 17:3797-3806.
41. Ireton K, Cossart P. Host-pathogen interactions during entry and actin-based movement of *Listeria monocytogenes*. Annu Rev Genet 1997; 31:113-138.
42. Mengaud J, Ohayon H, Gounon P et al. E-cadherin is the receptor for internalin, a surface protein required for entry of *L. monocytogenes* into epithelial cells. Cell 1996; 84:923-932.
43. Gaillard JL, Berche P, Frehel C et al. Entry of *L. monocytogenes* into cells is mediated by internalin, a repeat protein reminiscent of surface antigens from gram-positive cocci. Cell 1991; 65:1127-1141.
44. Lecuit M, Vandormael-Pournin S, Lefort J et al. A transgenic model for listeriosis: role of internalin in crossing the intestinal barrier. Science 2001; 292:1722-1725.
45. Braun L, Ghebrehiwet B, Cossart P. gC1q-R/p32, a C1q-binding protein, is a receptor for the InlB invasion protein of *Listeria monocytogenes*. Embo J 2000; 19:1458-1466.
46. Shen Y, Naujokas M, Park M et al. InlB-dependent internalization of *Listeria* is mediated by the Met receptor tyrosine kinase. Cell 2000; 103:501-510.
47. Cossart P, Bierne H. The use of host cell machinery in the pathogenesis of *Listeria monocytogenes*. Curr Opin Immunol 2001; 13:96-103.
48. Kolb-Maurer A, Gentschev I, Fries HW et al. *Listeria monocytogenes*-infected human dendritic cells: uptake and host cell response. Infect Immun 2000; 68:3680-3688.
49. Goebel W, Kuhn M. Bacterial replication in the host cell cytosol. Curr Opin Microbiol 2000; 3:49-53.
50. Cossart P. The listeriolysin O gene: a chromosomal locus crucial for the virulence of *Listeria monocytogenes*. Infection 1988; 16 Suppl 2:S157-S159.
51. Kayal S, Lilienbaum A, Poyart C et al. Listeriolysin O-dependent activation of endothelial cells during infection with *Listeria monocytogenes*: activation of NF-kappa B and upregulation of adhesion molecules and chemokines. Mol Microbiol 1999; 31:1709-1722.
52. Decatur AL, Portnoy DA. A PEST-like sequence in listeriolysin O essential for *Listeria monocytogenes* pathogenicity. Science 2000; 290:992-995.
53. Vazquez-Boland JA, Kocks C, Dramsi S et al. Nucleotide sequence of the lecithinase operon of *Listeria monocytogenes* and possible role of lecithinase in cell-to-cell spread. Infect Immun 1992; 60:219-230.
54. Gedde MM, Higgins DE, Tilney LG et al. Role of listeriolysin O in cell-to-cell spread of *Listeria monocytogenes*. Infect Immun 2000; 68:999-1003.
55. Rogers HW, Unanue ER. Neutrophils are involved in acute, nonspecific resistance to *Listeria monocytogenes* in mice. Infect Immun 1993; 61:5090-5096.
56. Rogers HW, Callery MP, Deck B et al. *Listeria monocytogenes* induces apoptosis of infected hepatocytes. J Immunol 1996; 156:679-684.

57. Czuprynski CJ, Brown JF, Maroushek N et al. Administration of anti-granulocyte mAb RB6-8C5 impairs the resistance of mice to *Listeria monocytogenes* infection. J Immunol 1994; 152:1836-1846.

58. Unanue ER. Studies in listeriosis show the strong symbiosis between the innate cellular system and the T-cell response. Immunol Rev 1997; 158:11-25.

59. Tripp CS, Wolf SF, Unanue ER. Interleukin 12 and tumor necrosis factor alpha are costimulators of interferon gamma production by natural killer cells in severe combined immunodeficiency mice with listeriosis, and interleukin 10 is a physiologic antagonist. Proc Natl Acad Sci USA 1993; 90:3725-3729.

60. Pasparakis M, Alexopoulou L, Episkopou V et al. Immune and inflammatory responses in TNF alpha-deficient mice: a critical requirement for TNF alpha in the formation of primary B cell follicles, follicular dendritic cell networks and germinal centers, and in the maturation of the humoral immune response. J Exp Med 1996; 184:1397-1411.

61. Rothe J, Lesslauer W, Lotscher H et al. Mice lacking the tumour necrosis factor receptor 1 are resistant to TNF-mediated toxicity but highly susceptible to infection by *Listeria monocytogenes*. Nature 1993; 364:798-802.

62. Kaplan MH, Sun YL, Hoey T et al. Impaired IL-12 responses and enhanced development of Th2 cells in Stat4-deficient mice. Nature 1996; 382:174-177.

63. Tripp CS, Gately MK, Hakimi J et al. Neutralization of IL-12 decreases resistance to *Listeria* in SCID and C. B-17 mice. Reversal by IFN-gamma. J Immunol 1994; 152:1883-1887.

64. Kaufmann SH, Hahn H, Berger R et al. Interferon-gamma production by *Listeria monocytogenes*-specific T cells active in cellular antibacterial immunity. Eur J Immunol 1983; 13:265-268.

65. Dai WJ, Bartens W, Kohler G et al. Impaired macrophage listericidal and cytokine activities are responsible for the rapid death of *Listeria monocytogenes*-infected IFN-gamma receptor-deficient mice. J Immunol 1997; 158:5297-5304.

66. Rogers HW, Tripp CS, Schreiber RD et al. Endogenous IL-1 is required for neutrophil recruitment and macrophage activation during murine listeriosis. J Immunol 1994; 153:2093-2101.

67. Czuprynski CJ, Brown JF. Purified human and recombinant murine interleukin-1 alpha induced accumulation of inflammatory peritoneal neutrophils and mononuclear phagocytes: possible contributions to antibacterial resistance. Microb Pathog 1987; 3:377-386.

68. Dalrymple SA, Lucian LA, Slattery R et al. Interleukin-6-deficient mice are highly susceptible to *Listeria monocytogenes* infection: correlation with inefficient neutrophilia. Infect Immun 1995; 63:2262-2268.

69. Ashkar S, Weber GF, Panoutsakopoulou V et al. Eta-1 (osteopontin): an early component of type-1 (cell-mediated) immunity. Science 2000; 287:860-864.

70. Patarca R, Freeman GJ, Singh RP et al. Structural and functional studies of the early T lymphocyte activation 1 (Eta-1) gene. Definition of a novel T cell-dependent response associated with genetic resistance to bacterial infection. J Exp Med 1989; 170:145-161.

71. Singh RP, Patarca R, Schwartz J et al. Definition of a specific interaction between the early T lymphocyte activation 1 (Eta-1) protein and murine macrophages in vitro and its effect upon macrophages in vivo. J Exp Med 1990; 171:1931-1942.

72. Weber GF, Cantor H. The immunology of Eta-1/osteopontin. Cytokine Growth Factor Rev 1996; 7:241-248.

73. Pamer EG, Harty JT, Bevan MJ. Precise prediction of a dominant class I MHC-restricted epitope of *Listeria monocytogenes*. Nature 1991; 353:852-855.

74. Sanderson S, Campbell DJ, Shastri N. Identification of a CD4+ T cell-stimulating antigen of pathogenic bacteria by expression cloning. J Exp Med 1995; 182:1751-1757.

75. Geginat G, Schenk S, Skoberne M et al. A novel approach of direct ex vivo epitope mapping identifies dominant and subdominant CD4 and CD8 T cell epitopes from *Listeria monocytogenes*. J Immunol 2001; 166:1877-1884.

76. Busch DH, Kerksiek K, Pamer EG. Processing of *Listeria monocytogenes* antigens and the in vivo T-cell response to bacterial infection. Immunol Rev 1999; 172:163-169.

77. Kaufmann SHE, Simon MM, Hahn H. Specific Lyt 123 cells are involved in protection against *Listeria monocytogenes* and in delayed-type hypersensitivity to listerial antigens. J Exp Med 1979; 150:1033-1038.

78. Kagi D, Ledermann B, Burki K et al. CD8+ T cell-mediated protection against an intracellular bacterium by perforin-dependent cytotoxicity. Eur J Immunol 1994; 24:3068-3072.

79. Harty JT, Bevan MJ. Specific immunity to *Listeria monocytogenes* in the absence of IFN gamma. Immunity 1995; 3:109-117.

80. White DW, Harty JT. Perforin-deficient CD8+ T cells provide immunity to *Listeria monocytogenes* by a mechanism that is independent of CD95 and IFN-gamma but requires TNF-alpha. J Immunol 1998; 160:898-905.

81. Altman JD, Moss PAH, Goulder PJR et al. Phenotypic analysis of antigen-specific T lymphocytes. Science 1996; 274:94-96.
82. Busch DH, Pilip IM, Vijh S. Coordinate regulation of complex T cell populations responding to bacterial infection. Immunity 1998; 8:353-362.
83. Kaufmann SHE, Hahn H. Biological functions of T cell lines with specificity for the intracellular bacterium Listeria monocytogenes in vitro and in vivo. J Exp Med 1982; 155:1754-1765.
84. Geginat G, Lalic M, Kretschmar M et al. Th1 cells specific for a secreted protein of *Listeria monocytogenes* are protective in vivo. J Immunol 1998; 160:6046-6055.
85. Edelson BT, Unanue ER. Immunity to *Listeria* infection. Curr Opin Immunol 2000; 12:425-431.
86. Kaisho T, Akira S. Dendritic-cell function in Toll-like receptor- and MyD88-knockout mice. Trends Immunol 2001; 22:78-83.
87. Reis e Sousa C, Sher A, Kaye P. The role of dendritic cells in the induction and regulation of immunity to microbial infection. Curr Opin Immunol 1999; 11:392-399.
88. Alexander JE, Andrew PW, Jones D et al. Characterization of an aromatic amino acid-dependent *Listeria monocytogenes* mutant: attenuation, persistence, and ability to induce protective immunity in mice. Infect Immun 1993; 61:2245-2248.
89. Rouquette C, Bolla JM, Berche P. An iron-dependent mutant of *Listeria monocytogenes* of attenuated virulence. FEMS Microbiol Lett 1995; 133:77-83.
90. Thompson RJ, Bouwer HG, Portnoy DA et al. Pathogenicity and immunogenicity of a *Listeria monocytogenes* strain that requires D-alanine for growth. Infect Immun 1998; 66:3552-3561.
91. Michel E, Reich KA, Favier R et al. Attenuated mutants of the intracellular bacterium *Listeria monocytogenes* obtained by single amino acid substitutions in listeriolysin O Mol Microbiol 1990; 4:2167-2178.
92. Goossens PL, Milon G. Induction of protective CD8+ T lymphocytes by an attenuated *Listeria monocytogenes* actA mutant. Int Immunol 1992; 4:1413-1418.
93. Guzman CA, Saverino D, Medina E et al. Attenuated *Listeria monocytogenes* carrier strains can deliver an HIV-1 gp120 T helper epitope to MHC class II-restricted human CD4+ T cells. Eur J Immunol 1998; 28:1807-1814.
94. Chakraborty T, Ebel F, Wehland J et al. Naturally occurring virulence-attenuated isolates of *Listeria monocytogenes* capable of inducing long term protection against infection by virulent strains of homologous and heterologous serotypes. FEMS Immunol Med Microbiol 1994; 10:1-9.
95. Kotloff KL, Winickoff JP, Ivanoff B et al. Global burden of *Shigella* infections: implications for vaccine development and implementation of control strategies. Bull. World Health Organ 1999; 77:651-666.
96. Sansonetti PJ. Slaying the Hydra all at once or head by head? Nat Med 1998; 4:499-500.
97. Parsot C, Sansonetti PJ. Invasion and the pathogenesis of *Shigella* infections. Curr Top Microbiol Immunol 1996; 209:25-42.
98. Dorman CJ, Porter ME. The *Shigella* virulence gene regulatory cascade: a paradigm of bacterial gene control mechanisms. Mol Microbiol 1998; 29:677-684.
99. Winstanley C, Hart CA. Type III secretion systems and pathogenicity islands. J Med Microbiol 2001; 50:116-126.
100. Cornelis GR, Van Gijsegem F. Assembly and function of type III secretory systems. Annu Rev Microbiol 2000; 54:735-774.
101. Blocker A, Gounon P, Larquet E et al. The tripartite type III secreton of *Shigella flexneri* inserts IpaB and IpaC into host membranes. J Cell Biol 1999; 147:683-693.
102. Blocker A, Jouihri N, Larquet E et al. Structure and composition of the *Shigella flexneri* "needle complex", a part of its type III secreton. Mol Microbiol 2001; 39:652-663.
103. Tran Van Nhieu G, Caron E, Hall A et al. IpaC induces actin polymerization and filopodia formation during *Shigella* entry into epithelial cells. Embo J 1999; 18:3249-3262.
104. Sansonetti PJ. Rupture, invasion and inflammatory destruction of the intestinal barrier by *Shigella*, making sense of prokaryote-eukaryote cross-talks. FEMS Microbiol Rev 2001; 25:3-14.
105. Tran Van Nhieu G, Ben-Ze'ev A, Sansonetti PJ. Modulation of bacterial entry into epithelial cells by association between vinculin and the *Shigella* IpaA invasin. Embo J 1997; 16:2717-2729.
106. Bourdet-Sicard R, Rudiger M, Jockusch BM et al. Binding of the *Shigella* protein IpaA to vinculin induces F-actin depolymerization. Embo J 1999; 18:5853-5862.
107. Page AL, Ohayon H, Sansonetti PJ et al. The secreted IpaB and IpaC invasins and their cytoplasmic chaperone IpgC are required for intercellular dissemination of *Shigella flexneri*. Cell Microbiol 1999; 1:183-193.
108. Barzu S, Benjelloun-Touimi Z, Phalipon A et al. Functional analysis of the *Shigella flexneri* IpaC invasin by insertional mutagenesis. Infect Immun 1997; 65:1599-1605.

109. Bernardini ML, Mounier J, d'Hauteville H et al. Identification of icsA, a plasmid locus of Shigella flexneri that governs bacterial intra- and intercellular spread through interaction with F-actin. Proc Natl Acad Sci USA 1989; 86:3867-3871.

110. Suzuki T, Miki H, Takenawa T. Neural Wiskott-Aldrich syndrome protein is implicated in the actin- based motility of Shigella flexneri. Embo J 1998; 17:2767-2776.

111. Suzuki T, Saga S, Sasakawa C. Functional analysis of Shigella VirG domains essential for interaction with vinculin and actin-based motility. J Biol Chem 1996; 271:21878-21885.

112. Egile C, Loisel TP, Laurent V et al. Activation of the CDC42 effector N-WASP by the Shigella flexneri IcsA protein promotes actin nucleation by Arp2/3 complex and bacterial actin- based motility. J Cell Biol 1999; 146:1319-1332.

113. Sansonetti PJ, Phalipon A. M cells as ports of entry for enteroinvasive pathogens: mechanisms of interaction, consequences for the disease process. Semin Immunol 1999; 11:193-203.

114. Zychlinsky A, Prevost MC, Sansonetti PJ. Shigella flexneri induces apoptosis in infected macrophages. Nature 1992; 358:167-169.

115. Zychlinsky A, Kenny B, Menard R et al. IpaB mediates macrophage apoptosis induced by Shigella flexneri. Mol Microbiol 1994; 11:619-627.

116. Chen Y, Smith MR, Thirumalai K et al. A bacterial invasin induces macrophage apoptosis by binding directly to ICE. Embo J 1996; 15:3853-3860.

117. Sansonetti PJ, Arondel J, Cavaillon JM et al. Role of interleukin-1 in the pathogenesis of experimental shigellosis. J Clin Invest 1995; 96:884-892.

118. Sansonetti PJ, Phalipon A, Arondel J et al. Caspase-1 activation of IL-1beta and IL-18 are essential for Shigella flexneri-induced inflammation. Immunity 2000; 12:581-590.

119. Way SS, Borczuk AC, Dominitz R et al. An essential role for gamma interferon in innate resistance to Shigella flexneri infection. Infect Immun 1998; 66:1342-1348.

120. Raqib R, Gustafsson A, Andersson J et al. A systemic downregulation of gamma interferon production is associated with acute shigellosis. Infect Immun 1997; 65:5338-5341.

121. Wenneras C, Ave P, Huerre M et al. Blockade of CD14 increases Shigella-mediated invasion and tissue destruction. J Immunol 2000; 164:3214-3221.

122. Arondel J, Singer M, Matsukawa A et al. Increased interleukin-1 (IL-1) and imbalance between IL-1 and IL-1 receptor antagonist during acute inflammation in experimental Shigellosis. Infect Immun 1999; 67:6056-6066.

123. Jung HC, Eckmann L, Yang SK et al. A distinct array of proinflammatory cytokines is expressed in human colon epithelial cells in response to bacterial invasion. J Clin Invest 1995; 95:55-65.

124. Perdomo JJ, Cavaillon JM, Huerre M et al. Acute inflammation causes epithelial invasion and mucosal destruction in experimental shigellosis. J Exp Med 1994; 180:1307-1319.

125. Cersini A, Salvia AM, Bernardini ML. Intracellular multiplication and virulence of Shigella flexneri auxotrophic mutants. Infect Immun 1998; 66:549-557.

126. Klee SR, Tzschaschel BD, Falt I et al. Construction and characterization of a live attenuated vaccine candidate against Shigella dysenteriae type 1. Infect Immun 1997; 65:2112-2118.

127. Kotloff KL, Noriega F, Losonsky GA et al. Safety, immunogenicity, and transmissibility in humans of CVD 1203, a live oral Shigella flexneri 2a vaccine candidate attenuated by deletions in aroA and virG. Infect Immun 1996; 64:4542-4548.

128. Noriega FR, Losonsky G, Lauderbaugh C et al. Engineered deltaguaB-A deltavirG Shigella flexneri 2a strain CVD 1205: construction, safety, immunogenicity, and potential efficacy as a mucosal vaccine. Infect Immun 1996; 64:3055-3061.

129. Noriega FR, Wang JY, Losonsky G et al. Construction and characterization of attenuated delta aroA delta virG Shigella flexneri 2a strain CVD 1203, a prototype live oral vaccine. Infect Immun 1994; 62:5168-5172.

130. Karnell A, Cam PD, Verma N et al. AroD deletion attenuates Shigella flexneri strain 2457T and makes it a safe and efficacious oral vaccine in monkeys. Vaccine 1993; 11:830-836.

131. Li A, Pal T, Forsum U et al. Safety and immunogenicity of the live oral auxotrophic Shigella flexneri SFL124 in volunteers. Vaccine 1992; 10:395-404.

132. Lindberg AA, Karnell A, Pal T et al. Construction of an auxotrophic Shigella flexneri strain for use as a live vaccine. Microb Pathog 1990; 8:433-440.

133. Lindberg AA, Karnell A, Stocker BA et al. Development of an auxotrophic oral live Shigella flexneri vaccine. Vaccine 1988; 6:146-150.

134. DuPont HL, Hornick RB, Dawkins AT et al. The response of man to virulent Shigella flexneri 2a. J Infect Dis 1969; 119:296-299.

135. Coster TS, Hoge CW, VanDeVerg LL et al. Vaccination against shigellosis with attenuated Shigella flexneri 2a strain SC602. Infect Immun 1999; 67:3437-3443.

136. Hartman AB, Venkatesan MM. Construction of a stable attenuated *Shigella* sonnei DeltavirG vaccine strain, WRSS1, and protective efficacy and immunogenicity in the guinea pig keratoconjunctivitis model. Infect Immun 1998; 66:4572-4576.

137. Yoshikawa M, Sasakawa C, Okada N et al. Construction and evaluation of a virG thyA double mutant of *Shigella flexneri* 2a as a candidate live-attenuated oral vaccine. Vaccine 1995; 13:1436-1440.

138. Sansonetti PJ, Arondel J, Fontaine A et al. OmpB (osmo-regulation) and icsA (cell-to-cell spread) mutants of *Shigella flexneri*: vaccine candidates and probes to study the pathogenesis of shigellosis. Vaccine 1991; 9:416-422.

139. Kotloff KL, Noriega FR, Samandari T et al. *Shigella flexneri* 2a strain CVD 1207, with specific deletions in virG, sen, set, and guaBA, is highly attenuated in humans. Infect Immun 2000; 68:1034-1039.

140. Le Minor L. Typing of *Salmonella* species. Eur J Clin Microbiol Infect Dis 1988; 7:214-218.

141. Edwards BH. *Salmonella* and *Shigella* species. Clin Lab Med 1999; 19:469-487.

142. Angulo FJ, Swerdlow DL. *Salmonella* enteritidis infections in the United States. J Am Vet Med Assoc 1998; 213:1729-1731.

143. Rowe B, Ward LR, Threlfall EJ. Multidrug-resistant *Salmonella* typhi: a worldwide epidemic. Clin Infect Dis 1997; 24 Suppl 1:S106-109.

144. Foster JW, Spector MP. How *Salmonella* survive against the odds. Annu Rev Microbiol 1995; 49:145-174.

145. Foster JW, Hall HK. Adaptive acidification tolerance response of *Salmonella typhimurium*. J Bacteriol 1990; 172:771-778.

146. Foster JW, Moreno M. Inducible acid tolerance mechanisms in enteric bacteria. Novartis Found Symp 1999; 221:55-69; discussion 70-54.

147. Wallis TS, Galyov EE. Molecular basis of *Salmonella*-induced enteritis. Mol Microbiol 2000; 36:997-1005.

148. Marcus SL, Brumell JH, Pfeifer CG et al. *Salmonella* pathogenicity islands: big virulence in small packages. Microbes Infect 2000; 2:145-156.

149. Hansen-Wester I, Hensel M. *Salmonella* pathogenicity islands encoding type III secretion systems. Microbes Infect 2001; 3:549-559.

150. Edwards RA, Schifferli DM, Maloy SR. A role for *Salmonella* fimbriae in intraperitoneal infections. Proc Natl Acad Sci USA 2000; 97:1258-1262.

151. van der Velden AW, Baumler AJ, Tsolis RM et al. Multiple fimbrial adhesins are required for full virulence of *Salmonella typhimurium* in mice. Infect Immun 1998; 66:2803-2808.

152. Baumler AJ, Tsolis RM, Heffron F. The lpf fimbrial operon mediates adhesion of *Salmonella typhimurium* to murine Peyer's patches. Proc Natl Acad Sci USA 1996; 93:279-283.

153. Lindquist BL, Lebenthal E, Lee PC et al. Adherence of *Salmonella typhimurium* to small-intestinal enterocytes of the rat. Infect Immun 1987; 55:3044-3050.

154. Galan JE. Interaction of *Salmonella* with host cells through the centisome 63 type III secretion system. Curr Opin Microbiol 1999; 2:46-50.

155. Kubori T, Matsushima Y, Nakamura D et al. Supramolecular structure of the *Salmonella typhimurium* type III protein secretion system. Science 1998; 280:602-605.

156. Kaniga K, Trollinger D, Galan JE. Identification of two targets of the type III protein secretion system encoded by the inv and spa loci of *Salmonella typhimurium* that have homology to the *Shigella* IpaD and IpaA proteins. J Bacteriol 1995; 177:7078-7085.

157. Collazo CM, Galan JE. Requirement for exported proteins in secretion through the invasion- associated type III system of *Salmonella typhimurium*. Infect Immun 1996; 64:3524-3531.

158. Fu Y, Galan JE. Identification of a specific chaperone for SptP, a substrate of the centisome 63 type III secretion system of *Salmonella typhimurium*. J Bacteriol 1998; 180:3393-3399.

159. Collazo CM, Galan JE. The invasion-associated type III system of *Salmonella typhimurium* directs the translocation of Sip proteins into the host cell. Mol Microbiol 1997; 24:747-756.

160. Wood MW, Rosqvist R, Mullan PB et al. SopE, a secreted protein of *Salmonella* dublin, is translocated into the target eukaryotic cell via a sip-dependent mechanism and promotes bacterial entry. Mol Microbiol 1996; 22:327-338.

161. Hardt WD, Chen LM, Schuebel KE et al. *S. typhimurium* encodes an activator of Rho GTPases that induces membrane ruffling and nuclear responses in host cells. Cell 1998; 93:815-826.

162. Hobbie S, Chen LM, Davis RJ et al. Involvement of mitogen-activated protein kinase pathways in the nuclear responses and cytokine production induced by *Salmonella typhimurium* in cultured intestinal epithelial cells. J Immunol 1997; 159:5550-5559.

163. Bakshi CS, Singh VP, Wood MW et al. Identification of SopE2, a *Salmonella* secreted protein which is highly homologous to SopE and involved in bacterial invasion of epithelial cells. J Bacteriol 2000; 182:2341-2344.

164. Norris FA, Wilson MP, Wallis TS et al. SopB, a protein required for virulence of *Salmonella dublin*, is an inositol phosphate phosphatase. Proc Natl Acad Sci USA 1998; 95:14057-14059.

165. Hong KH, Miller VL. Identification of a novel *Salmonella* invasion locus homologous to *Shigella* ipgDE. J Bacteriol 1998; 180:1793-1802.

166. Eckmann L, Rudolf MT, Ptasznik A et al. D-myo-Inositol 1,4,5,6-tetrakisphosphate produced in human intestinal epithelial cells in response to *Salmonella* invasion inhibits phosphoinositide 3-kinase signaling pathways. Proc Natl Acad Sci USA 1997; 94:14456-14460.

167. Galyov EE, Wood MW, Rosqvist R et al. A secreted effector protein of *Salmonella dublin* is translocated into eukaryotic cells and mediates inflammation and fluid secretion in infected ileal mucosa. Mol Microbiol 1997; 25:903-912.

168. Hersh D, Monack DM, Smith MR et al. The *Salmonella* invasin SipB induces macrophage apoptosis by binding to caspase-1. Proc Natl Acad Sci USA 1999; 96:2396-2401.

169. Hensel M. Salmonella pathogenicity island 2. Mol Microbiol 2000; 36:1015-1023.

170. Hensel M, Shea JE, Waterman SR et al. Genes encoding putative effector proteins of the type III secretion system of *Salmonella* pathogenicity island 2 are required for bacterial virulence and proliferation in macrophages. Mol Microbiol 1998; 30:163-174.

171. Uchiya K, Barbieri MA, Funato K et al. A *Salmonella* virulence protein that inhibits cellular trafficking. Embo J 1999; 18:3924-3933.

172. Groisman EA, Kayser J, Soncini FC. Regulation of polymyxin resistance and adaptation to low-Mg2+ environments. J Bacteriol 1997; 179:7040-7045.

173. Guiney DG, Libby S, Fang FC et al. Growth-phase regulation of plasmid virulence genes in *Salmonella*. Trends Microbiol 1995; 3:275-279.

174. Mastroeni P, Vazquez-Torres A, Fang FC et al. Antimicrobial actions of the NADPH phagocyte oxidase and inducible nitric oxide synthase in experimental salmonellosis. II. Effects on microbial proliferation and host survival in vivo. J Exp Med 2000; 192:237-248.

175. Vazquez-Torres A, Jones-Carson J, Mastroeni P et al. Antimicrobial actions of the NADPH phagocyte oxidase and inducible nitric oxide synthase in experimental salmonellosis. I. Effects on microbial killing by activated peritoneal macrophages in vitro. J Exp Med 2000; 192:227-236.

176. Vazquez-Torres A, Xu Y, Jones-Carson J et al. *Salmonella* pathogenicity island 2-dependent evasion of the phagocyte NADPH oxidase. Science 2000; 287:1655-1658.

177. Lucas RL, Lee CA. Unravelling the mysteries of virulence gene regulation in *Salmonella typhimurium*. Mol Microbiol 2000; 36:1024-1033.

178. Groisman EA. The pleiotropic two-component regulatory system PhoP-PhoQ. J Bacteriol 2001; 183:1835-1842.

179. Lucas RL, Lostroh CP, DiRusso CC et al. Multiple factors independently regulate hilA and invasion gene expression in *Salmonella* enterica serovar *typhimurium*. J Bacteriol 2000; 182:1872-1882.

180. Jones BD, Falkow S. Salmonellosis: host immune responses and bacterial virulence determinants. Annu Rev Immunol 1996; 14:533-561.

181. Nauciel C, Espinasse-Maes F. Role of gamma interferon and tumor necrosis factor alpha in resistance to *Salmonella typhimurium* infection. Infect Immun 1992; 60:450-454.

182. Gulig PA, Doyle TJ, Clare-Salzler MJ et al. Systemic infection of mice by wild-type but not Spv-*Salmonella typhimurium* is enhanced by neutralization of gamma interferon and tumor necrosis factor alpha. Infect Immun 1997; 65:5191-5197.

183. Vidal S, Tremblay ML, Govoni G et al. The Ity/Lsh/Bcg locus: natural resistance to infection with intracellular parasites is abrogated by disruption of the Nramp1 gene. J Exp Med 1995; 182:655-666.

184. Blackwell JM, Searle S, Goswami T et al. Understanding the multiple functions of Nramp1. Microbes Infect 2000; 2:317-321.

185. Jabado N, Jankowski A, Dougaparsad S et al. Natural resistance to intracellular infections: natural resistance-associated macrophage protein 1 (Nramp1) functions as a pH-dependent manganese transporter at the phagosomal membrane. J Exp Med 2000; 192:1237-1248.

186. Biggs TE, Baker ST, Botham MS et al. Nramp1 modulates iron homoeostasis in vivo and in vitro: evidence for a role in cellular iron release involving de-acidification of intracellular vesicles. Eur J Immunol 2001; 31:2060-2070.

187. Mayeux PR. Pathobiology of lipopolysaccharide. J Toxicol Environ Health 1997; 51:415-435.

188. Flad HD, Loppnow H, Rietschel ET et al. Agonists and antagonists for lipopolysaccharide-induced cytokines. Immunobiology 1993; 187:303-316.

189. Guha M, Mackman N. LPS induction of gene expression in human monocytes. Cell Signal 2001; 13:85-94.

190. Munder M, Mallo M, Eichmann K. Murine macrophages secrete interferon gamma upon combined stimulation with interleukin (IL)-12 and IL-18: A novel pathway of autocrine macrophage activation. J Exp Med 1998; 187:2103-2108.

191. Trinchieri G. Interleukin-12: a proinflammatory cytokine with immunoregulatory functions that bridge innate resistance and antigen-specific adaptive immunity. Annu Rev Immunol 1995; 13:251-276.
192. Emoto M, Emoto Y, Buchwalow IB et al. Induction of IFN-gamma-producing CD4+ natural killer T cells by *Mycobacterium* bovis bacillus Calmette Guerin. Eur J Immunol 1999; 29:650-659.
193. Mastroeni P, Harrison JA, Chabalgoity JA et al. Effect of interleukin 12 neutralization on host resistance and gamma interferon production in mouse typhoid. Infect Immun 1996; 64:189-196.
194. Mastroeni P, Harrison JA, Robinson JH et al. Interleukin-12 is required for control of the growth of attenuated aromatic-compound-dependent salmonellae in BALB/c mice: role of gamma interferon and macrophage activation. Infect Immun 1998; 66:4767-4776.
195. Ramarathinam L, Niesel DW, Klimpel GR. *Salmonella typhimurium* induces IFN-gamma production in murine splenocytes. Role of natural killer cells and macrophages. J Immunol 1993; 150:3973-3981.
196. Schafer R, Eisenstein TK. Natural killer cells mediate protection induced by a *Salmonella* aroA mutant. Infect Immun 1992; 60:791-797.
197. Aggarwal A, Kumar S, Jaffe R et al. Oral *Salmonella*: malaria circumsporozoite recombinants induce specific CD8+ cytotoxic T cells. J Exp Med 1990; 172:1083-1090.
198. Yrlid U, Wick MJ. *Salmonella*-induced apoptosis of infected macrophages results in presentation of a bacteria-encoded antigen after uptake by bystander dendritic cells. J Exp Med 2000; 191:613-624.
199. Pfeifer JD, Wick MJ, Roberts RL et al. Phagocytic processing of bacterial antigens for class I MHC presentation to T cells. Nature 1993; 361:359-362.
200. Mittrücker HW, Kaufmann SHE. Immune response to infection with *Salmonella typhimurium* in mice. J Leukoc Biol 2000; 67:457-463.
201. Bumann D, Hueck C, Aebischer T et al. Recombinant live *Salmonella* spp. for human vaccination against heterologous pathogens. FEMS Immunol Med Microbiol 2000; 27:357-364.
202. Garcia Vescovi E, Soncini FC, Groisman EA. The role of the PhoP/PhoQ regulon in *Salmonella* virulence. Res Microbiol 1994; 145:473-480.
203. McCormick BA, Miller SI, Carnes D et al. Transepithelial signaling to neutrophils by salmonellae: a novel virulence mechanism for gastroenteritis. Infect Immun 1995; 63:2302-2309.
204. Stocker BA. Aromatic-dependent salmonella as anti-bacterial vaccines and as presenters of heterologous antigens or of DNA encoding them. J Biotechnol 2000; 83:45-50.
205. Phalipon A, Sansonetti P. Live attenuated *Shigella flexneri* mutants as vaccine candidates against shigellosis and vectors for antigen delivery. Biologicals 1995; 23:125-134.
206. Levine MM, Galen J, Barry E et al. Attenuated *Salmonella* typhi and *Shigella* as live oral vaccines and as live vectors. Behring Inst Mitt 1997; 98:120-123.
207. Weiskirch LM, Paterson Y. *Listeria monocytogenes*: a potent vaccine vector for neoplastic and infectious disease. Immunol Rev 1997; 158:159-169.
208. Sizemore DR, Branstrom AA, Sadoff JC. Attenuated *Shigella* as a DNA delivery vehicle for DNA-mediated immunization. Science 1995; 270:299-302.
209. Nakayama KK, Curtiss R III. Construction of an Asd+ expression-cloning vector: Stable maintenance and high level expression of cloned genes in a *Salmonella* vaccine strain. Biotechnology 1988; 6:693-697.
210. Sizemore DR, Branstrom AA, Sadoff JC. Attenuated bacteria as a DNA delivery vehicle for DNA-mediated immunization. Vaccine 1997; 15:804-807.
211. Fennelly GJ, Khan SA, Abadi MA et al. Mucosal DNA vaccine immunization against measles with a highly attenuated *Shigella flexneri* vector. J Immunol 1999; 162:1603-1610.
212. Powell RJ, Lewis GK, Hone DM. Method for introducing and expressing genes in animal cells and live invasive vectors for use in the same. In: US Patent 5877159. University of Maryland at Baltimore, Baltimore, MD, United States of America. 1995.
213. Powell RJ, Lewis GK, Hone DM. Introduction of eukaryotic expression cassettes into animal cells using a bacterial vector delivery system. In: Brown F, Chanock RM, Ginsberg MS et al, eds. Vaccine 96: Molecular Approaches to the Control of Infectious Disease. Cold Spring Harbor: Cold Spring Harbor Laboratory Press, 1996:183-187.
214. Courvalin P, Goussard S, Grillot-Courvalin C. Gene transfer from bacteria to mammalian cells. C R Acad Sci III 1995; 318:1207-1212.
215. Anderson RJ, Pasetti MF, Sztein MB et al. DeltaguaBA attenuated *Shigella flexneri* 2a strain CVD 1204 as a *Shigella* vaccine and as a live mucosal delivery system for fragment C of tetanus toxin. Vaccine 2000; 18:2193-2202.
216. Spreng S, Dietrich G, Niewiesk S et al. Novel bacterial systems for the delivery of recombinant protein or DNA. FEMS Immunol Med Microbiol 2000; 27:299-304.
217. Gentschev I, Dietrich G, Spreng S et al. Delivery of protein antigens and DNA by virulence-attenuated strains of *Salmonella typhimurium* and *Listeria monocytogenes*. J Biotechnol 2000; 83:19-26.
218. Kolb-Mäurer A, Gentschev I, Fries HW et al. *Listeria monocytogenes*-infected human dendritic cells: uptake and host cell response. Infect Immun 2000; 68:3680-3688.

219. Darji A, Guzman CA, Gerstel B et al. Oral somatic transgene vaccination using attenuated *S. typhimurium.* Cell 1997; 91:765-775.
220. Vijh S, Pamer EG. Immunodominant and subdominant CTL responses to *Listeria monocytogenes* infection. J Immunol 1997; 158:3366-3371.
221. Darji A, Bruder D, zur Lage S et al. The role of the bacterial membrane protein ActA in immunity and protection against *Listeria monocytogenes.* J Immunol 1998; 161:2414-2420.
222. Galen JE, Nair J, Wang JY et al. Optimization of plasmid maintenance in the attenuated live vector vaccine strain *Salmonella* typhi CVD 908-htrA. Infect Immun 1999; 67:6424-6433.
223. Paglia P, Medina E, Arioli I et al. Gene transfer in dendritic cells, induced by oral DNA vaccination with *Salmonella typhimurium,* results in protective immunity against a murine fibrosarcoma. Blood 1998; 92:3172-3176.
224. Pasetti MF, Anderson RJ, Noriega FR et al. Attenuated deltaguaBA *Salmonella typhi* vaccine strain CVD 915 as a live vector utilizing prokaryotic or eukaryotic expression systems to deliver foreign antigens and elicit immune responses. Clin Immunol 1999; 92:76-89.
225. Brunham RC, Zhang D. Transgene as vaccine for chlamydia. Am Heart J 1999; 138:S519-522.
226. Shata MT, Stevceva L, Agwale S et al. Recent advances with recombinant bacterial vaccine vectors. Mol Med Today 2000; 6:66-71.
227. Cichutek K. Obtaining marketing authorization for nucleic acid vaccines in the European Union. Ann N Y Acad Sci 1995; 772:178-185.
228. Wolff JA, Ludtke JJ, Acsadi G et al. Long-term persistence of plasmid DNA and foreign gene expression in mouse muscle. Hum Mol Genet 1992; 1:363-369.
229. Manam S, Ledwith BJ, Barnum AB et al. Plasmid DNA vaccines: tissue distribution and effects of DNA sequence, adjuvants and delivery method on integration into host DNA. Intervirology 2000; 43:273-281.
230. Wagener S, Norley S, zur Megede J et al. Induction of antibodies against SIV antigens after intramuscular nucleic acid inoculation using complex expression constructs. J Biotechnol 1996; 44:59-65.
231. Nichols WW, Ledwith BJ, Manam SV et al. Potential DNA vaccine integration into host cell genome. Ann N Y Acad Sci 1995; 772:30-39.
232. Gerloni M, Billetta R, Xiong S et al. Somatic transgene immunization with DNA encoding an immunoglobulin heavy chain. DNA Cell Biol 1997; 16:611-625.
233. Dietrich G, Spreng S, Gentschev I et al. Bacterial systems for the delivery of eukaryotic antigen expression vectors. Antisense Nucleic Acid Drug Dev 2000; 10:391-399.
234. Catic A, Dietrich G, Gentschev I et al. Introduction of protein or DNA delivered via recombinant *Salmonella typhimurium* into the major histocompatibility complex class I presentation pathway of macrophages. Microbes Infect 1999; 1:113-121.
235. Hess J, Gentschev I, Miko D et al. Superior efficacy of secreted over somatic antigen display in recombinant *Salmonella* vaccine induced protection against listeriosis. Proc Natl Acad Sci USA 1996; 93:1458-1463.
236. Hess J, Grode L, Gentschev I et al. Secretion of different listeriolysin cognates by recombinant attenuated *Salmonella typhimurium*: superior efficacy of haemolytic over non- haemolytic constructs after oral vaccination. Microbes Infect 2000; 2:1799-1806.
237. Corbel MJ. Reasons for instability of bacterial vaccines. Dev Biol Stand 1996; 87:113-124.
238. Barbotin JN. Immobilization of recombinant bacteria. A strategy to improve plasmid stability. Ann N Y Acad Sci 1994; 721:303-309.
239. Gentschev I, Dietrich G, Spreng S et al. Recombinant attenuated bacteria for the delivery of subunit vaccines. Vaccine 2001; 19:2621-2628.
240. Jain V, Mekalanos JJ. Use of lambda phage S and R gene products in an inducible lysis system for Vibrio cholerae- and *Salmonella* enterica serovar *typhimurium*-based DNA vaccine delivery systems. Infect Immun 2000; 68:986-989.
241. Klenerman P, Hengartner H, Zinkernagel RM. A nonretroviral RNA virus persists in DNA form. Nature 1997; 390:298-301.
242. Ochsenbein AF, Karrer U, Klenerman P et al. A comparison of T cell memory against the same antigen induced by virus versus intracellular bacteria. Proc Natl Acad Sci USA 1999; 96:9293-9298.
243. Zinkernagel RM, Bachmann MF, Kundig TM et al. On immunological memory. Annu Rev Immunol 1996; 14:333-367.

CHAPTER 16

Cytokines and Immunomodulatory Ligands as Genetic Adjuvants

Hildegund C. J. Ertl

Genetic Adjuvants—Definition

Genetic adjuvant is a term for vaccine ingredients composed of expression vectors encoding biologically active molecules such as cytokines, chemokines (detailed in Chapter 15), soluble forms of cell ligands, adhesion proteins, or other immunoreactive molecules. These vectors added to a DNA vaccine produce the adjuvant upon in situ transfection and nuclear translocation. Compared to systemic or local application of immunomodulatory mediators the vector-expressed adjuvants given concomitantly with the antigen-encoding DNA vaccine provide the active factors at the site of antigen production for a prolonged, yet poorly defined period of time. The adjuvant may act either close to the site of inoculation or upon transfection of migratory cells at other locations such as the draining lymph nodes. The mediators are presumably secreted for an extended time until the transfected cells are eliminated or transcription is terminated by other means. The effect of the mediators is thus not limited by their generally short half-life in biological fluids.

Cytokines are a group of soluble polypeptides or proteins that exert their specific activity through interactions with cell surface receptors. One of their functions pertinent to vaccines is their ability to regulate cells involved in host defense mechanisms. Cytokines (excluding chemokines) include interleukins (IL) 1 through 23, interferons, colony stimulating factors (CSFs), tumor necrosis factors (TNFs) and growth factors (GFs) such as platelet derived growth factors (PDGFs), and transforming growth factors (TGF).

Cytokines due to their ability to regulate activation of different cell subsets have been used for medicinal purposes and as adjuvants in vaccine preparations. Other types of molecules such as co-stimulatory molecules, which provide a crucial activation signal to T and B lymphocytes, complement components involved in lymphocyte activation or adhesion molecules, which target cell to certain tissues, have been tested as genetic adjuvants and with the exception of chemokines are described in this chapter.

Classification of Cytokines

Their exceedingly pleoitropic and commonly redundant activities frustrate classification of cytokines according to biological functions. The term interleukins, coined in 1979, at the 2nd International Lymphokine Workshop in Interlaken, Switzerland, was originally thought to reflect cytokines that originated exclusively from leukocytes in contrast to lymphocyte derived lymphokines or monokines produced by monocytes. It is now appreciated that this subdivision is inaccurate. Alternatively, cytokines have been subdivided according to their biological functions into those that sponsor type 1 T helper (Th1) cell responses as opposed to those that promote type 2 T helper (Th2) cell responses. Again, depending on circumstances Th2 cytokines can promote Th1 responses and vice versa. Cytokines can be classified according to their structure

DNA Vaccines, edited by Hildegund C. J. Ertl. ©2003 Eurekah.com and Kluwer Academic / Plenum Publishers.

into alpha or beta spiral structure proteins. Structural subdivision has been useful for chemokines as a subgroup of cytokines that are detailed in the following chapter. Cytokines can be grouped according to their receptor usage. Many of the cytokine receptors are composed of several chains, some of which are used jointly by different receptors. Accordingly, downstream transduction signals eventually resulting in transcriptional regulation are shared by different receptors, which contributes to the redundancy of the cytokine system.

General Characteristics and Biological Functions of Cytokines, Co-Stimulatory Molecules and Other Immunomodulators

The function of cytokines as intercellular messengers within the context of antigen-induced immune responses varies depending on the micromileu, which in turn is affected by other cytokines and chemokines. The differentiation status of receptor-bearing cell subsets and the presence of different cell types that might respond differently to a given mediator influence the activity of cytokines. The concentration of cytokines can control their function. Different cytokines can act synergistically, in opposing or in a cascade-like mechanism. Cytokines are often used as adjuvants taking advantage of one function, ignoring others that may have counterproductive effects. For the intelligent design of cytokine adjuvants, it is crucial not only to consider the isolated functions of a given cytokine on the desired immune response, but also to understand the utterly complex and not yet fully unraveled interactions with other mediators and cell subsets. Cytokines and their receptors are described in more detail by the recently published and online cytokine reference books of Oppenheim and Feldmann[1] and by a cytokine website.[2] It is impossible to review this enormous field within this chapter which will focus on the immunologically relevant facts concerning those cytokines that may have adjuvant activity.

Interleukins

Interleukin 1

IL-1, is a proinflammatory cytokine, which is mainly produced by macrophages and neutrophils, but can also be secreted by endothelial cells, keratinocytes, fibroblasts, smooth muscle cells, some epithelial cells, cells within the brain such as microglia and astrocytes, T, B, and NK cells. There are two distinct IL-1s consisting of IL-1α, which is mostly cell-associated, and IL-1β, which is mainly secreted. These two forms of IL-1 have limited sequence homology, but closely related structures. IL-1 is produced as a precursor molecule that in the case of IL-1β gains functional activity upon proteolytic cleavage by the Th-1 convertase, an enzyme of the caspase family. IL-1α is cleaved by a cysteine protease. Infectious agents and other cytokines including granulocyte macrophage-colony stimulating factor (GM-CSF), interferons (IFN), TGF-β and IL-2 induce IL-1. Some cytokines such as IFN-γ and IL-4 can down-regulate IL-1 production. Prostaglandin E2 (PGE2) and several viral infections including those with human immunodeficiency virus (HIV), Epstein Barr virus (EBV), cytomegalovirus (CMV), and respiratory syncitial virus (RSV) inhibit production of IL-1. IL-1 binds to CD121a on T cells and CD121b on B cells, NK cells and macrophages. IL-1 has a positive feedback on IL-1 receptor expression. IL-1 induces T cells to express the IL-2 receptor and to secrete IL-2. IL-1 promotes B cell proliferation and induces their responsiveness to IL-5. IL-1 augments the expression of adhesion molecules on T and B cells, monocytes and neutrophils, induces secretion of chemokines and thus affects lymphocyte trafficking. IL-1 stimulates synthesis of GM-CSF and contributes to the activation of immature dendritic cells such as Langerhans cells. IL-1 contributes to inflammatory reactions by recruitment of neutrophils.[3-9]

Interleukin 2

IL-2 is a lymphoproliferative cytokine, which is mainly derived from activated CD4$^+$ T cells. The 15.4 kDa IL-2 protein, which has limited cross-reactivity between different species (i.e., human IL-2 acts on murine cells but not vice versa), binds to the IL-2 receptor, which is composed of three chains and depending on the composition of these chains modulates the function of IL-2. The high affinity IL-2 receptor contains the alpha chain (p55), the beta chain (p75) which together bind IL-2 and the gamma chain which provides optimal signaling activity. This latter chain is also a signaling component of the IL-4, IL-7, IL-9 and IL-15 receptors. Functional IL-2 receptors are mainly expressed on activated CD4$^+$ and CD8$^+$ T cells although the p75 chain is also found on resting T and NK cells. A soluble form of the IL-2 receptor is present in sera and is strongly augmented in certain diseases such as those caused by autoimmunity or viral infections. IL-2 stimulates secretion of INF-γ, and tumor necrosis factor (TNF)α and β. IL-2 as well as IL-4, IL-7 and IL-15 effects activation-induced apoptosis of T cells and thus presumably influences the magnitude of T cell memory.[10-17]

Interleukin 4

IL-4 is a lymphoproliferative cytokine, which is produced mainly by T helper cells of the Th2 subset. Synthesis of IL-4 is induced by IL-2 and inhibited by IFN-γ and TGF-β. The species-specific 20 kDa protein binds to the CD124 receptor, which is related to the IL-6 receptor and the beta chain of the IL-2 receptor. Signaling is provided through a gamma chain, which is identical to that of the IL-2 receptor. A variant of the IL-4 receptor is secreted and blocks the activity of IL-4. IL-4 promotes proliferation and differentiation of naïve B cells. In activated B cells, IL-4 promotes synthesis of IgG1 and IgE and inhibits synthesis of IgG2a, IgM, IgG2b and IgG3. Isotype switching by IL-4 is counterbalanced by INF-γ. IL-4 enhances expression of MHC class II molecules on B cells thus increasing their ability to serve as antigen-presenting cells to primed T cells. IL-4 inhibits IL-2 induced activation of NK cells. IL-4 suppresses production of inflammatory cytokines by monocytes and T cells and IFN-γ by Th1 cells [18-23]

Interleukin 5

Activated T cells of the Th2 subset secrete IL-5. The protein shows limited cross-reactivity between species. The two IL-5 receptors (high affinity and low affinity) are expressed on lymphoid and hematopoietic cells. IL-5 promotes proliferation and differentiation of eosinophils. In mice, IL-5 induces IL-2 receptor expression on B cells, promotes B cell proliferation and their differentiation into IgM and IgA secreting plasma cells. IL-5 together with IL-2 promotes the generation of cytolytic T lymphocytes. [24-28]

Interleukin 6

IL-6 is a 26 kDa protein that upon activation is produced by fibroblasts, endothelial cells, monocytes, macrophages, T and B cells, granulocytes, smooth muscle cells, glia cells, keratinocytes, chondrocytes and osteoblasts. Bacterial and viral products, interferons, TNF, IL-1 and IL-17 induce IL-6 production. IL-4 and TGF-β inhibit synthesis of IL-6. The IL-6 receptor (CD126) is expressed on T cells, activated B cells and some monocytes. IL-6 cooperates with IL-2 in driving differentiation and maturation of cytolytic T cells. IL-6 causes maturation of B cells that are pretreated with IL-4 and stimulates secretion of antibodies.[29-32]

Interleukin 7

IL-7 is involved in lymphocyte development and is secreted by stromal cells of the bone marrow and thymus, and by keratinocytes. The IL-7 receptor (CD127) is expressed on activated T cells, pre-B cells and bone marrow macrophages. IL-7 stimulates proliferation of pre-B cells. This activity is inhibited by TGF-β. IL-7, in synergy with IL-1, induces proliferation of activated T cells. It promotes generation of lymphokine activated killer (LAK) cells. IL-7 induces monocytes to secrete IL-1 and IL-6 and promotes production of IL-3 and GM-CSF by activated T cells. IL-7 inhibits production of TGF-β. [33-34]

Interleukin 9

Activated CD4⁺ T cells are the main source of IL-9. Synthesis of IL-9 is induced by IL-1 and inhibited by IL-2. IL-9 induces proliferation of a subpopulation of activated Th2 cells. IL-9 synergizes with IL-4 in promoting production of IgM, IgG and IgE. IL-9 induces IL-6 secretion by mast cells.[35,36]

Interleukin 10

IL-10 is produced by activated Th2 cells, B cells of the Ly-1 lineage and keratinocytes. IL-10 is largely but not exclusively an anti-inflammatory cytokine that inhibits synthesis of IL-12 and thus INF-γ. It also interferes with production of IL-2 and TNF-β by Th1 cells. IL-10 suppresses synthesis of IL-1, IL-6, and TNF-α by macrophages. IL-10 prevents T cell proliferation. It also inhibits activation of B cells by IL-5. It promotes differentiation of B cells and reduces MHC class II expression on monocytes. IL-10 inhibits activation of macrophages. On the other hand, IL-10 promotes the development of cytolytic T effector cells and can support induction of anti-tumor immunity. IL-10 can have chemotactic activity by recruiting CD8⁺ but not CD4⁺ T cells.[37-40]

Interleukin 11

Mesenchymal cells and bone marrow stromal cells secrete IL-11. IL-11 shares the signalling gp130 chain of its receptor with IL-6. IL-1 induces production of IL-11. IL-11 has hematopoietic activities and promotes differentiation of B cells into antibody secreting cells.[41,42]

Interleukin 12

IL-12 is mainly produced by phagocytes such as monocytes, macrophages, neutrophils and dendritic cells, and to a lesser extent by B cells. Bacterial products and some viruses such as lymphocytic choriomeningitis virus (LCMV) induce IL-12. Other viruses such as HIV or parasites such as Leishmania inhibit production of IL-12. This is based on the induction of Th2 linked cytokines most notably IL-4 and IL-10. IL-12 is composed of two chains, p40 and p35, which as heterodimers form biologically active molecules. Homodimers of p40 block the activity of heterodimers. The IL-12 receptor, which is composed of a b1 and b2 subunit, is expressed on activated NK cells and on activated T cells. The b1 subunit is expressed on resting human T and NK cells. Induction of the b2 subunit upon activation is inhibited by IL-4. Activated T cells from BALB/c mice, commonly used to test the effect of genetic adjuvants, unlike other strains of mice cease upon activation to express the IL-12 b2 receptor chain and thus become unresponsive to IL-12. This is presumably the basis for the Th2 prone immune responses of Balb/c mice to certain pathogens and vaccines.

IL-12 activates NK cells to secrete INF-γ and augments their lytic activity. It promotes polarization of T helper cells towards the Th1 pathways and inhibits polarization of an antigen driven immune response towards the Th2 pathway. IL-12 enhances the cytolytic activity of T cells by augmenting production of perforin. It inhibits synthesis of Th2 linked cytokines such as IL-4, IL-5 and IL-10. IL-12 has anti-angiogenic activity presumably through induction of INF-γ, which in turn promotes synthesis of IP-10 and MIG, two chemokines that inhibit blood vessel formation.[43-49]

Interleukin 13

Activated CD4⁺ T cells of the Th2 type and basophils produce IL-13. The high affinity IL-13 receptor is composed of an IL-13 specific alpha chain and the beta chain of the IL-4 receptor, both of which combine with the gamma chain of the IL-2 receptor. IL-13 shares several of the biological activities of IL-4, but unlike IL-4 has no direct effect on T cells. IL-13 inhibits production of proinflammatory cytokines by macrophages in response to INF-γ and bacterial products. It suppresses production of nitric oxide by macrophages. IL-13 induces differentiation and proliferation of activated B cells. It increases CD23, CD72 and MHC class II expression on B cells.[50,51]

Interleukin 15

IL-15 is produced by fibroblasts, epithelial cells, monocytes and muscle cells. IL-15 has similar biological activity to IL-2. It promotes proliferation of activated T and B cells and the lytic activity of NK cells. IL-15 protects activated T cells for fas-ligand induced apoptosis.[52-54]

Interleukin 16

IL-16 is produced mainly by CD8[+] T cells and to a lesser extent by CD4[+] T cells. The protein is synthesized as a precursor that requires cleavage by caspase 3. IL-16 binds to the CD4 receptor. IL-16 is chemotactic for CD4[+] T cells, eosinophils and monocytes. Binding to the CD4 receptor results in some cells in expression of IL-2 receptors and in production of GM-CSF, TNF-α and IL-6. IL-16 augments the anti-apoptotic effect of IL-2 on T cells. Binding of IL-16 to CD4 transiently inhibits activation of T cells through T cell receptor (TcR) or CD3 engagement.[55]

Interleukin 17

Activated CD4[+] T cells produce IL-17. IL-17 enhances expression of the ICAM-1 on fibroblasts. IL-17 has immunoenhancing properties and stimulates production of IL-6 and IL-8.

Interleukin 18

IL-18 is produced by a variety of cells including macrophages, dendritic cells, gut epithelial cells, keratinocytes, chondrocytes, osteoblasts and others. The 192 (mouse) or 193 (human) amino acid long protein induces production of IFN-γ by lymphocytes, including CD8[+] T cells, Th1 cells and B cells. Concomitant stimulation by IL-18 with IL-12 has a strongly synergistic effect in producing IFN-γ based on the induction of IL-18 receptors on T cells by IL-12. IL-18 is required for maturation and activation of NK cells; it augments the cytolytic activity of NK cells and induces them to secrete IFN-γ. IL-18 furthermore induces secretion of GM-CSF, IL-8 and TNF by T cells. [56]

Interleukin 21

IL-21 is a recently identified cytokine that is related to IL-2 and IL-15.[57] Activated T cells probably produce IL-21. IL-21 promotes maturation and proliferation of NK cells, induces B cell proliferation in the presence of CD40 and expansion of activated T cells. The IL-21 receptor is expressed in lymphoid organs and on NK cells.

Granulocyte-Macrophage Colony Stimulating Factor (GM-CSF)

Activated T cells, fibroblasts, endothelial cells, and macrophages produce GM-CSF. T cells produce GM-CSF in response to TcR engagement; other cells produce this cytokine upon encounter of LPS, proinflammatory cytokines and phorbol esters. Anti-inflammatory cytokines and glucocorticoids inhibit production of GM-CSF. The 23 kDa protein acts in a species-specific manner. GM-CSF is a survival factor for hematopoetic precursor cells. It stimulates proliferation and differentiation of myeloid dendritic cells. It activates neutrophils, eosinophils and macrophages. It induces secretion of other cytokines such as IL-1, IL-6 and TNF and it increases expression of MHC class II and Fc receptors on dendritic cells and macrophages.[58-61]

Interferons

Interferon α/β (IFN-1)

IFN-α and IFN-β are produced by a number of different cell types including monocytes/macrophages, fibroblasts and endothelial cells and type 2 dendritic cells in lymphatic tissues. Secretion is induced by pathogens such as viruses. The IFN-α protein has more than 20 different variants encoded by distinct genes. IFN-α/β have anti-viral and anti-proliferative activities.

IFN-α and β bind to the same receptor, which is expressed on a wide variety of cells. IFN-β increases the expression of MHC class I antigens. IFN-β stimulates the activity of NK cells. In mice, IFN-α contributes to activation/maturation of monocytes and dendritic cells. IFN-1s induce IL-15, which provides a proliferative signal to activated T cells. IFN-1s protect T cells from apoptosis, thus contributing to the persistence of memory T cells.[62-65]

Interferon (IFN)-γ

IFN-γ is produced by activated NK cells, B cells and activated T cells of the CD4⁺ Th1 subset and the CD8⁺ T cell subset. The synthesis of IFN-γ is induced by IL-2, IL-12 and IL-18.

IFN-γ receptors are expressed on a variety of cell types. IFN-γ similar to IFNα/β has anti-viral and anti-proliferative activities. By activating macrophages, IFN-γ also exhibits anti-bacterial activities. In addition, IFN-γ preferentially up-regulates expression of MHC class II determinants on a variety of cells. IFN-γ inhibits the effect of IL-4 on Th2 cells and on B cell switching towards the IgG1 and IgE isotypes. It is essential for production of IgG2α. IFN-γ stimulates secretion of TNF-α by macrophages. It furthermore instigates production of nitric oxide. IFN-γ induces a number of angiostatic chemokines such as MIG and IP-10. IFN-γ has a negative effect on protein expression driven by numerous mammalian or viral promoters.[66-70]

Tumor Necrosis Factor (TNF) Family

Tumor Necrosis Factor (TNF)-α

TNF-α is produced by most cells of the immune system including B, T and NK cells and macrophages. Substances derived from infectious agents, cytokines such as IL-1, IFN-γ, IL-2 and GM-CSF and cellular insults such as hypothermia or irradiation induce production of TNF-α, which is inhibited by IFN-α/β, and by anti-inflammatory cytokines such as IL-10 and TGF-β. Human TNF-α is active on murine cells with a slightly reduced specific activity. TNF-α is a pleiotropic cytokine that exhibits multiple functions on a variety of cell types. TNF-α induces production of chemokines such as MCP-1, MIP-1α and IP-10, augments expression of MHC determinants and promotes production of IL-1 and IFN-γ by lymphocytes, leukocytes and macrophages. TNF-α enhances the proliferation of T and B cells. TNF-α promotes the maturation of Langerhans cells into dendritic cells.

Tumor Necrosis Factor-β (TNF-β)/ Lymphotoxin (LT)

This factor is produced predominantly by activated T-lymphocytes of the Th1 subset. IFN-γ and IL-2 stimulate the synthesis of TNF-β. TNF-β is a protein of 171 amino acids that shows a high degree of homology between mice and humans. TNF-β acts on a plethora of different cells. It induces the synthesis of GM-CSF and IL-1, is mitogenic for B lymphocytes and is necessary for the development of lymph nodes and proper splenic organogenesis through the induction of chemokines such as SLC and BLC.[71-74]

CD40 Ligand (CD154)

CD40 ligand (CD40L) is a member of the TNF family. It is an unusual cytokine since it is a 39 kDa transmembrane protein, which is expressed by activated T cells, macrophages, dendritic cells, activated NK cells, eosinophils, smooth muscle cells, and other cell types. Expression of CD40L on T cells is induced by their activation through TcR interactions or through CD3 signaling. Expression is augmented by co-stimulation through CD28 or by co-activation with cytokines such as IL-2 or IL-12. Expression of CD40L by Th cells is inhibited by IFN-γ; expression on Th2 cells is blocked by TGF-β. Engagement of CD40L on T cells by CD40⁺ cells induces cytokine release and drives proliferation of the T cells. Conversely, T cells expressing CD40L activate CD40 expressing B cells, monocytes and dendritic cells. Dendritic cells upon engagement of CD40 with CD40L up-regulate expression of other costimulatory molecules.[75-78]

Co-Stimulatory Molecules

CD80/CD86

B7.1 known as CD80 and B7.2 called CD86 are co-stimulatory molecules. As described in more detail in the earlier chapters, activation of T cells requires two signals. Engagement of the TcRs with their corresponding targets, in most instances peptides displayed by MHC molecules serves as signal 1. Co-stimulatory molecules provide signal 2. Signal 1 in absence of signal 2 leads to T cell tolerance rather than to the induction of a productive immune response. B7.1 is a 262 amino acid protein which has ~ 25% sequence homology with the 306 amino acid long B7.2 molecule. B7.2 is constitutively expressed on dendritic cells, while expression of both B7.1 and B7.2 on other antigen presenting cells such as macrophages, keratinocytes and B cells is induced. B7 binds to CD28 and CTLA-4, which are both expressed on T cells. CD28 is expressed on most resting T cells, its density is increased upon their activation. Only activated T cells express CTLA-4. Following activation of T cells, engagement of CD28 by B7 stimulates production of IL-2, drives T cell proliferation, induces expression of CTLA-4 and the IL-2 receptor and up-regulates CD40L expression. It also induces production of anti-apoptotic proteins such as bcl-$_{xL}$, which protects T cells from fas-mediated cell death. Engagement of CD28 with B7.1 or B7.2 respectively influences the nature of the resulting immune response. Some evidence suggests that while both induce IL-2 and IFN-γ production B7.1 favors production of GM-CSF while B7.2 sponsors induction of IL-4, which may drive the immune response towards the Th2 pathway. Engagement of CTLA-4 with B7 down-regulates the immune response. [79-81]

Cytotoxic T lymphocyte Antigen-4 (CTLA-4)

As discussed above CTLA-4 is expressed on activated T cells and bind to the co-stimulatory molecules CD80 and CD86.

Immunomodulators

L-Selectin (CD62L)

CD62L is expressed on T and B cells, neutrophils, monocytes and NK cells. Upon activation, T cells shed CD62L. Expression levels of CD62 distinguish naïve (CD62L$_{high}$) T cells from effector/memory T cells (CD62L$_{low}$). CD62L binds to a sialated oligosaccharide expressed on high endothelial venules in lymph nodes. CD62L thus serves as an adherence receptor for lymphocytes. [82-84]

Complement Component C3d

C3 is part of the complement system. The 190 kDa molecule is composed of a 120 kDa alpha and a 75 kDa beta chain. Upon cleavage by the C3 convertase the molecule is digested into a large 185 kDa C3b fragment and a small 77 amino long C3a fragment. C3b, which is still composed of an alpha and a beta chain, is further cleaved into three fragments of 68, 43 and 3 kDa respectively with the 68 and 43 kDa fragment forming C3bi with the unaltered 75 kDa beta chain. The 68 kDa chain of this complex can be cleaved further releasing a 145 kDa complex (75 kDa beta chain + 43 kDa fragment + residual 27 kDa fragment of the 68 kDa protein) termed C3c and the 41 kDa C3dg fragment. Certain enzymes such as trypsin or neutrophil elastase can digest C3dg. By removing a 10 kDa fragment they create C3d. C3d and C3dg bind to CD21, which is the complement receptor (CR)2 that is expressed on B cells, granulocytes, follicular dendritic cells and a subset of macrophages. CD21 associates with CD19, a co-receptor for B cells, that facilitates T cell-mediated activation of B cell responses. [85]

The Effects of Cytokines and Immunostimulatory Molecules on Activation of an Immune Response

As detailed in the beginning chapters of this book, encounter of an antigen leads through a cascade of finely orchestrated events to activation of specific T and B cell responses followed by elimination of the antigen and then the development of immunological memory. In order to initiate a primary immune response, the antigen has to be taken up and processed by dendritic cells, which in an immature stage are dispersed throughout an organism. To efficiently present antigen to naive B and T cells, dendritic cells have to be activated by the so-called signal 0 also termed danger signal.[86] This danger signal causes up-regulation of MHC class II and co-stimulatory molecule expression. Maturation of dendritic cells also changes their chemokine receptor expression profile.[87] This equips them to migrate to draining lymph nodes which provide an optimized environment for induction of a primary immune response. Within lymphatic tissues, dendritic cells present the antigen, the so-called signal 1, in form of small peptides associate with MHC class II molecules together with co-stimulatory molecules such as B7.1 or B7.2 (signal 2) to CD4+ T cells. Activated CD4+ T cells in turn facilitate activation and proliferation of CD8+ T cells which recognize antigenic peptides associated with MHC class I molecules. CD4+ T cells furthermore provide help to B cells to respond to soluble antigen complexed with natural antibodies to the Fcγ receptor on follicular dendritic cells. Following induction by the antigen in association with a second costimulatory signal, CD8+ T cells and, to a lesser extent CD4+ T cells, proliferate.[88] Once fully matured, they leave the lymphatic tissue and migrate to the site of antigen invasion where they commence effector functions in form of cytolysis and cytokine release. Upon removal of the antigen most of the effector T cells undergo programmed cell death, the survivors form the pool of memory cells, which is maintained in absence of antigen with minimal turnover of the memory T cell populations.[89] B cells upon activation proliferate, their Ig genes rearrange to produce mature isotypes, such as IgGs and IgA and their hypervariable region genes hypermutate. B cells with mutations that result in improved affinity to the antigen are selected for, the others undergo apoptosis. Activated, antibody-producing B cells leave the lymph nodes and home mainly to the bone marrow and to a lesser extent to the spleen. Continued production of antibody may or may not require the presence of antigen.[90,91] Most of these events are influenced by cytokines, chemokines and immunostimulatory molecules, which can thus be used therapeutically to enhance or redirect an antigen-specific immune response.

Invasion by a pathogen or administration of a vaccine is commonly associated with an inflammatory reaction caused for example in the case of systemic vaccination by the minor injury at the injection site as well as by the adjuvants that are commonly included in the vaccines. In the case of DNA vaccines, the adjuvant effect is provided by the bacterial DNA which carries unmethylated CpG sequences. These sequences upon binding to Toll receptors (TLR-9) initiate an inflammatory reaction.[92] At the site of entry the antigens, or in case of a DNA vaccine, the plasmid vectors are in part taken up by resident immature dendritic cells through phagocytosis, pinocytosis or receptor-mediated endocytosis.[93] Other cells that become infected or transfected can provide additional antigen which is reprocessed by antigen presenting cells and can thus contribute to the initiation of an immune response. Immature dendritic cells express chemokine receptors such as CCR1, CCR2, CCR5, CCR6 and CXCR1. Expression of these receptors allows them to respond to inflammatory chemokine such as MIP-1α, RANTES, MIP-3α and MCP-1 that are secreted at sites of inflammations. Immature dendritic cells are also recruited by GM-CSF, a cytokine that provides a maturation signal to dendritic cells. Immature dendritic cells can thus be recruited to tissues subjected to infection by a pathogen or in response to injection of a vaccine.

Immature dendritic cells, present at or recruited to an area of inflammation, are able to take up the antigen. In the immature stage they are poorly equipped to initiate an immune response by presentation of the antigen to naïve T cells. Most of the MHC class II molecules needed for presentation of antigenic peptides to CD4+ T cells are located in intracellular vesicles and their expression level of co-stimulatory molecules is low. Following activation through the so-called danger signal, dendritic cells become partially mature. A pathogen or an adjuvant present in a vaccine formulation can provide the danger signal. Danger signals include bacterial products such as LPS, unmethgylated CpG motifs present in bacterial genome, heat shock proteins, cytokines such as IFN-1s or cocktails of cytokines including TNF-α, GM-CSF, IL-4 and IL-13.

Interactions of CD40 with CD40L expressed on activated T cells can drive maturation of dendritic cells. Initially after receiving a maturation signal, dendritic cells increase and then upon maturation decrease their antigen uptake. They down-regulate expression of CCR1, CCR5, CCR6 and CXCR1 and up-regulate expression of extracellular MHC class II determinants, co-stimulatory molecules and CCR7. Due to expression of CCR7, dendritic cells become responsive to chemokines such as SLC (secondary lymphoid tissue chemokine) and ELC (EBI1-ligand chemokine) also called MIP-3β (macrophage inflammatory protein), which are produced within lymphatic tissues. Dendritic cells migrate there through afferent lymph vessels and home to the T cell rich areas. Here they commence their function as professional antigen presenting cells.[94-96]

For activation of T cells the antigen has to be presented as of small peptides associated with MHC determinants. Activation of CD8[+] T cells requires presentation of peptides by MHC class I determinants. The peptides are generated by degradation of de novo synthesized proteins through a proteasome complex. The resulting peptides are translocated by a specialized transporter system called TAP into the endoplasmatic reticulum where they associate with an immature form of newly synthesized MHC class I molecules. Binding of peptides stabilizes the MHC class I molecules which are then transported to the cell surface where they engage the TcR of CD8[+] T cells. Both synthesis of the TAP molecules and MHC class I molecules are up-regulated by IFN-γ. MHC class I expression on certain cells is further increased by TNF-α.

The TcR of CD4[+] T cells recognizes peptides derived from proteins that are taken up by the cells and then enter the lysosomal pathway where they bind to the groove of MHC class II molecules. Again binding of the peptides stabilizes the MHC class II molecules, which are then translocated to the cell surface. Expression of MHC class II determinants is induced and increased by IFN- γ, IFN-β and IL-4. Expression of MHC class II on B cells is enhanced by IL-13 while GM-CSF enhances its expression on dendritic cells and macrophages. IL-10 reduces the expression of MHC class II determinants.

B cells recognize mainly conformational epitopes on soluble undigested antigen complexed by natural antibodies to Fcγ receptors on follicular dendritic cells. IL-4, IL-10, IL-11 and IL-6 increase differentiation of B cells in cooperation with IL-4.

In addition to recognition of the antigen, T and B cells require a second signal also called co-stimulatory signal for their activation.[97,98] Engagements of CD28 expressed on T cells with CD80 (B7.1) or CD86 (B7.2) on antigen presenting cells provide a costimulatory signal for T cells. Interactions between CD40 carried by B cells with CD40L expressed by activated T cells serves as a second signal for B cells. Certain cytokines such as IL-2 augment and amplify signal two. Other cytokines, such as IL-4 and IL-7 that signal through the common IL-2Rγ chain can have the same effect. The expression of co-stimulatory signals on antigen presenting cells is also enhanced by IFN-γ but inhibited by IL-10.

Upon activation T cells differentiate and proliferate in response to cytokines. In mice differentiation of CD8[+] T cells into cytolytic T cells is facilitated by cytokines including IL-5 and IL-6 which both cooperate with IL-2. IL-12 and IL-10 also promote maturation of CD8[+] T cells. Differentiation of CD4[+] T cells is influenced by IL-12 and IL-4, which have opposing signals as detailed below. Proliferation of T cells is mainly mediated by IL-2 but other cytokines that utilize the IL-2Rγ chain can also drive expansion of T cells. IL-7 together with IL-4 induce T cell proliferation, IL-1 indirectly increases proliferation of T cells by up-regulating expression of the IL-2 receptor. IL-15 and IL-21 serve as growth factors for activated T cells while IL-9 drives proliferation of a subset of Th2 cells. Proliferation of T cells is augmented in the presence of TNF-α and inhibited by TGF-β and interferons. B cells proliferate in response to a number of cytokines such as IL-1, IL-2, IL-4, IL-6, IL-10, IL-13, IL-15 and IL-21. IL-4 promotes B cell maturation, which is enhanced by ligation of CD40. IL-5 acts synergistically with IL-4 by enhancing CD40 expression. IL-13 has an effect similar to that of IL-4. Together these two cytokines also block programmed cell death of B cells. Proliferation of B cells is inhibited by TGF-β. Isotype switching of B cells towards IgE is mediated by IL-4 and IL-13, switching towards the IgG2a isotype is favored by IFN-γ and switching to IgA is facilitated by TGF-β.

T helper cells during the course of an immune response differentiate into either Th1 or Th2 type cells. These two subsets differ in their cytokine secretion profile and thus the type of the immune response they promote.[99] Th1 cells secrete IFN-γ and lymphotoxin and facilitate activation of CD8+ T cells and of B cells secreting IgG2a and IgG3. Th2 cells produce IL-4, IL-5 and IL-10. Most acute immune responses have a mixture of uncommitted Th0, Th1 and Th2 cells while chronic exposure to antigen commonly polarizes the responses towards either Th1 or Th2. IL-12 promotes differentiation of Th0 cells into Th1 cells, which promotes cellular immunity, while IL-4 promotes the development of Th2 cells, which drives humoral immunity and can induce allergies through the induction of IgE secreting B cells. IFN-γ inhibits differentiation into Th2 cells while IL-4 and IL-10 suppress differentiation into Th1 cells. The effect of IL-4 at the initiation of immune responses overrides the effects of cytokines that sponsor Th1 type immune responses.

The role of cytokines in establishing long-lasting immunological memory is not yet fully understood. Studies in knockout mice indicate that IL-2 or IL-4 is not essential for induction or maintenance of T or B cell memory. Cytokines, such as IL-7, IL-15 or TGF-β that provide survival signals to T cells by up-regulation of intracellular anti-apoptotic pathways or down-regulation of cell surface fas expression might play a role in increasing the number of lymphocytes that enter the memory pool. In contrast, cytokines that promote apoptosis of T cells such as IFN-γ may have a negative effect on the magnitude and in consequence the duration of immunological memory.

One should keep in mind that the induction of an immune response is not a single-file series of events. Antigen such as that encoded by DNA vaccines continues to be produced for an extended period of time and antigen processing and presentation and stimulation of T and B cells remain ongoing until the antigen is eventually eliminated. Antigen processing is at later stages influenced by infiltrating activated immune effector cells. These cells not only change the local cytokine milieu but also provide additional signals that affect antigen presentation.

Genetic adjuvants are generally directly added to the DNA vaccine. As they require initial transcription and translation of the vector encoded cytokine or immunomodulatory molecule they are unlikely to affect the very early stages of antigen processing, presentation and T and B cell activation. Although they might to a limited extend transfect dendritic cells and upon transport to draining lymph nodes may affect the initial immune responses, the bulk of cytokine encoding vectors will transfect stationary cells close to the inoculation site. Cytokines released by these cells are likely to influence the later stages of the immune response driven by antigen released from transfected tissue cells. They may act by recruiting antigen presenting cells, by causing maturation of dendritic cells or by preventing apoptosis of T effector cells. Whether or not they exert additional effects on activated lymphocytes by providing proliferation or differentiation signals remains to be elucidated.

The Effects of Genetic Adjuvants on the Immune Responses to DNA Vaccines

Many cytokines and other immunomodulators have been tested as genetic adjuvants in mice, rats as well as primates. Some have only been tested once or twice and it is premature to form conclusions concerning their usefulness as genetic adjuvants. Others such as IL-12 and GM-CSF upon extensive testing have yielded fairly consistent adjuvant effects and one of those, i.e., GM-CSF, which enhances B and T helper cell responses to vector-encoded antigen is now scheduled for a phase I clinical trial. Additional factors such as IL-4, IL-10 and IFN-γ have yielded sharply contrasting results in different systems and additional studies are needed to resolve these conflicts. Only a few studies have explored the use of cytokine mixture. Considering that cytokines commonly act in networks or cascades, affect each other synergistically, potentiate each other or reduce or subvert each others functions, additional studies with cocktails of cytokine adjuvants are warranted.

Interleukins

Interleukin 1

Both IL-1α and IL-1β have been tested for their ability to modulate the immune response to DNA vaccines. IL-1α when tested as a genetic adjuvant to an HIV-1 DNA vaccine had no effect on the ensuing immune response although previous results indicated that tumor cells transfected with this cytokine were rendered immunogenic.[100] A sequence encoding a nonapeptide of IL-1β with receptor binding activity fused to a model antigen, i.e., ovalbumin (OVA) exhibited disperate effects depending on the genetic make-up of the host. In C57Bl/6 mice addition of the IL-1β sequence had no effect on the B or T cell response to OVA. In BALB/c mice addition of the nonapeptide increased IFN-γ secretion by T cells and augmented the OVA-specific cytolytic T cell response while decreasing secretion of IL-4. Although the mechanism of action was not further elucidated, the authors speculated that IL-1β, which polarized the response towards the Th1 pathway in BALB/c mice, acted by inducing IL-12 production by antigen presenting cells. The IL 12 induced IFN-γ, which in turn inhibited Th2 development.[101] In a tumor model, a B cell idiotype expressed by a single variable chain fused to the IL-1β nonapeptide induced protection to challenge with the B cell lymphoma cell line expressing the same idiotype.[102]

Interleukin 2

IL-2 has been tested in different systems with variable results. In our model based on a DNA vaccine encoding the rabies virus glycoprotein under the control of the SV40 promoter, concomitant application of a plasmid for mouse IL-2 had in most experiments no effect although occasionally a slight increase in the ensuing B and T cell response was observed. Similar results were reported with an IL-2-OVA fusion protein.[101] In contrast, addition of an IL-2 encoding plasmid to a DNA vaccine expressing the hepatitis B virus (HBV) surface antigen augmented antibody responses as well as T helper and cytolytic T cell responses to the transgene product.[103] Addition of the IL-2 plasmid did not cause a clear switch in the nature of the immune response.[104] Addition of an IL-2 encoding vector to a DNA vaccines to the HIV Env and Gag antigens enhanced antibody production and T cell proliferation upon in vitro culture with antigen.[105] It also increased induction of cytolytic T cells[106] and stimulation of IFN-γ producing T cells thus indicating a switch towards a Th1 type immune response.[107] In another study, IL-2, when applied as a separate vector or linked to the transgene product as a fusion protein strongly increased antibody titers to the hepatitis B virus (HBV) surface antigen[104] and augmented T cell proliferation, IFN-γ and IL-2 secretion. IL-4 secretion was not increased. The effect was long lasting and could be observed for the duration of the experiment (~ 4 month post vaccination). In combination with a DNA vaccine encoding the herpes simplex virus (HSV) type 2 gD antigen, an IL-2 vector enhanced vaccine efficacy by increasing survival rates and reducing lesions upon intravaginal viral challenge.[108] Addition of an IL-2 vector to a cationic lipid formulation containing a DNA vaccine to HIV-1 increased upon intranasal immunization HIV-1 specific delayed-type hypersensitivity (DTH) reactions, T cell-mediated cytolysis and IFN-γ and IL-2 secretion by T cells. IL-4 secretion was inhibited. Antibody titers remained unchanged but showed an increase in the IgG2a/IgG1 ratio indicative for a bias towards a Th1 response.[109] Similarly, using DNA vaccines encoding HIV-1 gp120 or the influenza A virus nucleoprotein (NP), another group reported that gene gun delivery of a DNA vaccine without the addition of a cytokine adjuvant resulted in a Th2 biased immune response. Including a vector encoding IL-2 increased IFN-γ responses apparently redirecting the response towards the Th1 pathway. This effect was short-lived and not preserved when mice were boosted after a 3 month rest period.[110] In contrast, Barush et al reported that mice injected intramuscularly with a bicistronic vector for IL-2 and Env of HIV-1 showed decreased immune responses.[111] The same effect was seen if mice were injected with a DNA vaccine to HIV-1 given concomitantly with a vector encoding IL-2 or a fusion protein composed of IL-2 and the Fc portion of an immunoglobulin molecule (IL-2-Ig). Administration of the IL-2 encoding vector prior to injection of the DNA vaccine had the same effect. On the other hand, if the DNA vaccine was given first, followed 2 days later by administration of the IL-2-Ig construct, mice developed an increased antibody response. This increase was not observed if mice were injected with the vector encoding native IL-2. The difference in the efficacy of the two IL-2 constructs might reflect the extended half life of the IL-2-Ig construct. Furthermore, the bivalency of IL-2 linked to Ig might more readily cause activation due to receptor cross-linkage. The same group of investigators subsequently extended their studies to rhesus monkeys.[112, 113] The animals were immunized with a DNA vaccine to the simian immunodeficiency virus (SIV) Gag protein or to a chimeric simian human immunodeficiency virus (SHIV) envelope (Env 89.6P). Monkeys that received the IL-2-Ig plasmid together with the DNA vaccines (given together) developed 30-fold higher antibody titers to Env and 5-fold higher levels of CD8+ T cells to Gag as determined by staining with a MHC-Gag peptide tetramer. Upon challenge with the pathogenic SHIV 89.6P virus, sham-vaccinated monkeys generated marginal cytolytic T cell responses, showed a precipitous drop in CD4 counts, developed high viral titers and clinical evidence of progressive disease. Monkeys vaccinated with the DNA vaccines to Gag and Env together with the IL-2-Ig construct generated a robust cytolytic T cell response upon challenge indicating that the vaccine-induced immune response had not completely prevented the infection. Nevertheless, the animals developed low to undetectable set point viral titers, maintained normal CD4 counts and remained disease free.

The lack of consistency between the results obtained with IL-2 used as a genetic adjuvant can currently not be explained. The very promising results obtained both with the HIV and HBV antigens indicate that for some antigens IL-2 might be an appropriate genetic adjuvant. IL-2 most likely acts at a later stage of the induction phase of the immune response by promoting B and T cell proliferation thus causing an expansion of antigen-reactive clones. IL-2 per se does not polarize the immune response towards the Th1 pathway as was observed in several studies. IL-2 affects NK cells by up-regulating their IL-2 receptor expression and by inducing secretion of cytokines such as IFN-γ and TNF-α. This could indirectly augment the release of IL-12 from macrophages, which in turn would favor Th1 type immune responses.

Interleukin 4

Several investigators tested IL-4 as a genetic adjuvant to DNA vaccines with highly variable results. In our system, using an intramuscularly administered DNA vaccine to the rabies virus glycoprotein in mice we observed early after immunization a slight decrease in the antibody response of animals that had received the IL-4 encoding vector compared to the controls. In mice vaccinated with the DNA vaccine only, the antibody titers stabilized –10 weeks after vaccination, while titers in mice co-injected with the IL-4 expressing plasmid continued to rise and after 3 month eventually exceeded those of the control mice.[114] IL-4 did not polarize the response towards the Th2 pathway, as antibodies to rabies elicited in presence of the IL-4 vector remained predominantly of the IgG2a isotype. Furthermore, addition of the IL-4 vector resulted in an increase of the cytolytic T cell response to rabies. Mice immunized with a DNA vaccine to rabies in presence of an IL-4 vector developed upon intranasal immunization with an E1-deleted adenoviral recombinant expressing the same transgene product an enhanced IgG2a antibody response at the vaginal mucosa.[115] Another group of investigators, Kim et al, reported within 2 weeks after immunization an increased antibody response to Env and Gag of HIV-1 in presence of an IL-4 vector. Co-administration of the IL-4 vector also resulted in a slight increase in T cell proliferation to antigen and reduced the generation of cytolytic and IFN-γ secreting T cells.[106] A 3rd study reported a slight increase in the proliferative response of antigen-specific CD4+ T cells in mice injected with a DNA vaccine to HBV concomitantly with an IL-4 encoding vector. IL-2 and IFN-γ secretion by T cells was abrogated in presence of IL-4 while IL-4 secretion was increased. The activity of cytolytic T cells was strongly reduced. The overall production of antibodies was augmented; those of the IgG2a isotype were reduced while those of the IgG1 isotype were increased.[103] The efficacy of a DNA vaccine to the gD antigen of HSV type 2 was diminished upon addition of an IL-4 producing vector.[108] Using a

DNA vaccine encoding a fusion protein composed of OVA and IL-4, another group reported an increase in the antibody response to OVA without a shift towards isotypes indicative of a Th2 response. The same study reported a delay in the response of cytolytic T cells.[107] Using a tumor antigen, i.e., the human carcinoembryonic antigen (CEA), one group reported that addition of an IL-4 encoding vector reduced generation of cytolytic T cells and resistance to tumor challenge.[116] One study addressed the safety of an IL-4 encoding vector given to adult or neonatal mice. The vector sponsored the development of a Th2 type immune response to the simultaneously administered vaccine antigen. Long-term effects on subsequent immunizations were not observed nor did mice develop overt signs of autoimmune reactions.[117]

In most studies IL-4 switched the immune response to a DNA vaccine towards the Th2 pathway. CpG sequences present in the bacterial backbone of a plasmid vector induce by signalling through TLR-9 secretion of IL-12 which in turn promotes Th1 responses. An excess of IL-4 present at early stages of the immune response presumably inhibits the activity of IL-12 thus favoring generation of Th2 responses. The opposing results obtained by different investigators may reflect the balance between the adjuvanticity of the DNA vaccine favoring Th1 responses and the amount of vector-encoded IL-4 inhibiting such responses. The DNA vaccine we used in our studies in the rabies virus system is exceptionally rich in those CpG sequences that provide optimized stimulatory motifs for mice. Induction of a potent IL-12 response by such motifs stimulates production of IFN-γ which if present in high enough amounts can reduce the activity of the viral promoters used in DNA vaccines and thus result in a marked reduction of the antigenic load. Under such circumstances addition of a cytokine such as IL-4 that inhibits synthesis of IL-12 may reduce production of IFN-γ to levels that are not detrimental for vector-driven antigen production and thus increase the overall immune response that, due to sufficient levels of IL-12 preceding IL-4, remains Th 1 biased.

Interleukin 5

We tested IL-5 for its effect on a DNA vaccine to the rabies virus glycoprotein. The DNA vaccine was used for priming followed by an intranasal booster immunization with an E1-deleted adenoviral recombinant. Mice developed serum antibodies as well as antibodies that were secreted at the genital mucosa. IL-5 enhanced vaginal antibody titers by increasing antibodies of the IgA, IgG1 and IgG2b isotypes; antibodies of the IgG2a isotypes that dominated in mice injected with the DNA vaccine alone prior to the intranasal booster immunization were only marginally increased.[115] This effect of IL-5 presumably reflects its ability of this cytokine to promote B cell differentiation and in mice switching towards the IgA isotype.

Interleukin 6

IL-6 was tested as an adjuvant to a gene gun-delivered DNA vaccine expressing the hemagglutinin (HA) molecule of influenza A virus.[118] Mice that were vaccinated in the presence of the IL-6 vector developed reduced viral titers in lung lavage upon intranasal challenge with influenza A virus. They showed increased titers of HA-specific IgG antibodies in nasal wash before challenge. After challenge, they generated an accelerated serum antibody response to HA, which included antibodies of the IgA isotype absent in sera from mice vaccinated with the DNA vaccine alone. IL-6 causes maturation of IL-4 pre-treated B cells and stimulates secretion of antibodies, which might have facilitated the development of antibodies both in sera and at mucosal surfaces in this system.

Interleukin 7

IL-7 was used as a genetic adjuvant to a DNA vaccine expressing HIV gp120. Upon gene gun delivery of the DNA vaccine given alone, mice developed an immune response that was biased towards the Th2 pathway. Addition of the IL-7 encoding vector switched the immune response towards the Th1 pathway by favoring IFN-γ secreting T cells and reducing IL-4 secretion. These results were unexpected. None of the known functions of IL-7 suggest that this cytokine promotes Th1 responses. In contrast, IL-7 can prime naïve (human) T cells to secrete IL-4, which would favor the development of Th2 responses.[119]

Interleukin 10

We tested IL-10 as a genetic adjuvant to a DNA vaccine to the rabies virus glycoprotein and observed that addition of this cytokine unexpectedly caused a marked increase of a Th1 biased antibody response. IL-10 generally drives the immune response towards a Th2 pathway by inhibiting synthesis of IL-12. We assume that the unexpected augmentation of the Th1 response may be linked to the ability of IL-10 to down-regulate IL-12 production and thus output of IFN-γ, which can have negative effects on transgene expression driven by viral promoters. Another group obtained similar results upon testing IL-10 as a genetic adjuvant for an HIV-1 *gag/pol* vaccine. In contrast, in a study based on the bovine herpes virus gD protein co-administration of IL-10 reduced the antibody response.[120] In a mouse model for HSV type 2 addition of an IL-10 encoding vector reduced the efficacy of a DNA vaccine to the gD antigen and resulted in increased morbidity and mortality of the animals upon viral challenge.[108] In an HIV-1 model the same group of investigators reported that IL-10 used as a genetic adjuvant increased the antibody response to the vaccine antigen.[100]

Interleukin 12

This cytokine has been tested extensively for its ability to modulate immune responses to DNA vaccines and in general results were consistent in demonstrating that this cytokine enhances Th1 type immune responses to DNA vaccines. We tested IL-12 using two vectors each expressing one of the two chains, i.e., p40 and p35, which upon assembly into heterodimers form biologically active molecules. Addition of both vectors to a DNA vaccine expressing the rabies virus glycoprotein had no significant effect on the ensuing immune response. Another group tested the effect of one plasmid expressing in tandem the p35 and p40 chains of IL-12 on the immune response to an HIV-1 *env* vaccine. The IL-12 vector used at moderate doses significantly enhanced the DTH response to a peptide carrying an immunodominant epitope of the HIV-1 Env protein recognized in association with MHC class I determinants by CD8+ T cell. Accordingly, T cell-mediated cytolysis was also enhanced.[121] At higher doses[121] of the IL-12 vector the immune response decreased, which might have reflected formation of p40 homodimers, that can inhibit the biologically active heterodimers. Alternatively, addition of excessive amounts of the IL-12 vector might have resulted in high production of IFN-γ, which in turn might have effected vector-driven antigen expression. In another study, an IL-12 encoding vector was shown to enhance proliferation of T cells to HIV-1 or SIV Gag/Pol or Env proteins and to increase activation of cytolytic T cells. The antibody response was decreased upon co-administration of the IL-12 encoding vector.[100,106,122] A formulation of cationic lipids containing a DNA vaccine to HIV-1 gp160 together with an IL-12 encoding vector increased the induction of a cellular immune response.[123] In rhesus macaques addition of an IL-12 vector to an HIV-1 vaccine augmented the proliferative T cell response.[124] One group showed only a moderate effect of an IL-12 encoding vector on the primary immune response to a DNA vaccine encoding the HIV-1 Env antigen.[125] Nevertheless, upon booster immunization with a vaccinia virus recombinant expressing the same transgene product addition of IL-12 to the initial priming markedly increased the number of IFN-γ secreting T cells and decreased the number of T cells producing IL-4. This effect was not observed when mice were re-exposed to IL-12 at the time of the booster immunization with the vaccinia virus recombinant. The inhibition of the immune response by the second dose of IL-12 was linked to production of nitric oxide (NO). NO induced by cytokines such as IFN-γ, produced in response to IL-12 can be immunosuppressive. NO impairs T cell proliferation through inhibition of JAK 2 and 3 kinases needed to phosphorylate and activate members of the STAT family of transcription factors (STAT 1, 3 and 4), which serve as transactivators for a number of promoters including the one driving IFN-γ synthesis.

An IL-12 expressing vector was shown to increase the cytolytic T cell response to a DNA vaccine carrying a poorly immunogenic influenza A virus NP epitope. Two publications reported on the effect of IL-12 on DNA vaccines to HSV-2.[108,126] Addition of the cytokine-encoding vector enhanced vaccine efficacy by sponsoring activation of Th1 cells. IL-12 when added to a DNA vaccine for the surface antigen of HBV increased production of IgG2a antibodies and Th1 cytokines[121] while inhibiting the development of Th2 linked immune responses.[103] Upon co-injection with a DNA vaccine to the core protein of hepatitis C virus (HCV) IL-12 encoded by a vector increased the antigen-specific cytolytic T cell response and reduced generation of antibodies.[128]

An IL-12 vector given concomitantly with a protein vaccine to the Leishmania parasite increased protection against challenge with the pathogen. Protection was mediated by CD4 $^+$ T cell-derived IFN-γ. Addition of recombinant IL-12 protein to the vaccine was less efficacious, which may reflect the need for a prolonged presence of IL-12. Gene gun delivery of a DNA vaccine to *Plasmodium berghei* circumsporozoite protein together with a vector to IL-12 increased subsequent protection to parasite challenge through the induction of a Th1 response.[129] IL-12 also increased the efficacy of a DNA vaccine to a fungus, i.e., *arthoconidia* that is the causative agent for coccidioidomycosis. Protection was linked to an increased Th1 type immune response.[130]

In a murine cancer model for human papilloma virus type 16 associated malignancies, addition of an IL-12 encoding vector increased the cytolytic T cell response to a DNA vaccine encoding the E6 oncoprotein. This in turn increased the number of mice that remained tumor free upon challenge with an E6 expressing tumor cell line.[131] Similar results were obtained with a DNA vaccine to a mutant form of p53 where co-delivery of an IL-12 vector by gene gun delivery improved induction of protective immunity against challenge with a methylcholan-threne-induced fibrosarcoma cell line expressing the same mutation of p53.[132] Protection to tumor challenge was also augmented if IL-12 was co-injected with a DNA vaccine encoding CEA.[112] Again, improved vaccine efficacy was linked to an augmented Th1 response. In another tumor model addition of an IL-12 vector to a DNA vaccine to Her-2/neu, which is commonly over-expressed in breast cancer, reduced the incidence of spontaneously developing mammary carcinomas in FVB/N neu-transgenic mice.[133]

IL-12 sponsors development of CD8$^+$ T cell responses by driving the immune response towards a Th1 pathway. It can also play a role in generating helper cell independent cytolytic T cell responses to alloantigen.[134] IL-12 does not have a major role in the activation of anti-viral cytolytic T cell responses in vivo[135] although it was shown to enhance in vitro the ability of dendritic cells to induce this T cell subset.[136] This effect was not linked to IFN-γ. IL-12 may favor activation of cytolytic T cells to DNA vaccines by sponsoring the development of Th1 type T helper cells[137] and subsequently by driving proliferation of activated CD8$^+$ T cells. IL-12 can furthermore induce the expression of perforin and thus enhance the lytic capability of individual cytolytic T cells.[138] One study showed that the increased cytolytic T cell response to a DNA vaccine in the presence of IL-12 depended on IFN-γ as neutralization of this cytokine abrogated the IL-12 effect.[128] IFN-γ may increase production or longevity of cytolytic T cells by increasing MHC class I expression on transfected cells thus increasing the apparent antigenic load for this T cell subset. This effect is independent of IL-12, nor are the effects of IL-12 on cytolytic T cells linked to IFN-γ. Abrogation of the beneficial effect of the IL-12 genetic adjuvant through depletion of IFN-γ thus most likely does not reflect a causative linkage.

Interleukin 13

We tested an IL-13 vector in combination with a DNA vaccine to the rabies virus glycoprotein. The response to the DNA vaccine given at an optimal concentration was not altered upon addition of the cytokine-encoding vector (unpublished). Similar results were obtained upon addition of an IL-13 vector to DNA vaccines to HIV-1 Gag/Pol and Env.[106]

Interleukin 15

IL-15, a Th1 cytokine with an activity profile similar to IL-2, was found to augment the cytolytic T cell response to DNA vaccines expressing HIV Env or HIV-1 Gag/Pol. The cytokine did not affect T cell proliferation to antigen.[100] In a separate study IL-15 was administered intranasally together with a DNA vaccine to HIV-1. Addition of the cytokine increased generation of cytolytic T cells, of T cells that induced a DTH reaction and of T cells that secreted IFN-γ. Secretion of IL-4 was reduced. Further addition of vectors encoding IL-2 or IL-12 did not have a synergistic effect to the IL-15 producing plasmid.[139] In a murine HSV type 2 model, addition of an IL-15 encoding vector improved vaccine efficacy.[108]

Interleukin-18

IL-18 is a Th1 cytokine that similar to IL-12 promotes IFN-γ production. This cytokine was shown to increase proliferative T cell responses and to augment the induction of cytolytic T cells to DNA vaccines encoding HIV-1 antigens.[124] In a separate study, the effect of an IL-18 encoding vector on the immune response following intradermal injection of an HIV-1 Nef encoding construct was investigated.[140] Addition of the IL-18 vector reduced the antibody response following booster immunization with Nef protein and had no significant effect on the proliferative T cell response to the antigen. In a third report IL-18 used as a genetic adjuvant enhanced the efficacy of an HSV-2 DNA vaccine.[108]

Granulocyte Macrophage-Colony Stimulating Factor

GM-CSF similar to IL-12 has been tested in a wide range of systems and consistently augmented T helper and B cell responses. We tested the effect of a GM-CSF encoding vector on the immune response to an intramuscularly applied DNA vaccine to the rabies virus glycoprotein. GM-CSF markedly enhanced the antibody response and increased IL-2 release upon in vitro culture of splenocytes with the antigen. Addition of the GM-CSF encoding vector increased the efficacy of low doses of the DNA vaccine in providing protection to an otherwise lethal challenge with rabies virus.[141] Several groups tested the effect of GM-CSF as a genetic adjuvant to HIV-1 vaccines. Lee et al reported an increase in proliferative T cell responses and an augmented antibody response in rats.[105] Another group tested the effect of GM-CSF on the immune response to a DNA vaccine to the HIV-1 Env given intranasally in liposomes.[122] The vaccine given alone without adjuvant induced IgG, IgA and IgE antibodies to the V3 loop of HIV in serum, feces and vaginal lavage. Including a GM-CSF encoding vector in the vaccine formulation resulted in increased antibody titers. Mice that received the DNA vaccine together with vectors encoding GM-CSF and IL-12 also developed higher cell-mediated immune responses. The same group tested if timing influenced the effect of GM-CSF used as a genetic adjuvant.[142] The GM-CSF encoding vector given 3 days before the DNA vaccine augmented a Th2 type immune response, simultaneous administration enhanced a mixed immune response and administration 3 days after immunization favored a Th1 response. Conducting similar experiments we were unable to achieve an augmentation of the immune response unless the DNA vaccine and the cytokine-encoding vector were applied as a mixture. This suggests that the cytokine and the vaccine antigen had to be co-localized in vivo. In several experiments, mice injected first with the GM-CSF vector followed 48 hrs later with the DNA vaccine to the rabies virus glycoprotein developed an impaired immune response. We hypothesized that the initial injection of the cytokine-encoding DNA may have caused an excavation of the resident dendritic cells prior to their exposure to the DNA vaccine. This would have transiently depleted the inoculation site of antigen presenting cells.

In an influenza A virus system, one group reported an increase in the efficacy of a DNA vaccine to the viral HA in the presence of a GM-CSF encoding plasmid.[143] Another group reported augmentation of a cytolytic T cell response to a poorly immunogenic epitope of the influenza virus NP.[144] Mice immunized with a vaccine to the HBV surface antigen developed enhanced antibody titers upon addition of a GM-CSF encoding vector.[103] Similar results were obtained in rats immunized with a vaccine to the HCV envelope antigens.[145] Using a DNA vaccine to the HCV core antigen another group compared the effect of a GM-CSF encoding vector delivered by a separate plasmid or together with the antigen in a bicistronic vector. GM-CSF administered either way increased the lymphoproliferative response; nevertheless, the effect was more pronounced if the cytokine gene was expressed by a bicistronic vector.[146] In a mouse model to HSV type 2, addition of a GM-CSF vector to a gD expressing plasmid increased the antibody response favoring generation of antibodies of the IgG1 isotype indicative of a Th2 type response.[147] Another group using a similar antigen reported a marked enhancement of the number of cytokine secreting T cells. As both IL-4 and IFNγ producing cells were increased, these results did not confirm the other study's conclusion that GM-CSF polarizes the response towards the Th2 pathway. GM-CSF was not only effective in rodents but also in porcines. Pigs, upon immunization with DNA vaccines to the gB and gD antigen of pseudorabies virus, developed increased antibody titers and were protected against challenge with a highly virulent strain of pseudorabies virus. In the same study, vectors encoding IL-2 or IFN-γ failed to improve the efficacy of the DNA vaccine.[148] Combining a vector encoding GM-CSF with a DNA vaccine to encephalomyocarditis virus increased the sero-conversion rate of mice and improved survival upon viral challenge.[149]

Using a DNA vaccine to the *Plasmodium yolii* Py circumsporoite protein given twice, S. Hoffman's group demonstrated that addition of a GM-CSF encoding vector increased vaccine efficacy.[150] Mice injected with the DNA vaccine together with the GM-CSF vector developed an increased number of IFN-γ, IL-2 and IL-4 secreting T cells; serum antibody titers including IgG2a and IgG1 were augmented. Activation of cytolytic T cells was not affected. The effect was long lived and persisted for at least 12 weeks. The same group showed that priming of mice with a DNA vaccine in combination with a GM-CSF encoding vector significantly increased the immune response upon a booster immunization with a recombinant poxvirus expressing the same transgene product. This allowed for a significant reduction of the dose of the DNA vaccine without losing vaccine efficacy. Upon booster immunization GM-CSF increased antibody titers, the number of IFN-γ secreting T cells and the activity of cytolytic T cells.[150] The DNA vaccine to the circumsporoite protein of malaria had been shown previously to induce tolerance in neonatal mice, a finding that was not confirmed in other systems where DNA vaccines stimulated immune responses in newborn mice.[151,152] In the malaria model addition of a plasmid encoding GM-CSF to the DNA vaccine circumvented tolerization of neonates and instead primed them for a subsequent booster immunization.[153]

GM-CSF increased T cell responses to a DNA vaccine expressing the dense granular protein (GRA4) of *Toxoplasma gondii*[154] resulting in increased survival following challenge. Similarly T cell responses to secreted proteins (MPT64 and Ag85B) of *Mycobacterium tuberculosis* were enhanced although they failed to provide protection against subsequent challenge.[155] In rabbits vaccination with a DNA vaccine encoding the E6 oncoprotein of cottontail rabbit papilloma virus together with a GM-CSF encoding vector decreased papilloma formation upon viral challenge.[156] In another mouse tumor model the efficacy of a DNA vaccine expressing the melanoma associated antigen gp100/pmel17 was not improved upon addition of GM-CSF.[157] In contrast, a DNA tumor vaccine expressing the melanoma antigen gp75TRP-1 induced superior protection against tumor challenge upon addition of a GM-CSF encoding vector.[158]

One study addressed the mode of action of GM-CSF encoding vectors. Injection of a GM-CSF encoding vector into the skin of mice resulted in an increase in the numbers of epidermal dendritic cells close to the inoculation site as well as in draining lymph nodes. This effect was long lasting. Maximal numbers of dendritic cells accumulated in the skin by day 7 after injection.

The effect of GM-CSF presumably relates to the ability of this cytokine to recruit immature dendritic cells and to induce partial maturation of this cell subset. Like all genetic adjuvants given together with a DNA vaccine, GM-CSF given as a genetic adjuvant is not expected to affect the very early stages of the immune response to a DNA vaccine, which are initiated before transfected cells produce the cytokine. These very early events include transfection of dendritic cells by some of the injected plasmid vectors, their activation by the CpG motifs of the bacterial DNA followed by their trafficking to draining lymph nodes where they initiate an immune response. Considering that the number of dendritic cells that become directly transfected by vector DNA is small, the ensuing immune response is expected to be rather marginal. Indeed we showed that a vector encoding the vaccine antigen under the control of the MHC class II promoter that is inactive in most cells but drives expression of the transgene product in antigen presenting cells induced a far lower immune response compared to a plasmid that expressed the same transgene under the control of the comparably potent MHC class I promoter that is active in most cells. Most of the cells that become transfected upon DNA vaccination are resident cells such as muscle cells or keratinocytes in the skin. Antigen produced by these cells upon processing by antigen presenting cells followed by their activation and migration to draining lymph nodes is expected to initiate a second wave of T and B cell activation. We assume that GM-CSF affects this second wave by recruiting additional dendritic cells to the site of transfection. These dendritic cells upon engulfing the antigen secreted by the transfected cells or released upon their demise due to T cell-mediated cytolysis require activation prior to homing to lymph nodes. GM-CSF may in part together with other factors such as TNF-α contribute to the maturation of the dendritic cells. Infiltrating activated T cells through CD40-CD40L (CD40L) interactions may provide additional activation/maturation signal.

In most models, GM-CSF augmented the T helper and B cell response. The effect on antibody production might have been secondary due to improved T cell help. Only in a few studies was GM-CSF shown to increase generation of cytolytic T cell responses. This effect of GM-CSF may depend on the nature of the antigen. Direct transfection of dendritic cells results in production of antigen that enters the cytosolic pathway where degradation products can readily be transported to the endoplasmatic reticulum for association with MHC class I antigens. In contrast antigen that is taken up by pinocytosis, endocytosis or phagocytosis generally enters the lysosomal pathway which favors association of peptides with MHC class II determinants and thus activation of CD4⁺ T cells. Dendritic cells can under certain circumstances present lysosomaly processed antigen to CD8⁺ T cells. This type of cross priming is generally inefficient, and most soluble protein fail to induce detectable cytolytic T cell responses. It is therefore to be expected that GM-CSF primarily augments T helper cell responses and only rarely the induction of cytolytic T cell responses.

Interferons

Type 1 Interferons

Using a tumor antigen, co-delivery of a plasmid encoding IFN-α was shown to increase production of IgG2a antibodies upon gene gun immunization.[132] In our hands a plasmid encoding this cytokine failed to modulate the B cell response to an intramuscularly applied DNA vaccine to the rabies virus glycoprotein.

Interferon-γ

Several studies addressed the effect of IFN-γ as a genetic adjuvant. The results varied. We found that addition of an IFN-γ encoding vector strongly reduced the B and T helper cell response to a DNA vaccine encoding the rabies virus glycoprotein.[160] A negative effect of endogenous IFN-γ was also demonstrated by comparing GKO (IFN-γ knock-out) with C57Bl/6 mice. GKO mice generated superior antibody responses upon DNA vaccination compared to syngeneic mice carrying a functional IFN-γ gene. Others reported that addition of an IFN-γ encoding vector increased the T and B cell response to a DNA vaccine-encoded antigen and biased the response towards the Th1 pathway.[103] One study showed that the effect of IFN-γ was linked to its ability to promote IL-12 production by antigen-presenting cells.[161] Using OVA as a model antigen, another study failed to observe an effect upon addition of an IFN-γ encoding vector to the antigen-expressing vector on the ensuing immune response. Upon generating a hybrid gene in which the two cDNAs were covalently linked the authors described a marked increase in the CD4 and B cell response that was predominated by IFN-γ secretion and antibodies of the IgG2a isotypes.[161] Using a similar construct, another group reported that addition of IFN-γ had no effect on the overall antibody response to OVA but slightly favored induction of IgG2a antibody responses while the induction of cytolytic T cells was marginally decreased.[162] This observation is consistent with the ability of IFN-γ to counterbalance IL-4 mediated B cell switching and to favor production of IgG2a.

IFN-γ down-regulates protein production under the control of most viral promoters including the CMV promoter used in most DNA vaccines.[163] Depending on the amount of IFN-γ produced endogenously through IL-12 induction by the unmethylated CpG motifs of the DNA vaccines further addition of IFN-γ may profoundly reduce the antigenic load and thus the immune response.

TNF Family

Tumor Necrosis Factor α and β

TNF-α and β were tested in combination with DNA vaccines to HIV-1 antigens in mice. Both increased B cell responses. TNF-α and to a lesser extent TNF-LT augmented T cell proliferative responses and the activity of cytolytic T cells.[100]

CD40 Ligand

Addition of a vector encoding CD40L (CD154) increased both humoral and cell-mediated immune responses to a reporter protein (beta-gal).[164] Addition of a vector encoding a trimeric form of CD40L given together with a plasmid vector encoding ß-gal enhanced the IFN-γ response, production of IgG1 and IgG2a antibodies and generation of cytolytic T cells.[165] The induction of the cytolytic T cell response could be inhibited by treatment with CTLA4-Ig suggesting involvement of a B7-dependent pathway. In vivo injection of the CD40L vector augmented B7.1 and B7.2 expression on splenocytes suggesting that CD40L increased the immune response by activating dendritic cells.

Co-Stimulatory Molecules

CD80 and CD86

Testing vectors encoding either CD80 or CD86 we only observed a marginal increase in the resulting immune response if the antigen expressing DNA vaccine was delivered at sub-optimal doses. Addition of a CD86 vector to a DNA vaccine of HIV-1 antigens markedly enhanced T helper and cytolytic T cell responses; a less pronounced increase was observed upon co-administration of a vector encoding CD80. Neither of these two co-stimulatory factors affected the B cell response to HIV-1.[166] Another study of DNA vaccines to the Env protein of HIV-1 reported that CD80 had no effect on the immune response to the DNA vaccine. In contrast, CD86 strongly increased antibody titers, the cytolytic T cell response, and evoked a DTH reaction to antigen re-challenge. CD86 selectively increased antibodies of the IgG2a isotype suggesting a bias towards Th1 responses. Treatment of mice with CTLA-4Ig or an anti-interferon antibody abrogated the effect of CD86.[167] In an influenza A virus mouse model, induction of cytolytic T cells to an NP-expressing DNA vaccine was increased upon co-administration of a plasmid encoding CD86 while CD80 had no effect.[144] In contrast, another group reported that co-administration of a vector to CD80 with a DNA vaccine expressing a minigene encoding an allogeneic MHC class I epitope increased the cytolytic T cell response while addition of a vector to CD86 had no effect. The same study confirmed this finding with a DNA vaccine to OVA that induced protection to challenge with an OVA expressing tumor cell line if a vector to CD80 but not to CD86 was added to the vaccine.[168]

The reports that showed the CD80 or CD86 function as genetic adjuvants concluded that resident cells such as muscle cells upon transfection and expression of co-stimulatory molecules gained the ability to activate naive T cells to the vector-encoded antigen. T cell activation generally occurs in lymphatic tissues, which provide a suitable environment of ligands, cytokines and chemokines for activation of naïve T cells. Furthermore only lymphatic tissues have a sufficient density of naive T helper and cytolytic T cells to ensure that those with the correct TcR encounter their antigen and interact with other lymphocyte subsets for generation of a full-blown immune response. Naïve T cells reside in lymphatic tissues or circulate through the blood; they lack the adhesion molecules needed to extravasate into tissue. It is thus unlikely for naïve lymphocytes to reach cells expressing co-stimulatory molecules in non-lymphatic tissues. Granulomas can behave like tertiary lymphatic tissues and provide a suitable environment for activation of naive lymphocytes by antigen presented by dendritic cells. If such granulomas are formed upon DNA vaccination remains to be established.

Alternatively, activated NK cells can kill cells expressing CD80 or CD86 in part through interactions with CD28.[169-171] One report described that the adjuvant effect of co-stimulatory molecules could be abrogated by treatment of mice with antibodies to IFN-γ. This is compatible with an involvement of activated NK cells. One could envision that upon inoculation of plasmids to co-stimulatory molecules NK cells activated through IL-12 kill the transfected CD80 or CD86 expressing cells thus releasing antigen from those that were concomitantly transfected with the DNA vaccine. The increased amount of antigen available for processing by dendritic cells would shift the kinetics of the immune response and augment the immune response during the early phase after vaccination. One would expect that after vaccination, in the absence of vectors to CD80 or CD86, antigen released through T cell-mediated cytolysis of the transfected cells would compensate for the effect of the co-stimulatory molecules. Most studies on CD80 and CD86 as genetic adjuvants tested T and B cell responses within 4 weeks after immunization but failed to assess the effect at later time points. In one report, mice were boosted twice and then tested after the first and second boost, i.e., 5 and 8 weeks after the initial vaccination. In this study the effect of the co-stimulatory molecule was far more pronounced after the first boost than after the second boost which supports the above-proposed model.

CTLA-4

CTLA-4 is expressed by activated T cells and interacts with co-stimulatory molecules on dendritic cells. One report tested the effect of CTLA-4 linked in frame as a fusion protein to a DNA vaccine encoded antigen. Mice immunized with the vaccine encoding the fusion protein developed accelerated and enhanced antibody responses that were predominantly of the IgG1 isotype and increased T cell responses measured by a proliferation assay.[172] CTLA-4 presumably increased the immune response by selectively targeting antigen secreted from transfected cells towards antigen presenting cells.

Other Immunomodulators

L-selectin

L-selectin is expressed on high endothelial venules cells of lymph nodes. A DNA vaccine encoded antigen expressed as a fusion protein with L-selectin was shown to result in enhanced B and T cell responses presumably by directing the antigen towards lymphatic tissues.[172]

C3d

A DNA vaccine expressing a secreted form of the influenza virus HA molecule linked in frame to three in tandem copies of C3d accelerated affinity maturation of the anti-HA antibodies resulting in improved protection against viral challenge.[173] The underlying mechanism is presumably linked to binding of the HA portion of the protein to the Ig receptor on

B cells and concomitant binding of C3d to CD21 on the same B cells resulting in activation upon CD19 signaling.

Combinations of Cytokines

Only a limited number of studies have tested combinations of cytokines. IL-15 when tested in combination with either IL-2 or IL-12 was not found to have a synergistic effect on the activity of either genetic adjuvant.[139] A combination of GM-CSF and IL-12 encoding vectors to a DNA vaccine was shown to have an additive effect by both enhancing humoral responses through the activity of GM-CSF and cell-mediated responses through IL-12.[142] Additional studies are needed to establish if multiple cytokine adjuvants with known synergistic activity can further improve selected immune responses to DNA vaccines.

Mode of Delivery of Genetic Adjuvants

Route of Immunization

The route of immunization affects at least in some systems the type of the immune response generated upon DNA vaccination. Intramuscular immunization generally favors the development of Th1 responses while responses to some DNA vaccines given intradermally by gene gun immunization show a bias towards Th2 type immunity. This may be linked in part to the dose of the DNA vaccine required to drive an immune response. Intramuscular immunization requires 100-1000 times more DNA compared to gene gun immunization, which projects the DNA coated to gold beads through membranes and is thus far more efficient in transducing cells. This difference in dose may affect the magnitude of the DNA vaccine's adjuvanticity that favors the development of Th1 responses. Alternatively or in addition, differences in antigen presenting cells residing or recruited to the skin or muscle may also influence the resulting type of the immune response. Cytokines that drive immune responses towards the Th1 pathway, such as IL-12, would thus be expected to have a more profound effect on intradermal gene gun immunization while those that favor development of Th2 responses might be more effective upon intramuscular application. This has not been investigated in sufficient depth. Results with cytokines that have been studied extensively, most notably IL-12 and GM-CSF, yielded comparable results upon either route of immunization.

Mode of Delivery

Genetic adjuvants have been given either in separate vectors most commonly mixed with the DNA vaccines, as bicistronic vectors, which express the vaccine antigen together with the cytokine under separate promoters in the same plasmid, or as fusion products where the antigen is linked in frame to the cytokine or a biologically active fragment thereof. Fusion products or bicistronic vectors have the advantage that the antigen and the immunomodulator are expressed by the same cells thus providing a microenvironment of optimized cytokine levels at the site of antigen expression. DNA vaccines expressing fusion products might have the disadvantage of compromising the three-dimensional structure of the antigen, which in turn might have a negative impact on the generation of antibodies to the native protein. Bicistronic vectors have the potential disadvantage of lowered expression levels compared to those achieved by individual plasmids. In spite of these disadvantages, studies to date indicate that upon intramuscular immunization bicistronic vectors are more efficacious compared to individual vectors. Bicistronic vectors are not expected to offer an advantage to gene gun immunization where gold beads can be coated with multiple different vectors that deliver the antigen together with the genetic adjuvant to the same cells. Encapsulation of cytokine encoding vectors and DNA vaccines into lipid formulations to ensure their concomitant uptake has not been tested extensively but might provide an alternative to bicistronic vectors for intramuscular immunization.

Kinetics of Delivery

In most studies the DNA vaccines were given together with the genetic adjuvants. Most of those that tested the effect of sequential immunization, i.e., injection of the genetic adjuvant prior or following DNA vaccination, reported that this compromised the efficacy of the genetic adjuvant. One study reported that IL-2 was more efficacious as a genetic adjuvant when given 2 days after the DNA vaccine.[111] Another group of investigators described that the time of administration of a vector encoding GM-CSF relative to the DNA vaccine profoundly affected the type of the ensuing immune response as detailed above.[142]

The Effect of Genetic Adjuvants on Prime-Boost Regimens

DNA vaccines in spite of all their advantages have the clear disadvantage of eliciting far lower immune responses compared to traditional vaccines. Using DNA vaccines for priming followed by a booster immunization with a conventional vaccine can circumvent this. This vaccine regimen not only results in potent humoral and cell-mediated immune responses but also counterbalances the effect of pre-existing immunity to viral vectors such as replication-defective adenoviral recombinants.[141] Several studies have addressed the effect of genetic adjuvants on such prime-boost regimens. Results showed that the adjuvant enhances the immune response following the booster immunization. We demonstrated this for IL-4 and IL-5 which when mixed with a DNA vaccine increased vaginal antibody secretion following an intranasal booster immunization with an adenoviral recombinant expressing the same antigen.[141] S. Hoffman showed that a combination of a DNA vaccine to a malaria antigen mixed with a vector encoding GM-CSF dramatically increased the antibody response upon a booster immunization with a recombinant poxvirus.[150] An IL-12 encoding vector increased cell-mediated immune responses to HIV-1 Env following booster immunization with a vaccinia virus recombinant.[124] An IL-18 vector in contrast decreased the antibody response to HIV-1 Nef following a second immunization with the protein. Again, additional studies are needed to assess the viability of this approach.

Summary

DNA vaccines first described in 1992 have by now been tested in a variety of species, including humans, for the induction of immune responses to viral, bacterial or parasitic pathogens, cancer-associated antigens, allergens, or auto-antigens. They induce a full spectrum of adaptive immune responses including B, T helper and cytolytic T cells and in experimental rodents provide protection to challenge with a variety of different microbes or tumor cells. Nevertheless, in general the magnitude of the immune response that can be achieved with DNA vaccines falls short of that induced with traditional attenuated vaccines. Genetic adjuvants provide one of many avenues to improve or alternatively redirect the immune response to DNA vaccines. Adjuvants that lead to an increased immune response would be desirable for most vaccines that aim at protecting against potential encounters with a pathogenic microorganism or that are designed for immunotherapy of cancer or chronic infections. Redirection of immune responses by vaccines would serve treatment or prevention of allergies or autoimmune reactions. Vaccines that present the antigen without co-stimulatory signals to a naïve immune system can theoretically achieve inhibition of the immune response to a given antigen by induction of tolerance. This would serve patients that require gene therapy for replacement of missing or faulty genes or transplants for the replacement of an organ.

Cytokines and other immunomodulators have been used extensively in experimental animal models mainly with the goal of enhancing immune responses to pathogens or cancer cells. At least two of the cytokines, i.e., GM-CSF and IL-12, have yielded highly reproducible results in different systems evaluated by different groups of investigators. Other cytokines, most notably those that affect the development of Th2 type immune responses gave inconsistent or even contradictory results. Additional studies are warranted to resolve these conflicts. The effects of many immunomodulators were marginal and often only demonstrable in certain mouse strains or under conditions where the dose of the DNA vaccine was suboptimal. In general, concomitant application of the DNA vaccine together with the genetic adjuvant gave superior results to staggered injections. Bicistronic vectors where tested were more efficacious compared to individual plasmids expressing either the antigen or the adjuvant. Only a limited number of studies addressed the use of cytokine cocktails and additional experiments are needed to explore this approach.

DNA vaccines have been shown to prime the immune response to a subsequent booster immunization with a traditional vaccine. This type of vaccine regimen that has yielded highly promising results in experimental animals, including primates, will soon be tested in human volunteers. Addition of genetic adjuvants to the DNA vaccine was shown by several investigators to further improve the immune response following the booster immunization and might thus provide an avenue to combat diseases which have resisted prevention or treatment with current vaccines.

Safety studies for the use of genetic adjuvants have not yet been performed in depth. Immunomodulators, especially when delivered by plasmids separate from the antigen-encoding vectors, may be produced in vivo for extended periods of time. This could readily result in adverse reactions or even effect serendipitous infections encountered by the patient upon vaccination. Modified vectors that allow drug-induced termination of production of the immunomodulator should be developed to reduce anticipated side effects of genetic adjuvants.

References

1. Oppenheim JJ, Feldmann M, eds. Cytokine Reference Volumes 1 and 2. New York: Academic Press, 2000 and at www. academicpress.com/cytokine reference.
2. www.copewithcytokines.de.
3. Rosenwasser LJ. Biologic activities of IL-1 and its role in human disease. J Allergy Clin Immunol 1998; 102(3):344-350.
4. Mantovani A, Locati M, Allavena P et al. The chemokine superfamily:crosstalk with the IL-1 system. Immunobiology 1996; 195(4-5):522-549.
5. Dinarello CA. The role of interleukin-1 in host responses to infectious diseases. Infect Agents Dis 1992; 1(5):227-236.
6. Dinarello CA. Role of interleukin-1 in infectious diseases. Immunol Rev 1992; 127:119-146.
7. Platanias LC, Vogelzang NJ. Interleukin-1:biology, pathophysiology, and clinical prospects. Am J Med 1990; 89(5):621-629.
8. Caussy D, Sauder DN. The role of interleukin-1 in the immunological response. Transfus Med Rev 1989; 3(3):194-205.
9. Fuhlbrigge RC, Hogquist KA, Unanue ER et al. Molecular biology and genetics of interleukin-1. Year Immunol 1989; 5:21-37.
10. Powell JD, Ragheb JA., Kitagawa-Sakakida S et al. Molecular regulation of interleukin-2 expression by CD28 co-stimulation and anergy. Immunol Rev 1998; 165:287-300.
11. Gesbert F, Delespine-Carmagnat M, Bertoglio J. Recent advances in the understanding of interleukin-2 signal transduction. J Clin Immunol 1998; 18(5):307-20.
12. Theze J, Alzari PM, Bertoglio J. Interleukin 2 and its receptors:recent advances and new immunological functions. Immunol Today 1996; 17(10):481-486.
13. Semenzato G, Pizzolo G, Zambello R. The interleukin-2/interleukin-2 receptor system:structural, immunological, and clinical features. Int J Clin Lab Res 1992; 22(3):133-142.
14. Swain SL. Lymphokines and the immune response:the central role of interleukin-2. Curr Opin Immunol 1991; 3(3):304-310.
15. Gillis S. Interleukin 2:biology and biochemistry. J Clin Immunol 1983; 3(1):1-13.
16. Gillis S, Mochizuki DY, Conlon PJ et al. Molecular characterization of interleukin 2. Immunol Rev 1982; 63:167-209.
17. Farrar JJ, Benjamin WR, Hilfiker Ml et al. The biochemistry, biology, and role of interleukin 2 in the induction of cytotoxic T cell and antibody-forming B cell responses. Immunol Rev 1982; 63:129-66.
18. Nelms K, Huang H, Ryan J et al. Interleukin-4 receptor signalling mechanisms and their biological significance. Adv Exp Med Biol 1998; 452:37-43.

19. Paludan SR. Interleukin-4 and interferon-gamma:the quintessence of a mutual antagonistic relationship. Scand J Immunol 1998; 48(5):459-68.

20. Szabo SJ, Glimcher LH, Ho IC. Genes that regulate interleukin-4 expression in T cells. Curr Opin Immunol 1997; 9(6):776-781.

21. Brown MA, Hural J. Functions of IL-4 and control of its expression. Crit Rev Immunol 1997; 17(1):1-32.

22. Keegan AD, Nelms K, Wang LM et al . Interleukin 4 receptor:signaling mechanisms. Immunol Today1994; 15(9):423-432.

3. Banchereau J, Rousset F. Functions of interleukin-4 on human B lymphocytes. Immunol Res 1991; 10(3-4):423-427.

24. Takatsu K, Tominaga A, Harada N et al. T cell-replacing factor (TRF)/interleukin 5 (IL-5):molecular and functional properties. Immunol Rev 1988; 102:107-135.

25. Kopf M, Le Gros G, Coyle AJ et al. Immune responses of IL-4, IL-5, IL-6 deficient mice. Immunol Rev 1995; 148:45-69.

26. Koike M, Takatsu K. IL-5 and its receptor:which role do they play in the immune response? Int Arch Allergy Immunol 1994; 104(1):1-9.

27. Lopez AF, Shannon MF, Chia MM et al. Regulation of human eosinophil production and function by interleukin-5. Immunol Ser 1992; 57:549-571.

28. Takatsu K, Takaki S, Hitoshi Y et al. Cytokine receptors on Ly-1 B cells. IL-5 and its receptor system. Annu NY Acad Sci 1992; 651:241-58.

29. Taga T, Kishimoto T. Gp130 and the interleukin-6 family of cytokines. Annu Rev Immunol 1997; 15:797-819.

30. Hibi M, Nakajima K, Hirano T. IL-6 cytokine family and signal transduction:a model of the cytokine system. J Mol Med 1996; 74(1):1-12.

31. Akira S, Kishimoto T. IL-6 and NF-IL6 in acute-phase response and viral infection. Immunol Rev 1992; 127:25-50.

32. Hirano T. The biology of interleukin-6. Chem Immunol 1992; 51:153-80.

33. Or R, Abdul-Hai A, Ben-Yehuda A. Reviewing the potential utility of interleukin-7 as a promoter of thymopoiesis and immune recovery. Cytokine Cell Mol Ther 1998; 4(4):287-294.

34. Costello R, Duffaud F, The pleiotropic effects of interleukin 7 and their pathologic and therapeutic implications. Eur J Med 1992;1(2):119-121.

35. Temann UA, Geba GP, Rankin JA et al. Expression of interleukin 9 in the lungs of transgenic mice causes airway inflammation, mast cell hyperplasia, and bronchial hyperresponsiveness. J Exp Med 1998; 188(7):1307-1320.

36. Petit-Frere C, Dugas B, Braquet P et al. Interleukin-9 potentiates the interleukin-4-induced IgE and IgG1 release from murine B lymphocytes. Immunology 1993; 79(1):146-51.

37. Rennick DM, Fort MM, Davidson NJ. Studies with IL-10-/- mice:an overview. J Leukoc Biol 1997; 61(4):389-396.

38. Rennick D, Davidson N, Berg D. Interleukin-10 gene knock-out mice:a model of chronic inflammation. Clin Immunol Immunopathol 1995; 76(3 Pt 2):S174-178.

39. Moore KW, O'Garra A, de Waal Malefyt R et al. Interleukin-10. Annu Rev Immunol 1993; 11:165-190.

39. Howard M, O'Garra A, Ishida H et al. Biological properties of interleukin 10. J Clin Immunol 1992; 12(4):239-247.

41. Jacques Y, Minvielle S, Muller-Newen G et al. The interleukin-11/receptor complex:rational design of agonists/antagonists and immunoassays potentially useful in human therapy. Res Immunol 1998; 149(7-8):737-740.

42. Hermann J, Walmsley M, Brennan FM. Cytokine therapy in rheumatoid arthritis. Springer Semin Immunopathol 1998; 20(1-2):275-88.

43. Komastu T, Ireland DD, Reiss CS. IL-12 and viral infections. Cytokine Growth Factor Rev 1998; 9(3-4):277-285.

44. Trinchieri G. Interleukin-12: a cytokine at the interface of inflammation and immunity. Adv Immunol 1998; 70:83-243.

45. Gately MK, Renzetti LM, Magram J et al. The interleukin-12/interleukin-12-receptor system:role in normal and pathologic immune responses. Annu Rev Immunol 1998; 16:495-521.

46. Trinchieri G. Proinflammatory and immunoregulatory functions of interleukin-12. Int Rev Immunol 1998; 16(3-4):365-396.

47. Trinchieri G. Immunobiology of interleukin-12. Immunol Res 1998; 17(1-2):269-78.

48. Jacobson NG, Szabo SJ, Guler ML et al. Regulation of interleukin-12 signalling during T helper phenotype development. Adv Exp Med Biol 1996; 409:61-73.

49. Trinchieri G, Gerosa F. Immunoregulation by interleukin-12. J Leukoc Biol 1996; 59(4):505-511.

50. Corry DB. IL-13 in allergy: home at last. Curr Opin Immunol 1999; 11(6):610-614.
51. Hart PH, Bonder CS, Balogh J et al. Differential responses of human monocytes and macrophages to IL-4 and IL-13. J Leukoc Biol 1999; 66(4):575-578.
52. Yoshikai Y, Nishimura H. The role of interleukin 15 in mounting an immune response against microbial infections. Microbes & Infection 2000; 2(4):381-389.
53. Sprent J, Zhang X, Sun S et al. T-cell proliferation in vivo and the role of cytokines. Philos Trans R Soc Lond B Biol Sci 2000; 355(1395):317-322.
54. Carson W, Caligiuri MA. Interleukin-15 as a potential regulator of the innate immune response. Braz J Med Biol Res 1998; 31(1):1-9.
55. Cruikshank WW, Kornfeld H. Center DM. Interleukin-16. J Leukoc Biol 2000; 67(6):757-766.
56. Okamura H, Tsutsui H, Kashiwamura S et al. Interleukin-18:a novel cytokine that augments both innate and acquired immunity. Adv Immunol 1998; 70:281-312.
57. Parrish-Novak J, Dillon SR, Nelson A et al. Interleukin 21 and its receptor are involved in NK cell expansion and regulation of lymphocyte function. Nature 2000; 408(6808):57-63.
58. Storozynsky E, Woodward JG, Frelinger JG et al. Interleukin-3 and granulocyte-macrophage colony-stimulating factor enhance the generation and function of dendritic cells. Immunology 1999; 97(1):138-149.
59. Gaudernack G, Gjertsen MK. Combination of GM-CSF with antitumour vaccine strategies. Eur J Cancer 1999; 35 Suppl 3:S33-35.
60. Zecher R, Scheicher C, Wagener S et al. Modulation of accessory cell function of immortalized bone marrow-derived macrophages by granulocyte/macrophage colony-stimulating factor. Med Microbiol Immunol 1993; 182(3):153-166.
61. Fabian I, Kletter Y, Mor S et al. Activation of human eosinophil and neutrophil functions by haematopoietic growth factors:comparisons of IL-1, IL-3, IL-5 and GM-CSF. Br J Haematol 1992; 80(2):137-143.
62. Bogdan C. The function of type I interferons in antimicrobial immunity. Curr Opin Immunol 2000; 12(4):419-424.
63. Sun S, Sprent J. Role of type I interferons in T cell activation induced by CpG DNA. Curr Top Microbiol Immunol 2000; 247:107-117.
64. Sinigaglia F, D'Ambrosio D, Rogge L. Type I interferons and the Th1/Th2 paradigm. Dev Comp Immunol 1999; 23(7-8):657-663.
65. Akbar AN, Lord JM, Salmon M. IFN-alpha and IFN-beta: a link between immune memory and chronic inflammation. Immunol Today 2000; 21(7):337-342.
66. Paludan SR. Interleukin-4 and interferon-gamma:the quintessence of a mutual antagonistic relationship. Scand J Immunol 1998; 48(5):459-468.
67. Murray HW. Current and future clinical applications of interferon-gamma in host antimicrobial defense. Intensive Care Med 1996; 22 Suppl 4:S456-61.
68. Young HA, Hardy KJ. Role of interferon-gamma in immune cell regulation. J Leukoc Biol 1995; 58(4):373-381.
69. Billiau A, Dijkmans R. Interferon-gamma:mechanism of action and therapeutic potential. Biochem Pharmacol 1990; 40(7):1433-1439.
70. Mond JJ, Brunswick M. A role for IFN-gamma and NK cells in immune responses to T cell-regulated antigens types 1 and 2. Immunol Rev 1987; 99:105-118.
71. Makhatadze NJ. Tumor necrosis factor locus:genetic organisation and biological implications. Hum Immunol 1998; 59(9):571-579.
72. Pasparakis M, Alexopoulou L, Douni E et al. Tumour necrosis factors in immune regulation: everything that's interesting is...new. Cytokine Growth Factor Rev 1996; 7(3):223-229.
73. Tewari M, Dixit VM. Recent advances in tumor necrosis factor and CD40 signaling. Curr Opin Genet Dev 1996; 6(1):39-44.
74. Lynch DH. The role of FasL and TNF in the homeostatic regulation of immune responses. Adv Exp Med Biol 1996; 406:135-138.
75. Clarke SR. The critical role of CD40/CD40L in the CD4-dependent generation of CD8+ T cell immunity. J Leuko Biol 2000; 67(5):607-614.
76. Laman JD, Claassen E, Noelle RJ. Functions of CD40 and its ligand, gp39 (CD40L). Crit Rev Immunol 1996; 16(1):59-108.
77. Noelle RJ. The role of gp39 (CD40L) in immunity. Clin Immunol Immunopathol 1995; 76(3 Pt 2):S203-7.
78. Armitage RJ, Maliszewski CR, Alderson MR et al. CD40L:a multi-functional ligand. Semin Immunol 1993; 5(6):401-412.
79. Slavik JM, Hutchcroft JE, Bierer BE. CD28/CTLA-4 and CD80/CD86 families:signaling and function. Immunol Res 1999; 19(1):1-24.

80. Greenfield EA, Nguyen KA, Kuchroo VK. CD28/B7 costimulation:a review. Crit Rev Immunol 1998; 18(5):389-418.

81. Lu P, Wang YL, Linsley PS. Regulation of self-tolerance by CD80/CD86 interactions. Curr Opin Immunol 1997; 9(6):858-862.

82. Swarte VV, Mebius RE, Joziasse DH et al. Lymphocyte triggering via L-selectin leads to enhanced galectin-3-mediated binding to dendritic cells. Eur J Immunol 1998; 28(9):2864-2871.

83. Giblin PA, Hwang ST, Katsumoto TR et al. Ligation of L-selectin on T lymphocytes activates beta1 integrins and promotes adhesion to fibronectin. J Immunol 1997; 159(7):3498-3507.

84. Tang ML, Hale LP, Steeber DA et al. L-selectin is involved in lymphocyte migration to sites of inflammation in the skin:delayed rejection of allografts in L-selectin-deficient mice. J Immunol 1997; 158(11):5191-5199.

85. Dempsey PW, Allison ME, Akkaraju S et al. C3d of complement as a molecular adjuvant:bridging innate and acquired immunity. Science. 1996; 271(5247):348-350.

86. Anderson CC, Matzinger P. Danger:the view from the bottom of the cliff Semin Immunol 2000; 12(3):231-238.

87. Sallusto F, Palermo B, Lenig D et al. Distinct patterns and kinetics of chemokine production regulate dendritic cell function. Eur J Immunol 1999; 29(5):1617-1625.

88. Whitmire JK, Murali-Krishna K, Altman J et al. Antiviral CD4 and CD8 T-cell memory:differences in the size of the response and activation requirements. Philos Trans R Soc Lond B Biol Sci 2000; 355(1395):373-379.

89. Lau LL, Jamieson BD, Somasundaram T et al. Cytotoxic T-cell memory without antigen Nature1994; 369(6482):648-652.

90. Slifka MK, Ahmed R. Long-lived plasma cells:a mechanism for maintaining persistent antibody production. Curr Opin Immunol 1998; 10(3):252-258.

91. Maruyama M, Lam KP, Rajewsky K. Memory B-cell persistence is independent of persisting immunizing antigen. Nature 2000; 407(6804):636-642.

92. Krieg AM. The role of CpG motifs in innate immunity. Curr Opin Immunol 2000; 12(1):35-43.

93. Bot A, Stan AC. Inaba K et al. Dendritic cells at a DNA vaccination site express the encoded influenza nucleoprotein and prime MHC class I-restricted cytolytic lymphocytes upon adoptive transfer. Int Immunol 2000; 12(6):825-832.

94. Banchereau J, Steinman RM. Dendritic cells and the control of immunity. Nature 1998, 392(6673):245-252.

95. Lanzavecchia A. Dendritic cell maturation and generation of immune responses. Haematologica 1999; 84:23-25.

96. Sallusto F, Mackay CR, Lanzavecchia A. The role of chemokine receptors in primary, effector, and memory immune responses. Annu Rev Immunol 2000; 18:593-620.

97. Schwartz RH. Costimulation of T lymphocytes:the role of CD28, CTLA-4, and B7/BB1 in interleukin-2 production and immunotherapy. Cell 1992; 71(7):1065-1068.

98. Wegmann DR, Shehadeh N, Lafferty KJ et al. Establishment of islet-specific T-cell lines and clones from islet isografts placed in spontaneously diabetic NOD mice. J Autoimmun 1993; 6(5):517-527.

99. O'Garra A. Immunology. Commit ye helpers. Nature 2000, 404(6779):719-720.

100. Kim JJ, Trivedi NN, Nottingham LK et al. Modulation of amplitude and direction of in vivo immune responses by co-administration of cytokine gene expression cassettes with DNA immunogens. Eur J Immunol 1998; 28(3):1089-1103.

101. Maecker HT, Umetsu DT, DeKruyff RH et al. DNA vaccination with cytokine fusion constructs biases the immune response to ovalbumin. Vaccine 1997; 15(15):1687-1696.

102. Hakim I, Levy S, Levy R. A nine-amino acid peptide from IL-1beta augments antitumor immune responses induced by protein and DNA vaccines. J Immunol 1996; 157(12):5503-5511.

103. Chow YH, Chiang BL, Lee YL et al. Development of Th1 and Th2 populations and the nature of immune responses to hepatitis B virus DNA vaccines can be modulated by codelivery of various cytokine genes. J Immunol 1998; 160(3):1320-1329.

104. Chow YH, Huang WL, Chi WK et al. Improvement of hepatitis B virus DNA vaccines by plasmids coexpressing hepatitis B surface antigen and interleukin-2. J Virol 1997; 71(1):169-178.

105. Lee AH, Suh YS, Sung YC. DNA inoculations with HIV-1 recombinant genomes that express cytokine genes enhance HIV-1 specific immune responses. Vaccine 1999; 17(5):473-479.

106. Kim JJ, Simbiri KA, Sin JI et al. Cytokine molecular adjuvants modulate immune responses induced by DNA vaccine constructs for HIV-1 and SIV. J Interferon Cytokine Res 1999; 19(1):77-84.

107. Kim JJ, Yang JS, Montaner L et al. Coimmunization with IFN-gamma or IL-2, but not IL-13 or IL-4 cDNA can enhance Th1-type DNA vaccine-induced immune responses in vivo. J Interferon Cytokine Res 2000; 20(5):311-319.

108. Sin JI, Kim JJ, Boyer JD et al. In vivo modulation of vaccine-induced immune responses toward a Th1 phenotype increases potency and vaccine effectiveness in a herpes simplex virus type 2 mouse model. J Virol 1999; 73(1):501-509.
109. Xin KQ, Hamajima K, Sasaki S et al. Intranasal administration of human immunodeficiency virus type-1 (HIV-1) DNA vaccine with interleukin-2 expression plasmid enhances cell-mediated immunity against HIV-1. Immunology 1998; 94(3):438-444.
110. Prayaga SK, Ford MJ, Haynes JR. Manipulation of HIV-1 gp120-specific immune responses elicited via gene gun-based DNA immunization. Vaccine 1997; 15(12-13):1349-1352.
111. Barouch DH. Santra S. Steenbeke TD et al. Augmentation and suppression of immune responses to an HIV-1 DNA vaccine by plasmid cytokine/Ig administration. J Immunol 1998; 161(4):1875-1882.
112. Barouch DH, Craiu A, Kuroda MJ et al. Augmentation of immune responses to HIV-1 and simian immunodeficiency virus DNA vaccines by IL-2/Ig plasmid administration in rhesus monkeys. Proc Natl Acad Sci USA 2000; 97(8):4192-4197.
113. Barouch DH, Santra S, Schmitz JE et al. Control of viremia and prevention of clinical AIDS in rhesus monkeys by cytokine-augmented DNA vaccination. Science 2000; 290(5491):486-492.
114. Deng H, Pasquini S, Wang Y et al. Augmentation of immune responses by DNA vaccines that contain cytokine genes. Proc 1998 Natl Immunol Meeting 1998; 26-35.
115. Xiang ZQ, Pasquini S, Ertl HC. Induction of genital immunity by DNA priming and intranasal booster immunization with a replication-defective adenoviral recombinant. J Immunol 1999; 162(11):6716-6723.
116. Song K, Chang Y, Prud'homme GJ. Regulation of T-helper-1 versus T-helper-2 activity and enhancement of tumor immunity by combined DNA-based vaccination and nonviral cytokine gene transfer. Gene Ther 2000; 7(6):481-492.
117. Ishii KJ, Weiss WR, Ichino M et al. Activity and safety of DNA plasmids encoding IL-4 and IFN gamma. Gene Ther 1999; 6(2):237-244.
118. Lee SW, Youn JW, Seong BL et al. IL-6 induces long-term protective immunity against a lethal challenge of influenza virus. Vaccine 1999; 17(5):490-496.
119. Webb LM, Foxwell BM, Feldmann M. Interleukin 7 activates human naïve [CD4+] cells and primes for interleukin 4 production. Eur J Immunol 1997; 27(3):633-640.
120. Lewis PJ, van Drunen Littel-Van Den Hurk, Babiuk LA. Induction of immune responses to bovine herpesvirus type 1 gD in passively immune mice after immunization with a DNA-based vaccine. J Gen Virol 1999; 80 (11):2829-2837.
121. Tsuji T, Hamajima K, Fukushima J et al. Enhancement of cell-mediated immunity against HIV-1 induced by co-inoculation of plasmid-encoded HIV-1 antigen with plasmid expressing IL-12. J Immunol 1997; 158(8):4008-4013.
122. Kim JJ, Ayyavoo V, Bagarazzi ML et al. In vivo engineering of a cellular immune response by coadministration of IL-12 expression vector with a DNA immunogen. J Immunol 1997; 158(8):816-826.
123. Okada E, Sasaki S, Ishii N et al. Intranasal immunization of a DNA vaccine with IL-12- and granulocyte-macrophage colony-stimulating factor (GM-CSF)-expressing plasmids in liposomes induces strong mucosal and cell-mediated immune responses against HIV-1 antigens. J Immunol 1997; 159(7):3638-3647.
124. Kim JJ, Nottingham LK, Tsai A et al. Antigen-specific humoral and cellular immune responses can be modulated in rhesus macaques through the use of IFN-gamma, IL-12, or IL-18 gene adjuvants. J Med Primatol 1999; 28(4-5):214-223.
125. Gherardi MM, Ramirez JC, Esteban M. Interleukin-12 (IL-12) enhancement of the cellular immune response against human immunodeficiency virus type 1 env antigen in a DNA prime/vaccinia virus boost vaccine regimen is time and dose dependent:suppressive effects of IL-12 boost are mediated by nitric oxide. J Virol 2000; 74(14):6278-6286.
126. Sin JI, Kim JJ, Arnold RL et al. IL-12 gene as a DNA vaccine adjuvant in a herpes mouse model:IL-12 enhances Th1-type [CD4+] T cell-mediated protective immunity against herpes simplex virus-2 challenge. J Immunol 1999; 162(5):2912-21.
127. Lee YL, Tao MH, Chow YH et al. Construction of vectors expressing bioactive heterodimeric and single-chain murine interleukin-12 for gene therapy. Hum Gene Ther 1998; 9(4):457-465.
128. Shan M, Liu K, Fang H. [DNA vaccination of the induction of immune responses by codelivery of IL-12 expression vector with hepatitis C structural antigens]. [Chinese] Chung Hua Kan Tsang Ping Tsa Chih 1999; 7(4):236-9.
129. Yoshida S, Kashiwamura SI, Hosoya Y et al. Direct immunization of malaria DNA vaccine into the liver by gene gun protects against lethal challenge of Plasmodium berghei sporozoite. Biochem Biophys Res Commun 2000; 271(1):107-115

130. Jiang C, Magee DM, Cox RA. Coadministration of interleukin 12 expression vector with antigen 2 cDNA enhances induction of protective immunity against Coccidioides immitis. Infect Immun 1999; 67(6):5848-5853.

131. Tan J, Yang NS, Turner JG et al. Interleukin-12 cDNA skin transfection potentiates human papillomavirus E6 DNA vaccine-induced antitumor immune response. Cancer Gene Ther 1999; 6(4):331-339.

132. Tuting T, Gambotto A, Robbins PD et al. Co-delivery of T helper 1-biasing cytokine genes enhances the efficacy of gene gun immunization of mice:studies with the model tumor antigen beta-galactosidase and the BALB/c Meth A p53 tumor-specific antigen. Gene Ther 1999; 6(4):629-636.

133. Amici A, Smorlesi A, Noce G et al. DNA vaccination with full-length or truncated neu induces protective immunity against the development of spontaneous mammary tumors in HER-2/neu transgenic mice. Gene Ther 2000; 7(8):703-706.

134. Gately MK, Wolitzky AG, Quinn PM et al. 1992. Regulation of human cytolytic lymphocyte responses by interleukin 12. Cell Immunol 1992; 143(1):127-142.

135. Biron CA, and Garzinelly RT. Effect of IL-12 on microbial immune responses:a key mediator in regulating disease outcome. Curr Opin Immunol 1995; 7(4):485-496.

136. Bhardwaj N, Seder R, Reddy A et al. IL-12 in conjunction with dendritic cells enhances anti-viral CD8+ CTL responses in vitro. J Clin Invest 1996; 98(3):715-722.

137. Trinchieri, G. Interleukin-12 and its role in the generation of Th1 cells. Immunol. Today 1993; 14(7):335-338.

138. Chouaib S, Chehimi J, Bani L et al. Interleukin 12 induces differentiation of major histocompatibility complex class I-primed cytotoxic T-lymphocyte precursors into allospecific cytotoxic effectors. Proc Natl Acad Sci USA 1994; 91(26):12659-12663.

139. Xin KQ, Hamajima K, Sasaki S et al. IL-15 expression plasmid enhances cell-mediated immunity induced by an HIV-1 DNA vaccine. Vaccine 1999; 17(7-8):858-866.

140. Billaut-Mulot O, Idziorek T, Ban E et al. Interleukin-18 modulates immune responses induced by HIV-1 Nef DNA prime/protein boost vaccine. Vaccine 2000; 19(1):95-102.

141. Xiang Z, Ertl HC. Manipulation of the immune response to a plasmid-encoded viral antigen by coinoculation with plasmids expressing cytokines. Immunity 1995; 2(2):129-135.

142. Kusakabe K, Xin KQ, Katoh H et al. The timing of GM-CSF expression plasmid administration influences the Th1/Th2 response induced by an HIV-1-specific DNA vaccine. J Immunol 2000; 164(6):3102-3111.

143. Operschall E, Schuh T, Heinzerling L et al. Enhanced protection against viral infection by co-administration of plasmid DNA coding for viral antigen and cytokines in mice. J Clin Virol 1999; 13(1-2):17-27.

144. Iwasaki A, Stiernholm BJ, Chan AK et al. Enhanced CTL responses mediated by plasmid DNA immunogens encoding costimulatory molecules and cytokines. J Immunol 1997; 158(10):4591-4601.

145. Lee SW, Cho JH, Sung YC. Optimal induction of hepatitis C virus envelope-specific immunity by bicistronic plasmid DNA inoculation with the granulocyte-macrophage colony-stimulating factor gene. J Virol 1998; 72(10):8430-8436.

146. Cho JH, Lee SW, Sung YC. Enhanced cellular immunity to hepatitis C virus nonstructural proteins by codelivery of granulocyte macrophage-colony stimulating factor gene in intramuscular DNA immunization. Vaccine1999; 17(9-10):1136-1144.

147. Sin JI, Kim JJ, Ugen KE et al. Enhancement of protective humoral (Th2) and cell-mediated (Th1) immune responses against herpes simplex virus-2 through co-delivery of granulocyte-macrophage colony-stimulating factor expression cassettes. Eur J Immunol 1998; 28(11):3530-3540.

148. Somasundaram C, Takamatsu H, Andreoni C et al. Enhanced protective response and immuno-adjuvant effects of porcine GM-CSF on DNA vaccination of pigs against Aujeszky's disease virus. Vet Immunol Immunopath 1999; 70(3-4):277-287.

149. Sin JI, Sung JH, Suh YS et al. Protective immunity against heterologous challenge with encephalomyocarditis virus by VP1 DNA vaccination:effect of coinjection with a granulocyte-macrophage colony stimulating factor gene. Vaccine 1997;15(17-18):1827-1833.

150. Weiss WR, Ishii KJ, Hedstrom RC et al. A plasmid encoding murine granulocyte-macrophage colony-stimulating factor increases protection conferred by a malaria DNA vaccine. J Immunol 1998; 161(5):2325-2332.

151. Bot A, Shearer M, Bot S et al. Induction of antibody response by DNA immunization of newborn baboons against influenza virus. Viral Immunol 1999; 12(2):91-96.

152. Wang Y, Xiang Z, Pasquini S. Ertl HC. Immune response to neonatal genetic immunization. Virology 1997; 228(2):278-284.

153. Ishii KJ, Weiss WR, Klinman DM. Prevention of neonatal tolerance by a plasmid encoding granulocyte-macrophage colony stimulating factor. Vaccine1999; 18(7-8):703-710.
154. Desolme B, Mevelec MN, Buzoni-Gatel D et al. Induction of protective immunity against toxoplasmosis in mice by DNA immunization with a plasmid encoding Toxoplasma gondii GRA4 gene. Vaccine 2000; 18(23):2512-2521.
155. Kamath AT, Hanke T, Briscoe H et al. Co-immunization with DNA vaccines expressing granulocyte-macrophage colony-stimulating factor and mycobacterial secreted proteins enhances T-cell immunity, but not protective efficacy against Mycobacterium tuberculosis. Immunology 1999; 96(4):511-6.
156. Leachman SA, Tigelaar RE, Shlyankevich M et al. Granulocyte-macrophage colony-stimulating factor priming plus papillomavirus E6 DNA vaccination:effects on papilloma formation and regression in the cottontail rabbit papillomavirus—rabbit model. J Virol 2000; 74(18):8700-8708.
157. Nawrath M, Pavlovic J, Dummet R et al. Reduced melanoma tumor formation in mice immunized with DNA expressing the melanoma-specific antigen gp100/pmel17. Leukemia 1999; 13 Suppl 1:S48-51.
158. Bowne WB, Wolchok JD, Hawkins WG et al. Injection of DNA encoding granulocyte-macrophage colony-stimulating factor recruits dendritic cells for immune adjuvant effects. Cytokines Cell Mol Ther 1999; 5(4):217-225.
159. Xiang ZQ, He Z, Wang Y. et al. The effect of interferon-gamma on genetic immunization. Vaccine 1997; 15(8):896-898.
160. Asakura Y, Liu LJ, Shono N et al. Th1-biased immune responses induced by DNA-based immunizations are mediated via action on professional antigen-presenting cells to up-regulate IL-12 production. Clin Exp Immunol 2000; 119(1):130-139.
161. Lim YS, Kang BY, Kim EJ et al. Potentiation of antigen-specific, Th1 immune responses by multiple DNA vaccination with an ovalbumin/interferon-gamma hybrid construct. Immunology 1998; 94(2):135-141.
162. Harms JS, Splitter GA. Interferon-γ inhibits transgene expression driven by SV40 or CMV promoters but augments expression driven by the MHC class I promoter. Hum Gene Ther 1995; 6(10):1291-1297.
163. Qin L, Ding Y, Pahud DR et al. Promoter attenuation in gene therapy:interferon-gamma and tumor necrosis factor-alpha inhibit transgene expression. Hum Gene Ther 1997; 8(17):2019-2029.
164. Mendoza RB, Cantwell MJ, Kipps TJ. Immunostimulatory effects of a plasmid expressing CD40 ligand (CD154) on gene immunization. J Immunol 1997; 159(12):5777-5781.
165. Gurunathan S, Irvine KR, Wu C-Y et al. CD40 ligand/trimer DNA enhances both humoral and cellular immune responses and induces protective immunity to infectious and tumor challenge, J Immunol 1998; 161(9):4563-4571.
166. Kim JJ, Bagarazzi ML, Trivedi N et al. Engineering of in vivo immune responses to DNA immunization via codelivery of costimulatory molecule genes. Nat Biotechnol 1997; 15(7):641-646.
167. Tsuji T, Hamajima K, Ishii N et al. Immunomodulatory effects of a plasmid expressing B7-2 on human immunodeficiency virus-1-specific cell-mediated immunity induced by a plasmid encoding the viral antigen. Eur J Immunol 1997; 27(3):782-787.
168. Corr M, Tighe H, Lee D et al. Costimulation provided by DNA immunization enhances antitumor immunity. J Immunol 1997; 159(10):4999-5004.
169. Geldhof AB, Raes G, Bakkus M et al. Expression of B7-1 by highly metastatic mouse T lymphomas induces optimal natural killer cell-mediated cytotoxicity. Cancer Res 1995; 55(13):2730-2733.
170. Geldhof AB, Moser M, Lespagnard L, et al. Interleukin-12-activated natural killer cells recognize B7 costimulatory molecules on tumor cells and autologous dendritic cells. Blood 1998; 91(1):196-206.
171. Chambers BJ, Salcedo M, Ljunggren HG. Triggering of natural killer cells by the costimulatory molecule CD80 (B7-1). Immunity 1996; 5(4):311-317.
172. Boyle JS, Brady JL, Lew AM. 1998. Enhanced responses to a DNA vaccine encoding a fusion antigen that is directed to sites of immune induction. Nature 1998; 392(6674):408-411.
173. Ross TM, Xu Y, Bright A et al. C3d enhancement of antibodies to hemagglutinin accelerates protection against influenza virus challenge. Nat Immunol 2000; 1:127-131.

CHAPTER 17

Chemokines: Role as Immunomodulators and Potential as Adjuvants for DNA Vaccines

Philip M. Murphy

Abstract

Adaptive immune responses require proper positioning of antigen-presenting cells (APCs) and antigen-specific lymphocytes in specific microdomains of secondary lymphoid tissue. This process is guided in part by members of the chemokine family of leukocyte chemoattractants. In addition, chemokines may use mechanisms independent of cell attraction to modulate the magnitude and cytokine polarity of the immune response. Together, these properties have suggested that exogenous chemokines given as adjuvants might be able to boost the immune response to vaccines, especially DNA vaccines where immune responses in large animals are often weak. Evidence, reviewed here, is now emerging to support this approach.

Introduction

A major problem confronting clinical development of DNA vaccines is that it is hard to elicit an effective immune response in large animals. One potential solution to the problem is to deliver chemoattractants specific for antigen presenting cells (APC) into the injection site, with the aim of boosting antigen uptake by increasing APC trafficking through the site. The chemoattractants that are most likely to be useful in this regard are members of the chemokine family, a large group of chemotactic cytokines which are now recognized as the main chemoattractants for leukocytes in vivo.[1-4]

Chemokines work by triggering specific Gi-coupled 7-transmembrane domain receptors on the cell surface.[5-7] In vivo, this induces leukocyte extravasation from blood to tissue through a sequential multistep process. Chemokines first activate leukocyte β2 integrins, which mediate firm adhesion of the cell to activated endothelial cells in post-capillary venules. Next, remodeling of the cytoskeleton takes place, which causes the cell to become elongated and polarized, and the cell then crawls towards progressively higher chemokine concentrations in the tissues. In this way, chemokines, chemokine receptors and leukocyte adhesion molecules together form a combinatorial network that controls both tissue specific homing of leukocytes and navigation of leukocytes within tissue.[8,9] In addition, some chemokines can regulate differentiation of effector T cells into Th1 and Th2 polarized phenotypes.[10] Thus, chemokines can modulate both the nature and magnitude of the immune response. These properties imply that blockade of endogenous chemokines may be effective in preventing or treating immunologically mediated disease, and conversely that exogenous chemokines may be effective as adjuvants to amplify and shape the immune response to vaccines. This chapter will review recent data that support these concepts.

Background

Classification of the Chemokine System

Chemokines are defined by structure, not function, whereas, conversely, chemokine receptors are defined by function, not structure. Nevertheless, almost all chemokines are leukocyte chemoattractants and all known chemokine receptors are 7-transmembrane domain (7TM) proteins. The chemokine system also includes chemokine binding proteins and chemokine mimics. Many examples of mimicry have been reported in viruses, but so far only a few have been identified in man.[11]

Chemokine Classification

Chemokines occupy a sector of sequence space bounded by ~20% amino acid identity among all paired comparisons. To date 43 human chemokines have been identified (see Table 1). They can be divided into four main subclasses, designated C, CC, CXC and CX3C, based on conserved cysteine motifs. CC, CXC and CX3C chemokines have at least four conserved cysteines, but can be distinguished by the number of amino acids separating the two cysteines located near the N-terminus. For CC chemokines the first two cysteines are adjacent, whereas they are separated by one and three amino acids for CXC and CX3C chemokines, respectively. The C chemokines are atypical in that only the first and third cysteines found in other chemokines are conserved. Two smaller subgroups have also been identified: one in the CC subclass, in which the members have two additional cysteines located in the C-terminal region, and a second in the CXC subclass, which is defined by a conserved glutamic acid-leucine-arginine (ELR) motif that lies N-terminal to the first cysteine.

Recently a systematic nomenclature has been developed to provide order to the chaotic lists of synonyms which cropped up in cases where multiple groups had independently identified the same chemokine.[3] Each name is formed by a subclass root followed by the letter "L" signifying "ligand", followed by a number which matches the number used in the name of the corresponding gene. Chemokines are unevenly distributed among the four principal subclasses. At one extreme, there are only two known C chemokines and one CX3C chemokine, while at the other extreme there are 25 known human CC chemokines. The CXC subclass is intermediate in size with 16 members. The chemokine repertoire is very similar but not identical in different mammalian species.[12] In addition, the tissue distribution and regulation of expression can vary by species, which complicates interpretation of animal models of chemokine biology.

Chemokine Receptor Classification

Chemokine receptors are defined by the ability to activate a signal transduction pathway upon binding a chemokine. Eighteen receptor subtypes have been identified that meet these criteria, all of which are members of the rhodopsin family of 7-transmembrane domain G protein-coupled receptors. The main G protein specificity is Gi (pertussis toxin-sensitive), but there is also some evidence for coupling to Gs, G16 and other Gq proteins in cotransfection systems. Chemokine-receptor pairs can be monogamous or promiscuous, however promiscuous receptors are generally restricted to binding chemokines from a single subclass. Accordingly, the receptors are named by simply adding the letter "R" for receptor after the appropriate ligand subclass root followed by a number, which generally corresponds to the chronological order in which the receptor was discovered.[7] An exception is the C chemokine receptor XCR1, in which the X is used to distinguish it from CR1, the previously assigned acronym for complement receptor 1.

At the time of this writing, there are 10, 6, 1 and 1 receptor subtypes in the CC, CXC, C and CX3C subclasses, respectively, which is proportional to the sizes of the corresponding ligand classes. There is also a separate category of cellular 7TM proteins which bind chemokines but do not transmit a signal. Three proteins have been identified that have this profile: Duffy, D6 and a molecule named inappropriately by one group as CCR10 and by another as CCR11.

Table 1. Chemokine targets and immune functions

Systematic Name	Nonstandard Names	Main Regulation and Source	Chemokine Receptor(s)	Main Leukocyte Targets	Functional Classification
CXC Chemokine/Receptors					
CXCL1	GROα/MGSA-α	Inducible: Leukocytes and tissue cells	CXCR2	Neutrophil	Inflammatory, innate
CXCL2	GROβ/MGSA-β	Inducible: Leukocytes and tissue cells	CXCR2	Neutrophil	Inflammatory, innate
CXCL3	GROγ/MGSA-γ	Inducible: Leukocytes and tissue cells	CXCR2	Neutrophil	Inflammatory, innate
CXCL4	PF4	Constitutive: platelets	Unknown		
CXCL5	ENA-78	Inducible: Leukocytes and tissue cells	CXCR2	Neutrophil	Inflammatory, innate
CXCL6	GCP-2	Inducible: Leukocytes and tissue cells	CXCR1, CXCR2	Neutrophil	Inflammatory, innate
CXCL7	NAP-2	Constitutive: platelets	CXCR2	Neutrophil	Inflammatory, innate
CXCL8	IL-8	Inducible: Leukocytes and tissue cells	CXCR1, CXCR2	Neutrophil	Inflammatory, innate
CXCL9	Mig	Inducible, especially by IFN-γ: tissue cells	CXCR3	Monocyte, Th1 T cells	Inflammatory, adaptive, Th1
CXCL10	IP-10	Inducible, especially by IFN-γ: tissue cells	CXCR3	Monocyte, Th1 T cells	Inflammatory, adaptive, Th1
CXCL11	I-TAC	Inducible, especially by IFN-γ: tissue cells	CXCR3	Monocyte, Th1 T cells	Inflammatory, adaptive, Th1
CXCL12	SDF-1 α/β	Constitutive: most tissues	CXCR4	Most leukocyte subsets	Homeostatic
CXCL13	BCA-1	Constitutive: secondary lymphoid tissue	CXCR5	B cells, CD4+ T$_{FH}$ cells	Homeostatic
CXCL14 (CXCL15)	BRAK/bolekine	Constitutive: breast and kidney	Unknown		
CXCL 16	Unknown		Unknown CXCR6	NK-T cells, CD4+ and CD8+ T cells	Inflammatory, adaptive, Th1
C Chemokine/Receptors					
XCL1	Lymphotactin/ SCM-1α/ATAC		XCR1	CD4+ and CD8+ T cells	Inflammatory, adaptive
XCL2	SCM-1β		XCR1	CD4+ and CD8+ T cells	Inflammatory, adaptive

continued on next page

Table 1. Chemokine targets and immune functions (con'd)

Systematic Name	Nonstandard Names	Main Regulation and Source	Chemokine Receptor(s)	Main Leukocyte Targets	Functional Classification
CX3C Chemokine/Receptor					
CX3CL1	Fractalkine	Inducible: Endothelial cells and neurons	CX3CR1	Monocyte, Th1 T cells	Inflammatory, adaptive, Th1; Adhesion
CC Chemokine/Receptors					
CCL1	I-309		CCR8	Thymocytes, Th2 T cells	Inflammatory, adaptive, Th2
CCL2	MCP-1/MCAF/TDCF	Inducible: many sources	CCR2	Monocytes, Th1 and Th2 T cells	Inflammatory, innate & adaptive, Th1 and Th2
CCL3	MIP-1α/LD78α	Inducible: many sources	CCR1, CCR5	Monocytes, Th1 T cells (CCR1 & 5); Th2 T cells (CCR1)	Inflammatory, innate & adaptive, mainly Th1
CCL3L1	LD78β	Inducible: many sources	CCR1, CCR5	Monocytes, Th1 T cells (CCR1 & 5); Th2 T cells (CCR1)	Inflammatory, innate & adaptive, mainly Th1
CCL4	MIP-1β	Inducible: many sources	CCR5	Monocytes, Th1 T cells	Inflammatory, innate & adaptive, Th1
CCL5	RANTES	Inducible: many sources	CCR1, CCR3, CCR5	Monocytes, eosinophils, Th1 T cells (CCR1 & 5); Th2 T cells (CCR1 & 3)	Inflammatory, innate & adaptive, Th1 (CCR1 & 5); Th2 (CCR1 & 3)
(CCL6)	Unknown		Unknown		
CCL7	MCP-3	Inducible: many sources	CCR1, CCR2, CCR3	Monocytes, eosinophils, Th1 T cells (CCR1 & 2); Th2 T cells (CCR1 & 3)	Inflammatory, innate & adaptive, Th1 (CCR1 & 2) and Th2 (CCR1 & 3)

continued on next page

Table 1. Chemokine targets and immune functions (con'd)

Systematic Name	Nonstandard Names	Main Regulation and Source	Chemokine Receptor(s)	Main Leukocyte Targets	Functional Classification
CC Chemokine/Receptors (con'd)					
CCL8	MCP-2	Inducible: many sources	CCR3, CCR5	Monocytes, eosinophils, Th1 T cells (CCR5); Th2 T cells (CCR3)	Inflammatory, innate & adaptive, Th1 (CCR5) and Th2 (CCR3)
(CCL9/10)	Unknown				
CCL11	Eotaxin	Inducible: many sources	CCR3	Eosinophils, basophils, Th2 T cells	Inflammatory, innate & adaptive, Th2
(CCL12)	Unknown				
CCL13	MCP-4	Inducible: many sources	CCR2, CCR3	Monocytes & Th1 T cells (CCR2); Eosinophils & Th2 T cells (CCR3)	Inflammatory, innate & adaptive, Th1 (CCR2) and Th2 (CCR3)
CCL14	HCC-1	Constitutive: plasma	CCR1, CCR5	Monocyte, Th1 T cells	Inflammatory, adaptive, Th1
CCL15	HCC-2/Lkn-1/ MIP-1δ	Inducible: many sources	CCR1, CCR3	Monocytes & Th1 & Th2 T cells (CCR1); Eosinophils & Th2 T cells (CCR3)	Inflammatory, innate & adaptive, Th1 (CCR1 & 3) and Th2 (CCR1)
CCL16	HCC-4/LEC/ LCC-1	Inducible: many sources	CCR1, CCR2	Monocytes, Th1 T cells (CCR1 & 2); Th2 T cells (CCR1)	Inflammatory, innate & adaptive, Th1 and Th2
CCL17	TARC	Constitutive: thymus, secondary lymphoid tissue, skin	CCR4	Transitional thymocytes, Th2 T cells, memory T cells (CLA+)	Inflammatory, innate & adaptive, Th2
CCL18	DC-CK1/PARC/ AMAC-1	Constitutive: dendritic cells	Unknown	Dendritic cells	
CCL19	MIP-3β/ELC/ exodus-3	Constitutive: thymus, secondary lymphoid tissue, skin	CCR7	Naïve T cells, T$_{CM}$ T cells, B cells Mature DC, medullary thymocytes	Homeostatic, adaptive: Ag recognition in lymph node

continued on next page

Table 1. Chemokine targets and immune functions (con'd)

Systematic Name	Nonstandard Names	Main Regulation and Source	Chemokine Receptor(s)	Main Leukocyte Targets	Functional Classification
CC Chemokine/Receptors (con'd)					
CCL20	MIP-3α/LARC/ exodus-1	Constitutive: secondary lymphoid tissue	CCR6	Immature DC, B cell, Effector T cells, memory T cells (CLA+)	Inflammatory & Homeostatic
CCL21	6Ckine/SLC/ exodus-2	Constitutive: thymus, secondary lymphoid tissue	CCR7	Naïve T cells, T_{CM} T cells, B cells Mature DC, medullary thymocytes	Homeostatic, adaptive: Ag recognition in lymph node
CCL22	MDC/STCP-1	Constitutive: thymus, secondary lymphoid tissue	CCR4	Transitional thymocytes, Th2 T cells, memory T cells (CLA+)	Inflammatory, innate & adaptive, Th2
CCL23	MPIF-1/CKβ8/ CKβ8-1	Inducible: many sources	CCR1	Monocytes, Th1 and Th2 cells	Inflammatory, innate and adaptive
CCL24	Eotaxin-2/ MPIF-2	Inducible: many sources	CCR3	Eosinophils, basophils, Th2 T cells	Inflammatory, innate & adaptive, Th2
CCL25	TECK	Constitutive: mucosa, thymus	CCR9	Immature thymocytes, memory T cells (α4β7+), B cells	Homeostatic, adaptive, immune surveillance
CCL26	Eotaxin-3	Inducible: many sources	CCR3	Eosinophils, basophils, Th2 T cells	Inflammatory, innate & adaptive, Th2
CCL27	CTACK/ILC	Constitutive: skin	CCR10	Memory T cells (CLA+)	Homeostatic, adaptive, immune surveillance
CCL28	MEC	Constitutive: mucosa, skin	CCR3, CCR10	Eosinophils, basophils, Th2 T cells; Memory T cells (CLA+)	Homeostatic, innate, immune surveillance; Inflammatory, innate & adaptive, Th2

Chemokine Structure and Presentation

Direct structural studies of CC, CXC and CX3C chemokines have revealed a common fold consisting of three anti-parallel β sheets connected by loops and constrained by disulfide bonds between cysteines 1 and 3 and cysteines 2 and 4; an unordered domain N-terminal to the first cysteine; and a C-terminal α helical domain.[13,14] Although chemokines have signal peptides and are released from cells, they are probably presented to leukocytes in vivo as tethered ligands, attached either to endothelial cells or extracellular matrix by glycosaminoglycan adaptors.[15,16] CXCL16 and CX3CL1 (also known as fractalkine) use a different plasma membrane attachment mechanism which involves an intrinsic transmembrane domain connected to the chemokine domain by a mucin-like stalk. A proteolytically cleaved form of fractalkine has also been identified. Tethered full-length fractalkine can mediate direct adhesion of monocytes, T cells and NK cells, whereas cleaved fractalkine is chemotactic. Both activities are mediated through the same receptor CX3CR1.[17,18]

Patterns of Regulation of the Chemokine System

Chemokine expression is subject to strong temporal and spatial regulation and each chemokine has a unique distribution pattern and set of regulators. Although there is a very large literature for this subject, it can be reduced to a simple and useful organizing concept: chemokines can be subclassified into two main groups based on whether they are produced constitutively or whether they are produced only in cells that have been stimulated by an inducing agent. The reality is a bit more complicated since most chemokines which are constitutively expressed can be further induced, and chemokines which are highly inducible in some cell types may be constitutively expressed in others. Moreover, the number of cell types capable of expressing a given chemokine is highly variable, whether one considers the inducible or constitutive types. Some can be expressed by most if not all cell types, whereas others are limited to specific microenvironments and are produced by one or only a few cell types.

Even if one just considers ligands that bind to the same receptor, the patterns of expression can vary substantially. Take for example the two ELR+ CXC chemokines NAP-2 and IL-8, which bind to the same neutrophil receptor, CXCR2. IL-8, but not NAP-2, also triggers a second neutrophil receptor named CXCR1. NAP-2 is made constitutively in platelets and is released from storage sites in platelet granules by proinflammatory stimuli such as thrombin. IL-8, in contrast, can be made by all cell types, but generally requires induction at the transcriptional level by proinflammatory stimuli. Thus in this case inflamed tissue sites can produce both immediate and delayed signals arising either from vessel wall or more deeply in inflamed tissue, which allows for flexibility in the timing and extent of neutrophil recruitment. More generally, the chemokine system's diverse and flexible mechanisms of regulation are well-suited to the diverse homeostatic and emergency functions of the immune system.

Chemokine expression can also be regulated post-transcriptionally, through adenine-thymidine-rich elements in untranslated regions, and post-translationally by glycosaminoglycans, proteases, and scavengers. Scavengers include autoantibodies, the 7TM chemokine binding proteins and even signaling chemokine receptors, which can be coupled or uncoupled to chemotactic signal transduction pathways in a dynamic fashion in response to specific cytokines.[19,20]

Like chemokines, the temporal expression of chemokine receptors can be constitutive or inducible, and the spatial distribution can be broad or narrow. Different receptors that bind many of the same ligands can vary substantially in distribution, as in the case of CCR1 and CCR5 on T cells, or they can be tightly coexpressed, as in the case of CXCR1 and CXCR2 on neutrophils. In addition to transcriptional and post-transcriptional mechanisms of regulation, receptors may also be regulated by priming agents, homologous and heterologous desensitization processes and the balance of endogenous agonists versus antagonists.[21] Endogenous chemokine antagonists of chemokine receptors have only recently been identified and their specific roles in vivo are not yet established.[22]

Chemokine Regulation of the Immune System

Functional Classification of the Chemokine System

The structural classification of chemokines has two strong functional correlates: ELR$^+$ CXC chemokines are the major neutrophil-targeted chemokines and also have angiogenic activity, whereas ELR$^-$ CXC chemokines lack neutrophil activity and are angiostatic.[23] A more comprehensive functional classification recognizes two major groups, "inflammatory" and "homeostatic", based on where and when a chemokine and its receptor are expressed and how they are regulated.[24] An idealized inflammatory chemokine is not produced constitutively but can be induced by many tissue cells and leukocytes by pro-inflammatory and infectious stimuli. Receptors for inflammatory chemokines are expressed constitutively on effector cells of the innate immune system (neutrophils, monocytes, macrophages, eosinophils, NK cells), and on activated effector T cells and immature dendritic cells. In contrast, homeostatic chemokines, also referred to as lymphoid chemokines, are produced constitutively in specific microenvironments of primary, secondary and peripheral lymphoid tissue, and their receptors are expressed on mature dendritic cells, and naïve and memory T and B lymphocytes.

Some chemokines do not fit these polarized descriptions perfectly, but are skewed more or less to one type. Inflammatory chemokines mediate effector responses of the innate and adaptive arms of the immune system, whereas homeostatic chemokines support lymphopoiesis and immune surveillance by regulating the localization of dendritic cells and lymphocytes in lymphoid and nonlymphoid tissue. Dynamic shifts in chemokine receptor expression on dendritic cells and T and B lymphocytes have been associated with progression of the adaptive immune response.[25]

Chemokine Regulation of Immune System Development

Gene knockout studies have demonstrated specific roles for the receptors CXCR2, CXCR4, CXCR5 and CCR7 and their ligands in immune system development. CXCR2 knockout mice exhibit overdevelopment of the immune system, whereas CXCR4, CXCR5 and CCR7 knockouts have severe underdevelopment of the immune system.

CXCR2 knockouts have massive expansion of neutrophils and B lymphocytes in the circulation and greatly enlarged lymph nodes and spleen.[26] Despite the increased number of neutrophils, mobilization to sites of inflammation and infection is severely impaired which increases the susceptibility to infection in challenge experiments (see below). Leukocytosis does not occur when CXCR2 knockout mice are derived in germ-free conditions, suggesting a potential interaction between microbe-stimulated cytokines and CXCR2 signaling. CXCR2 has been shown in vitro to be a negative regulator of hematopoiesis and a positive regulator of leukocyte trafficking, and both of these mechanisms may contribute to leukocyte expansion in CXCR2 knockout mice.[26,27] Curiously, MIP-1α, which was the first chemokine reported to negatively regulate hematopoiesis in vitro, is dispensable for normal hematopoiesis in vivo, as revealed in MIP-1α knockout mice. These mice do have impaired inflammatory responses however when challenged with microbes such as coxsackie A and influenza A viruses and *Listeria monocytogenes*.[28,29] Many other chemokines have also been shown in vitro to function as negative regulators of hematopoiesis,[30] however evidence that this is important in vivo is currently lacking.

Of the three receptor knockouts associated with an underdeveloped immune system, CXCR4 knockouts are the most severely affected and have defects in primary hematopoietic tissue.[31,32] These mice die in the perinatal period and lack B cell lymphopoiesis and bone marrow myelopoiesis, but have normal thymi and thymocyte maturation. This is consistent with the pre B cell-stimulatory activity of the CXCR4 ligand CXCL12 (SDF-1), expression of CXCR4 on CD34$^+$ myeloid progenitor cells, and expression of SDF-1 in bone marrow.[33] The mechanism may involve loss of the ability of SDF-1 to retain hematopoietic precursors in the bone marrow. CXCL12 and CXCR4 knockout mice have the same phenotype, which is consistent with the notion that they function as a monogamous ligand-receptor pair. It is important to note that CXCL12 and CXCR4 are unique in that no other chemokine or chemokine receptor gene knockouts reported so far have been associated with early mortality or spontaneous disease. Defective vascular and cerebellar development has also been documented in CXCL12 and CXCR4 knockout mice.[31,32] Still a mystery is the exact role of CXCR4 in mature leukocyte migration. This receptor is by far the most broadly and highly expressed of the chemokine receptors. Moreover it is efficiently coupled to chemotaxis as assessed by SDF-1 stimulation of leukocytes in vitro. Its functionality on T cells is especially noteworthy in that other chemokine receptors expressed on these cells typically require activation of the cells to transduce chemoattractant signals.

Both CCR7 and CXCR5 deficient mice have defects in development of and trafficking to secondary lymphoid tissue.[34,35] CCR7-deficient mice have atrophic T cell zones populated by a paucity of naïve T cells; moreover, adoptively transferred T cells from wild type mice fail to traffic to lymph nodes in CCR7 knockout mice. Consistent with this the CCR7 ligand SLC is expressed on high endothelial venules of lymph node and Peyer's patches and on stromal cells in T cell zones of lymph node, Peyer's patches and spleen, and CCR7 is expressed on naïve T cells.[36] A second CCR7 ligand ELC is expressed on interdigitating dendritic cells in T cell zones and attracts naïve T cells and activated B cells.[37] Moreover, the mouse strain *plt* (acronym for paucity of lymphoid tissue) has a phenotype similar to the CCR7 knockout mouse and is genetically deficient in SLC production.[38] CCR7 knockout mice and *plt* mice have not been reported to develop spontaneous infections, however *plt* mice do have increased susceptibility to challenge with mouse hepatitis virus. Delayed type hypersensitivity reactions and antibody production are severely impaired in these mice. Taken together, these results suggest that the SLC/ELC-CCR7 axis is critical for development of a functional microenvironment for priming naïve T cells by directing DC and T cells to T cell zones of secondary lymphoid tissue.

Mice deficient in CXCR5 also have aberrant development of lymphoid organs, including absent inguinal lymph nodes, absent or phenotypically abnormal Peyer's patches and ectopic development of germinal centers in T cell zones of the spleen.[34,39] This is consistent with expression of CXCR5 on B cells and expression of its only known ligand BLC in secondary lymphoid tissue.[40]

To date, a dominant chemokine has not been found that regulates thymocyte development, perhaps because chemokine function is redundant in this organ. CCR5, CXCR4, CCR4, CCR8 and CCR9 are differentially expressed on subsets of developing thymocytes; and their ligands are expressed in thymus and chemoattract appropriate thymocyte subsets in vitro.[41] However CXCR4, CCR5, CCR4 and CCR8 knockout mice appear to have normal thymic development and T cell maturation.

Chemokine Regulation of the Innate Immune Response

There is abundant direct and indirect evidence from animal models that specific chemokines regulate innate immunity in vivo (see Table 2). At the level of expression, neutrophil-targeted CXC chemokines such as IL-8 and mouse KC and MIP-2 are highly induced in macrophages, endothelial cells and epithelial cells by inflammatory cytokines, as well as by bacterial products such as lipopolysaccharides (LPS). Further, live LPS-containing Gram negative pathogens such as *Shigella flexneri*, which are controlled predominantly by innate immune mechanisms, induce IL-8 production in an LPS-dependent manner.[42] The neutrophil itself can produce chemokines, particularly IL-8, in response to bacterial formylpeptides and primary cytokines or during phagocytosis of microbial agents.[43-45] This suggests that autocrine and paracrine pathways involving chemokines may be important to the evolution of the innate immune response to bacterial infection. NK cells also produce significant amounts of inflammatory chemokines.[46]

At the functional level, chemokines have been shown in vitro and in some cases in vivo to be powerful chemoattractants for phagocytic cells and NK cells.[1,47,48] There are also reports that IL-8 can directly induce antimicrobial effector functions of neutrophils,[49] although additional work is needed to catalogue which other chemokines are capable of these actions, which pathogens are susceptible to chemokine-stimulated killing, and whether this is significant in vivo.

Table 2. Chemokine receptor-disease associations based on phenotypes of natural and targeted inactivating mutations of the corresponding gene

Molecule	Human Disease	Rodent Disease Model
CXCR2		Acute pyelonephritis
CXCR3		Cardiac allograft rejection
CXCR4	HIV/AIDS	Perinatal lethality
CCR1		Cardiac allograft rejection
		EAE
CCR2		*L. monocytogenes*
		L. major
		M. tuberculosis
		Atherosclerosis
		EAE
		Cardiac allograft rejection
CCR5	HIV/AIDS	*L. monocytogenes*
		C. neoformans
		T. gondii
		Influenza A
		L. major
		Cardiac allograft rejection
CCR6		Allergic pulmonary inflammation
CCR8		Allergic pulmonary inflammation
CX3CR1		Cardiac allograft rejection

With the development of neutralizing antibodies to chemokines and mice lacking chemokines and chemokine receptors, tools now exist for evaluating normal chemokine action in vivo. Comprehensive analysis of the vast and rapidly growing literature on this subject is beyond the scope of this review, however I will cite a number of compelling examples focused on the area of infectious disease susceptibility, since this will be a major target of DNA vaccines.

Mouse models have confirmed that inflammatory chemokines do regulate innate immune mechanisms and oppose infectious agents by their effects on leukocyte recruitment. Studies have revealed a critical role for CXCR2 in mice challenged with *Pseudomonas aeruginosa* in the lung,[50] *E. coli* in the urinary tract,[51] and *O. volvulus* in the cornea,[52] as well as for CXCR2 ligands in mice challenged in the lung with *Aspergillus fumigatus*[53] or *Nocardia asteroides*,[54] or in the brain with Gram negative bacteria.[55] Inflammatory CXC chemokines have also been linked to wound healing and angiogenesis.[56] CC chemokines are also important in innate immunity in vivo. For example, mice lacking CCR1 are more susceptible to systemic challenge with Respiratory Syncytial Virus.[57] and *Aspergillus fumigatus*. Consistent with this, the CCR1 ligand MIP-1α has been shown to be a critical component of innate defense against invasive pulmonary aspergillosis in neutropenic hosts.[58] CCR2 knockout mice are more susceptible to *Listeria monocytogenes*[59] and *Mycobacterium tuberculosis*,[60] in both cases apparently through innate immune deficiency. CXCL15 (lungkine) also plays a critical role in pulmonary host defense to *Klebsiella* infection.[61] Several examples have been reported in which chemokine injection has resulted in reduced susceptibility to infectious agents in animal models. For example, the CC chemokine C10 promotes disease resolution and survival in an experimental model of bacterial sepsis,[62] and treatment of mice with the CCR2 ligand MCP-1 protects mice against lethal bacterial infection.[63] The mechanism for these effects is unclear but cannot involve establishment of chemotactic gradients of the administered chemokine.

These results show that despite the overlapping specificities of inflammatory chemokines for receptors and leukocytes, the system is not completely redundant. It is important to note however that chemokine and chemokine receptor deficiency, whether it is induced experimentally in animals or inherited in humans, has never been associated with increased susceptibility to naturally acquired infections. This implies a substantial degree of functional redundancy does occur in vivo. Two examples of complete chemokine receptor deficiency have been identified in man, which are the result of inactivating mutations in CCR5 and the Duffy antigen receptor for chemokines. Counterintuitively, in both cases affected individuals are highly resistant to specific infectious agents: HIV-1 in the case of CCR5[64] and *Plasmodium vivax*, which causes a form of malaria, in the case of Duffy.[65] This is the result of chemokine mimicry by the surface proteins of these organisms, which interact with the respective chemokine receptors to facilitate infection of target cells.

Chemokine Regulation of the Adaptive Immune Response

Role of Chemokine-Directed Trafficking in Immune Recognition

Dendritic Cells
Trafficking of T cells and mature DCs from the periphery to secondary lymphoid tissue as well as their exact positioning within lymphoid microenvironments is critical for adaptive immunity and is thought to be regulated by specific sets of chemokines and chemokine receptors. In particular, the SLC/ELC-CCR7 axis has emerged as key coordinator for bringing T cells and DCs together in lymphoid tissue. A model has emerged in which DC trafficking is regulated by a switch in chemokine receptor expression that is tightly coupled to cell maturation. In this way, immature DCs which express the inflammatory chemokine receptors CCR1, CCR2, CCR5 and CCR6 are able to migrate into inflamed sites in response to cognate chemokines, where they then take up antigen and begin to develop into mature DCs capable of presenting antigen to specific T cells. During maturation, inflammatory chemokine receptors are downregulated while CCR4 and most importantly CCR7 are upregulated, thereby sensitizing the cells to the CCR7 ligands SLC and ELC which are produced constitutively in secondary lymphoid tissue. This provides the directional cues for migration of DC to secondary lymphoid organs. Consistent with this model, *plt* mice, which are genetically deficient in SLC production, have defects not only in T cell but also DC localization in secondary lymphoid organs.[38]

This model has also been tested in CCR6 knockout mice, which revealed a deficiency of myeloid DCs in the subepithelial dome of Peyer's patches, implying that CCR6 and its ligand MIP-3α control gut-specific trafficking of these cells. These mice have deficient humoral immune responses within the gut mucosa but normal skin responses. Restriction to a gut DC phenotype implies that other chemokine receptors control DC trafficking in other organs. In this regard, CD8α⁺ DC trafficking in the spleen in mice injected with Stag, a soluble *Toxoplasma gondii* antigen preparation, has been shown to be dependent on CCR5.[66] Moreover, there is evidence that the chemokine receptor that mediates DC trafficking in any given microenvironment can differ according to the inflammatory stimulus.[67] Finally, chemokine receptor functionality can differ dramatically on different types of immature DCs. In particular, blood derived plasmacytoid and myeloid DCs express a similar repertoire of inflammatory chemoattractant receptors but they are functional only on myeloid DCs.[68]

Lymphocytes
The phenotype of CCR7 knockout and *plt* mice described above provides strong genetic evidence for the importance of the SLC/ELC-CCR7 axis for trafficking of naïve T lymphocytes to secondary lymphoid tissue. The importance of CCR7 in memory cell trafficking in

vivo is less clear. An intriguing recent development is the separation of memory cells based on CCR7 phenotyping into 2 distinct subsets with distinct homing potential and effector function.[69] Central memory T cells (T_{CM}) express CCR7, traffic between the blood and secondary lymphoid organs, but are not Th1 or Th2 polarized. Effector memory T cells (T_{EM}) lack CCR7, traffic through peripheral tissues as immune surveillance cells, and rapidly release cytokines in response to activation. Both populations can be activated by APCs and antigen.

A third subset of CD4[+] memory cells, named follicular help T cells (T_{FH}), is defined by CXCR5 expression and absence of CCR7.[70] These cells do not produce Th1 or Th2 cytokines upon activation but are able to provide help for B cell maturation and antibody production. Consistent with this, they constitute the majority of CD4[+] memory cells in follicular zones of inflamed tonsils, whereas in contrast CXCR5 is present on only a small subset of T_{CM} cells in peripheral blood. The CXCR5 ligand BLC is expressed on follicular HEV and therefore is properly positioned to focus CXCR5 positive B and T cells from the blood into follicles. Thus the BLC-CXCR5 axis may play an analogous trafficking role for T-B interaction in follicles as SLC/ELC-CCR7 does for T-DC interaction in T cell zones. The phenotype of the CXCR5 knockout mouse, as described above, has validated the importance of this axis for germinal center development and B cell homing to secondary lymphoid tissue.[34]

The chemoattractants that regulate leukocyte homing to spleen are less well-defined. Migration to splenic red pulp may involve CXCL16 and its specific receptor CXCR6.[71] CXCL16 is made by cells in the splenic red pulp and is upregulated after exposure to inflammatory stimuli. NK T cells, and activated CD4[+] and CD8[+] T cells are found in this area and express CXCR6. CXCL16 is also made by T cell zone DCs, and CXCR6 is also found on naïve CD8[+] T cells, and intraepithelial lymphocytes. Thus, CXCL16 may function in T cell–DC interactions and in regulating movements of activated T cells in the splenic red pulp and in peripheral tissues.

The chemokines that control steady state trafficking of T lymphocytes in nonlymphoid tissue are not defined yet, although good candidates have recently been identified. CCR9 appears to mark a subset of T lymphocytes that home preferentially to small intestine.[72] Consistent with this, small intestine expresses high levels of the CCR9 ligand CCL25/TECK. Similarly, CLA[+] (cutaneous lymphocyte antigens) T lymphocytes that home to skin preferentially express CCR4 and CCR10, and their ligands CCL17/TARC and CCL27/CTACK are expressed in skin.[73]

Role of Chemokines in T Effector Cell Differentiation and Trafficking

After T cell receptor (TCR) activation and IL-2 stimulated proliferation, T lymphocytes downregulate the homeostatic chemokine receptors CXCR5 and CCR7, and upregulate inflammatory chemokine receptors. This switch renders the cells chemotactically responsive to cognate inflammatory chemokines and is thought to facilitate exit from lymph node and homing to peripheral sites of inflammation and infection.[74] Moreover, inflammatory chemokine receptors are differentially expressed on T effector cell subsets, and there is increasing evidence that this specialization not only regulates the trafficking of these cells but also controls differentiation of their cytokine phenotype.[10,75] Th1 cells, which secrete IFN-γ but not IL-4 and control cell-mediated immunity, selectively express CXCR3, CXCR6, CCR5 and CX3CR1, whereas CCR3, CCR4 and CCR8 are preferentially expressed on subsets of Th2 cells, which express IL-4 but not IFN-γ and are associated with humoral immunity and allergic inflammation.[71,76,77] Expression of chemokine ligands for these receptors correlates well with Th1 and Th2 polarized inflammatory processes in vivo.[78]

Effector T cell differentiation and function have been most extensively studied for MCP-1 and its receptor CCR2, but the results are complex and at first glance contradictory: CCR2, which is expressed on both Th1 and Th2 subsets, is strongly associated with Th1 polarization, whereas MCP-1 promotes Th2 polarization. The following are a few examples that illustrate this dichotomy. MCP-1 transgenic mice that express plasma concentrations of MCP-1 sufficient to desensitize CCR2 have increased susceptibility to infection with the intracellular pathogens *M. tuberculosis* and *L. monocytogenes*, suggesting that MCP-1 inhibits Th1 responses.[79] Consistent with this, MCP-1 deficient mice are reported to have enhanced Th1 responses in *M. tuberculosis* infection and impaired Th2 responses in *Leishmania* infection.[80,81] However, CCR2 knockout mice have the opposite phenotype: markedly reduced Th1 responses and increased susceptibility to *M. tuberculosis* and markedly increased Th2 responses and increased susceptibility to *L. major*.[82,83]

MCP-1 appears to promote Th2 polarization directly, by inhibiting IL-12 production in monocytes, and by enhancing IL-4 but not IFN-γ production in memory and activated T cells.[81] Thus MCP-1 may be important both in recruiting monocytes to sites of antigen deposition as well as in shaping the immune response to antigen by direct effects on T effector cell polarization. CCR5 has also been shown to regulate IL-12 production: in the *Toxoplasma* model referred to previously IL-12 production is CCR5 dependent and associated with DC migration in spleen.[66]

Why MCP-1 and CCR2 are polar opposites in regulating effector T cell polarization is not yet resolved, although recent data indicating that CCR2 knockout mice challenged with *M. tuberculosis* have markedly impaired DC migration suggest that the problem may be a profound defect in antigen detection and presentation that preempts any direct effects of MCP-1 on T cell differentiation.[60,82] Consideration of the redundancy of chemokine signaling suggests other potential explanations. For example, CCR2 can still be triggered by other CCR2 ligands in MCP-1 knockout mice, but none of these ligands can activate CCR2 pathways in CCR2 knockout mice. Also, CCR2 ligands that are normally depleted by CCR2 ligation may accumulate in CCR2 knockout mice, thereby allowing triggering of alternative receptors such as CCR3 that, unlike CCR2, could be coupled to Th2 differentiation. Lastly, MCP-1 could have a second receptor besides CCR2, although so far there is no evidence for this in hematopoietic cells.[10]

The role of CCR1 and its ligands on T cell polarization is also complex. MIP-1α can directly enhance IFN-γ production in activated T cells, and MIP-1α neutralization attenuates Th1-driven experimental autoimmune encephalomyelitis and Th1-dependent granuloma formation in mice.[84,85] Nevertheless, mice lacking the MIP-1α receptor CCR1 typically have reduced Th2 responses.[86] This suggests a role for CCR5 in mediating Th1 polarizing effects of MIP-1α.

CXCR3 and CX3CR1 expression on Th1 cells also appears to be functionally important in effector T cell polarization. When treated with low doses of cyclosporine, mice lacking CXCR3 or its ligand IP-10 are completely protected from acute and chronic cardiac allograft rejection, a Th1 driven process.[87] CX3CR1 deficiency also reduces the risk of rejection in this model, but it is not completely protective.[88] CXCR3 is also highly associated with Th1 cells present in inflamed tissues in patients, for example in psoriasis, a Th1-polarized disease.[89]

CCR8 expression on Th2 cells is also functionally significant in vivo. CCR8 knockout mice have reduced allergic airway inflammation in response to three different Th2 polarizing antigens: *Schistosoma mansoni* soluble egg antigen, ovalbumin and cockroach antigen.[90] In contrast, the response of CCR8 knockout mice to purified protein derivative (PPD), a prototypical Th1 antigen, is normal. Not all inflammatory chemokine receptors expressed on effector T cells, however, have been shown to affect function. For example, CCR4 knockout mice do not reveal any defect in Th2 cell polarization or Th2-dependent inflammation, but instead unexpectedly have increased resistance to endotoxin, a classic stress of the innate immune response.[91]

Use of Chemokines as Adjuvants for DNA Vaccines

Many studies have shown that vaccines and traditional adjuvants can induce endogenous local production of inflammatory chemokines.[92-94] In the case of DNA vaccines injected into muscle, the mechanism appears to involve myocyte stimulation by CpG motifs in the plasmid DNA.[95] Studies with knockout mice have indicated that endogenous chemokines can modulate vaccine responses. But can this activity be translated into new and better vaccines for the treatment or prevention of clinical disease?[93,96,97] There are now proof of concept studies in small animal models indicating that chemokines can be engineered as adjuvants, delivered either as protein or as immunomodulatory plasmids. Although the mechanisms are not fully defined, they appear to involve chemokine action at multiple different steps in the immune response, for example, to strengthen trafficking of specific classes of immune cells through the site of vaccination, or to enhance uptake of antigen by APCs, or to tilt the Th1/Th2 balance to a position that is optimal for a particular disease.

Chemokines were first evaluated as therapeutic agents in cancer.[98] One of the first papers showed that administration of the inflammatory Th1 and monocyte directed chemokine IP-10 could prevent tumor growth in mice in a T cell dependent manner, indicating that the chemokine was acting as an adjuvant to endogenous tumor antigens.[99] Subsequently many other chemokines, from both inflammatory and homeostatic subclasses, have also been reported to have antitumor effects.[100-102] Several delivery methods have been used successfully, including chemokine gene transfection of cultured tumor cells, and direct injection of chemokine protein or recombinant chemokine encoding adenoviruses into established tumor in vivo.[103] The mechanism appears to involve recruitment of monocytes, NK cells and CD8$^+$ cytotoxic T cells to the tumor.[104,105] Synergistic effects with cytokines such as IL-2 and IL-12 have been observed.[103]

More recently, chemokines have been validated as potent adjuvants for DNA vaccines in infectious disease models, the HSV-2 model being the most extensively studied.[106-108] Sin et al found that the inflammatory chemokines IL-8 and RANTES encoded on plasmids drive antigen-specific Th1 responses, and reduce morbidity and mortality due to HSV-2. In contrast, plasmids encoding IP-10 and MIP-1α, which also promote Th1 polarization, or MCP-1, which promotes Th2 polarization, increased mortality in this study.[106] This contrasted with a second study using plasmids encoding MIP-1α and HSV gB protein as a mucosal vaccine, in which vaccinated animals, while also exhibiting Th1 responses, had reduced susceptibility to vaginal infection with HSV. MIP-1α appeared to act by upregulating costimulatory molecules B7-1 and B7-2 on APCs, which was associated with a Th1 effector T cell response. The CXC chemokine MIP-2 was also active in this model, whereas MCP-1 and MIP-1β were not.[107] MIP-1α expression plasmid has also been used in mice to enhance Th1 type responses to a DNA vaccine constructed from pCMV160IIIB and pcREV.[109] Both humoral and cytotoxic T lymphocyte responses were enhanced; mucosal administration increased antigen-specific mucosal IgA production.

Somewhat surprisingly, plasmids encoding the homeostatic chemokines SLC and ELC, which normally are constitutively produced in lymphoid tissue, have also been shown to enhance immune responses to an HSV gB DNA vaccine.[108] Since CCR7 is the only signaling receptor identified for these chemokines, it is also surprising they promoted somewhat different immune responses: SLC induced a Th1 response whereas ELC promoted both Th1 and Th2 responses. With regard to humoral immunity, mucosal administration enhanced responses in both proximal and distal mucosal sites, but did not affect the systemic response. Conversely, intramuscular administration of chemokine plasmid enhanced the systemic but not the mucosal humoral response.

Together these results highlight several important points about chemokine adjuvants. First, the normal immunoregulatory functions of endogenous chemokines may not accurately predict their pharmacological or adjuvant effects. Therefore choosing chemokines for evaluation in particular vaccine applications will have to be done empirically. Second, the effects of a chemokine plasmid may be different, even opposite, depending on whether an antigen plasmid is coinjected and depending on the route of administration. Third, chemokine adjuvants may work by strengthening the functional link between innate and adaptive immune responses to antigens.

Biragyn et al have developed a chemokine adjuvant that works independently of effects on cell trafficking and differentiation.[110] This group has genetically engineered chemokine-antigen fusion proteins in which the chemokine domain acts as a Trojan horse to facilitate uptake of the antigen via the normal process of ligand-receptor internalization. This approach has been used to render a poorly immunogenic and nonprotective mouse B cell lymphoma tumor antigen immunogenic and partially protective. The fusion protein is immunogenic whether it is delivered as a protein or as a DNA plasmid. In contrast, a vaccine composed of the antigen and the chemokine on separate plasmids or as separate proteins is neither immunogenic nor protective suggesting that cell trafficking cannot be involved in the effect.

In a completely unexpected development, chemokine gene administration has also been shown to induce neutralizing antibody against the encoded chemokine. This has been used to block immune responses and to treat experimental autoimmune encephalitis and experimental arthritis in rodent models.[111-113]

Conclusions

The word chemokine is a contraction of chemotactic cytokine. While early research in this field repeatedly documented the powerful chemoattractant properties of these molecules, their cytokine-like effects on the immune responses have only recently come into focus. Together these activities may account for the adjuvant effects of chemokines revealed in vaccine studies in small animals. The impressive results obtained so far, which include achieving protective immune responses to lethal infectious challenge in mice, clearly justify further research to test this approach in large animals and ultimately in the clinic. However, it is important to keep in mind that chemokine adjuvants have been harmful in some contexts, and that the dose, route of administration, timing of administration and nature of the vaccine antigen may all unpredictably affect safety and efficacy.

References

1. Baggiolini M, Dewald B, Moser B. Human chemokines: an update. Annu Rev Immunol 1997; 15:675-705.
2. Mackay CR. Chemokines: immunology's high impact factors. Nat Immunol 2001; 2:95-101.
3. Zlotnik A, Yoshie O. Chemokines: a new classification system and their role in immunity. Immunity 2000; 12:121-7.
4. Rollins BJ. Chemokines. Blood 1997; 90:909-28.
5. Murphy PM. The molecular biology of leukocyte chemoattractant receptors. Annu Rev Immunol 1994; 12:593-633.
6. Wells TN, Power CA, Proudfoot AE. Definition, function and pathophysiological significance of chemokine receptors. Trends Pharmacol Sci 1998; 19:376-80.
7. Murphy PM, Baggiolini M, Charo IF et al. International union of pharmacology. XXII. Nomenclature for chemokine receptors. Pharmacol Rev 2000; 52:145-76.
8. Springer TA. Traffic signals for lymphocyte recirculation and leukocyte emigration: the multistep paradigm. Cell 1994; 76:301-14.
9. Foxman EF, Campbell JJ, Butcher EC. Multistep navigation and the combinatorial control of leukocyte chemotaxis. J Cell Biol 1997; 139:1349-60.
10. Luther SA, Cyster JG. Chemokines as regulators of T cell differentiation. Nat Immunol 2001; 2:102-7.
11. Murphy PM. Viral exploitation and subversion of the immune system through chemokine mimicry. Nat Immunol 2001; 2:116-22.
12. Modi WS, Yoshimura T. Isolation of novel GRO genes and a phylogenetic analysis of the CXC chemokine subfamily in mammals. Mol Biol Evol 1999; 16:180-93.
13. Mizoue LS, Bazan JF, Johnson EC et al. Solution structure and dynamics of the CX3C chemokine domain of fractalkine and its interaction with an N-terminal fragment of CX3CR1. Biochemistry 1999; 38:1402-14.
14. Clark-Lewis I, Kim KS, Rajarathnam K et al. Structure-activity relationships of chemokines. J Leukoc Biol 1995; 57:703-11.
15. Middleton J, Neil S, Wintle J et al. Transcytosis and surface presentation of IL-8 by venular endothelial cells. Cell 1997; 91:385-95.
16. Tanaka Y, Adams DH, Hubscher S et al. T-cell adhesion induced by proteoglycan-immobilized cytokine MIP-1 beta. Nature 1993; 361:79-82.
17. Imai T, Hieshima K, Haskell C et al. Identification and molecular characterization of fractalkine receptor CX3CR1, which mediates both leukocyte migration and adhesion. Cell 1997; 91:521-30.

18. Matloubian M, David A, Engel S et al. A transmembrane CXC chemokine is a ligand for HIV-coreceptor Bonzo. Nat Immunol 2000; 1:298-304.

19. Rabin RL, Park MK, Liao F et al. Chemokine receptor responses on T cells are achieved through regulation of both receptor expression and signaling. J Immunol 1999; 162:3840-50.

20. D'Amico G, Frascaroli G, Bianchi G et al. Uncoupling of inflammatory chemokine receptors by IL-10: generation of functional decoys. Nat Immunol 2000; 1:387-91.

21. Ali H, Richardson RM, Haribabu B et al. Chemoattractant receptor cross-desensitization. J Biol Chem 1999; 274:6027-30.

22. Loetscher P, Pellegrino A, Gong JH et al. The ligands of CXC chemokine receptor 3, I-TAC, Mig, and IP10, are natural antagonists for CCR3. J Biol Chem 2001; 276:2986-91.

23. Belperio JA, Keane MP, Arenberg DA et al. CXC chemokines in angiogenesis. J Leukoc Biol 2000; 68:1-8.

24. Moser B, Loetscher P. Lymphocyte traffic control by chemokines. Nat Immunol 2001; 2:123-8.

25. Sozzani S, Allavena P, Vecchi A et al. The role of chemokines in the regulation of dendritic cell trafficking. J Leukoc Biol 1999; 66:1-9.

26. Cacalano G, Lee J, Kikly K et al. Neutrophil and B cell expansion in mice that lack the murine IL-8 receptor homolog. Science 1994; 265:682-4.

27. Broxmeyer HE, Cooper S, Cacalano G et al. Involvement of Interleukin (IL) 8 receptor in negative regulation of myeloid progenitor cells in vivo: evidence from mice lacking the murine IL-8 receptor homologue. J Exp Med 1996; 184:1825-32.

28. Cook DN, Beck MA, Coffman TM et al. Requirement of MIP-1 alpha for an inflammatory response to viral infection. Science 1995; 269:1583-5.

29. Cook DN, Smithies O, Strieter RM et al. CD8+ T cells are a biologically relevant source of macrophage inflammatory protein-1 alpha in vivo. J Immunol 1999; 162:5423-8.

30. Youn BS, Mantel C, Broxmeyer HE. Chemokines, chemokine receptors and hematopoiesis. Immunol Rev 2000; 177:150-74.

31. Tachibana K, Hirota S, Iizasa H et al. The chemokine receptor CXCR4 is essential for vascularization of the gastrointestinal tract. Nature 1998; 393:591-4.

32. Zou YR, Kottmann AH, Kuroda M et al. Function of the chemokine receptor CXCR4 in haematopoiesis and in cerebellar development. Nature 1998; 393:595-9.

33. Nagasawa T, Tachibana K, Kishimoto T. A novel CXC chemokine PBSF/SDF-1 and its receptor CXCR4: their functions in development, hematopoiesis and HIV infection. Semin Immunol 1998; 10:179-85.

34. Forster R, Mattis AE, Kremmer E et al. A putative chemokine receptor, BLR1, directs B cell migration to defined lymphoid organs and specific anatomic compartments of the spleen. Cell 1996; 87:1037-47.

35. Forster R, Schubel A, Breitfeld D et al. CCR7 coordinates the primary immune response by establishing functional microenvironments in secondary lymphoid organs. Cell 1999; 99:23-33.

36. Gunn MD, Tangemann K, Tam C et al. A chemokine expressed in lymphoid high endothelial venules promotes the adhesion and chemotaxis of naive T lymphocytes. Proc Natl Acad Sci USA 1998; 95:258-63.

37. Ngo VN, Tang HL, Cyster JG. Epstein-Barr virus-induced molecule 1 ligand chemokine is expressed by dendritic cells in lymphoid tissues and strongly attracts naive T cells and activated B cells. J Exp Med 1998; 188:181-91.

38. Gunn MD, Kyuwa S, Tam C et al. Mice lacking expression of secondary lymphoid organ chemokine have defects in lymphocyte homing and dendritic cell localization. J Exp Med 1999; 189:451-60.

39. Voigt I, Camacho SA, de Boer BA et al. CXCR5-deficient mice develop functional germinal centers in the splenic T cell zone. Eur J Immunol 2000; 30:560-7.

40. Gunn MD, Ngo VN, Ansel KM et al. A B-cell-homing chemokine made in lymphoid follicles activates Burkitt's lymphoma receptor-1. Nature 1998; 391:799-803.

41. Campbell JJ, Pan J, Butcher EC. Cutting edge: developmental switches in chemokine responses during T cell maturation. J Immunol 1999; 163:2353-7.

42. Philpott DJ, Yamaoka S, Israel A et al. Invasive Shigella flexneri activates NF-kappa B through a lipopolysaccharide-dependent innate intracellular response and leads to IL-8 expression in epithelial cells. J Immunol 2000; 165:903-14.

43. Hachicha M, Rathanaswami P, Naccache PH et al. Regulation of chemokine gene expression in human peripheral blood neutrophils phagocytosing microbial pathogens. J Immunol 1998; 160:449-54.

44. Friedland JS, Constantin D, Shaw TC et al. Regulation of interleukin-8 gene expression after phagocytosis of zymosan by human monocytic cells. J Leukoc Biol 2001; 70:447-54.

45. Kuhns DB, Nelson EL, Alvord WG et al. Fibrinogen Induces IL-8 Synthesis in Human Neutrophils Stimulated with Formyl-Methionyl-Leucyl-Phenylalanine or Leukotriene B(4). J Immunol 2001; 167:2869-78.

46. Fehniger TA, Shah MH, Turner MJ et al. Differential cytokine and chemokine gene expression by human NK cells following activation with IL-18 or IL-15 in combination with IL-12: implications for the innate immune response. J Immunol 1999; 162:4511-20.

47. Campbell JJ, Qin S, Unutmaz D et al. Unique subpopulations of CD56+ NK and NK-T peripheral blood lymphocytes identified by chemokine receptor expression repertoire. J Immunol 2001; 166:6477-82.

48. Inngjerdingen M, Damaj B, Maghazachi AA. Expression and regulation of chemokine receptors in human natural killer cells. Blood 2001; 97:367-75.

49. Thelen M, Peveri P, Kernen P et al. Mechanism of neutrophil activation by NAF, a novel monocyte-derived peptide agonist. Faseb J 1988; 2:2702-6.

50. Tsai WC, Strieter RM, Mehrad B et al. CXC chemokine receptor CXCR2 is essential for protective innate host response in murine Pseudomonas aeruginosa pneumonia. Infect Immun 2000; 68:4289-96.

51. Olszyna DP, Florquin S, Sewnath M et al. CXC chemokine receptor 2 contributes to host defense in murine urinary tract infection. J Infect Dis 2001; 184:301-7.

52. Hall LR, Diaconu E, Patel R et al. CXC chemokine receptor 2 but not C-C chemokine receptor 1 expression is essential for neutrophil recruitment to the cornea in helminth-mediated keratitis (river blindness). J Immunol 2001; 166:4035-41.

53. Mehrad B, Strieter RM, Moore TA et al. CXC chemokine receptor-2 ligands are necessary components of neutrophil- mediated host defense in invasive pulmonary aspergillosis. J Immunol 1999; 163:6086-94.

54. Moore TA, Newstead MW, Strieter RM et al. Bacterial clearance and survival are dependent on CXC chemokine receptor-2 ligands in a murine model of pulmonary Nocardia asteroides infection. J Immunol 2000; 164:908-15.

55. Kielian T, Barry B, Hickey WF. CXC chemokine receptor-2 ligands are required for neutrophil-mediated host defense in experimental brain abscesses. Journal of Immunology 2001; 166:4634-43.

56. Devalaraja RM, Nanney LB, Du J et al. Delayed wound healing in CXCR2 knockout mice. J Invest Dermatol 2000; 115:234-44.

57. Domachowske JB, Bonville CA, Gao JL et al. The chemokine macrophage-inflammatory protein-1 alpha and its receptor CCR1 control pulmonary inflammation and antiviral host defense in paramyxovirus infection. J Immunol 2000; 165:2677-82.

58. Mehrad B, Moore TA, Standiford TJ. Macrophage inflammatory protein-1 alpha is a critical mediator of host defense against invasive pulmonary aspergillosis in neutropenic hosts. J Immunol 2000; 165:962-8.

59. Kurihara T, Warr G, Loy J et al. Defects in macrophage recruitment and host defense in mice lacking the CCR2 chemokine receptor. J Exp Med 1997; 186:1757-62.

60. Peters W, Dupuis M, Charo IF. A mechanism for the impaired IFN-gamma production in C-C chemokine receptor 2 (CCR2) knockout mice: role of CCR2 in linking the innate and adaptive immune responses. J Immunol 2000; 165:7072-7.

61. Chen SC, Mehrad B, Deng JC et al. Impaired pulmonary host defense in mice lacking expression of the CXC chemokine lungkine. J Immunol 2001; 166:3362-8.

62. Steinhauser ML, Hogaboam CM, Matsukawa A et al. Chemokine C10 promotes disease resolution and survival in an experimental model of bacterial sepsis. Infect Immun 2000; 68:6108-14.

63. Nakano Y, Kasahara T, Mukaida N et al. Protection against lethal bacterial infection in mice by monocyte- chemotactic and -activating factor. Infect Immun 1994; 62:377-83.

64. Berger EA, Murphy PM, Farber JM. Chemokine receptors as HIV-1 coreceptors: roles in viral entry, tropism, and disease. Annu Rev Immunol 1999; 17:657-700.

65. Horuk R, Chitnis CE, Darbonne WC et al. A receptor for the malarial parasite Plasmodium vivax: the erythrocyte chemokine receptor. Science 1993; 261:1182-4.

66. Aliberti J, Reis e Sousa C, Schito M et al. CCR5 provides a signal for microbial induced production of IL-12 by CD8 alpha+ dendritic cells. Nat Immunol 2000; 1:83-7.

67. Stumbles PA, Strickland DH, Pimm CL et al. Regulation of dendritic cell recruitment into resting and inflamed airway epithelium: Use of alternative chemokine receptors as a function of inducing stimulus. Journal of Immunology 2001; 167:228-34.

68. Penna G, Sozzani S, Adorini L. Cutting edge: selective usage of chemokine receptors by plasmacytoid dendritic cells. J Immunol 2001; 167:1862-6.

69. Sallusto F, Lenig D, Forster R et al. Two subsets of memory T lymphocytes with distinct homing potentials and effector functions. Nature 1999; 401:708-12.

70. Schaerli P, Willimann K, Lang AB et al. CXC chemokine receptor 5 expression defines follicular homing T cells with B cell helper function. J Exp Med 2000; 192:1553-62.
71. Kim CH, Kunkel EJ, Boisvert J et al. Bonzo/CXCR6 expression defines type 1-polarized T-cell subsets with extralymphoid tissue homing potential. J Clin Invest 2001; 107:595-601.
72. Zabel BA, Agace WW, Campbell JJ et al. Human G protein-coupled receptor GPR-9-6/CC chemokine receptor 9 is selectively expressed on intestinal homing T lymphocytes, mucosal lymphocytes, and thymocytes and is required for thymus-expressed chemokine-mediated chemotaxis. J Exp Med 1999; 190:1241-56.
73. Homey B, Wang W, Soto H et al. Cutting edge: the orphan chemokine receptor G protein-coupled receptor- 2 (GPR-2, CCR10) binds the skin-associated chemokine CCL27 (CTACK/ALP/ILC). J Immunol 2000; 164:3465-70.
74. Loetscher P, Seitz M, Baggiolini M et al. Interleukin-2 regulates CC chemokine receptor expression and chemotactic responsiveness in T lymphocytes. J Exp Med 1996; 184:569-77.
75. Sallusto F, Mackay CR, Lanzavecchia A. The role of chemokine receptors in primary, effector, and memory immune responses. Annu Rev Immunol 2000; 18:593-620
76. Bonecchi R, Bianchi G, Bordignon PP et al. Differential expression of chemokine receptors and chemotactic responsiveness of type 1 T helper cells (Th1s) and Th2s. J Exp Med 1998; 187:129-34.
77. Sallusto F, Lenig D, Mackay CR et al. Flexible programs of chemokine receptor expression on human polarized T helper 1 and 2 lymphocytes. J Exp Med 1998; 187:875-83.
78. Qiu B, Frait KA, Reich F et al. Chemokine expression dynamics in mycobacterial (type-1) and schistosomal (type-2) antigen-elicited pulmonary granuloma formation. Am J Pathol 2001; 158:1503-15.
79. Rutledge BJ, Rayburn H, Rosenberg R et al. High level monocyte chemoattractant protein-1 expression in transgenic mice increases their susceptibility to intracellular pathogens. J Immunol 1995; 155:4838-43.
80. Chensue SW, Warmington KS, Ruth JH et al. Role of monocyte chemoattractant protein-1 (MCP-1) in Th1 (mycobacterial) and Th2 (schistosomal) antigen-induced granuloma formation: relationship to local inflammation, Th cell expression, and IL-12 production. J Immunol 1996; 157:4602-8.
81. Gu L, Tseng S, Horner RM et al. Control of TH2 polarization by the chemokine monocyte chemoattractant protein-1. Nature 2000; 404:407-11.
82. Peters W, Scott HM, Chambers HF et al. Chemokine receptor 2 serves an early and essential role in resistance to Mycobacterium tuberculosis. Proc Natl Acad Sci USA 2001; 98:7958-63.
83. Sato N, Ahuja SK, Quinones M et al. CC chemokine receptor (CCR)2 is required for langerhans cell migration and localization of T helper cell type 1 (Th1)-inducing dendritic cells. Absence of CCR2 shifts the Leishmania major-resistant phenotype to a susceptible state dominated by Th2 cytokines, b cell outgrowth, and sustained neutrophilic inflammation. J Exp Med 2000; 192:205-18.
84. Karpus WJ, Lukacs NW, McRae BL et al. An important role for the chemokine macrophage inflammatory protein-1 alpha in the pathogenesis of the T cell-mediated autoimmune disease, experimental autoimmune encephalomyelitis. J Immunol 1995; 155:5003-10.
85. Lukacs NW, Kunkel SL, Strieter RM et al. The role of macrophage inflammatory protein 1 alpha in Schistosoma mansoni egg-induced granulomatous inflammation. J Exp Med 1993; 177:1551-9.
86. Gao JL, Wynn TA, Chang Y et al. Impaired host defense, hematopoiesis, granulomatous inflammation and type 1-type 2 cytokine balance in mice lacking CC chemokine receptor 1. J Exp Med 1997; 185:1959-68.
87. Hancock WW, Lu B, Gao W et al. Requirement of the chemokine receptor CXCR3 for acute allograft rejection. J Exp Med 2000; 192:1515-20.
88. Haskell CA, Hancock WW, Salant DJ et al. Targeted deletion of CX(3)CR1 reveals a role for fractalkine in cardiac allograft rejection. J Clin Invest 2001; 108:679-88.
89. Flier J, Boorsma DM, van Beek PJ et al. Differential expression of CXCR3 targeting chemokines CXCL10, CXCL9, and CXCL11 in different types of skin inflammation. J Pathol 2001; 194:398-405.
90. Chensue SW, Lukacs NW, Yang TY et al. Aberrant in vivo T helper type 2 cell response and impaired eosinophil recruitment in CC chemokine receptor 8 knockout mice. J Exp Med 2001; 193:573-84.
91. Chvatchko Y, Hoogewerf AJ, Meyer A et al. A key role for CC chemokine receptor 4 in lipopolysaccharide-induced endotoxic shock. J Exp Med 2000; 191:1755-64.
92. Zaitseva M, King LR, Manischewitz J et al. Human peripheral blood T cells, monocytes, and macrophages secrete macrophage inflammatory proteins 1alpha and 1beta following stimulation with heat-inactivated Brucella abortus. Infect Immun 2001; 69:3817-26.
93. Boyer JD, Kim J, Ugen K et al. HIV-1 DNA vaccines and chemokines. Vaccine 1999; 17:S53-64.
94. Boyer JD, Cohen AD, Vogt S et al. Vaccination of seronegative volunteers with a human immunodeficiency virus type 1 env/rev DNA vaccine induces antigen-specific proliferation and lymphocyte production of beta-chemokines. J Infect Dis 2000; 181:476-83.

95. Stan AC, Casares S, Brumeanu TD et al. CpG motifs of DNA vaccines induce the expression of chemokines and MHC class II molecules on myocytes. Eur J Immunol 2001; 31:301-10.
96. Kipps T, Mendoza R. Extending genetic vaccines with chemokines. Nat Biotechnol 1999; 17:226-7.
97. Scheerlinck JY. Genetic adjuvants for DNA vaccines. Vaccine 2001; 19:2647-56.
98. Rollins BJ, Sunday ME. Suppression of tumor formation in vivo by expression of the JE gene in malignant cells. Mol Cell Biol 1991; 11:3125-31.
99. Luster AD, Leder P. IP-10, a -C-X-C- chemokine, elicits a potent thymus-dependent antitumor response in vivo. J Exp Med 1993; 178:1057-65.
100. Nomura T, Hasegawa H, Kohno M et al. Enhancement of anti-tumor immunity by tumor cells transfected with the secondary lymphoid tissue chemokine EBI-1-ligand chemokine and stromal cell-derived factor-1alpha chemokine genes. Int J Cancer 2001; 91:597-606.
101. Narvaiza I, Mazzolini G, Barajas M et al. Intratumoral coinjection of two adenoviruses, one encoding the chemokine IFN-gamma-inducible protein-10 and another encoding IL-12, results in marked antitumoral synergy. J Immunol 2000; 164:3112-22.
102. Braun SE, Chen K, Foster RG et al. The CC chemokine CK beta-11/MIP-3 beta/ELC/Exodus 3 mediates tumor rejection of murine breast cancer cells through NK cells. J Immunol 2000; 164:4025-31.
103. Emtage PC, Wan Y, Hitt M et al. Adenoviral vectors expressing lymphotactin and interleukin 2 or lymphotactin and interleukin 12 synergize to facilitate tumor regression in murine breast cancer models. Hum Gene Ther 1999; 10:697-709.
104. Pertl U, Luster AD, Varki NM et al. IFN-gamma-inducible protein-10 is essential for the generation of a protective tumor-specific CD8 T cell response induced by single-chain IL-12 gene therapy. J Immunol 2001; 166:6944-51.
105. Nagai M, Masuzawa T. Vaccination with MCP-1 cDNA transfectant on human malignant glioma in nude mice induces migration of monocytes and NK cells to the tumor. Int Immunopharmacol 2001; 1:657-64.
106. Sin J, Kim JJ, Pachuk C et al. DNA vaccines encoding interleukin-8 and RANTES enhance antigen-specific Th1-type CD4(+) T-cell-mediated protective immunity against herpes simplex virus type 2 in vivo. J Virol 2000; 74:11173-80.
107. Eo SK, Lee S, Chun S et al. Modulation of immunity against herpes simplex virus infection via mucosal genetic transfer of plasmid DNA encoding chemokines. J Virol 2001; 75:569-78.
108. Eo SK, Lee S, Kumaraguru U et al. Immunopotentiation of DNA vaccine against herpes simplex virus via co-delivery of plasmid DNA expressing CCR7 ligands. Vaccine 2001; 19:4685-93.
109. Lu Y, Xin KQ, Hamajima K et al. Macrophage inflammatory protein-1alpha (MIP-1alpha) expression plasmid enhances DNA vaccine-induced immune response against HIV-1. Clin Exp Immunol 1999; 115:335-41.
110. Biragyn A, Tani K, Grimm MC et al. Genetic fusion of chemokines to a self tumor antigen induces protective, T-cell dependent antitumor immunity. Nat Biotechnol 1999; 17:253-8.
111. Youssef S, Maor G, Wildbaum G et al. C-C chemokine-encoding DNA vaccines enhance breakdown of tolerance to their gene products and treat ongoing adjuvant arthritis. J Clin Invest 2000; 106:361-71.
112. Youssef S, Wildbaum G, Karin N. Prevention of experimental autoimmune encephalomyelitis by MIP-1alpha and MCP-1 naked DNA vaccines. J Autoimmun 1999; 13:21-9.
113. Youssef S, Wildbaum G, Maor G et al. Long-lasting protective immunity to experimental autoimmune encephalomyelitis following vaccination with naked DNA encoding C-C chemokines. J Immunol 1998; 161:3870-9.

CHAPTER 18

DNA Vaccines: Safety and Regulatory Issues

Dennis M. Klinman and Herbert A. Smith

Summary

Plans to initiate clinical trials involving DNA vaccines prompted the US Food and Drug Administration to examine the safety of this form of vaccination. Concerns were raised that DNA vaccines might possibly:

1. integrate into the host genome, thereby increasing the risk of oncogenesis;
2. stimulate the production of autoantibodies, thereby inducing or accelerating the development of autoimmune disease;
3. induce tolerance rather than immunity to the encoded antigen; and/or
4. alter the host's immune milieu, thereby reducing responsiveness to subsequent vaccination and/or increasing susceptibility to subsequent infection.

Pre-clinical studies evaluated the likelihood of these adverse effects occurring, and showed that the risks of DNA vaccination were of insufficient magnitude to prevent the initiation of phase I safety trials. Results from ongoing clinical trials indicate that "first generation" DNA vaccines are safe in humans. While mild-moderate pain, swelling and redness at the site of vaccination have been reported, no major systemic toxicity has been reported. Efforts continue to improve the immunogenicity of the "next generation" of DNA vaccines by modifying the plasmid backbone, its method of delivery, or by co-administering various immune adjuvants. These efforts may increase the risk of adverse side effects, reinforcing the need for continued pre-clinical and clinical safety studies of novel vaccine formulations.

Risks from Plasmid Integration

Small amounts of foreign DNA exist as contaminants in "conventional" vaccines. Historically, vaccine manufacturers sought to minimize human exposure to this material in order to reduce concern that foreign DNA might integrate into the host genome, thereby increasing the likelihood of malignant transformation, genomic instability, or cell growth dysregulation.[1] Not surprisingly, these concerns were raised when purified DNA plasmids were considered for use as human vaccines.[2,3]

Initial studies determined the tissue distribution patterns and persistence of DNA plasmids in vivo. Results showed that plasmids were carried through the blood and lymph to all highly vascularized organs. Yet high levels of plasmid DNA persisted long term (weeks to months) only at the site of injection.[4,7] Efforts were made to determine whether these plasmids had integrated into the host genome. To date, very few sponsors have published the results of their integration studies, making it difficult to reach a consensus concerning the frequency of plasmid integration. Indeed, there is concern that findings are published only by those sponsors who detect little or no evidence of integration, leading to a "reporting bias" in the literature. In one of the studies on this issue, the behavior of a DNA vaccine encoding the circumsporozoite protein of malaria was examined. Results indicate that 3-30 copies of this plasmid remained associated with genomic DNA two months after intramuscular injection in mice.[4,6] This presumably reflects the level of plasmid integration at the site of vaccine administration.

The risk posed by such integration is difficult to assess. An integration frequency of 3 30 copies per 10^5 cells is 3,000 times lower than a spontaneous mutation rate of 1×10^5 per gene.[6] Yet the effect of a single base mutation differs from the impact of inserting long stretches of plasmid DNA (containing strong promoter regions) into the mammalian genome. Moreover, the technology used to monitor plasmid persistence examines sections of DNA hundreds of base pairs in length. Since DNA fragments as short as 7 bp can integrate and affect recombination rates, the precise frequency (and therefore the risk) posed by DNA vaccination remains uncertain.[4]

The issue of plasmid integration and biodistribution also impacts the need to perform reproductive toxicity studies. If plasmids reach gonadal tissue and integrate into germinal cells, heritable genetic defects could result.[8] If plasmid does not reach gonadal tissue, there is no need to perform reproductive toxicity studies. While pre-clinical studies involving mice and rabbits indicate that intramuscular injected DNA vaccines do not persist in the testes or ovaries,[6] equivalent studies investigating the effect of administering plasmids by other methods (gene gun, biojector, electroporation, liposome encapsulation) have not been published.

Available evidence suggests that the DNA vaccines undergoing phase I human trials rarely integrate. However, efforts to increase vaccine immunogenicity by modifying the vector, co-injecting agents that increase cellular uptake, or altering the site and/or method of plasmid delivery, may increase the likelihood of integration. Recent findings document that intramuscular administration of plasmid DNA encapsulated in poly-lactide-co-glycolide microspheresincreases uptake by dendritic cells in the primary lymphoid organs.[9] Plasmids expressing self genes (such as cytokines) may integrate and persist more frequently, due to homologous recombination and absence of an immune response against transfected cells.[10,11] For these reasons, regulatory authorities advise that plasmid distribution/persistence studies continue to be performed on new vaccines. Should these yield evidence that plasmids are persisting in vivo at levels exceeding those previously described, thorough integration studies should be performed.[8,12,13]

Autoimmunity

It has been known for decades that bacterial DNA can induce the production of anti-double-stranded DNA autoantibodies in normal mice and accelerate the development of autoimmune disease in lupus-prone animals.[14,16] These effects have now been attributed to the presence of "CpG motifs" in DNA plasmids of bacterial origin. Such motifs stimulate polyclonal B cell activation, trigger the production of proinflammatory cytokines, and prevent the apoptotic death of activated immune cells.[14,20] In various model systems, each of these effects has been shown to contribute to the development of autoimmune disease. For example, polyclonal B cell activation is an early manifestation of systemic lupus erythematosus in (NZB/NZW) F1 (hereafter NZB/W) mice, over-production of IL-6 causes lupus-like disease in patients with cardiac myxoma, and defects in apoptosis contribute to the development of autoimmune disease in *lpr* and *gld* mice.[21 23]

Since DNA vaccines contain CpG motifs, it was important to determine whether vaccination induced or accelerated the development of systemic autoimmunity. Towards that end, plasmids and/or CpG oligonucleotides (ODN) were administered to BALB/c and NZB/W mice. The latter animals were selected because they spontaneously develop a disease similar to human SLE. Results indicate that repeated DNA vaccination of normal mice triggers up to a 4-fold increase in the number of B cells activated to secrete IgG anti-DNA autoantibodies.[17] This immune stimulation returns to normal within two weeks after the cessation of therapy. When compared to IgG anti-DNA levels in mice with lupus like disease, the amount of autoantibody present in DNA vaccinated normal mice was well below the amount capable of causing disease.[17,24] No immune response was detected in vaccinated mice against cells expressing the vaccine encoded antigen (such as muscle or dendritic cells), suggesting that these tissues were not targeted for elimination by the immune system.[17]

DNA vaccination of young NZB/W mice accelerated the production of anti-DNA autoantibodies (these were predominantly of the IgG2a isotype) and stimulated a marked increase in the ratio of IFNγ : IL 4 secreting cells in vivo.[25] However, these changes delayed rather than accelerated the development of autoimmune disease.[26] Presumably, changes in the cytokine milieu induced by CpG motifs interfered with lupus pathogenesis. Taken together, available evidence indicates that systemic autoimmunity is unlikely to result from DNA vaccination. Consistent with this conclusion, there have been no clinical reports of systemic autoimmune disease arising as a consequence of DNA vaccination.

The situation is somewhat more complex for organ specific autoimmune diseases, whose induction can be promoted by Th1 mediated immunity. Since CpG motifs boost the production of Th1 cytokines, there is concern that DNA vaccination might induce or worsen such diseases. One well studied example of Th1 dependent organ specific autoimmunity is murine experimental allergic encephalomyelitis (EAE), which serves as a model of human multiple sclerosis.[27] To examine whether CpG DNA can contribute to the development of EAE, SJL mice were immunized with myelin basic protein$_{87-106}$ (MBP) plus ODN. Disease inducing effector cells were not elicited when this EAE resistant strain was treated with MBP alone. However, the combination of MBP co-administered with CpG DNA caused disease, manifest by severe muscle weakness. This effect was linked to the up-regulation of IL-12 production.[27,28]

In another model system, CpG DNA potentiated the development of a different type of organ specific autoimmunity. Chlamydia infection can cause autoimmune myocarditis, due to molecular mimicry between bacterial antigens and epitopes expressed by heart muscle myosin.[29] When normal mice were co-immunized with chlamydia antigens plus CpG (but not control) DNA, the production of heart muscle specific autoantibodies and the severity of autoimmune disease increased.[29] In this system, CpG DNA improved the presentation of cross-reactive antigen while promoting the production of proinflammatory cytokines.

Results from both EAE and myositis studies suggest that the adjuvant effect of CpG DNA may contribute to the development of organ specific autoimmunity, both by changing the cytokine milieu and by enhancing the presentation of self antigens.[30,31] Yet the level of risk posed by DNA vaccination is difficult to quantitate. To date, autoimmunity was detected only when CpG DNA was administered in conjunction with disease inducing antigens. Since there is no evidence of systemic or organ specific autoimmunity developing in healthy animals treated with CpG DNA alone, clinical trials have proceeded with appropriate safeguards.

Tolerance

Most vaccines intended for human use are administered to infants and children. Due to the immaturity of their immune systems, very young animals exposed to foreign antigens are at risk for developing tolerance rather than immunity.[32] Since the protein encoded by a DNA vaccine is produced endogenously and expressed in the context of self MHC, the potential exists for the neonatal immune system to recognize it as "self", triggering the development of long lasting immune tolerance. Moreover, studies indicate that DNA vaccination during pregnancy can result in transplacental transfer of the plasmid to the fetus, where the encoded protein is expressed.[33]

Studies addressing the ability of DNA vaccines to induce tolerance yielded divergent results. When neonatal mice were vaccinated with a plasmid encoding the circumsporozoite protein of malaria (CSP), long lasting tolerance was observed.[34,35] The likelihood of inducing tolerance increased with vaccine dose, decreased with recipient age, and was MHC independent.[34,36] Yet DNA vaccines encoding antigens from influenza, rabies, hepatitis B, and lymphocytic choriomeningitis virus, as well as other malaria proteins, failed to induce neonatal tolerance.[36-40] Differences in the age of initial vaccination, dose of plasmid, and nature of the encoded antigen make it difficult to compare these studies. Fortunately, tolerance has never been observed when immunologically mature animals were immunized with a DNA vaccine. Experience with the circumsporozoite vaccine suggests that careful preclinical studies in appropriate models be performed prior to the use of DNA vaccines in children or newborns.

Alteration of the Immune Milieu

The vector backbone of DNA plasmids contains CpG motifs that preferentially stimulate the production of pro-inflammatory and Th1 cytokines.[18,41 43] This raises the possibility that DNA vaccination might bias the host's cytokine profile, interfering with immune homeostasis or increasing susceptibility to pathogens whose elimination requires the host to mount a vigorous Th2 response.[10,27,44,45]

Immune homeostasis involves maintaining a balance between Th1 : Th2 cytokine secreting cells. These form a dynamic and mutually inhibitory network. Th1 cytokines can block the maturation of Th0 into Th2 type cells, and vice versa.[46] To determine whether DNA vaccines alter this balance, normal mice were injected with CpG DNA. As expected, a short term increase in the ratio of Th1 : Th2 secreting cells was detected in the recipient's spleen and lymph nodes. However, no evidence for long term skewing of the immune repertoire or increased susceptibility to infection was observed when young mice were treated repeatedly with CpG DNA.[44]

There is also concern about the effect of co-administering plasmids that encode cytokines or co-stimulatory molecules with DNA vaccines. Vaccine developers hope that these novel adjuvants will boost immunogenicity and promote the development of optimally protective responses.[47] Plasmids encoding a variety of cytokines have been evaluated for use as vaccine adjuvants. These plasmids have the potential of persisting for prolonged periods in vivo, since the self protein they encode does not elicit an immune response against the transfected cell.[11] In one study, plasmids encoding IL-4 and IFN-γ were repeatedly administered to newborn and adult BALB/c mice. As expected, these plasmids altered the immune milieu of recipient animals, with IL-4 inducing Th0 cells to mature into Th2 cells while IFN-γ promoted maturation into Th1 cells.[10] Other studies confirm that cytokine encoding plasmids can alter the response to a co administered DNA vaccine.[48] Yet observed effects on cytokine balance have been short lived and have not been associated with altered immune responses to unrelated antigens or with the production of pathogenic autoantibodies (although it is unclear whether systematic efforts to detect such events were widely undertaken). A recent publication suggests that cytokine encoding plasmids can induce autoantibodies against the encoded cytokine.[49] Whether this has long term deleterious side effects requires further study. No evidence of human toxicity has been reported from a clinical trial involving a plasmid encoding GM-CSF. Thus, although cytokine encoding plasmids may alter immune responsiveness locally, long term systemic effects have not been reported.

Human Clinical Studies

Phase I clinical studies designed to assess the safety of plasmid DNA vaccines for the prevention and/or treatment of HIV, malaria, and hepatitis B have been conducted.[50 54] Single doses ranging from 250 ng to 2,500 μg were administered up to three times, with some vaccinees receiving a cumulative dose of 7.5 mg.[50 54] All DNA vaccines have been delivered intramuscularly or intradermally by syringe injection, needleless injector, or gene gun.[50-54] Vaccine safety has been monitored by following clinical symptomatology and by analyzing the hematologic, serologic and biochemical profiles of vaccinees. There have been no major adverse events reported following DNA vaccination. Vaccinees describe a variety of minor complaints, primarily involving transient pain, swelling and/or redness at the injection site.[50-54] Of note, there was no clear trend towards a worsening of clinical complaints with the administration of higher or more frequent doses of vaccine.[52]

A secondary goal of ongoing phase I trials involves determining whether DNA vaccines are immunogenic in humans. Various parameters have been monitored, including the amount and isotype of antibody produced against the vaccine encoded protein, the activation and/or proliferation of periferal blood mononuclear cells (PBMC), the production of cytokines, and/or the development of antigen specific cytotoxic activity. While evidence of both B- and T-cell stimulation has been reported, the magnitude of these responses was modest.[50-54] Efforts are therefore underway to improve immunogenicity by boosting the uptake, expression and/or persistence of plasmids in vivo, co-administering immune adjuvants, or boosting with protein or viral vectors. In some cases, these efforts are likely to alter the safety profile of the DNA vaccine. For example, animal studies indicate that the use of DL-lactide co-glycolide micoparticles for intramuscular injection can increase plasmid persistence in vivo,[55,56] while the use of electroporation to increase cellular uptake of the vaccine may increase the likelihood of plasmid integration.

Conclusions

Local injection of DNA plasmids results in their widespread distribution throughout the body. However, the vast majority of plasmid persists long-term primarily at the site of injection.Available evidence suggests that the frequency of integration of current DNA vaccines is low. DNA vaccination does induce non-specific immune activation, manifest by increased autoantibody production. Yet pre-clinical and clinical data indicate that the level of such activation is insufficient to induce systemic autoimmune disease. There is concern that certain Th1 dependent organ-specific diseases may be induced or worsened by DNA vaccination. In particular, co-administration of CpG DNA with specific self antigens or antigens cross-reactive with self may be deleterious. Less likely is the potential that long-term changes in the immune milieu will result from DNA vaccination. Tolerance has not been observed in adults, but may result from neonatal immunization with vaccines encoding certain antigens.

Clinical studies indicate that DNA vaccines are well tolerated. Only mild to moderate adverse events have been reported, and these were primarily local. Yet these same studies indicate that DNA vaccines are only modestly immunogenic in humans. Sponsors are therefore pursuing strategies to increase vaccine immunogenicity by co-delivering various adjuvants, modifying plasmids and/or changing their method of delivery. These efforts may alter the safety profile of the resultant vaccines, requiring that diligent pre clinical safety evaluations continue.

References

1. Acceptability of Cell Substrates for Production of Biogicals. World Health Organization Technical Report. 1987, Series 747.
2. Klinman DM, Takeno M, Ichino M et al. DNA vaccines: safety and efficacy issues. Springer Semin Immunopathol 1997; 19:245-256.
3. Robertson JS. Safety considerations for nucleic acid vaccines. Vaccine 1994; 12:1526-1528.
4. Ledwith BJ, Manam S, Troilo PJ et al. Plasmid DNa vaccines: Assay for integration into host genomic DNA. Dev Biol 2000; 104:33-43.
5. Martin T, Parker SE, Hedstrom R et al. Plasmid DNA malaria vaccine: the potential for genomic integration after intramuscular injection. Hum Gene Ther 1999; 10:759-768.
6. Parker SE, Borellini F, Wenk ML et al. Plasmid DNA malaria vaccine: Tissue distribution and safety studies in mice and rabbits. Hum Gene Ther 1999; 10:741-758.
7. Mor G, Klinman DM, Shapiro S et al. Complexity of the cytokine and antibody response elicited by immunizing mice with Plasmodium yoelii circumsporozoite protein plasmid DNA. J Immunol 1995; 155:2039-2046.
8. Points to consider on Plasmid DNA Vaccines for Preventive Infectious Disease Indications Food and Drug Administration, Center for Biologics Evaluation and Research, Office of Vaccine Research and Review, Docket No. 96 N 0400, 1996.
9. Lunsford L, McKeever U, Eckstein V et al. Tissue distribution and persistence in mice of plasmid DNA encapsulated in a PLGA based microsphere. J Drug Target 2000; 8:39-50.
10. Ishii KJ, Weiss WR, Ichino M et al. Activity and safety of DNA plasmids encoding IL-4 and IFN gamma. Gene Ther 1999; 6:237-244.
11. Pasquini S, Xiang ZQ, Wang Y et al. Cytokines and costimulatory molecules as genetic adjuvants. Immunolgy and Cell Biology 1997; 75:397-401.
12. The International Association of Biological Standardization Biologicals 1998; 26:205-212.
13. Sanceau J, Kaisho T, Hirano T et al. Triggering of the human interleukin-6 gene by interferon gamma and tumor necrosis factor alpha in monocytic cells involves cooperation between interferon reguratory factor 1, NFκB, and Sp1 transcription factors. J Biol Chem 1995; 270:27920-27931.

14. Steinberg AD, Krieg AM, Gourley MF, Klinman DM. Theoretical and experimental approaches to generalized autoimmunity. Immunol Rev 1990; 118:129-163.
15. Gilkeson GS, Pippen AM, Pisetsky DS. Induction of cross-reactive anti-dsDNA antibodies in preautoimmune NZB/NZW mice by immunization with bacterial DNA. J Clin Invest 1995; 95:1398-1402.
16. Gilkeson GS, Riuz P, Howell D et al. Induction of immune-mediated glomerulonephritis in normal mice immunized with bacterial DNA. Clin.Immunol and Immunopathol 1993; 68:283-292.
17. Mor G, Singla M, Steinberg Adet al. Do DNA vaccines induce autoimmune disease? Hum Gene Ther 1997; 8:293-300.
18. Klinman DM, Yi A, Beaucage SL et al. CpG motifs expressed by bacterial DNA rapidly induce lymphocytes to secrete IL-6, IL-12 and IFNγ. Proc Natl Acad Sci USA 1996; 93:2879-2883.
19. Krieg AM. CpG DNA : A pathogenic factor in systemic lupus erythematosus? J Clin Immunol 1995; 15:284-292.
20. Yi A, Hornbeck P, Lafrenz DE, Krieg, AM CpG DNA rescue of murine B lymphoma cells from anti-IgM induced growth arrest and programmed cell death is associated with increased expression of c myc and bcl xl. J Immunology 1996; 157:4918-4925.
21. Klinman D., Steinberg AD. Systemic autoimmune disease arises from polyclonal B cell activation. J Exp Med 1987; 165:1755-1760.
22. Klinman DM. Polyclonal B-cell activation in lupus prone mice precedes and predicts the development of autoimmune disease. J Clin Invest 1990; 86:1249-1254.
23. Watanabe Fukunaga R, Brannan CI, Copeland NG et al. Lymphoproliferation disorder in mice explained by defects in Fas antigen that mediates apoptosis. Nature 1992; 356:314-317.
24. Klinman DM, Takeshita F, Kamstrup S et al. DNA vaccines: Capacity to induce autoimmunity and tolerance. Dev Biol 2000; 104:45-51.
25. Gilkeson GS, Conover JS, Halpern Met al. Effects of bacterial DNA on cytokine production by (NZB/NZW)F1 mice. J Immunol 1998;161:3890-3895.
26. Gilkeson GS, Ruiz P, Pippen AM et al. Modulation of renal disease in autoimmune NZB/NZW mice by immunization with bacterial DNA. J Exp Med 1996; 183:1389-1397.
27. Segal BM, Klinman DM, Shevach EM. Microbial products induce autoimmune disease by an IL-12 dependent process. J Immunol 1997; 158:5087-5091.
28 Segal BM, Chang JT, Shevach EM. CpG oligonucleotides are potent adjuvants for the activation of autoreactive encephalotogenic T cells in vivo. J Immunol 2000; 164:5683-5688.
29. Bachmaier, K., Meu, N., Maza, L.M., Pal, S., Nessel, A., & Penninger, J.M. (1999). Chlamydia infections and heart disease linked through antigenic mimicry. Science, 283:1335-1339.
30. Cowdery JS, Chace JH, Yi A, Krieg AM. Bacterial DNA induces NK cells to produce IFNgamma in vivo and increases the toxicity of lipopolysaccharides. J Immunol 1996; 156:4570-4575.
31. Sparwasser T, Meithke T, Lipford G et al. Bacterial DNA causes septic shock. Nature 1997; 386:336-338.
32. Silverstein AM, Segal S. The ontogeny of antigen specific T cells. J Exp Med 1975; 142:802-804.
33. Tsukamoto M, Ochiya T, Yoshida S et al. Gene transfer and expression in progeny after intravenous DNA injection into pregnant mice. Nat Genet 1995; 9:243-248.
34. Ichino M, Mor G, Conover J et al. Factors associated with the development of neonatal tolerance after the administration of a plasmid DNA vaccine. J Immunol 1999; 162:3814-3818.
35. Mor G, Yamshchikov G, Sedega, M et al. Induction of neonatal tolerance by plasmid DNA vaccination of mice. J Clin Invest 1997; 98:2700-2705.
36. Pertmer T, Robinson HL. Studies of antibody responses following neonatal immunization with influenza hemagluttinin DNA or protein. Virology 1999; 257:406-414.
37. Hassett D, Zhang J, Slifka M, Whitton L. Immune responses following neonatal DNA vaccination are long lived, abundant and qualitatively similar to those induced by conventional immunization. J Virol 2000; 74:2620-2627.
38. Bot A, Bot S, Bona C. Enhanced protection against influenza virus of mice immunized as newborns with a mixture of plasmids expressing hemagglutinin and nucleoprotein. Vaccine 1998; 16:1675-1682.
39. Prince AM, Whalen R, Brotman B. Successful nucleic acid based immunization of newborn chimpanzees against hepatitis B virus. Vaccine 1997; 15:916-919.
40. Sarzotti M, Dean TA, Remington MP et al. Induction of CTl responses in newborn mice by DNA immunization. Vaccine 1997; 15:795-781. (abstract)
41. Sato Y, Roman M, Tighe H et al. Immunostimulatory DNA sequences necessary for effective intradermal gene immunization. Science 1996; 273:352-354.
42. Klinman DM, Yamshchikov G, Ishigatsubo Y. Contribution of CpG motifs to the immunogenicity of DNA vaccines. J Immunol 1997; 158:3635-3642.

43. Krieg AM, Yi A, Matson S et al. CpG motifs in bacterial DNA trigger direct B cell activation. Nature 1995; 374:546-548.
44. Klinman DM, Conover J, Coban C. Repeated administration of synthetic oligodeoxynucleotides expressing CpG motifs provides long term protection against bacterial infection. Infect Immun 1999; 67:5658-5663.
45. Kovarik J, Siegrist CA. Immunity in early life. Immunol Today 1998; 19:150-152.
46. Mond JJ, Lees A, Snapper CM. T cell independent antigens Type 2. Annu Rev Immunol 1995; 13:655-692.
47. Kim JJ, Bagarazzi ML, Trivedi N et al. Engineering of in vivo immune responses to DNA immunization via codelivery of costimulatory molecule genes. Nature Biotechnology 1997; 15:641-646.
48. Kim JJ, Nottingham LK, Sin JI et al. CD8⁺ T cells influence antigen specific immune responses through the expression of chemokines. J Clin Invest 1998; 102:1112-1124.
49. Quintana FJ, Rotem A, Carmi P, Cohen IR. Vaccination with empty plasmid DNA or CpG ODN inhibits diabetes in NOD mice. J Immunol 2000; 165:6148-6155.
50. Tacket CO, Roy MJ, Widera G et al. Phase 1 safety and immune response studies of a DNA vaccine encoding hepatitis B surface antigen delievered by a gene delivery device. Vaccine 1999; 17:2826-2829.
51. Wang R, Doolan DL, Le TP et al. Induction of antigen specific cytotoxic T lymphocytes in humans by a malaria DNA vaccine. Science 1998; 282:476-480.
52. Le TP, Noonon KM, Hedstrom RC et al. Safety, tolerability, and humoral immune responses after intramuscular administratrion of a malaria DNA vaccine to healthy adult volunteers. Vaccine 2000; 18:1893-1901.
53. MacGregor RR, Boyer JD, Ugen, KE et al. First human trial of a DNA based vaccine for treatment of HIV infection: Safety and host response. J Infect Dis 1998; 178:92-100.
54. Calarota S, Bratt G, Nordlund S et al. Cellular cytotoxic response induced by DNA vaccination in HIV-1 infected patients. Lancet 1998; 351:1320-1325.
55. Singh M, Briones M, Ott G, O'Hagan D. Cationic microparticles: a potent delivery system for DNA vaccines. Proc Natl Acad Sci USA 2000; 97:811-816.
56. Jones DH, Clegg JCS, Farrar GH. Oral delivery of micro-encapsulated DNA vaccines. Dev Biol Stand 1998; 92:149-155.

CHAPTER 19

The Introduction of New DNA Vaccines into Developing Countries

Richard T. Mahoney, Yu-Mei Wen, Henry Wilde and Zhi-Yi Xu

Abstract

The development and introduction of new vaccines is a costly and time-consuming process. Unfortunately, those most in need—individuals in developing countries—are the last to receive these powerful disease-preventing products. From the time a vaccine is first licensed in a developed country to the time most of the poor in developing countries have access to the vaccine can be 20-30 years. This delay is unacceptable. There is a great need to reduce this time span. This paper examines five ways of reducing the time span with special reference to DNA vaccines. Each of the five is essential and achieving success on all five will require a heightened level of international effort and coordination.

Introduction

During the last decade, naked DNA immunization, an entirely novel technology to induce humoral and cellular immune responses in hosts has been employed to develop vaccines against a number of pathogens. Basically, the recombinant DNA used for immunization consists of a plasmid vector backbone which has adjuvant activity, a transcription unit which includes a strong viral promoter/enhancer sequence, the gene of interest (usually the gene encoding the protective antigen), polyadenylation sequences, and an antibiotic resistance gene for in vitro manipulation and selection. Because this new approach for vaccine development offers many advantages over the traditional vaccines, DNA immunization has attracted much interest in the developing world. The most appealing advantage of using simply plasmid DNAs as vaccines for developing countries is the relative ease and speed of production and manipulation. In addition, DNA vaccine is stable and can be stored and delivered without requiring a cold chain, which is much preferred to distribute vaccines efficiently in rural areas.

Much has been written about the length of time and financial requirements to develop a new vaccine. Authors usually address the time and costs of vaccine research and development from first proof of principle through to first licensure. The times cited are often 10 to 15 years and the costs projected are cited as between $200 and $300 million. These are impressive figures and point to the need for finding ways to accelerate the development of new vaccines and to reduce development costs where possible. The figures do not, however, take into account the time and costs involved in bringing these vaccines to the majority of the world's populations; those who live in developing countries under economically disadvantaged conditions. Introduction there can take an additional 15-20 years. As noted, DNA vaccines have certain potential advantages of manufacture that may ultimately lead to lower prices and ease of distribution and use. (A recent review of these issues with respect to orphan vaccines has been published by one of us.)[1]

DNA Vaccines, edited by Hildegund C. J. Ertl. ©2003 Eurekah.com and Kluwer Academic / Plenum Publishers.

Vaccine Research and Development

Most basic childhood vaccines (DTP, BCG, and OPV) research, development and production scale up took place in public-funded vaccine institutes in developed countries and the development costs were recovered from sales in Europe and North America. Technology transfer to production facilities in developing countries took place largely as a public sector exercise financially supported by the recipient country and by multilateral and bilateral donors. These transfers were mostly devoid of intellectual property constraints because there were no patents on the vaccines and because the production know-how was in the public domain. However, the creation of vaccine research and development capability in developing countries was largely overlooked and R&D departments in developing country vaccine production facilities were either absent or grossly under-funded and neglected.

The development of vaccines against diseases that predominantly affect humans in developing countries has been neglected by the modern biotechnology industry. Vaccines such as those against cholera, *Shigella*, Japanese encephalitis and dengue have received little attention from the large international research-based firms. This situation will affect the development of DNA vaccines against similar diseases. This is especially troubling because it is believed that DNA vaccines will be particularly easy to produce and thus potentially have relatively low cost. For example, Prince has developed a prototype DNA vaccine against hepatitis C virus,[2] but further development of this vaccine has been impeded by lack of interest on the part of industry and therefore lack of resources.

To consider developing DNA vaccines in developing countries, several additional issues are to be considered:

1. The immunogenicity of the gene(s) of interest:

   To date, most DNA immunization studies have been carried out in mice. Results obtained were promising, as both humoral and cellular immune responses were induced, as well as protection against challenge with a variety of infectious agents.[3-7] However, the immunogenicity of different genes from different pathogens varies, and it is necessary to study and develop DNA vaccines for each pathogen individually. For instance, with a single injection in mice of 100 μg of DNA for the S gene of hepatitis B virus, peak antibody titers could be maintained for at least 74 weeks.[8] On the other hand, rabies DNA immunization requires three injections to induce protective immunity.[9] Clearly, it is preferable to develop DNA vaccines that can induce effective immune responses by only one injection. The possible boosting effect of subsequent subclinical infection in developing countries may contribute to secondary responses.

2. The delivery system and route of delivery:

   An important problem for developing DNA vaccine is the huge quantity of DNA needed for inducing immune responses in large animals and human. The production and purification of large quantity of plasmid DNA not only would be a financial burden but would increase the risks of side effects of DNA immunization, especially causing autoimmune responses or plasmid DNA integration into host cells. Though several delivery systems have been reported including gene guns, microparticle formulation, cationic lipids, and bacterial delivery system,[10] it is more practical to develop systems that are simple and can be manipulated by local health workers in developing countries. Some studies attempted to prime with DNA immunization followed by boosting with protein immunization to induce adequate immune responses. Though this combination vaccination protocol yielded better immune responses, it may be impractical for a vaccine producer to install two completely different types of equipment and producing pipelines. Besides, it will also be difficult for local health workers to carry out vaccination using two different vaccines. Regarding the route of delivery, oral or mucosal delivery are simple and will be better accepted; however, to monitor and control the dosage of DNA being successfully delivered via these routes would be a major problem.

3. Preventive and therapeutic DNA vaccines:

For preventive vaccines, the vaccinees are mainly children or infants, therefore, DNA vaccines should be carefully studied in young or newborn animals. Some studies have presented data that DNA vaccines could induce immune responses in newborn animals,[11] which suggest potential effectiveness in children. More studies should be carried out, not only for the efficacy of DNA vaccination, but also for examining the possible integration of DNA into the host genome of young and newborn animals. As for therapeutic vaccines, because DNA immunization can induce both cellular and humoral immune responses, it is predicted to be effective for treatment of persistent virus or bacterial infections. In developing countries, tuberculosis, AIDS, and viral hepatitis B and C are still prevalent, and because antibiotics or antiviral drugs are expensive or sometimes not available in the rural areas, the need for a DNA therapeutic vaccine is urgent and would be better accepted than antiviral drugs.

Recently there are promising developments that point toward better prospects for the development and use of vaccines for developing countries. There are new efforts and international arrangements; new funding; and new vaccine producers in developing countries who are struggling to overcome barriers; and there are new organizations participating in the process.

New International Arrangements

With the advent of the World Trade Organization (WTO) and its mandate to oversee the Agreement on Trade Related Aspects of Intellectual Property Rights (TRIPS), a fundamental change is taking place that will have far reaching impact for vaccine research, development, manufacture, and sales. It is not yet known what impact this will have on vaccines. WTO will be a means to try to encourage, and enforce, global respect for intellectual property rights. The day when manufacturers in developing countries could simply reverse-engineer a product will pass. Reverse engineering and outright expropriation of know-how and patents have been significant deterrents to major manufacturers working with their counterparts in poor countries. Developing country manufacturers do offer certain advantages including lower land, building, and labor costs. There are, however, burgeoning middle classes in many developing countries that are able to pay full market prices for new vaccines. These populations offer enticing markets to the major international manufacturers and may slowly overcome the reluctance of international vaccine manufacturers to work in the "Third World." On the other hand, some fear that WTO and TRIPS will lead to the formation of monopolies by developed country vaccine developers who will focus on the wealth markets of developed countries and will neglect the poor in developing countries. Their monopoly positions will prevent others from making efforts to supply the poor.

New Funding

The promise of major strides in the control of morbidity and mortality with vaccines has attracted important new donors. One of particular note is the Bill and Melinda Gates Foundation. The United States National Institutes of Health is also receiving substantial additional funds for vaccine research and development and has even established a new Vaccine Research Center. A new international coalition is also stimulating new allocations for vaccines. The Global Alliance for Vaccines and Immunization (GAVI) has engendered new funding from the Netherlands, Norway, the United Kingdom and the United States. To date, these funds are largely for vaccine procurement and for upgrading immunization delivery infrastructure. GAVI has formed a Task Force for Research and Development, but it is not clear how much funding will flow through this task force, if any. Nevertheless, GAVI represents yet another significant initiative enlarging the vaccine effort. One implication of these new undertakings is that the introduction of new vaccines to the developing world should proceed at a much faster pace than previously.

Newly Invigorated Vaccine Producers in Developing Countries

Some developing country vaccine manufacturers have made major strides in technology. For example, one Indian manufacturer (Serum Institute of India) now is the world's single largest supplier of vaccines used in national immunization programs around the world. Another manufacturer, Bio Farma, located in Indonesia has recently begun to supply DTP and polio vaccine to UNICEF. At least three manufacturers in India are building and staffing state-of-the art production and research facilities. Similar developments are underway in China and Latin America. The lack of capability in developing countries until recently has resulted in stagnation of effort either for vaccine improvement or new vaccine development in these facilities. A center, the International Vaccine Institute in Seoul, Korea, aimed at promoting public sector vaccine research and development, targeted to developing countries, has been created at the initiative of the United Nations Development Program. Because of these recent developments and initiatives, there are increasing opportunities for developing countries to advance in applied vaccine research and development.

The Introduction Process

The global introduction of hepatitis B vaccine as the 7th immunogen in the WHO Expanded Program on Immunization (EPI) has resulted in the definition of 5 key elements[12] necessary for such acceleration. These elements are:

1. Establishment and dissemination of disease burden data and of cost effectiveness computations.
2. Vaccine introduction trials and effectiveness evaluations.
3. Establishment of an international consensus on recommendations for vaccine use.
4. Assurance of adequate, sustainable and competitive vaccine supply.
5. Creation of funding mechanisms to supply vaccine to countries unable to finance their own procurement.

These activities overlap with each other and are carried out in parallel. They are listed in a rough order of sequence but the sequence may vary depending on the vaccine and the country.

Establishment and Dissemination of Disease Burden Data and of Cost Effectiveness Computations

It is not enough to know that a vaccine preventable disease exists in developing countries. To justify the cost and effort of introducing a vaccine and of sustaining its use for decades, it is also necessary to quantify the disease burden. Disease burden data allow for rational priority setting through the development of believable models that elucidate cost effectiveness.

Vaccine Introduction Trials and Effectiveness Evaluations

Clemens[13] described the usefulness of carrying out an introductory effectiveness trial for a new vaccine. These trials evaluate newly licensed vaccines to determine how best to provide the vaccine cost effectively and to assess its impact on disease. These projects allow national health authorities to gain experience with a new vaccine before making a full commitment to incorporate it into public health immunization programs. The projects are not clinical safety or efficacy trials and do not utilize placebos, strict randomization, or blinding. Data from control populations may be derived from historical information or concurrent data derived from populations not receiving vaccine. Vaccines to be used have already been licensed in countries of origin. Vaccines used for such introductory trials may be donated by manufacturers or paid for by donors.

Establishment of International Consensus and Recommendations for Vaccine Use

The application of political, social, and economic pressure in support of introducing new vaccines to developing countries requires vigorous effort and innovative thought. The decision to introduce a new vaccine into wide scale public health use requires consensus among ministries of health, international agencies such as WHO and UNICEF, national regulatory bodies, academic and medical experts, and others. This consensus is achieved through the weight of data about the vaccines being brought to the attention of key decision makers in these crucial groups.

Assurance of Adequate and Cost Competitive Vaccine Supply

Since the inception of the WHO/EPI program, adequate supplies of vaccines have been provided through both private and public sector producers in industrialized and developing countries. For example, over half of the currently produced DTP for global needs is produced in developing country facilities. The experience of UNICEF, in which vaccines have been provided through the tender and bid process at or near marginal cost, has yielded a very low price for developing world public sector immunization programs. This will not be easily emulated in the future and certainly not with newly developed vaccines. Newer vaccines will be considerably more expensive. It is thus imperative that ways are found to 1) encourage competitiveness in the vaccine production market, 2) develop strategies which encourage availability of vaccines to poorest countries at low price, 3) retain and strengthen some vaccine producers in the developing world, and 4) find inducements to encourage major vaccine developers to distribute new vaccines to poor countries earlier than in the past.

Encouragement of Competitiveness

In the current world with an economic environment that favors open markets and globalization, it is essential that competitiveness in the vaccine production market be maintained. Under these conditions, the existence of two or more manufacturers for any given vaccine destined for global use is important. Intellectual property agreements that severely limit access to know-how through restrictive sublicensing might contain exceptions for supply to the public sector of developing countries. Several major vaccine manufacturers have recently merged, leading to a reduction in competition. Some lifesaving vaccines such as, for example, tissue cultured rabies vaccines are now in the hands of only two major producers that are financially related. The prices of these two vaccines are almost identical but, fortunately, the companies offer two tier pricing: one for developed countries and one for poorer developing countries.

Retention and Strengthening of Manufacturers in Developing Countries

The role of developing country vaccine manufacturers in the establishment and maintenance of the WHO/EPI program has been crucial. Although of variable quality and quantity, the DTP vaccines produced in developing country facilities have been an invaluable backbone of the EPI system. Recently, production capabilities in several developing countries have advanced rapidly through the emergence of private sector producers or the privatization of formerly government owned facilities. It is important to maintain and strengthen these and other developing country manufacturers because they are more likely to accord priority to vaccines of special concern in their countries (that may not be high priorities in developed countries), and because their prices are likely to be set to reach a broader market than what large multinationals may choose to pursue.

Though the technology for producing modern, safe and effective vaccines against several major endemic and epidemic diseases such as Japanese encephalitis are known, no such vaccine has as yet appeared on the international market. One major producer has gone as far as pilot plant production but discontinued the project since it was not considered a profitable venture in view of the ever-increasing costs of development and licensing. These costs are often governed by regulations for production and licensing and by the litigious environment in highly developed countries. Average length of time needed to bring a new vaccine from the research bench to the international market is 15-18 years with costs in the hundreds of millions of dollars. Thus, it is not surprising that a manufacturer will only consider developing and making vaccines that can first sell at high cost in the West (often 10-15 times the later-achieved public sector price in a developing country). It is only after most, if not all, development costs are recovered that the product is ready for EPI use in the poor world. These facts bode ill for development of DNA vaccines for tropical and poor country users since validation and licensing costs for these products are going to be high as well.

At present, there are many problems to be solved before DNA vaccines can be used for global immunization (as discussed in other chapters). The many potential advantages that DNA vaccine can provide will suit the demanding need in developing countries. It seems important and appropriate to set up joint research groups of scientists from developed and developing countries to study these problems and to develop the vaccines. Joint research groups can select genes of interest from the microbial isolates of developing countries and local scientists can provide information to choose the first few DNA vaccines to be developed. In addition, large animal studies in DNA immunization experiments will be less expensive in the developing countries, and the fund for field experiments can be significantly decreased. Since ultimately, these vaccines will be mainly used in people of the developing countries, an early involvement of local workers will build up mutual understanding and confidence, which may bypass some barriers in future vaccine production and clinical trials. Besides, early involvement would also shorten the time for the transfer of technology know-how in DNA vaccine production, which will bring DNA vaccines more rapidly to those who need these vaccines the most.

Creation of Funding Mechanisms

The single most important impediment to early introduction of new vaccines in the developing world is lack of funds to buy the new vaccines that are usually at prices higher than the EPI vaccines. Hepatitis B vaccine introduction is the model for understanding problems of lack of procurement funds. This vaccine, even under lowest price conditions, costs two to six times the amount per dose of the most expensive of the EPI vaccines, measles vaccine. It is unlikely that Hib vaccine, the next likely candidate for expanded introduction in the developing world, will achieve prices as low as hepatitis B vaccine. Conjugated pneumococcal vaccine, because of its complicated production process, will likely be significantly higher priced than Hib vaccine. These facts do not bode well for early introduction of newer vaccines into developing world public sector immunization programs. However, important progress is being made. The Bill and Melinda Gates Foundation has donated $750 million for the creation of the Global Fund for Children's Vaccine. Several European countries have announced their intention to provide funds for this Fund. The Fund will be used to support the procurement of new vaccines. The success of this Fund is essential for the long-term success in introducing new and improved vaccines, and the development of effective policies and management systems that are required.[14] However, innovative ways to encourage major manufacturers to produce vaccines that cannot be marketed profitably in developed countries but would be lifesaving in poor countries, must be found.

Conclusion

Accelerating the time from first licensure of newer vaccines to actual use in public sector immunization programs in the developing world is a commitment easily urged but not easily accomplished. We have described a set of elements that we believe must be addressed to accelerate the process.

References

1. Wilde H. What are today's orphaned vaccines? Clinical Infectious Diseases; in press.
2. Prince A. New York Blood Center. personal communication.
3. Wang B, Ugen KE, Srikantan et al. Gene inoculation generates immune responses against human immunodeficiency virus type 1. Proc Natl Acad Sci USA 1993; 90:4156-4160.
4. Chang GJ, Hunt AR, Davis B. A single intramuscular injection of recombinant plasmid DNA induces protective immunity and prevents Japanese encephalitis in mice. J Virol 2000; 74:4244-4252.
5. Major ME, Vitvitski L, Mink MA et al. DNA-based immunization with chimeric vectors for the induction of immune responses against the hepatitis C virus nucleocapsid. J Virol 1995; 69:5798-5805.

6. Montgomery DL, Shiver JW, Karen R et al. Heterologous and homologous protection against influenza A by DNA vaccination: optimization of DNA vectors. DNA Cell Biol 1993; 12:777-783.

7. Lin YL, Chen LK, Liao CL et al. DNA immunization with Japanese encephalitis virus nonstrucural protein NS1 elicits protective immunity in mice. J Virol 1998; 72:191-200.

8. Davis HL, Mancini M, Michel M-L et al. DNA-mediated immunization to hepatitis B surface antigen: longevity of primary response and effect of boost. Vaccine 1996; 14:910-915.

9. Xiang ZQ, Spitalink S, Tran M et al. Vaccination with a plasmid vector carrying the rabies virus glycoprotein gene induces protective immunity against rabies virus. Virology 1994; 199:132-140.

10. Pachuck CJ, McCallus, Weiner BD et al. DNA vaccines-challenges in delivery Curr Opi Mol Ther 2000; 2:188-198.

11. Bot A, Bot S, Garcia-sastre A et al. DNA immunization of newborn mice with a plasmid-expressing nucleoprotein of influenza virus. Viral Immunol 1996; 9:207-210.

12. Mahoney R, Maynard J. The Introduction of New Vaccines into Developing countries. Vaccine 1999; 17:646-652.

13. Clemens J, Brenner R, Rao M et al. Evaluating New Vaccine for Developing countries: Efficacy or Effectiveness? JAMA 1996; 275(5).

14. Mahoney R, Ramachandran S, Xu Z. The Introduction of New Vaccines into Developing countries II: Vaccine Financing. Vaccine 2000; 18:2625-2635.

Index

The manufacturer's authorised representative in the EU is Springer
Nature Customer Service Centre GmbH, Europaplatz 3, 69115 Heidelberg,
Germany. If you have any concerns regarding our products, please
contact ProductSafety@springernature.com

Printed and bound by CPI Group (UK) Ltd, Croydon, CR0 4YY
23/04/2026
02095625-0007